Chemical Hazard Communication Guidebook

OSHA, EPA and DOT Requirements

2nd Edition

Andrew B. Waldo
and
Richard deC. Hinds

McGraw-Hill, Inc.
New York St. Louis San Francisco Auckland Bogotá
Caracas Lisbon London Madrid Mexico Milan
Montreal New Delhi Paris San Juan São Paulo
Singapore Sydney Tokyo Toronto

Executive Enterprises Publications Co., Inc., New York, New York

Copyright © 1993 by Executive Enterprises, Inc. All rights reserved. Printed in the United States of America. Except as permitted under the United States Copyright Act of 1976, no part of this publication may be reproduced or distributed in any form or by any means, or stored in a data base or retrieval system, without the prior written permission of the publisher.

1 2 3 4 5 6 7 8 9 0   HAL/HAL   9 8 7 6 5 4 3 2

ISBN 0-07-067755-7

The sponsoring editor for this book was Gail F. Nalven, and the production supervisor was Thomas G. Kowalczyk.

Printed and bound by Arcata Graphics/Halliday.

Cover photograph courtesy of S.C. Delaney/U.S. EPA.

This publication is designed to provide accurate and authoritative information regarding its subject matter. It is sold with the understanding that the publisher is not engaged in rendering legal, accounting, or other professional service. If legal advice or other expert assistance is required, the services of a competent professional person should be sought.—*From a Declaration of Principles Jointly Adopted by a Committee of the American Bar Association and a Committee of Publishers.*

TABLE OF CONTENTS

	Page
CHAPTER ONE: INTRODUCTION	1
ORGANIZATION	1
THE OSHA HAZARD COMMUNICATION STANDARD	3
SARA TITLE III EMERGENCY PLANNING AND COMMUNITY RIGHT-TO-KNOW REQUIREMENTS	6
HAZARDOUS MATERIALS TRANSPORTATION	6
COMPLIANCE PLANNING	7
CHAPTER TWO: THE OSHA HAZARD COMMUNICATION STANDARD	9
INTRODUCTION	9
WHO MUST COMPLY WITH THE HCS	11
CHEMICALS COVERED	12
OBLIGATIONS OF THE EMPLOYER	13
Hazard Determination	13
Hazard Determination Procedure	16
MATERIAL SAFETY DATA SHEETS	18
LABELS	20
EMPLOYEE INFORMATION AND TRAINING	22
THE WRITTEN HAZARD COMMUNICATION PROGRAM	24
TRADE SECRETS	26
ENFORCEMENT	29
RELATIONSHIP TO STATE AND LOCAL "RIGHT-TO-KNOW" LAWS	30
HAZARD COMMUNICATION STANDARD COMPLIANCE CHECKLIST FOR CHEMICAL USERS	32
CHAPTER THREE: SARA TITLE III EMERGENCY PLANNING AND COMMUNITY RIGHT-TO-KNOW REQUIREMENTS	33
INTRODUCTION	33
Preemption	34
EMERGENCY PLANNING REQUIREMENTS	34
EMERGENCY RELEASE REPORTING REQUIREMENTS	36
COVERED FACILITIES, RELEASES, AND SUBSTANCES	36
Covered Facilities	36
Covered Releases	37
Covered Substances	37
EXEMPTED RELEASES	38

	Page
Federally Permitted Releases	38
Other Exempted Releases	38
THE NOTIFICATION PROCESS	39
Immediate Notification Requirements	39
CONTENTS OF IMMEDIATE REPORT	39
FOLLOW-UP WRITTEN RELEASE REPORTS	40
COMPLIANCE PLANNING	40
HAZARDOUS CHEMICAL "RIGHT-TO-KNOW"	
REPORTING REQUIREMENTS	41
Basic Requirements	41
Submission of MSDSs or Lists	42
The List Option	43
Mixtures	44
Emergency and Hazardous Chemical Inventory	44
Content of Forms	45
Request for Tier 2 Information	47
Confidentiality	48
TOXIC CHEMICAL RELEASE REPORTING	48
Introduction	48
Toxic Chemicals Covered	48
Facilities Covered	50
Thresholds for Reporting	51
Toxic Chemical Release Reporting	52
Supplier Notification	54
PROTECTING TRADE SECRETS	55
Introduction	55
Procedure for Making Trade Secret Claims Under SARA	55
SUSBSTANTIATION REQUIREMENTS	56
Petition for Disclosure or Review of Claim	57
Disclosure to Health Professionals	58
ENFORCEMENT	59
Emergency Planning Notification Requirements	59
Emergency Release Notification Requirements	59
Reporting Requirements	60
Trade Secrets	60
Enforcement of Section 323 by Health Professionals	61
Enforcement by Citizen Suits and State or Local Suits	61
State Enforcement	61
COMPLIANCE PLANNING	62

	Page
COMPLIANCE CHECKLIST	62

CHAPTER FOUR: HAZARDOUS MATERIALS TRANSPORTATION ... 65
THE HAZARDOUS MATERIALS TRANSPORTATION ACT AND
    GENERAL STRUCTURE OF THE REGULATIONS ... 66
HAZARDOUS MATERIALS COVERED ... 68
GENERAL OVERVIEW OF THE SHIPPER'S RESPONSIBILITIES ... 71
SHIPMENT OF EPA HAZARDOUS SUBSTANCES ... 76
    Definition of Hazardous Substances ... 76
DOT REGULATION OF HAZARDOUS SUBSTANCES ... 78
INCIDENT REPORTING FOR SPILLS IN TRANSPORTATION ... 79
PROPOSED CHANGES TO THE DOT SYSTEM OF REGULATION ... 81
COMPLIANCE PLANNING ... 84

APPENDIX A: MODEL COMPANY PROGRAMS:
HAZARD COMMUNICATION AND EMPLOYEE TRAINING ... 91
INTRODUCTION ... 91
ABC COMPANY HAZARD COMMUNICATION PROGRAM ... 93
  1. Policy Statement ... 93
  2. Program Responsibilities ... 94
  3. Employee Notification and Training ... 97
  4. Location and Availability of Written Hazard
     Communication Program ... 98
  5. Sample Bulletin Board Notification ... 99
  6. Hazardous Material List ... 100
  7. Chemical Name Index ... 101
  8. Hazard Determinations ... 102
  9. Material Safety Data Sheets (MSDSs) ... 103
 10. Sample MSDS Request Letter ... 104
 11. Guide for Reviewing MSDS Completeness ... 105
 12. Labeling ... 107
 13. Employee Information/Training ... 108
 14. Outside Contractors ... 109
 15. Letter of Notification to Contractor Employers ... 111
HAZARDOUS CHEMICAL EMPLOYEE TRAINING ... 113
  Introduction ... 113
  Outline of a Model Hazardous Chemical
    Employee Training Program ... 115
  General Information ... 116
  Site Specific Information ... 123

Page

APPENDIX B: REFERENCE GUIDE TO THE REGULATIONS 125
   OSHA HAZARD COMMUNICATION STANDARD 127
     A. Scope of the Hazard Communication Standard 127
     B. Performing a Hazard Determination 129
     C. Preparing Material Safety Data Sheets 132
     D. Maintaining Material Safety Data Sheets 135
     E. Preparing and Maintaining Labels and Other Warnings 137
     F. Training Employees about Chemical Hazards 140
     G. Preparing and Providing a Written Hazard
        Communication Program 143
     H. Requirements for Those Handling Hazardous
        Chemicals in Sealed Containers 144
     I. Special Requirements for Laboratories 145
     J. Requirements for Disclosing and
        Withholding Trade Secret Information 146
   OSHA's Hazard Communication Standard 151
   OSHA Material Safety Data Sheet Form 179
   EMERGENCY PLANNING AND COMMUNITY RIGHT-TO-KNOW 181
     A. Emergency Planning Notice 181
     B. Emergency Release Reporting 183
     C. MSDS/List Submission 184
     D. Hazardous Chemical Inventory Reporting 186
     E. Submission of Toxic Chemical Release Reports 188
     F. Disclosure of Information 191
   EPA's Emergency Planning Rule 193
   EPA's Release Reporting Regulations Under CERCLA 199
   EPA's Regulation Establishing Emergency and Hazardous
       Chemical Inventory Forms and Community Right-to-Know Reporting
       Requirements 205
   EPA's Regulation Establishing Toxic Chemical
       Release Reporting 221
   EPA's Toxic Chemical Release Inventory Reporting Form R
       and Instructions 243
   EPA's Trade Secrecy Claims for Emergency Planning and
       Community Right-to-Know Information: and Trade Secret
       Disclosures to Health Professionals 279
   DOT REQUIREMENTS APPLICABLE TO SHIPPERS OF
   HAZARDOUS MATERIALS 305
     A. Determining the Proper Shipping Name and Classification 305

	Page
B. Complying with General Restrictions on Transporting Hazardous Materials	306
C. Determining the Proper Packaging	308
D. Selecting the Proper Label(s)	313
E. Marking the Packaging in the Manner Required	318
F. Determining the Proper Placard(s)	325
G. Preparing Shipping Papers	329
DOT Hazardous Materials Table	335
Appendix to § 172.101-List of Hazardous Substances and Reportable Quantities	426
APPENDIX C: DEFINITIONS OF HAZARDS	**461**
OSHA HAZARD COMMUNICATION STANDARD	463
SARA RIGHT-TO-KNOW REPORTING	476
HAZARDOUS MATERIALS TRANSPORTATION	482
APPENDIX D: LIST OF EXTREMELY HAZARDOUS SUBSTANCES (ALPHABETICAL ORDER)	**493**
APPENDIX E: LIST OF EXTREMELY HAZARDOUS SUBSTANCES (CAS ORDER)	**505**
APPENDIX F: LIST OF HAZARDOUS SUBSTANCES (ALPHABETICAL ORDER)	**517**
APPENDIX G: LIST OF HAZARDOUS SUBSTANCES (CAS ORDER)	**551**

About the Authors

Andrew B. (Pat) Waldo is president of Ariel Research Corporation, Bethesda, MD. He has served as a consultant on health, safety, and environmental issues to industry, arms of the U.S. Congress, Presidential committees, and federal and state agencies. He advises industrial clients on compliance management regarding state, federal, and international regulation of the manufacture and use of chemicals. He established Ariel's International Chemical Regulatory Monitoring System, a tool used in major corporate programs for tracking compliance with a diverse range of requirements. Mr. Waldo's recent books on regulatory compliance include: *Consolidated Chemical Regulation Guidebook* (Ariel Research Corp., 1990); *The Community Right-to-Know Manual* (Thompson Publications, 1987); *The Community Right-to-Know Handbook* (Thompson Publications, 1986); and, "A Review of U.S. and International Restrictions on Exports of Hazardous Substances" in *The Export of Hazard* (Routledge, Kegan, Paul, 1985).

Richard deC. Hinds is a partner with the law firm of Cleary, Gottlieb, Steen, and Hamilton in Washington, DC specializing in environmental regulation and litigation. He has written extensively on the regulation of toxic substances and is a frequent speaker at seminars in this rapidly developing area of law. He is an author of *Complying with OSHA's Hazard Communication Standard* (BNA Chemical Regulation Reporter, April 26, 1984) and a co-author of *The Community Right-To-Know Manual*. He received his B.A. degree from Tufts University in 1964 and his L.L.B. degree from Columbia University in 1967. He was an editor of the *Columbia Law Review* from 1966 to 1967.

Chapter One

INTRODUCTION

Organization

This book is written to assist the reader in complying with hazard communication requirements applicable to hazardous chemicals in the workplace, in meeting reporting responsibilities imposed by emergency planning and community right-to-know requirements, and in understanding restrictions on the transportation of hazardous materials. Chapter Two contains a detailed summary of the U.S. Occupational Safety and Health Administration's (OSHA) Hazard Communication Standard (HCS) which requires all employers to provide information to their employees on the hazards of the chemicals with which they work. Chapter Three details the obligations of employers established by the Environmental Protection Agency (EPA) under Title III of the Superfund Amendment and Reauthorization Act (SARA) to provide the community with information on the identity and amount of hazardous chemicals present at a facility and information on releases of hazardous chemicals to the environment. Finally, Chapter Four summarizes the major requirements for shipments of hazardous materials under the U.S. Department of Transportation's (DOT) regulations.

The book is designed to provide the reader with tools to determine what the regulations require and to develop a plan to meet these requirements. The chapters provide a summary of the requirements that apply in each area and provide practical suggestions for a cost-effective compliance program.

Appendix A includes a model hazard communication program that the reader may use as the basis for developing a company program to comply with OSHA's requirements.

Chemical Hazard Communication Guidebook

In addition, Appendix A provides an outline of an employee training course which can be used to prepare a training program that meets the HCS.

Appendix B is a reference guide to the OSHA, EPA, and DOT regulations covered. It contains an outline of the principal requirements in each area followed by the actual text of the regulations that apply. The outlines in the reference guide summarize the requirements, briefly describe its application, and provide a reference to the pertinent regulation. Although the regulations reprinted here include the principal ones of concern, space limitations precluded the inclusion of the entire text of DOT's regulations, which occupy several volumes.

Appendix C contains the definitions of hazard-related terms used in the regulations. Each of the regulations promulgated by OSHA, EPA, and DOT defines the types of hazards covered. In some cases, the same term may be defined differently. In other cases, the definition in one regulation may refer to the definition in another regulation. To clarify how these terms relate to one another, the appendix provides each definition and compares the terms used in the three program areas.

The appendices that follow provide the lists of chemicals subject to emergency planning, emergency release reporting, and toxic chemical release reporting in alphabetic and CAS order. The DOT Hazardous Materials Table is included as part of Appendix B. With the passage of SARA and its reference to various lists of chemicals, each defined differently, an important element in compliance is the need to compare the chemicals present at a facility with the lists adopted as part of the regulations. Although the OSHA Hazard Communication Standard established floor lists of chemicals which, at a minimum, are to be considered hazardous, the HCS is broad in scope and employs performance criteria in defining hazardous chemicals subject to regulation. The HCS applies to many chemicals not specifically included on any list. Similarly, the spill reporting requirements imposed by EPA incorporate by reference both listed and unlisted hazardous wastes. In developing a compliance program, close attention should be given to any performance criteria that may apply, as well as to any specific lists of chemicals to which the regulation may refer.

Introduction

The OSHA Hazard Communication Standard

The purpose of the OSHA Hazard Communication Standard is to ensure that the hazards of all chemicals produced in or imported into the United States are identified and that this information, together with information on protective measures, is provided to employees who are exposed or potentially exposed to hazardous chemicals in the workplace. Although the original HCS applied only to employers in the manufacturing sectors included in Standard Industrial Classification (SIC) codes 20-39, OSHA published a final rule on August 24, 1987, extending the scope of the OSHA HCS to cover all employers. The revised HCS first required manufacturers, importers, and distributors to provide a material safety data sheet (MSDS) to commercial customers in the non-manufacturing sector with the first shipment of a hazardous chemical after September 23, 1987. By May 23, 1988, all firms that ship, store, sell, or use chemicals were to have prepared and implemented a comprehensive hazard communication program for their employees. The HCS was recodified as 29 CFR 1910.1200 for general industry, 29 CFR 1915.99 for shipyard employment, 29 CFR 1917.28 for marine terminals, 29 CFR 1918.20 for longshoring, 29 CFR 1926.59 for construction, and 29 CFR 1928.21 for agriculture.

Due to court challenges and administrative actions, OSHA was prevented from enforcing the HCS in all industries, including MSDS provisions on multi-employer worksites, coverage of consumer products, and the coverage of drugs in the non-manufacturing sector. As a result of the February 21, 1990 Supreme Court decision in *Dole, Secretary of Labor et al. v. United Steelworkers of America et al.*, 110 S. Ct. 929 (1990), all provisions of the rule are in effect for all industrial segments, including the three areas for which a stay had been granted.

The original HCS required manufacturers and importers to prepare MSDSs and appropriate labels for each container of chemicals shipped after November 25, 1985. After that date, manufacturers, importers, and distributors of chemicals had to provide MSDSs to their manufacturing customers. Employers in the manufacturing sector were also required to provide appropriate training to employees and to develop a written hazard communication program by May 25, 1986.

Chemical Hazard Communication Guidebook

The following lists the SIC codes which became subject to the OSHA Standard with its revisions:

Division A. Agriculture, Forestry, and Fishing
    Major Group 01. Agricultural production—crops
    Major Group 02. Agricultural production—livestock
    Major Group 07. Agricultural services
    Major Group 08. Forestry
    Major Group 02. Fishing, hunting, and trapping

Division B. Mining
    Major Group 13. Oil and gas extraction

Division C. Construction
    Major Group 15. Building construction—general contractors and operative builders
    Major Group 16. Construction other than building construction—general contractors
    Major Group 17. Construction—special trade contractors

Division E. Transportation, Communication, Electric, Gas, and Sanitary Services
    Major Group 40. Railroad transportation
    Major Group 41. Local and suburban transit and interurban highway passenger transportation
    Major Group 42. Motor freight transportation and warehousing
    Major Group 44. Water transportation
    Major Group 45. Transportation by air
    Major Group 46. Pipe lines, except natural gas
    Major Group 47. Transportation services
    Major Group 48. Communication
    Major Group 49. Electric, gas, and sanitary services

Division F. Wholesale Trade
    Major Group 50. Wholesale trade—durable goods

Introduction

    Major Group 51. Wholesale trade—non-durable goods

Division G. Retail Trade
    Major Group 53. General merchandise stores
    Major Group 54. Food stores
    Major Group 55. Automotive dealers and gasoline service stations
    Major Group 56. Apparel and accessory stores
    Major Group 57. Furniture, home furnishing, and equipment stores
    Major Group 58. Eating and drinking places
    Major Group 59. Miscellaneous retail

Division H. Finance, Insurance, and Real Estate
    Major Group 60. Banking
    Major Group 61. Credit agencies other than banks
    Major Group 62. Security and commodity brokers, dealers, exchanges, and services
    Major Group 63. Insurance
    Major Group 64. Insurance agents, brokers, and service
    Major Group 65. Real estate
    Major Group 66. Combinations of real estate, insurance, loans, law office
    Major Group 67. Holding and other investment offices

Division I. Services
    Major Group 70. Hotels, rooming houses, camps, and other lodging places
    Major Group 72. Personal services
    Major Group 73. Business services
    Major Group 75. Automotive repair, services and garages
    Major Group 76. Miscellaneous repair services
    Major Group 78. Motion pictures
    Major Group 79. Amusement and recreation services, except motion pictures
    Major Group 80. Health services
    Major Group 81. Legal services
    Major Group 82. Education services
    Major Group 83. Social services

Chemical Hazard Communication Guidebook

> Major Group 84. Museums, art galleries, botanical and zoological gardens
> Major Group 86. Membership organizations
> Major Group 89. Miscellaneous services

SARA Title III Emergency Planning and Community Right-To-Know Requirements

On October 17, 1986, the President signed into law the Emergency Planning and Community Right-To-Know Act, which was included as Title III of SARA. It establishes a nationwide program of emergency planning involving extremely hazardous substances. Any facility handling extremely hazardous substances above thresholds set by EPA must comply with the notification provisions described in EPA's regulations. The Act also expands the spill reporting program established by the Comprehensive Environmental Response, Compensation, and Liability Act (CERCLA). These reporting obligations apply to any facility that has a covered spill of extremely hazardous substances or a hazardous substance as defined in CERCLA. It mandates the submission of MSDSs or lists of hazardous chemicals, to state and local entities as well as the preparation and submission of annual hazardous chemical inventories by facilities required to prepare or have available MSDSs under the OSHA hazard communication standard. Finally, it obligates covered facilities in SIC 20-39 to prepare and submit toxic chemical release reports.

Any facility required by OSHA to prepare or have available an MSDS for a hazardous chemical had to submit an MSDS or alternative list of substances, as required by SARA right-to-know regulations, on or before October 17, 1990, or within three months after the facility first becomes subject to SARA MSDS reporting. The facility had to submit its SARA hazardous chemical inventories by March 1, 1991, or March 1 of the first year after the facility becomes subject to the requirement, and annually thereafter. The toxic chemical release report applicable to facilities in SIC 20-39 must be submitted by July 1 annually.

Hazardous Materials Transportation

Under the Hazardous Materials Transportation Act, DOT has broad authority to regulate the shipment of hazardous materials and may designate as hazardous any material or class of materials that may pose an unreasonable risk to health, safety, or

Introduction

property. DOT's Hazardous Materials Transportation regulations govern the classification, packaging, labeling, marking, placarding, and documentation of hazardous materials being shipped.

DOT has proposed to modify the current system in order to conform with international systems of regulation. On May 5, 1987, and in a revision on November 6, 1987, DOT issued a major notice of proposed rulemaking (HM-181) revamping the current system to establish performance-oriented packaging standards. It is likely that DOT will issue a final rule soon. The shipper of hazardous materials faces the necessity of understanding not only the present system, but also how the proposals may change current practices.

Compliance Planning

The passage of SARA and the extension of the HCS to all employers brought about a major expansion in the number of facilities subject to federal health, safety, and environmental regulations and greatly increased the complexity of these regulations. These changes require an integrated approach to compliance with the HCS, "right-to-know" reporting obligations, and restrictions on the shipment of hazardous materials. The task of compliance necessitates an understanding not only of a particular regulation, but also of the cross-references that have been established between these areas of regulation. The following examples illustrate this need:

1. In SARA, the MSDS and hazardous chemical inventory reporting requirements apply to facilities required to prepare or have available an MSDS under the OSHA HCS.
2. In the HCS, OSHA requires that chemical manufacturers, importers, and distributors must ensure that labels, tags, or markings for hazardous materials leaving the workplace must not conflict with DOT regulations.
3. The emergency release reporting provisions established in SARA build upon the existing spill reporting program in CERCLA. Spill reporting in SARA/CERCLA applies both to covered releases occurring at a facility, and those in transportation. In the event of an accidental release in transportation, DOT's incident reporting requirements may also apply.

4. SARA also requires DOT to list and regulate as hazardous materials those hazardous substances defined by EPA in CERCLA. DOT has issued regulations establishing the transportation requirements applicable to such materials.

A management plan for compliance must address the specific steps necessary to comply with the regulatory programs in a coordinated manner. However, a compliance plan must do more than ensure that the actions required by the OSHA HCS, SARA Title III, and DOT have been taken. It must also provide for continuing compliance as new information becomes available on chemical health effects and safe handling techniques. The plan should not be limited to meeting the minimum requirements of the HCS, SARA Title III, and DOT, but should also take into account potential product liability, workmen's compensation, tort liability to employees, and employer-employee relations in general. Finally, the plan should deal with compliance issues raised by any applicable state and local right-to-know laws. The most cost-effective, efficient approach to employee and community hazard communication is to design a comprehensive program that incorporates all of these elements.

Chapter Two

THE OSHA HAZARD COMMUNICATION STANDARD

Introduction

The U.S. Occupational Safety and Health Administration (OSHA) has developed a comprehensive regulation concerning the communication to employees of the hazards of chemicals in the workplace. The OSHA Hazard Communication Standard (HCS) seeks to ensure that the hazards of all chemicals produced—or imported into the United States—are identified, and that this information, along with information on protective measures, is provided to employees exposed, or potentially exposed, to those chemicals in the workplace.

The original HCS applied only to employers in the manufacturing sectors identified in Standard Industrial Classification Codes 20-39. It required manufacturers and importers to prepare material safety data sheets (MSDS) and appropriate labels for each container of chemicals shipped after November 25, 1985. After that date manufacturers, importers, and distributors of chemicals had to provide an MSDS with the first shipment of a hazardous chemical to purchasers and users in the manufacturing sector. The HCS also required all employers in the manufacturing sector to provide appropriate training to employees exposed to hazardous chemicals and to develop a written hazard communication program by May 25, 1986.

Following a court challenge to the original HCS, the U.S. Court of Appeals for the

Chemical Hazard Communication Guidebook

Third Circuit issued an order requiring OSHA to extend the HCS to employers outside the manufacturing sector. Accordingly, on August 24, 1987, OSHA published a final rule extending the scope of the HCS to cover all employers (52 Fed. Reg. 31852). The revised HCS provided that, effective September 23, 1987, manufacturers, importers, and distributors had to provide an MSDS with the first shipment of a hazardous chemical to commercial customers in the non-manufacturing sector. By May 23, 1988, all firms that ship, store, sell, or use chemicals were to prepare and implement a comprehensive hazard communication program for their employees.

Due to court challenges and administrative actions, OSHA was stayed from enforcing certain requirements of the HCS in all industries, including MSDS provisions on multi-employer worksites, coverage of consumer products, and the coverage of drugs in the non-manufacturing sector. As a result of the February 21, 1990 Supreme Court decision in *Dole, Secretary of Labor et al. v. United Steelworkers of America et al.* (No. 88-1434), all provisions of the rule are in effect for all industrial segments, including the three areas for which a stay had been granted.

The HCS is a performance-oriented rule. Most requirements are stated in terms of objectives to be achieved rather than in terms of the methods employers must use to achieve those objectives. Employers thus have considerable flexibility to design hazard communication programs appropriate for their own workplaces.

This performance orientation also means that employers will have questions on how to comply with the standard. In seeking answers, employers should refer to the standard's requirements, and to the extensive preambles to the original and revised standard published in the *Federal Register* (48 Fed. Reg. 53280, November 25, 1983; 52 Fed. Reg. 31852, August 24, 1987). Employers may also consult OSHA's Inspection Procedures for the Hazard Communication Standard for interpretations of the standard as well as information on OSHA's plans for enforcement. (OSHA Instruction 2-2.38 C, October 22, 1990). OSHA regional offices and the headquarters office in Washington also can provide informal advice to employers. However, companies should not rely on OSHA advice that is not put in writing. Employers with remaining questions should consult with counsel familiar with the terms and objectives of the HCS.

OSHA Hazard Communication Standard

Who Must Comply with the HCS

All employers that manufacture, import, distribute, or use hazardous chemicals in the United States are subject to the HCS, but the specific obligations imposed will vary depending on which of those activities the employer engages in.

Chemical manufacturers and importers have special responsibilities under the standard including the obligation to make hazard determinations and prepare MSDSs and container labels. "Manufacturing" includes processing, formulating, and repackaging. Thus, a company that blends or formulates chemicals would be a chemical manufacturer, as would a company that repackages chemicals for sale in smaller quantities. An "importer" is the first business within the United States that receives hazardous chemicals produced in other countries for commercial sale in the United States.

Distributors of hazardous chemicals must pass on to other distributors and to purchasers the hazard information (MSDSs, labels) they receive from their suppliers.

All employers with workplaces containing hazardous chemicals must maintain an MSDS file and provide employees access to it. In addition, all employers must train their employees concerning the workplace hazards of the chemicals to which they are exposed and appropriate precautions to take in handling such chemicals. Employers that do not manufacture or import a chemical may rely on the hazard determination, labeling, and MSDS obtained from the manufacturer or importer of that chemical. However, if the employer receives no MSDS or label with appropriate hazard warnings for a purchased chemical, the employer may have an obligation to obtain or develop an MSDS for that chemical and appropriate labeling for each container.

Laboratories. The coverage of laboratories is limited under the HCS. It would include research facilities as well as quality control laboratory operations located within manufacturing facilities. OSHA promulgated a separate rule covering exposure to hazardous chemicals in laboratories (29 CFR 1910.1450, 55 Fed. Reg., January 31, 1990). The laboratories standard is consistent with the HCS, but has some additional requirements that must be applied in laboratories covered by that rule.

Chemical Hazard Communication Guidebook

It should be noted that the definition of a laboratory in the final rule for exposures to hazardous chemicals in laboratories is narrow, and specifies criteria for "laboratory use" and "laboratory scale" activities. The rule excludes procedures that are part of a production process. OSHA considers a laboratory to be a facility where relatively small quantities of hazardous chemicals are used on a non-production basis.

The laboratory rule is found in Appendix B.

Facilities Handling Sealed Containers. OSHA has indicated in its inspection procedures that employees who work in operations where they handle only sealed containers, such as in a warehouse, must receive an MSDS for chemicals in sealed containers and also training in their handling, if requested. The training required for employees who handle sealed containers would depend on the type of chemicals involved, the potential size of any spills or leaks, the type of work performed and what actions employees are expected to take when a spill or leak occurs.

Chemicals Covered

The HCS is extremely broad in its chemical coverage. It is not limited to chemicals included on any specific lists. Instead, the HCS generally applies to any hazardous chemicals known to be present in the workplace to which employees may be exposed under normal conditions of use or in a foreseeable emergency.

A number of chemical substances are exempt from the HCS, including all foods, drugs, and cosmetics in a retail store or in the workplace for personal consumption by employees; alcoholic beverages in a retail store; hazardous wastes regulated by the Environmental Protection Agency; tobacco; and wood products. In addition, all drugs regulated by the Food and Drug Administration are exempt.

Articles and consumer products are also exempt, but only if they meet certain conditions. Articles are manufactured items of specific shape that do not result in exposure to hazardous chemicals under normal use (*e.g.*, furniture, typewriter ribbons). If the use of the article results in exposure to hazardous substances, the article is subject to the HCS and must have an MSDS and be appropriately labeled. For example, textile products may be covered if an employee pressing the textile product could be exposed to hazardous formaldehyde fumes. Consumer products (*e.g.*, detergents) are exempt if used in the workplace in the same general form and concentration as a product packaged for household use by the general public. In the HCS, OSHA added the further condition that employee exposure must not be greater in duration or frequency than consumer exposure. Thus, for example, the use of cans of spray paint during production runs rather than for occasional, short, one-time applications that typify consumer use would be an example of hazardous chemical use that OSHA has indicated would not qualify as consumer product use.

Chemicals regulated by other federal agencies (except for hazardous wastes) are covered by the HCS, but they are exempt from its labeling provisions if subject to the labeling regulations of the other agency. Regulated hazardous wastes are exempt from the HCS.

Obligations of the Employer

Hazard Determination

The requirements in the HCS apply only to hazardous chemicals. Manufacturers and importers must identify those chemicals which are hazardous and determine what hazards they pose. (Importers will normally need to have their foreign suppliers make hazard determinations for them. As with all other requirements applicable to importers, however, the importers are responsible for their own compliance with the standard.) Other employers need only determine the hazards posed by a chemical if they choose not to rely on the hazard determinations made by the manufacturer or importer of the chemical.

The HCS defines a hazardous chemical as any chemical which is either a physical hazard, or a health hazard, or both. A chemical is considered to be a physical hazard only if there is scientifically valid evidence that it is a combustible liquid, a compressed gas, explosive, flammable, an organic peroxide, an oxidizer, pyrophoric, unstable (reactive), or water-reactive.

Chemical Hazard Communication Guidebook

A combustible liquid is a liquid having a flash point at or above 100 degrees F (37.8 degrees C) but below 200 degrees F (93.3 degrees C).

A compressed gas is (l) a gas or mixture of gases in a container which has an absolute pressure greater than 40 psi at 70 degrees F (21.1 degrees C) or an absolute pressure greater than 104 psi at 130 degrees F (54.4 degrees C), or (2) a liquid having a vapor pressure exceeding 40 psi at 100 degrees F (37.8 degrees C) (determined in accordance with the ASTM D-323-72).

An explosive is a chemical that causes a sudden, almost instantaneous release of pressure, gas, and heat when subjected to sudden shock, pressure, or high temperature.

A chemical is flammable if it is a flammable aerosol, gas, liquid, or solid under the conditions specified in the HCS. For example, a liquid is flammable if the flash point is below 100 degrees F (37.8 degrees C).

An organic peroxide is a chemical that contains the bivalent -0-0- structure and is structurally similar to hydrogen peroxide. Such chemicals typically undergo auto-accelerating thermal decomposition.

An oxidizer is a chemical other than a blasting agent or explosive that ignites or promotes combustion in other materials, thereby causing fire either of itself or through the release of oxygen or other gases.

A chemical is pyrophoric if it will ignite spontaneously in air at a temperature of 130 degrees F (54.4 degrees C) or below.

An unstable (reactive) chemical will vigorously polymerize, decompose, condense, or become self-reactive under conditions of shock, pressure, or temperature. A water-reactive chemical will react with water to release a gas that is either flammable or presents a health hazard.

A health hazard is defined as any chemical for which there is statistically significant evidence based on at least one study conducted in accordance with established scientific principles that acute or chronic health effects may occur in exposed employees. Appendix A to the HCS defines certain kinds of effects deemed to be acute or chronic health hazards. Acute effects include irritation, corrosivity, sensitization, and toxicity. Criteria are given for determining if a substance presents those hazards. For example, a chemical is deemed acutely toxic if it exhibits any of the following characteristics:

1. Oral median lethal dose (LD50) less than 500 mg/kg in albino rats.
2. A dermal LD50 of less than 1,000 mg/kg in albino rabbits.
3. An inhalation lethal concentration (LC50) in air of less than 2,000 parts per million in albino rats.

The term chronic effect includes carcinogenicity, teratogenicity, mutagenicity, and other chronic adverse health effects. Appendix A of the HCS gives target organ effects of certain chemicals to illustrate the range and diversity of the health hazards that may be found in the workplace.

The HCS deems hazardous some 2,400 chemicals which are included on one or more of the following four lists (the lists overlap but no single one includes all the chemicals): (1) OSHA's health standards, appearing in 29 CFR Part 1910, Subpart Z (the Z-tables of 29 CFR 1910.1000 and other standards which appear elsewhere in Subpart Z); (2) the latest edition (currently 1990-91) of the American Conference of Governmental Industrial Hygienists (ACGIH) publication, "Threshold Limit Values for Chemical Substances and Physical Agents in the Work Environment"; (3) the latest edition (currently the fifth, dated 1989) of the Annual Report on Carcinogens of the National Toxicology Program (NTP); and (4) the latest editions of the Monographs published by the International Agency for Research on Cancer (IARC). It should be emphasized that these are base or floor lists that only include a small percentage of the chemicals that are hazardous under the HCS criteria.

Chemicals listed in the NTP Annual Report or in the IARC Monographs as actual or potential carcinogens, and chemicals regulated by OSHA as carcinogens, are deemed by the HCS to be carcinogens. OSHA considers a chemical to be a carcinogen on the basis of an IARC report if the report finds that there is "limited" or "sufficient" evidence, by either animal or human data, of carcinogenicity, but not if the report finds that there is "inadequate evidence" or "no data" suggesting carcinogenicity. There are approximately 525 chemicals deemed to be carcinogens.

Hazard Determination Procedure

A person making hazard determinations must describe in writing the procedures used. These procedures may vary according to the circumstances, particularly for health hazards. Thus, for well-studied chemicals the person may base a hazard determination on a review of selected basic reference works, such as those listed in Appendix C to the HCS. However, if other information (such as information in the company's own files) indicates that additional effects are possible, they must also be taken into account. For less well-known chemicals, a person may need to consult computer data bases, information in the company's files, industry-sponsored studies, or other sources. Testing is not required by the HCS.

The findings of a hazard determination must be reported on an MSDS. The HCS requires an MSDS to list all health effects of a hazardous chemical, not just those which pose a significant risk of harm to employees in the workplace. In the information and training given to employees, an employer may explain the expected effects at actual workplace exposure levels.

To resolve controversy over whether a chemical poses a particular health effect, the HCS sets a low threshold favoring disclosure. A health effect must be reported on an MSDS if supported by statistically significant evidence based on at least one study conducted in accordance with established scientific principles. In reporting the effect, however, the person need not endorse it. Indeed, the person may note any limitations of the study and may include the results of other studies tending to show that the chemical does not cause that effect.

The requirements for statistical significance and compliance with established scientific principles mean that a person making a hazard determination need not report a health effect on an MSDS simply because it is reported in a study. The person may evaluate the study for compliance with these requirements by using professional judgment.

Special rules govern the determination of hazards posed by mixtures. If a mixture has been tested as a whole for hazards, those results may be used to determine the mixture's hazards. If the mixture has not been tested as a whole, the person making the hazard determination may use whatever scientifically valid information is available, if any, to determine the mixture's physical hazards. The standard irrefutably presumes the health hazards of an untested mixture to be those of the hazardous chemicals present in the mixture in concentrations of 1.0 percent or greater (as determined by weight or volume), except that an untested mixture is irrefutably presumed to present a risk of cancer if it contains a component deemed by the standard to be a carcinogen in a concentration of 0.1 percent or more (*i.e.*, a chemical included in the latest NTP Annual Report or listed as a potential or actual carcinogen in the latest edition of an IARC Monograph, or regulated by OSHA as a carcinogen).

To exclude a hazardous component from an MSDS for a mixture, the person preparing the MSDS must ensure that the hazardous component is present in the mixture only in concentrations below the thresholds mentioned above. Furthermore, if the person has evidence indicating that, even at concentrations below those thresholds, the hazardous component could be released in concentrations that would exceed an OSHA permissible exposure limit (PEL), or an ACGIH threshold limit value (TLV), or could pose a health hazard to employees, the person must still report the hazardous component on the MSDS.

Persons making hazard determinations are required to document the steps taken to evaluate the hazards of each chemical. This documentation may have to be produced to demonstrate to an OSHA inspector that the person performed an adequate hazard determination. It may also be relevant in defending a personal injury suit brought by someone claiming to have been injured by an unreasonable failure to disclose particular hazards.

The HCS does not specifically require a person to repeat hazard determinations from time to time. However, it does require that MSDSs and labels be updated to reflect what is currently known about a substance. Accordingly, as new information comes in, the

Chemical Hazard Communication Guidebook

hazard determination needs to be reviewed.

Material Safety Data Sheets

All employers must obtain from their suppliers, and make available to their employees, an MSDS for each hazardous chemical used in their workplace. Manufacturers, importers, and distributors were required to provide MSDSs to customers outside the manufacturing sector with the first shipment of hazardous chemicals after September 23, 1987.

If an employer does not receive an MSDS for a chemical it purchases, the employer should not assume that the chemical has been determined to be non-hazardous. The employer should examine the label to determine whether it provides any hazard information. If the label indicates the product poses any hazards, the employer has the obligation under the HCS to have an MSDS available for that product. As a practical matter, an employer who does not have an MSDS for a chemical should: (1) require a supplier to confirm in writing that the chemical has been determined to be non-hazardous unless the label specifically so states; (2) if the chemical is hazardous, obtain an MSDS for the chemical from the supplier or from another source; or (3) prepare an MSDS itself. Because most chemicals pose some hazards, employers should insist that suppliers of all chemicals provide MSDSs. Most chemical suppliers have MSDSs for all chemicals they sell, including chemicals that pose no hazards, and can easily comply with such a request.

Distributors and other employers who only handle chemicals in sealed containers that are not opened under normal conditions must maintain a file of MSDSs received, but only have to obtain an MSDS for a hazardous chemical received without one if an employee specifically requests it.

Each MSDS must contain in English the following information about a hazardous chemical: (l) its specific chemical identity (unless claimed a trade secret) and synonyms, or if it is a mixture, its components which contribute to the hazard of the mixture (as determined through testing or through the presumptions discussed previously in the hazard determination section); (2) its physical and chemical characteristics; (3) its physical hazards, including the potential for fire, explosion, and reactivity; (4) its health hazards, including signs and symptoms of exposure and any medical conditions generally recognized as being aggravated by exposure to the chemical; (5) primary routes of exposure; (6) the OSHA PEL, ACGIH TLV, and any other exposure limit used or recommended by the person preparing the MSDS; (7) whether it is deemed by the standard to be a carcinogen; (8) any generally applicable precautions for safe handling and use known to the person preparing the MSDS; (9) any generally applicable control measures known to the person preparing the MSDS; (10) emergency and first aid procedures; (11) date of preparation or latest revision of the MSDS; and (12) the name, address, and telephone number of a responsible party who can provide additional information on the chemical and appropriate emergency procedures, if necessary. If some required information is unavailable, the MSDS must so indicate; blanks are not allowed.

OSHA has prepared a suggested form for an MSDS containing the above information. (See Appendix B.) Note that the form is for guidance purposes and does not have to be followed precisely. It does, however, provide a good model to follow in preparing an MSDS that complies with the HCS.

Employers receiving inconsistent MSDSs should review each carefully. The HCS contemplates that an employer may rely on any MSDS it receives for a given hazardous chemical. Tort liability considerations, however, suggest that an employer should seek to have the most complete and accurate MSDS available. The employer may want to question the firms that prepared the inconsistent MSDSs to evaluate which is the more accurate MSDS. An employer may prepare its own MSDS by incorporating elements from the MSDSs of its suppliers. If it does so, however, it must ensure that the new MSDS accurately reflects the scientific evidence. The employer also becomes responsible for updating the MSDS with any significant new information regarding the hazards of the chemical, or ways to protect against the hazards posed by it, within three months of becoming aware of such information.

Since MSDSs must be readily accessible to employees, the MSDSs should not be locked in an office overnight if the night shift needs to have access to them. Similarly, the employer may not keep the MSDSs in a location away from the work areas. When

Chemical Hazard Communication Guidebook

employees' work is carried out at more than one geographical location, the MSDSs may be kept at a central location so long as the employer ensures that the information can be obtained immediately in an emergency. An employer may computerize the MSDSs so long as a terminal is available to its employees.

Labels

A non-manufacturing user of hazardous chemicals must ensure that each container of hazardous chemicals in the workplace is properly labeled, unless the container is exempted. Exempted containers include pipes and containers into which an employee transfers hazardous chemicals for that employee's use during that work shift. Where the container is a tank truck, rail car, or the like, the label may either be posted on the tank or attached to the shipping papers.

The labels must contain the identity of the hazardous chemical, and "appropriate hazard warnings." It will not necessarily be "appropriate" to warn on the label about every hazard listed in the MSDS. The MSDS is supposed to contain a comprehensive description of the hazards posed by the chemical. The hazards highlighted on the label should be those that are most significant to employees under foreseeable conditions of exposure.

OSHA has provided very specific guidance on the labeling of carcinogens in OSHA Instruction CPL 2-2.38c (October 22, 1990), featured in **Table 1** on the facing page. In that instruction OSHA states that chemicals that have been classified by IARC as human carcinogens, or chemicals for which there is at least limited evidence of carcinogenicity to humans, and those chemicals identified by NTP as known to be, or reasonably anticipated to be, carcinogenic must have carcinogen warnings on the label as well as information on the MSDS. In addition, any chemical for which there is positive human evidence that it is a carcinogen must be labeled as such. The instruction contains the following table setting forth OSHA's conclusions regarding MSDS and label notations for carcinogens.

TABLE 1

OSHA GUIDANCE FOR MSDS AND LABEL NOTATIONS FOR CARCINOGENS

Source	MSDS	Label
Listed on NTP Annual Report on Carcinogens	X	X
IARC—Group 1A	X	X
IARC—Group 1B	X	X
IARC—Group 2A	X	X
IARC—Group 2B	X	Not Required
IARC—Group 3	Not Required	Not Required
One Positive Study—Animal Only	X	Not Required
One Positive Study—Some Human Evidence	X	X
Multiple Animal Studies	X	Depends on weight of evidence; review necessary

Chemical Hazard Communication Guidebook

It should be noted that the OSHA Instruction itself acknowledges that deviations from this guidance may be appropriate when, for example, recent evidence disputes the conclusions of past evaluations by IARC or NTP.

The nature of the hazards posed by a chemical should be specified with particularity. According to OSHA, if the human organs affected by a chemical are known, the label should so indicate. For example, if a chemical is known to cause bladder cancer in humans, the label should so state. A label that stated such a chemical was "harmful if ingested" or even a "cancer hazard" would not be specific enough according to OSHA.

In the case of containers of hazardous chemicals leaving the workplace, the labels must contain the name and address of a responsible party to contact for more information if necessary. For in-plant stationary containers, such as reactor vessels, labels may be in the form of batch tickets, process sheets, signs, or operating procedures. In addition, standardized in-plant labeling systems (*e.g.*, the HMIS system) can be used as long as training regarding the use of these systems, together with the MSDSs, provides the required information.

While it is the responsibility of the manufacturer or importer to prepare the label, all employers must check that the containers they receive are properly labeled. If a container arrives without a proper label, the supplier should be contacted and a label or labeling information obtained and placed on the container.

Employee Information and Training

All employers covered by the HCS must provide their employees, including their laboratory employees, with information and training on the hazards of chemicals in their work areas and the means to avoid those hazards. Employers must provide their employees with information about the requirements of the standard and how they may obtain access to MSDSs and the written hazard communication program. New hires and employees reassigned to a different work area must also receive such training.

The HCS provides that, at a minimum, the training program must cover the following four topics:

1. How to detect the presence or release of a hazardous chemical in the workplace (*e.g.*, monitoring devices, visual appearance or odor of chemicals when released, etc.);
2. The physical and health hazards of the chemicals in the workplace;
3. The measures employees can take to protect themselves from these hazards including information on appropriate work practices, emergency procedures and personal protective equipment; and
4. An explanation of the hazard communication program developed by the employer including explanation of the labeling system, the material safety data sheets, and the list of hazardous chemicals in the workplace and how employees can obtain and use this information.

The first step in complying with this requirement is to gather the necessary hazard information. Much of it should be contained in MSDSs and container labels. Other information will relate to the particular workplace and its hazard control program. A model training program is included in Appendix A.

Employers may train their employees with regard to general classes of hazards, rather than with respect to hundreds or thousands of different chemicals. For example, generic training films are available from BNA Communications, Inc., Rockville, Maryland (301-948-6540) on handling solvents, corrosives, oxidizers, poisons, gases, explosives, carcinogens, etc. Substance-specific information should be added to the training program as necessary. It is also available to employees in the MSDSs that must be maintained in the workplace.

Distributors and other facilities that only handle hazardous chemicals in sealed containers not opened under normal conditions of use need only provide information and training to the extent necessary to protect employees in the event of a spill or leak.

Employers should document employee training sessions to enable them to demonstrate compliance with the HCS and to defend against possible lawsuits. They should keep records of subjects covered and materials distributed. They should also require employees to sign attendance lists and to state in writing that they understood the material presented

Chemical Hazard Communication Guidebook

at the training sessions.

By its terms, the HCS does not require periodic refresher training once employees have been trained, except to the extent that chemical hazards in the workplace change or a new chemical is introduced. Refresher training would be helpful, however, to protect against possible worker compensation claims and personal injury lawsuits.

The Written Hazard Communication Program

Employers using hazardous chemicals must develop, implement, and maintain a written hazard communication program. Distributors and other facilities that only handle hazardous chemicals in sealed containers that are not opened under normal conditions of use are not required to prepare a written program. However, all covered facilities will find it useful to prepare such a program as a means of ensuring compliance with the other applicable requirements of the HCS.

The hazard communication program must include a list of all hazardous chemicals in the workplace to which employees may be exposed. The employer should ensure that the list is updated as new chemicals enter the workplace. The list need not contain the specific chemical identities of hazardous chemicals, but it must give names that are keyed to MSDSs that provide specific chemical identities, except to the extent that those identities are trade secrets. In that event, the MSDS need only provide a generic or trade name for the chemical.

In workplaces with several work areas, such as a large plant, the employer should compile the list for each work area. Since employees will have access to the list, it is preferable to specify those hazardous chemicals to which an employee may be exposed in that employee's work area, instead of creating the impression that each employee may be exposed to every hazardous chemical in a facility.

The hazard communication program must also include the procedures to be used in complying with the MSDS, labeling, and training requirements of the HCS. In addition, the program must specify the methods to be used to inform employees of the hazards of non-routine tasks (such as cleaning reactor vessels), and the hazards of chemicals in unlabeled pipes.

Finally, the program must explain the methods to be used to inform employees of another employer (*e.g.*, contractors) of the hazards that they may face in the workplace. Employer A need not train the employees of employer B working in the same workplace. However, employer A must provide employer B with sufficient information to permit it to train its employees about the hazards of the chemicals used by employer A. Sometimes it may be simpler for an employer itself to provide the required training for employees of maintenance and other contractors who are regularly in that employer's workplace. In addition, the plan must describe the methods the employer will use to provide the other employer(s) with information on the labeling system in use in the workplace and any precautionary measures that need to be taken to protect its employees.

The written program need not be lengthy or complicated, but it should adequately address each of the required components in the program. The following checklist is a guide to the elements that any program should take into consideration in an appropriate manner for the specific facility.

1. Labels and other Forms of Warning
    a. Designation of person(s) responsible for ensuring labeling of in-plant containers.
    b. Designation of person(s) responsible for ensuring labeling on shipped containers.
    c. Description of labeling system(s) used.
    d. Description of written alternatives to labeling of in-plant containers, if used.
    e. Procedures to review and update label information when necessary.

2. Material Safety Data Sheets
    a. Designation of person(s) responsible for obtaining/maintaining the MSDS.
    b. How such sheets are to be maintained (*e.g.*, in notebooks in the work area(s)), and how employees can obtain access to them.
    c. Procedure for updating the MSDS when new and significant health information

Chemical Hazard Communication Guidebook

   is found.
   d. Description of alternatives to actual data sheets in the workplace, if used.

3. Training
   a. Designation of person(s) responsible for conducting training.
   b. Format of the program to be used (audiovisuals, classroom instruction, etc.).
   c. Elements of the training program—compare to the elements required by the HCS. (See 29 CFR §1910.1200(h).)
   d. Procedure to train new employees at the time of their initial assignment to work with a hazardous chemical, and to train employees when a new hazard is introduced into the workplace.
   e. Guidelines on training programs prepared by the Office of Training and Education (49 FR 30290; July 27, 1984) provide general information on what constitutes a good training program.
4. Does a list of the hazardous chemicals exist in each work area or at a central location?
5. Are methods the employer will use to inform employees of the hazards of nonroutine tasks outlined?
6. Are employees informed of the hazards associated with chemicals contained in unlabeled pipes in their work areas?
7. Does the plan include methods employers will use to inform contractors of the hazards to which their employees may be exposed?
8. Is the written program made available to employees and their designated representatives?
9. Does the plan include the methods the employer will use at multi-employers? Worksites to inform other employers of any precautionary measures that need to be taken?
10. For multi-employer workplaces, are the methods the employer will use to inform other employers about the labeling systems being used adequately described?

A model hazard communication program is attached in Appendix A.

Trade Secrets

The trade secret provisions in the HCS balance an employer's need to protect trade secrets and the need of health professionals to know the specific identity of chemicals to which employees may be exposed. The HCS explicitly provides that it does not require the disclosure under any circumstances of process or percentage-of-mixture information that is a trade secret. The only trade secret information which must be disclosed, and then only under certain conditions, is the specific identity of a hazardous chemical.

The HCS permits health professionals (*i.e.*, a physician, industrial hygienist, toxicologist, epidemiologist, or occupational health nurse), employees, and their designated representatives to obtain access to trade secret specific chemical identity information, provided they demonstrate a need for the information and agree to sign a confidentiality agreement. In emergency situations, health professionals must be given trade secret chemical identity information immediately.

Manufacturers, importers, and users of hazardous chemicals may determine that the specific identity of any of the hazardous chemicals they produce or import is a trade secret. (Importers will normally obtain this information from their foreign suppliers.) The MSDSs prepared or obtained for these chemicals may withhold the specific identities of those chemicals as trade secrets, but they must so indicate.

The HCS permits a confidential chemical identity to be claimed a trade secret if maintaining its secrecy gives an employer an opportunity to obtain an advantage over competitors who do not know it or use the chemical in the same way. Under general trade secret law, from which the standard's definition of "trade secret" is derived, the basic requirements for a trade secret are secrecy and value. An identity may be a trade secret even if a competitor could discover it through reverse engineering. Questions about whether a chemical identity qualifies as a trade secret should be resolved by reference both to the HCS and to the law of trade secrets.

The employer should consider preparing a written justification for each specific identity classified as a trade secret. In some circumstances, the employer may need to present a written justification for a trade secret classification within a short period of time (within thirty days if the employer denies a health professional's written request for access to a specific identity withheld as a trade secret). In such a situation, advance preparation may prove useful. A written justification may also be required to comply with SARA and state right-to-know laws that require employers to submit a justification for a claim of trade secrecy to a government agency in order to be able to withhold specific chemical identity from an MSDS or required report.

To prove secrecy, the justification should describe the methods the employer uses to keep the chemical identity from its competitors and to limit internal access to the identity to those who need to know it. To establish value, the justification might show that the chemical identity permits the production of a better product than that of competitors, production of the same product at lower cost, or that the employer's competitors do not use the chemical in the novel way in which the employer uses it. Note that a firm's use of a chemical in manufacturing a product may be a trade secret even though the supplier of that chemical makes no trade secret claims for its identity.

The HCS does require limited disclosure of trade secrets. In a medical emergency, the employer must disclose a trade secret identity to a treating physician or nurse if that health professional deems disclosure necessary for emergency or first aid treatment. In a non-emergency situation, the employer must disclose such information to a health professional providing occupational health services to employees, including downstream employees and their designated representatives, but only if certain requirements are met including a written request demonstrating need. Employers should establish appropriate procedures for complying with requests for disclosure of trade secret information in both emergency and non-emergency contexts.

In both an emergency and a non-emergency situation, the employer may condition disclosure upon entry into a confidentiality agreement (as soon as circumstances permit in an emergency situation; prior to disclosure in a non-emergency situation). Accordingly, the employer should obtain or prepare a model confidentiality agreement governing access to trade secret identities. The model agreement may include a liquidated damages clause setting forth the reasonable damages that the employer would suffer in the event of unauthorized disclosure, and that the person requesting disclosure (or his employer) would have to agree to pay.

Enforcement

OSHA has had an active enforcement program for the HCS for years. Under the OSH Act, OSHA inspectors can issue citations for violations of any aspect of the HCS and each separate violation can result in a separate penalty. The specific violations that OSHA inspectors are instructed to look for include:

1. Failure by a manufacturer or importer to perform a hazard determination.
2. Failure to have a written hazard communication program or, if a plan exists, failure to include a required element. (Even if you are otherwise in full compliance with the HCS you must have a written program.)
3. Failure to have a complete list of hazardous chemicals in the workplace. (This is a common citation. Every hazardous substance in the workplace must be listed.)
4. Failure to have a system to inform employees of the hazards associated with non-routine tasks, and with chemicals in unlabeled pipes.
5. Failure to have containers labeled or labels that fail to include required information. (All labels must disclose the identity of the chemical and provide appropriate hazard warnings. The chemical identity shown on the label must cross-reference the identity shown on the MSDS and on the list of hazardous chemicals.)
6. Failure to have an MSDS for a hazardous chemical. (Documentation of efforts to obtain an MSDS from your supplier is essential to avoid a citation.)
7. Failure to have an MSDS file accessible to employees during work hours.
8. Failure of a distributor to transmit an MSDS with the first shipment of a hazardous chemical to a customer.
9. Failure to train exposed employees concerning chemical hazards and precautions.
10. Failure to have an adequate training program.
11. Failure to provide trade secret chemical identity information to a doctor or nurse in an emergency.

Chemical Hazard Communication Guidebook

A citation is required to be in writing and to describe with particularity the nature of the violation, including a reference to the provisions of the HCS alleged to be violated. In addition, the citation is required to set a "reasonable time" for the abatement of a violation.

OSHA has the authority only to propose penalties. The actual penalty paid will depend on whether the employer contests the citation. If the citation is not contested within fifteen days, the amount proposed must be paid. Otherwise, the amount of the penalty will be set by the OSH Review Commission, an independent tribunal, following a hearing, unless the case is settled previously. Most citations are not contested (generally a mistake) and most contested citations are resolved through informal settlement negotiations.

OSHA defines the type of violation in the citation as willful, repeat, serious, failure-to-abate, or nonserious (or "other"). An employer who receives a citation will be advised on the form itself of OSHA's classification of the violation(s). A violation of the HCS will be considered "serious" whenever a deficiency on an MSDS or label, or a violation of any other element of the HCS could result in or contribute to a potential exposure capable of producing serious physical harm or death.

Any employer found to have willfully or repeatedly violated the HCS can be assessed a civil penalty of not more than $10,000 for each violation. A willful citation will be issued if an employer refuses to provide specific chemical identity information to a health professional in a medical emergency. Any employer cited for serious or other violations of the HCS can be assessed a civil penalty of up to $1,000 for each violation. "Serious violations" are penalized more heavily than nonserious violations. Failure to correct a violation within the abatement period prescribed by OSHA can lead to a further penalty of not more than $1,000 for each day during which such failure continues.

In addition, criminal penalties can be imposed for willful violations that cause death to any employee. Upon conviction, an employer can be fined up to $10,000 ($20,000 if a repeat conviction) and/or imprisoned up to six months.

Relationship to State and Local "Right-To-Know" Laws

Right-to-know laws have been adopted by over thirty states and many localities. Their provisions differ considerably. Many involve aspects of hazard communication not included in the OSHA standard, such as providing information on chemical hazards to fire departments, and to the community at large. A major practical consideration for employers is whether the OSHA standard preempts state and local right-to-know laws, or whether they have to comply with both the federal standard and applicable state and local laws.

The OSHA standard provides that it "is intended to address comprehensively the issue of evaluating the potential hazards of chemicals, and communicating information concerning hazards and appropriate protective measures to employees, and to preempt any legal requirements of a state, or political subdivision of a state, pertaining to the subject." 29 CFR §1910.1200(a)(2). The preemptive effect of the HCS has been upheld in several cases. Accordingly, a state (or locality) should be able to regulate with respect to an issue covered by the standard only through the adoption and approval by OSHA of a state plan or a modification to an existing state plan.

OSHA's position is that state standards can be enforced only under the auspices of an OSHA-approved state plan. States without state plans are preempted from addressing the issue of hazard communication. Community right-to-know standards are outside the jurisdiction of OSHA and are not affected by this position.

Unfortunately, it is not simple. Although the Congress has in certain specific instances expressed an intention to preempt particular kinds of state and local legislation, it has not done so in any broad way. The courts, in consequence, have addressed conflicts between national and state or local legislation on a statute-by-statute basis. Cases include *Associated Industries of Massachusetts v. Snow*, 898 F. 2d 274 (1st Cir. 1990); *New Jersey State Chamber of Commerce v. Hughey*, 774 F. 2d 587 (3d Cir. 1985) ("Hughey I"), appeal after remand, 868 F. 2d 621 (3d Cir.) ("Hughey II"), cert. denied, 109 S. Ct. 3246 (1989); and *Manufacturers Ass'n of Tri-County v. Knepper*, 801 F. 2d 130 (3d Cir. 1986), cert. denied 484 U.S. 815 (1987). The OSHA standard has been held to have no preemptive effect on state environmental or right-to-know laws that seek to protect the health or safety of the general

Chemical Hazard Communication Guidebook

public because the standard is limited to workplace hazards to employees.

Community right-to-know laws enacted by the states have become a fixed part of regulatory compliance planning. These laws will, however, continue to be the subject of court challenges until more precise boundaries between federal and state regulation are defined.

Hazard Communication Standard Compliance Checklist for Chemical Users

The following checklist summarizes steps an employer using hazardous chemicals should take:

1. Review the OSHA standard and determine whether you are covered. Virtually every employer in the United States is covered.
2. Compile a list of the chemicals in each workplace.
3. Review the labels on chemical containers and any MSDSs you have received and compile a list of hazardous chemicals in each workplace. Check applicable exceptions.
4. Develop procedures for training and providing information to employees on chemical hazards; informing employees of the hazards of non-routine tasks; and informing other employers in the same workplace (*e.g.*, maintenance contractors, construction subcontractors) of chemical hazards to which their employees may be exposed.
5. Have MSDSs available in an easily accessible location for employee inspection and appropriate labels on every non-exempt container of hazardous chemicals in the workplace. (Complete by May 23, 1988.)
6. Unless you only handle hazardous chemicals in sealed containers that are not normally opened by your employees, prepare a written hazard communication program which includes the list from step 3 and a description of your compliance program for steps 4 and 5.

Chapter Three

SARA TITLE III EMERGENCY PLANNING AND COMMUNITY RIGHT-TO-KNOW REQUIREMENTS

Introduction

This chapter provides a summary of the emergency planning and community right-to-know requirements imposed by SARA Title III. These provisions have four major aspects: (1) emergency planning; (2) emergency release reporting; (3) MSDS and inventory reporting for hazardous chemicals; and (4) toxic chemical release reporting.

All users of extremely hazardous substances are potentially subject to the emergency planning program of Title III. All facilities are potentially subject to emergency release reporting in the event of a covered spill of an extremely hazardous substance or hazardous substance defined under CERCLA. Hazardous chemical MSDS and inventory reporting applies to facilities required to maintain MSDSs for hazardous chemicals by the OSHA Hazard Communication Standard. Facilities in SIC 20-39 are also potentially subject to the toxic chemical release reporting provisions if they use more than 10,000 pounds of a listed toxic chemical.

Title III establishes a framework of state commissions and local emergency planning committees, which are responsible for developing emergency plans to address the potential release of extremely hazardous substances. Facilities having more than a threshold planning quantity of a listed extremely hazardous chemical on hand must comply with notification requirements and participate in emergency planning.

Chemical Hazard Communication Guidebook

Any facility that has a release to the environment of a reportable quantity of an extremely hazardous or hazardous substance must immediately notify federal, state, and local emergency response authorities. The spill reporting provisions of SARA build upon the existing requirement of the Comprehensive Environmental Response Compensation and Liability Act (CERCLA) to report releases of hazardous substances exceeding reportable quantities set by the Environmental Protection Agency or by statute.

The community right-to-know provisions require the submission of material safety data sheets (MSDSs) or, alternatively, lists of hazardous chemicals present above certain thresholds. Moreover, the provisions require inventory reporting on the identity and amount of hazardous chemicals present above the specified thresholds at a facility subject to the Hazard Communication Standard (HCS). The MSDSs (or lists) and hazardous chemical inventories are provided to the state commission, the local committee, and the fire department with jurisdiction.

Certain manufacturing facilities in SIC 20-39 must also report on the amounts of toxic chemicals released to the environment. These annual reports must be provided to EPA and the states by July 1 of each year. The provision applies to facilities within SIC 20-39 that have ten or more full-time employees, which manufacture, process, or otherwise use a listed toxic chemical in excess of specified threshold quantities.

Preemption

Employers subject to existing state and local right-to-know laws may question whether those laws have been superseded by the right-to-know provisions of SARA Title III. SARA specifically addresses this issue and expressly provides that nothing in Title III preempts any state or local law except laws passed after August 1, 1985, that set forth a format for the MSDS that conflicts with the OSHA HCS.

Emergency Planning Requirements

Title III creates a framework for emergency planning in which the governor of each state appointed a state emergency response commission, which in turn was responsible for

SARA Title III Requirements

establishing emergency planning districts served by local emergency planning committees. Each committee was responsible for developing emergency response plans to address the potential for emergencies involving extremely hazardous substances.

The state commission and the local committee not only serve in the development of emergency plans, but also process public requests for information submitted by covered facilities under Title III, which includes not only the emergency notification, but also MSDSs or lists, hazardous chemical inventory reports, and toxic chemical release reports. Except for trade secret chemical identity and the location of hazardous chemicals claimed confidential, all information submitted under Title III is available to the public.

On April 22, 1987, EPA promulgated its final rule establishing the list of extremely hazardous substances, associated threshold planning quantities, and its requirements for emergency planning notification. The rule applies to any facility at which there is present an amount of any extremely hazardous substance equal to or in excess of the threshold planning quantity. The rule requires the owner or operator of such a facility to notify the appropriate state commission that it is subject to the emergency planning provisions of the Act before May 17, 1987, or within sixty days of first becoming subject to this reporting requirement.

A "facility" is broadly defined in 40 CFR § 355.20 of EPA's rule to include all buildings, equipment, and structures owned or operated by the same person that are located on a single site or contiguous sites. However, Title III excludes the transportation (or storage incident to transportation) of covered substances from its requirements, except the spill reporting requirements described below. The exemption relating to storage incident to transportation is limited to the storage of materials which are still moving under active shipping papers and which have not reached the ultimate consignee.

In addition, a newly covered facility must appoint a facility emergency coordinator who will participate in the emergency planning process of the local committees. Facilities subject to emergency planning were to notify their local committees of the facility representative by September 17, 1987 or, for a newly-covered facility, within sixty days of becoming subject to this planning requirement.

Chemical Hazard Communication Guidebook

The state commission may designate additional facilities that become subject to the emergency response planning and notification requirements of Title III, but only after notice and comment. If the facility becomes covered in this manner, it must be notified by the state commission. Moreover, facilities may become subject to emergency planning notification through a revision of the list of extremely hazardous substances by EPA.

Emergency Release Reporting Requirements

Emergency notification of leaks, spills, and other releases of specified chemicals into the environment is required under CERCLA and SARA Title III, and a number of other federal and state laws. Under CERCLA, persons in charge of a facility (including transporters) must report to the National Response Center (NRC) any spill or release of a specified "hazardous substance" in an amount equal to or greater than the reportable quantity (RQ) specified by EPA.

SARA significantly expanded the existing CERCLA notification requirement by adding a list of "extremely hazardous substances" (see Appendices D and E) to the list of "hazardous substances" (see Appendices F and G) for which reporting is mandatory, and by requiring that all reports of releases must be submitted immediately to the appropriate state emergency response commission and local emergency planning committee as well as to the NRC.

In most situations, initial and follow-up reporting under the broad CERCLA/SARA requirements will satisfy all federal emergency release notification obligations. However, other federal statutes may impose additional reporting requirements for certain chemicals, and many states have reporting requirements that demand notification of specified state agencies in the event of a release of specified substances.

Covered Facilities, Releases and Substances

Covered Facilities

The emergency notification requirements of CERCLA/SARA apply to a "release" from a "facility" of specified hazardous substances into any environmental medium: air, water, or land. Any landfill, building, or physical structure where hazardous substances are produced, used, or stored is a covered facility, including motor vehicles, rolling stock, aircraft, and pipelines.

Covered Releases

A "release" is defined in SARA as any "spilling, leaking, pumping, pouring, emitting, emptying, discharging, injecting, escaping, leaching, dumping, or disposing into the environment" of any covered substances. The definition of release also includes the discarding of barrels or other receptacles containing, or which once contained, such substances. The definition of release in CERCLA is essentially the same. Therefore if a spilled substance remains within an enclosed manufacturing plant and does not enter the environment, a reportable release has not occurred.

A spill outside the same manufacturing plant, even if contained within the boundaries of the facility, is a release to the environment and would thus be reportable to the NRC under CERCLA. Although the reporting requirements under SARA specifically include an exemption for releases that "result in exposure to persons solely within the boundaries of the facility" (i.e., the plant boundaries), such releases (especially those to the air) have the potential to result in exposure to persons offsite. Reporting under both CERCLA and SARA of all releases that escape from a plant other than into a secure, diked area is therefore strongly recommended.

Covered Substances

Reporting to the NRC is required under CERCLA for any "hazardous substance" released in an amount equal to or greater than the RQ for the substance. The list of hazardous substances is set forth at 40 CFR § 302.4 (see Appendices F and G). The substances covered under the release reporting requirements of SARA include all substances on the list of hazardous substances as well as those on a list of "extremely hazardous substances" set forth at 40 CFR Part 355, Appendix A (see Appendices D and E of this book).

Chemical Hazard Communication Guidebook

Exempted Releases

Federally Permitted Releases

Certain releases do not have to be reported because they are specifically permitted. Reporting is not required under CERCLA/SARA for "federally permitted releases." CERCLA Section 101(10) narrowly defines a federally permitted release to include, among others, a discharge authorized in a final RCRA permit, an underground injection authorized under federal or state programs, an air emission subject to a federal or state permit or control requirement, an effluent discharge subject to a Clean Water Act permit or identified in the permit application, or a discharge to a public sewer system if the discharge is in compliance with applicable general or categorical pretreatment standards.

Companies cannot determine whether or not they are covered by this "federally permitted release" exemption without careful review. Depending on the type of permit, the substance being released may have to be subject to a specific limit or order for the release to be exempted. For example, discharges to POTWs are exempt only if the released substance is specified in an applicable pretreatment standard and in a pretreatment program submitted to EPA under Section 402 of the Clean Water Act.

Other Exempted Releases

Also exempt from CERCLA/SARA notification requirements are:

1. Releases which only result in exposure within a facility (for purposes of CERCLA reporting) or only to persons within the facility boundaries (for purposes of SARA reporting);
2. Those releases which are in amounts below the RQ (or, in the absence of an RQ, below one pound);
3. "Continuous releases" that require annual reporting under CERCLA (unless there is a "statistically significant increase" in the release of hazardous substances), rather than immediate reporting; and
4. Releases of a pesticide being applied consistent with its FIFRA registration and

SARA Title III Requirements

labeling requirements. (Note that this exemption does not include accidental spills or other releases of pesticides.)

The Notification Process

Immediate Notification Requirements

In the event of a release of a hazardous substance or an extremely hazardous substance in excess of the applicable RQ, the owner/operator of the facility (or the person in charge of the facility) must immediately notify:

1. The National Response Center (800-424-8802 or 202-267-2675);
2. The state emergency response commission of any state that may be affected by the release; and
3. The community emergency coordinator for the local emergency committee for any area likely to be affected by the release.

Some states may require that additional local officials also be notified.

Contents of Immediate Report

Emergency notification to the NRC, the state commission, and the local committee of releases into the environment of hazardous substances for which there is an RQ, or releases of extremely hazardous substances, must contain:

1. The name of the chemical(s) or identity of the substance(s) involved. (SARA provides that the name of the chemical cannot be withheld as a trade secret);
2. An indication of whether the substance is on the list of extremely hazardous substances;
3. An estimate of the quantity of the substance released;
4. The location, time, and duration of the release;
5. The medium or media into which the release occurred;
6. Any known or anticipated acute or chronic risks associated with the release and,

where appropriate, advice regarding medical attention necessary for exposed individuals;
7. Proper precautions to take as a result of the release, including evacuation (unless such information is readily available to the community emergency coordinator pursuant to the community emergency plan); and
8. The name and telephone number of the person or persons to be contacted for further information.

The initial notification can be made orally by telephone or radio, or in person.

Follow-up Written Release Reports

As soon as practicable after a release which requires emergency notification, SARA requires the facility owner/operator to provide to the same entities that received the initial oral notification a written follow-up emergency notice setting forth and updating the initial information. In addition, the follow-up notice is required to include the following information:

1. Actions taken to respond to and contain the release;
2. Any known or anticipated acute or chronic health risks associated with the release; and
3. Where appropriate, advice regarding medical attention necessary for exposed individuals. State and local governments may require the written follow-up notification to include the results of a facility's inspection and the measures to be adopted to prevent future releases.

Compliance Planning

Facilities that have on hand hazardous or extremely hazardous substances should assemble in advance the information that must be reported in the event of a spill or other release. Facilities should also have determined precisely who must be notified at the state and local level of such releases.

Facilities that have recurring releases should review carefully the scope of the exemptions from reporting. Facilities that determine that CERCLA/SARA reporting is or is not required for a specific release should carefully review whether reporting is required under other federal or state laws under the specific circumstances of the release in question.

Hazardous Chemical "Right-to-Know" Reporting Requirements

Basic Requirements

Under SARA, employers required to prepare or have available MSDSs for hazardous chemicals in their workplace ("covered employers") must submit an MSDS for those chemicals or a list of them to the state commission, the designated local committee, and the fire department with jurisdiction. The facility must follow that up with annual inventory reporting on the amount of those chemicals present at the facility. As a result of the extension of the OSHA HCS to employers outside the manufacturing sector, those employers also become subject automatically to this federally mandated community right-to-know program.

SARA provides a number of exemptions from hazardous chemical reporting requirements, which largely duplicate the exemptions provided in the OSHA Hazard Communication Standard. These exemptions are for:

1. Any food, food additive, color additive, drug, or cosmetic regulated by the Food and Drug Administration;
2. Any substance present as a solid in any manufactured item if exposure to the substance does not occur under normal conditions of use;
3. Any consumer product used for personal, family, or household purposes, and substances present in the same form and concentration as a product packaged for distribution and use by the general public;
4. Any substance used in a research laboratory, hospital or other medical facility under the direct supervision of a technically qualified individual;
5. Any substance used in routine agricultural operations, or as a fertilizer held for sale by a retailer to the ultimate customer.

Chemical Hazard Communication Guidebook

Solids in a manufactured item are exempt from hazardous chemical inventory reporting to the extent exposure does not occur in normal use. Non-reactive solids may be required to have an accompanying MSDS because of potential hazards of the solid's use in the workplace, but these solids would be exempt from SARA notification.

The consumer product exemption applies if the chemical is in the same form and concentration as a product packaged for distribution and use by the general public. EPA includes within this exemption not only products purchased by industry from the retailer's shelf, but also containers of products in the same form (*i.e.,* packaging and physical state) and concentration as products intended for use by the general public. The threshold issues are whether the product is packaged in the same way as a consumer product and whether the substance is in the same concentration as one intended for consumer use.

The exemption for research laboratories applies to certain chemicals used in the facility, not to the facility as a whole. It applies to those substances used in the lab under the direct supervision of a technically qualified individual. The exemption includes research facilities as well as quality control laboratories located within manufacturing facilities.

The agricultural exemption applies both to the application of a chemical in routine agricultural operations and to fertilizers in retail sales. Agricultural operations may include a wide range of growing activities such as nurseries.

Submission of MSDSs or Lists

Under Section 311 of SARA, covered employers must provide an MSDS for each hazardous chemical present at the facility above thresholds set by EPA to the state emergency response commission, the local emergency planning committee for the planning district in which the facility is located, and the fire department with jurisdiction over the facility. Alternatively, the facility can provide a list of those chemicals, grouped by hazard category. The facility must submit an MSDS before October 17, 1990, or within three months after the facility first becomes subject to the requirement, for all hazardous chemicals present at any one time in amounts equal to or greater than 10,000 pounds and for all extremely hazardous substances present at the facility in an amount greater than or equal to 500 pounds or the threshold planning quantity (TPW) whichever is lower.

If an employer chooses the MSDS rather than the list compliance option, these same state and local entities must be provided with a revised MSDS or an MSDS for a new chemical within three months of receipt of same by the employer.

The List Option

Section 311 of SARA provides that in lieu of providing copies of MSDSs, a facility may provide a list of the hazardous chemicals present at the facility, grouped by categories of hazard, and including the chemical or common name of each chemical as set forth in the MSDS. If a chemical user does not know the chemical identities of some or all of the hazardous ingredients in a product because they are claimed as trade secrets under the OSHA HCS, the facility is only required to provide the information known to it, which is typically limited to what is contained in the MSDS. If the facility chooses the list approach, the facility does not need to provide updated MSDSs to state and local authorities; it only needs to update the list with the names of new chemicals used at the facility.

The hazard categories referred to above have been defined by EPA as follows:

A. Health Hazards

1. Immediate (acute) hazards (including irritants, sensitizers, corrosive chemicals, and other hazardous chemicals that cause an adverse effect to a target organ, occurring rapidly from short-term exposure)
2. Delayed (chronic) hazards (including carcinogens and other hazardous chemicals that cause an adverse effect to a target organ, occurring as a result of long-term exposure)

B. Physical Hazards

1. Fire hazards (including flammable, combustible, pyrophoric, and oxidizing chemicals)

2. Sudden release of pressure hazards (including explosives and compressed gasses)
3. Reactive hazards (including organic peroxides)

If a chemical falls in more than one of these five categories it would have to be listed under each category. If it is not clear from a review of the MSDS in which hazard category or categories a chemical should be placed, purchasers will need to contact their suppliers for this information. Appendix C contains a comparison of these EPA hazard categories with OSHA and DOT hazard categories to facilitate the correct categorization of a chemical.

Mixtures

SARA gives a facility two options with respect to hazardous mixtures. The first option is to submit an MSDS for each mixture or to list each mixture. Most companies will find this approach to compliance to be the least burdensome. The second option is to submit an MSDS for each chemical substance in each mixture which is a hazardous chemical or to identify the common hazardous components of the mixtures on a list.

Emergency and Hazardous Chemical Inventory

Section 312 of SARA requires a covered facility to prepare an annual Emergency and Hazardous Chemical Inventory Form annually on or before March 1, with respect to the preceding calendar year for each category of hazardous substance present at the facility during the year which is in excess of the thresholds established by the Administrator. The inventory form must be sent to the appropriate local emergency planning committee, the state emergency response commission, and the fire department with jurisdiction over the facility.

The annual inventory form contains basic "Tier 1" information on the amount and general location of hazardous chemicals present at a facility, aggregated by EPA hazard category. Upon the specific request of the state commission, local committee, or fire department, a facility must also supply more detailed "Tier 2" information on specific hazardous chemicals.

SARA Title III Requirements

Content of Forms

(a) Tier 1 Information

The Tier 1 form contains the following information:

(i) An estimate (in ranges) of the maximum amount of the hazardous chemicals in each hazard category present at a facility at any time during the preceding calendar year;
(ii) An estimate (in ranges) of the average daily amount of the hazardous chemicals in each category present at a facility during the preceding calendar year; and
(iii) The general location of the hazardous chemicals in each category. To protect confidential information, Section 312 permits facilities to report inventory information by broad hazard categories and to use ranges for the quantity information. The categories are the EPA categories referred to above.

EPA has published a rule that contains the forms and instructions for reporting (CFR 40 Part 370, attached as part of Appendix B). While the agency has published a uniform format for the inventory, the rule permits a state or locality to use its own form. Since SARA has no preemptive effect, so far governments impose supplemental requirements. Sections 370.40 and 370.41 of 40 CFR state that facilities will meet the Section 312 requirements if they submit the published form, or any state or local form that contains identical information.

(b) Tier 2 Information

Under the two-tier process for reporting established by SARA, a summary of the information on covered chemicals is provided in the annual Tier 1 report "with information on specific chemicals available upon subsequent request made on a *facility-by-facility* basis." (Conference Report 99-962, 99th Cong., 2nd Sess. at 288, emphasis added). However, a facility may, if it prefers, submit Tier 2 information on specific chemicals instead of the Tier 1 report. EPA has designed the Tier 2 form so that it can serve as a worksheet for the Tier 1 form. Some states relying upon independent state regulatory authority require Tier 2 information from all covered facilities.

Chemical Hazard Communication Guidebook

If a facility receives a request for Tier 2 information, it must prepare a form for specific hazardous chemicals present at the facility containing the following information:

(i) Chemical or common name of the hazardous chemical;

(ii) An estimate (in ranges) of the maximum amount of the hazardous chemical present at the facility at any time during the preceding calendar year;

(iii) An estimate (in ranges) of the average daily amount of the hazardous chemical present at the facility during the preceding calendar year (based on the total of all daily weights divided by the number of days present);

(iv) A brief description of the manner of storage of the chemical;

(v) The location at the facility of the hazardous chemical; and

(vi) An indication of whether the facility elects to have specific location information not disclosed to the public.

The inventory information submitted under Section 312 must correspond with the MSDS information or list submitted under Section 311. Thus, if a facility elects to submit an MSDS for the hazardous components of a mixture (or a list thereof) under Section 311, it must submit inventory information for those components (to the extent such information is ascertainable) rather than for the mixtures. Alternatively, if a facility submitted an MSDS for each mixture (or a list thereof) under Section 311, it should include the total quantity of such mixtures in its Tier 1 and Tier 2 reports.

In filling out the forms for submission, whether for Tier 1 or Tier 2, the process of hazard categorization does not entail mutually exclusive classification. Each hazardous chemical is evaluated against the categories and may be listed in more than one.

To determine the maximum amount present, in the case of Tier 1, the peak weights (*i.e.*, the greatest single-day weights) of each of the hazardous chemicals falling into the category are added together, whereas in the case of Tier 2, the peak weight of the individual chemical is used. The final result is then associated with a weight range, which is entered on the form.

The average daily amount present is computed by totalling the amount of each

SARA Title III Requirements

chemical present each day and dividing by the number of days that the chemical is on-site. The number of days on site must be reported separately. For Tier 1 reporting, these average weights for chemicals in a common hazard category are then added together, whereas in Tier 2 reporting the average weight of the particular chemical is used. Reporting of average weight is also in ranges.

The general location must be provided for Tier 1, whereas in Tier 2 reporting it must be the specific location. The general location information should include the names or identifications of buildings, tank fields, lots, sheds, or other such areas. In contrast, location information on the Tier 2 form is more specific, is included on a separate form, and may be claimed confidential. The Tier 2 location information reported must be "precise," enabling emergency responders to locate the area easily. This would be the building or lot, and where practical, the room or area.

Request for Tier 2 Information

Any state or local official can request the state emergency response commission or the local emergency planning committee to obtain and make available to the official Tier 2 information for a facility. In addition, any member of the public can request Tier 2 information that is in the possession of the state commission or local committee for the preceding calendar year with respect to a specific facility. If the commission or committee has the material in its possession it must make the information available to the requesting party.

If the commission or committee does not have available the Tier 2 information requested by a member of the public for a particular facility, it must request the facility to provide such information with respect to all hazardous chemicals the facility stored in an amount in excess of 10,000 pounds at any time during the preceding calendar year. In addition, the commission or committee has the discretion to ask a facility to provide Tier 2 information on substances stored in an amount less than 10,000 pounds. However, it need not ask for information on such small quantity chemicals if the person requesting it does not establish a legitimate need for it. (A state or local official acting in an official capacity can request information on low volume chemicals without a showing of need.) Facilities are given only thirty days to respond to a request for Tier 2 information and as noted above such information can be requested for even low volume chemicals under certain circumstances.

Confidentiality

With the exception of chemical identity and precise location information, the MSDSs, lists, and Tier 1 and Tier 2 forms obtained by the state commission or local committee must be made available to the public. To claim the specific chemical identity of a chemical substance to be confidential, it is necessary to follow the procedures set forth in EPA regulations that have been issued thereunder (40 CFR Part 350, see Appendix B). If a covered facility subject to these reporting requirements does not know the specific chemical identity of the chemical, it need only provide the information it has in its possession, *i.e.*, the name of the chemical found on the MSDS.

Toxic Chemical Release Reporting

Introduction

On February 16, 1988, EPA issued its final rule establishing the regulations for toxic chemical release reporting and the uniform format for submitting the required information (40 CFR, Part 372, included in the appendix). Toxic chemical release reporting applies to owners and operators of facilities that have ten or more full-time employees, that manufacture, import, process, or otherwise use a toxic chemical in excess of certain thresholds, and that are in SIC 20-39. The thresholds have changed over the first three years of the program. For firms manufacturing, importing, or processing a listed toxic chemical, the thresholds have 75,000 pounds for the 1987 reporting period, 50,000 for the 1988 reporting period, and 25,000 for the 1989 reporting period and thereafter. The threshold for persons who use a toxic chemical at a facility is permanently set at 10,000 pounds per chemical per year. The toxic chemical release reporting forms are due annually on July 1.

Toxic Chemicals Covered

The initial list of toxic chemicals was established by SARA on the basis of chemicals

SARA Title III Requirements

included in two state programs already in effect in New Jersey and Maryland. The list constitutes over 300 chemicals and categories. A chemical may be added to the list of toxic chemicals by rule if the Administrator determines that there is sufficient evidence that the chemical is known to cause or can reasonably be anticipated to cause significant adverse acute human health effects; cancer, teratogenicity, or other serious chronic health effects; or a significant adverse effect on the environment of sufficient seriousness to warrant reporting. A chemical may be removed from the list if the Administrator determines that there is not sufficient evidence that the chemical meets any of the criteria for inclusion on the list.

Any person may petition the EPA to add or delete a toxic chemical. The agency has acted on a number of these petitions, which have resulted in a modification of the toxic chemicals subject to reporting.

A number of exemptions have been established from reporting.

First, the agency has created a *de minimis* concentration for toxic chemicals in mixtures. A facility need not consider for purposes of reporting a toxic chemical which is present in a mixture in a concentration below 1 percent, or 0.1 percent in the case of a carcinogen as defined by the OSHA HCS. However, the exemption only applies to the quantity present in the mixture. If the toxic chemical is manufactured, processed, or used other than as part of the mixture or in concentrations higher than the *de minimis* level, reporting is required if the total amount manufactured, imported, processed, or used exceeds the applicable threshold.

Second, toxic chemicals present in articles are exempt from reporting. However, the exemption does not apply to items from which there are releases of toxic chemicals. For example, the milling of a metal normally generates fume or dust containing the toxic chemical that would disqualify it from being considered an article. In contrast, a battery containing lead does not release the metal during normal use.

Third, a number of specific uses are exempt, including:

a. Use as a structural component of the facility;
b. Use of products for routine janitorial or facility grounds maintenance;
c. Personal use of items containing toxic chemicals, together with supplies of such products in a facility operated cafeteria, store, or infirmary;
d. Use of products to maintain motor vehicles operated by the facility; and
e. Use of toxic chemicals present in process water and non-contact cooling water drawn from the environment or from municipal sources, or toxic chemicals present in the air used as compressed air or for combustion.

Fourth, toxic chemicals manufactured, processed, or used at laboratories under the direct supervision of a technically qualified individual are exempt.

Facilities Covered

In its rule, EPA describes a number of different situations which can exist at a facility and defines their reporting responsibilities. The agency's definition addresses compliance by owners and operators of facilities at which there are a number of establishments under common ownership or control. If such a facility meets one of the following criteria and exceeds the employment and threshold criteria, it is subject to reporting:

a. A facility containing one or more establishments with a primary SIC within 20-39;
b. A facility containing a number of different establishments where either the total value of products shipped or produced within SIC 20-39 from the establishments is more than 50 percent of all products shipped or produced; or one establishment within the facility has a primary SIC of 20-39 and contributes more in terms of value of products shipped than any other establishment within the facility.

Once a facility is covered, all releases of listed toxic chemicals must be accounted for, even from individual establishments in the facility that fall outside of SIC 20-39.

The agency's approach in the rule is to require a determination of coverage for the whole facility, but to permit separate reporting by the separate establishments as long as all releases and waste treatment methods are accounted for. The facility must maintain consistency in whatever method is selected. Thus, if one establishment at the facility uses 5000 pounds of benzene and another uses 8000 pounds within the same facility, the facility as a whole is subject to reporting because the combined total of benzene used is over 10,000 pounds, but the establishments have the option of combined or separate reporting.

However, where the owner has only a real estate interest in the facility, the owner is exempt from reporting. This exemption applies to owners of facilities such as industrial parks, all or part of which are leased to persons who operate establishments within the covered SICs. The owner or operator of an establishment within a facility such as an industrial park does not have to concern himself with toxic chemical reporting responsibilities for the amounts of toxic chemicals present or released at other unrelated establishments.

EPA has published a list of questions and answers, regarding compliance that may be obtained from the Office of Toxic Substances (*Toxic Chemical Release Inventory Questions and Answers*, EPA 560/4-90-003).

Thresholds for Reporting

As discussed above, the thresholds for reporting with respect to manufacturing (including importing) and processing are 25,000 for 1989 and any year following. The threshold for chemical use is 10,000 pounds per chemical for the applicable calendar year.

Manufacture means to produce, prepare, import, or compound a toxic chemical. It also includes the production of coincidental toxic chemical by-products during the manufacture of another chemical. Process means the preparation of a toxic chemical after manufacture for distribution in commerce either in the same form or physical state, or in a different form or physical state from that in which it was received by the person preparing the substance, or as part of an article containing the toxic chemical. Use is defined as any use that is not manufacturing or processing, and includes use of a toxic chemical contained in a mixture or trade name product. It does not include relabeling or redistributing a container of a toxic chemical where no repackaging of the toxic chemical occurs.

Chemical Hazard Communication Guidebook

The facility must meet any applicable threshold. When the facility engages in both manufacturing and use of the chemical, the lower use threshold applies. With regard to listed categories of toxic chemicals, the threshold applies to the total volume of all chemicals falling within the category. In the case of reuse or recycling operations, the threshold applies to the quantities of toxic chemicals added to the operation during the calendar year. Finally, in some cases, the list of toxic chemicals limits reporting to a particular form or operation, such as the strong acid manufacturing process for isopropyl alcohol. The determination of the threshold applies only to those described forms of the chemical.

Toxic Chemical Release Reporting

For each toxic chemical known to be present, a covered facility must submit to EPA and to the State a completed Toxic Chemical Release Inventory Reporting Form on EPA "Form R" that contains:

a. The name and location of the facility;
b. Information regarding the use of the toxic chemical;
c. An estimate of the maximum amounts of the toxic chemical present at the facility at any time during the preceding calendar year;
d. For each wastestream, the waste treatment or disposal methods employed and an estimate of the treatment efficiency typically achieved by such methods for that wastestream;
e. The annual quantity of the toxic chemical entering each wastestream; and
f. A certification regarding the accuracy of the report.

To provide the release information, the facility must use all readily available data (including relevant monitoring data and emissions measurements) collected at the facility under other applicable regulations or as part of routine plant operations. When monitoring information is not readily available, reasonable estimates of the amounts released must be made, but no additional monitoring or measurement is required.

There are four principal methods by which the estimates may be made. The first is

SARA Title III Requirements

by the use of monitoring data, which is the most accurate, but such data will rarely be available. The second is by a mass balance calculation. A mass balance depends on a knowledge of the amounts of the chemical entering and leaving a process. The error of a mass balance estimate may be large because the amounts being "lost" or released are likely to be small in relation to the amounts of the material going in or leaving the process. Thus, a small error in the figures for input and outgo may have a large effect on the estimate of the loss. The third method of estimation depends upon the use of published emission factors relating to the type of plant and equipment being used. These factors are only accurate to the extent that the model plant or equipment from which they are drawn resembles the facility at which the estimate is being made. The final method is by engineering estimates, which are nothing more than reasonable, educated guesses about the amount of toxic chemicals being released.

The estimates are entered on the form for fugitive and stack air emissions, discharges to receiving water bodies, underground injection, and releases to land. Stormwater discharges are included if the facility has monitoring data for the chemical. In the absence of such data, the facility indicates that it has no data. In addition, estimates of the amount of the toxic chemical in waste to off-site locations (*e.g.*, a POTW) must be provided for each off-site method of disposal. Locations to which the waste is sent for recovery, recycling, or reuse of the toxic chemical are not subject to this requirement.

In addition to the information about estimated releases, the facility must indicate the waste treatment methods used on-site and estimate their efficiency. This includes for each toxic chemical: the types of wastestreams containing the chemical being reported; the waste treatment methods being used; the range of concentrations of the chemical in the influent to the treatment method; whether sequential treatment is used; the efficiency or effectiveness of each treatment method in removing the chemical; and whether the treatment efficiency was based on actual operating data.

A great number of toxic chemicals will be in the form of mixtures. In an effort to achieve the most accurate estimates from submitters, the agency has created a hierarchy of requirements based upon the specificity of information available to the facility.

a. If the facility is aware of the identity and concentration of the toxic chemical in a mixture, it must calculate the weight of the chemical in the mixture in reporting;
b. If the facility is aware of the identity of the chemical and a maximum concentration in the mixture, it must calculate the weight of the chemical assuming the maximum concentration is present;
c. If the facility knows the identity but has no information about the concentration, it is not required to factor the chemical into its threshold or release calculations;
d. If the facility does not know the specific identity of the toxic chemical, but has information on its concentration in a mixture, it is required to report the generic chemical name or trade name (if the generic name is not known) and all releases of the toxic chemical;
e. If the facility has a maximum concentration for a chemical for which it has no specific chemical identity, it must report as described above; or
f. If the facility does not know the identity or the maximum concentration of the toxic chemical, it is not required to report.

To address the difficulties of estimating the amount of a toxic chemical present in a mixture, EPA has established a supplier notification program that took effect with the first shipment in 1989 of a mixture containing a toxic chemical.

Supplier Notification

The supplier notification program applies to facilities within SIC 20-39, that manufacture or process a toxic chemical, or sell, or otherwise distribute a mixture or trade name product containing a toxic chemical. The required notification is a statement that shows: the mixture or trade name product contains a toxic chemical subject to SARA; the name of each toxic chemical and the associated CAS number; and the percent by weight of each toxic chemical in the mixture. The notification was to be sent with the first shipment of the toxic chemical in 1989. If an MSDS must be provided with such a shipment, the notification should be attached to the MSDS.

Supplier notification is not required for mixtures which are articles; foods, drugs, or cosmetics packaged for distribution to the public; or any consumer product. If the identity is considered a trade secret, a generic name is to be included; if the concentration is considered a trade secret, a maximum concentration is to be supplied.

Protecting Trade Secrets

Introduction

Although SARA requires facilities to provide large amounts of information to federal, state, and local authorities, specific chemical identity and specific chemical location are the only pieces of information which SARA permits to be withheld as confidential or proprietary business information. Even chemical identity cannot be withheld in the case of a leak or spill subject to emergency notification requirements. Furthermore, to discourage the making of baseless or frivolous claims of trade secrecy for chemical identity, SARA Section 322 requires that such claims be substantiated in some detail at the time they are made. EPA has published a rule defining the procedures for claims of trade secrecy made by facilities reporting under SARA (40 CFR Part 350, included in Appendix B).

Procedure for Making Trade Secret Claims Under SARA

(a) Emergency Planning Communications

Under the emergency planning provisions of section 303, the local committee may request information from facilities "*necessary* for developing and implementing the emergency plans" (emphasis added). Moreover, facilities that make an emergency planning notification must "promptly inform the emergency planning committee of any relevant changes occurring at such facility as such changes occur or are expected to occur."

The facility may withhold only trade secret chemical identity information in responding to the local committee. If the chemical identity is withheld, a generic hazard category must be supplied. The principal EPA health or physical hazard category in which the chemical falls (*e.g.*, fire hazard) should be stated.

(b) Emergency Release Notification

No trade secret claim can be made with respect to information submitted as part of an emergency release notification.

(c) Hazardous Chemical Reporting

SARA, like the OSHA Hazard Communication Standard, permits only chemical identity to be withheld. The procedure is the same as that described for emergency planning communications. With regard to MSDSs and lists submitted under section 311 of SARA, the separate trade secret claim, together with sanitized and unsanitized versions of the substantiation, must be submitted simultaneously to EPA when a MSDS is submitted withholding a chemical identity claimed as a trade secret under the OSHA HCS.

Because the Tier 1 report contains general information, none of the information submitted may be withheld. With regard to Tier 2 information, the specific chemical name can be withheld as a trade secret by checking the trade secret box and some or all of the location information can be claimed confidential at the discretion of the facility. If a specific chemical name is withheld, a generic hazard category should be used.

(d) Toxic Chemical Release Reporting

Only the specific chemical identity can be claimed a trade secret by checking the trade secret box on the Form R. If the identity is withheld, a generic name (that is structurally descriptive of the material) must be provided.

Substantiation Requirements

Under Section 322 of SARA, all claims of trade secrecy for chemical identity must be substantiated. A sanitized and an unsanitized version (containing the trade secret chemical identity) of the submission giving rise to the trade secret claim must be submitted to EPA together with factual information that substantiates the claim that the withheld chemical identity is a trade secret. EPA has prepared a form containing a number of questions which must be answered to document the trade secret claim. (See Appendix B.)

The form requires that the following criteria be addressed:

1. Has the chemical identity been disclosed to any person outside the company (other than a government official or person subject to a confidentiality agreement)?
2. What measures have been taken to safeguard the confidentiality of the chemical identity?
3. Has the chemical identity been disclosed or made available to the public under any federal or state law? (*e.g.*, an air or water permit application)
4. Why is disclosure of the identity of the chemical (together with other information required to be submitted) likely to cause substantial harm to the company's competitive position?
5. Is the identity of the chemical "readily discoverable" through analysis or "reverse engineering" of the product produced by the company or wastes discharged by the company?

The information in the substantiation form can itself be claimed to be confidential or proprietary business information and such a claim does not need to be substantiated. However, in that case, a sanitized version of the substantiation form must be prepared and submitted to EPA, as well as to the local committee, state commissioner, fire department, or designated state official to whom the report or submission is being sent.

Petition for Disclosure or Review of Claim

Any person may petition EPA for the disclosure of the specific chemical identity of a chemical claimed as a trade secret, and the agency may itself initiate a review.

EPA's trade secret regulations require a petition for disclosure to be accompanied by a copy of the submission in which the submitter claimed chemical identity as a trade secret. No grounds for seeking disclosure need to be given in the petition. The agency must then initiate a review of the substantiation filed by the trade secret claimant and determine within 30 days whether the substantiation presents claims that, if true, are sufficient to support a finding that the specific chemical identity is a trade secret.

Chemical Hazard Communication Guidebook

If the assertions are found to be insufficient, the trade secret claimant must be given notice and 30 days within which to supplement its assertion or appeal the determination to the EPA General Counsel. Even if the initial assertions are found by EPA to be sufficient, the trade secret claimant will be notified that he has thirty days to supplement the explanation with detailed information to support the assertions. Failure to document the assertions could lead to a determination that the trade secret claim is not adequately supported. If EPA determines that the substantiation provided is sufficient, it will notify the petitioner of that decision, and the petitioner may seek judicial review.

If EPA determines that the trade secret claimant's substantiation is not sufficient to establish that a specific chemical identity is a trade secret, EPA must notify the trade secret claimant of that determination. The trade secret claimant has thirty days within which to take an administrative appeal to EPA's General Counsel. If the General Counsel does not reverse the determination, the trade secret claimant has 30 days within which to seek judicial review.

Disclosure to Health Professionals

The specific chemical identity of a hazardous chemical, extremely hazardous substance, or toxic chemical must be provided promptly to any health professional who requests such information in writing if the health professional provides a written statement of need and a written confidentiality agreement. EPA's regulations set forth criteria that the statement of need and confidentiality agreement must meet. The confidentiality agreement may restrict the use of the information to the health purposes indicated in the written statement of need. It may also provide for appropriate legal remedies in the event of a breach of the agreement, including stipulation of a reasonable estimate of likely damages. However, the agreement may not include requirements for the posting of a penalty bond.

A facility subject to SARA must immediately provide a MSDS, inventory form, or toxic chemical release form, including the specific chemical identity, if known, of the chemical, to any treating physician or nurse who requests such information if the physician or nurse determines that:

a. A medical emergency exists;
b. The specific chemical identity of the chemical concerned is necessary for or will assist in emergency or first aid diagnosis or treatment; and
c. The individual(s) being diagnosed or treated have been exposed to the chemical concerned.

While the requested information must be provided immediately, the owner or operator disclosing the information may require the doctor or nurse to sign a written confidentiality agreement as soon as circumstances permit.

An owner or operator of a facility must also provide the specific chemical identity, if known, of a hazardous chemical, extremely hazardous substance, or toxic chemical to any health professional (such as a physician, toxicologist, or epidemiologist) who is a local government employee, or under contract with a local government, and who provides a written statement of need and a written confidentiality agreement.

Enforcement

Emergency Planning Notification Requirements

To enforce the emergency planning notification requirements of SARA, the Administrator of EPA may issue a compliance order to a facility owner or operator. If a facility fails to obey a compliance order EPA can bring suit in federal district court for a civil penalty of not more than $25,000 for each day in which such failure to comply continues.

Emergency Release Notification Requirements

(a) Civil

Failure to make a timely notification of the release of an extremely hazardous substance, or hazardous substance requiring an emergency notification under SARA, renders a facility subject to a civil penalty to be assessed by the Administrator of EPA of not more than $25,000 per day for each day during which the violation continues. In the case of a second or subsequent violation, the amount of the penalty escalates to not more than $75,000 for each day the violation continues.

Chemical Hazard Communication Guidebook

(b) Criminal

Any person who "knowingly and willfully" fails to provide an emergency release notification can, in addition, be fined not more than $25,000 or imprisoned for not more than two years, or both. In the case of a second or subsequent conviction, the penalty escalates to a fine of up to $50,000, imprisonment for not more than five years, or both.

Reporting Requirements

Any facility owner who violates any requirements of Section 311 of SARA (concerning reporting of MSDSs or lists of hazardous chemicals) can be assessed a civil penalty not to exceed $10,000 for each violation. Any covered facility that fails to comply with the reporting requirements of SARA Section 312 (concerning emergency and hazardous chemical inventory forms) or Section 313 (concerning toxic chemical release forms) is liable for a civil penalty not to exceed $25,000 for each such violation. Each day any of the above violations continues constitutes a separate violation.

Trade Secrets

If the EPA Administrator determines that a trade secret claimant has made a frivolous trade secret claim, the trade secret claimant can be held liable for a civil penalty of $25,000 per claim. Failure to submit substantiation for a trade secret claim to EPA and failure to disclose specific chemical identity to a health professional in an emergency can result in a civil penalty of up to $10,000 per day of violation being imposed.

Any person who "knowingly and willfully" divulges or discloses any trade secret information entitled to protection under Section 322 shall, upon conviction, be subject to a fine of not more than $20,000, or to imprisonment not to exceed one year, or both. Any EPA employee or other federal official who discloses or "makes known in any manner" any information that concerns or relates to trade secrets shall be fined not more than $1,000, or imprisoned for not more than one year, or both, and shall be removed from federal employment (18 USC Section 1905).

SARA Title III Requirements

Enforcement of Section 323 by Health Professionals

Whenever any facility owner or operator required to provide information to a health professional fails or refuses to provide such information, the health professional may bring an action in federal court to require the facility owner or operator to provide such information.

Enforcement by Citizen Suits and State or Local Suits

SARA authorizes citizen suits in federal court against owners or operators of a facility for failure to comply with specified provisions of the Act. It also authorizes a state or local government to commence a civil action in federal district court against an owner or operator of a facility for failure to comply with the Act. In any actions against an owner or operator of a facility, the district court has jurisdiction to enforce the requirement concerned and to impose the civil penalty provided for violation of that requirement. SARA provides a significant incentive for citizens to bring such suits by authorizing the court to award costs and attorney's fees to a substantially prevailing party.

State Enforcement

As discussed above, many states have their own emergency planning and right-to-know requirements that also include enforcement and penalty provisions. In New Jersey, for example, failure to notify the state Department of Environmental Protection of an emergency release can result in civil penalties of not more than $25,000 per day. (NJSA 58:10-23.11e, .11u(a).) It is therefore important for each facility to determine if there are state and/or local reporting requirements in addition to those mandated by CERCLA/SARA.

Compliance Planning

Enforcement proceedings can be expensive to defend. The potential monetary penalties are high and any penalties assessed cannot be deducted as a business expense for tax purposes. It is therefore essential for any firm subject to SARA's requirements to devise and implement a carefully thought-out compliance program. Merely adopting such a

Chemical Hazard Communication Guidebook

program is not sufficient, however, to guarantee compliance. It is also necessary to conduct periodic audits to verify that the compliance program is actually being followed and not simply gathering dust in someone's drawer. Periodic training of appropriate employees on the requirements of SARA and the firm's compliance program is also important. Finally, consult promptly with counsel knowledgeable about administrative enforcement proceedings if you encounter a compliance problem.

Compliance Checklist

SARA TITLE III

1. Immediately determine whether any chemicals present at the facility are extremely hazardous or hazardous chemicals subject to reporting if released to the environment, and determine for each chemical the Reportable Quantity that triggers such a reporting requirement. In the event of a leak or spill of more than the Reportable Quantity of such a chemical into the environment, immediately contact the National Response Center (800-424-8802 or 202-267-2675) and the appropriate state commission and local committee emergency response coordinators. As soon as practicable after emergency telephone notification, provide a written follow-up notice to the state commission and local committee.

2. Determine if the amount of any extremely hazardous substance present at the facility exceeds the Threshold Planning Quantity for that substance. If any chemical is above the threshold, notify the state emergency response commission and the local emergency planning committee that the facility is subject to the emergency planning requirements.

3. Set up a procedure to obtain the quantity and location data required to complete a Hazardous Chemical Inventory Form for hazardous chemicals present at the facility in greater than applicable threshold amounts.

4. By October 17, 1990, or within three months of first becoming subject to the requirement, submit an MSDS for each hazardous chemical present at the facility in greater than applicable threshold amounts to the state emergency response commission, the local emergency planning committee, and the fire department with jurisdiction over the facility. Alternatively, submit by that date

SARA Title III Requirements

a list of the hazardous chemicals present at the facility in greater than applicable threshold amounts, grouped by EPA hazard category.

5. Set up a procedure to ensure that any updated MSDSs or new MSDSs are submitted to the appropriate state and local entities within three months of receipt of the new or revised MSDS.
6. Categorize the chemicals present at the facility using the EPA hazard categories. Suppliers may need to be contacted for assistance.
7. By March 1 of each year, prepare the Hazardous Chemical Inventory form by filling in the requested information for each hazard category and submit it to the state commission, local committee, and fire department. Contact the applicable state office for special state requirements or instructions.
8. By March 1 of each year, develop for each trade secret chemical, the identity of which the company desires to keep confidential, the necessary substantiation required by Section 322 of SARA for submission to EPA. (Note that only trade secret chemical identity information can be withheld from the state commission and local committee.)
9. Determine if toxic chemical release reporting applies to your facility by reviewing whether your facility, or an establishment within your facility, falls within SIC 20-39 with more than ten full-time employees, and determine whether you have present toxic chemicals above the specified thresholds. If so, the facility must prepare a toxic chemical release report and submit it to EPA and a designated state official by July 1 of each year.
10. Establish a procedure to monitor changes to the list of toxic chemicals, which may be modified by agency action.
11. Establish a procedure to comply with the toxic chemical supplier notification program applicable to companies manufacturing or distributing toxic chemicals. Covered suppliers must notify customers within SIC 20-39.

Chapter Four

HAZARDOUS MATERIALS TRANSPORTATION

This chapter provides a general overview of the regulation of hazardous materials in domestic transportation by the Department of Transportation (DOT) and outlines the significant changes which have been proposed to bring the U.S. system into conformance with international systems of regulation. The chapter addresses:

a. The Hazardous Materials Transportation Act and general structure of the regulations;
b. Hazardous materials covered;
c. General overview of the shipper's responsibilities;
d. Shipment of EPA hazardous substances;
e. Incident reporting for spills in transportation; and
f. Proposed changes to the DOT system of regulation.

The discussion of DOT's existing regulations in the first three sections is intended to provide a summary of the key features of these regulations for a reader unfamiliar with the agency's requirements.

The shipment of hazardous substances has been the subject of significant change and some confusion as DOT has carried out the provision, adopted as part of SARA, requiring DOT to list and regulate as "hazardous materials" the transportation of "hazardous substances" defined by EPA under CERCLA. Appendix B lists the EPA hazardous substances and their reportable quantities, which have been incorporated as an appendix to the Hazardous Materials Table by DOT.

Chemical Hazard Communication Guidebook

The discussion of incident reporting for spills in transportation in Section E focuses on the requirements imposed by DOT. The reader may wish to refer to the related topic of spill reporting under CERCLA/SARA, which is considered in Chapter 3.

The final section outlines the principal changes to the existing regulations soon to be finalized by DOT in order to put into effect performance-oriented packaging standards that have been developed as Recommendations by the United Nations Committee of Experts on the Transport of Dangerous Goods, which, in this area, is one of the primary specialized groups working on hazardous materials standards. If the proposed rules are finalized in 1991, which is likely, the regulation of the transportation of hazardous materials will undergo significant changes the new system comes into force.

The Hazardous Materials Transportation Act and General Structure of the Regulations

The Hazardous Materials Transportation Act (HMTA) is the principal statute governing interstate transportation of hazardous materials. Passed in 1975, the Act provides the Secretary of Transportation broad authority to regulate hazardous materials. In particular, the Act:

a. Gives DOT jurisdiction over any traffic of hazardous materials affecting interstate commerce;
b. Authorizes the designation of hazardous materials as substances or materials in a quantity and form which may pose an unreasonable risk to health and safety or property when transported in commerce;
c. Authorizes DOT to issue regulations related to packing, repacking, handling, labeling, marking, placarding, and routing as well as the manufacture, testing, maintenance, and reconditioning of containers or packages used to transport hazardous materials;
d. Authorizes the establishment of a registration program for shippers, carriers, and container manufacturers and reconditioners;
e. Establishes a procedure for granting exemptions where it can be demonstrated that a level of safety will be achieved that equals or exceeds that achieved by regulation;

f. Allows DOT to assess civil and criminal penalties for violations; and

g. Preempts any state or local law inconsistent with the Act, and establishes a procedure allowing DOT to waive preemption.

It should be noted that DOT also specifically regulates intrastate commerce in several respects through the regulation of hazardous wastes and hazardous substances defined by the Environmental Protection Agency, and flammable cryogenic liquids in portable tanks and cargo tanks.

DOT has promulgated voluminous regulations covering the transportation of hazardous materials that implement the HMTA. The regulations are contained in Title 49 of the Code of Federal Regulations.

a. Part 106 establishes the procedures for rulemaking employed by DOT's Research and Special Programs Administration (RSPA) that has primary responsibility for the development of hazardous materials transportation regulations. The Office of Hazardous Materials Transportation has the central responsibility within RSPA. These functions are coordinated with other "modal" administrations, which include the Federal Aviation Administration, the Federal Highway Administration, the Federal Railroad Administration, and the U.S. Coast Guard;

b. Part 107 contains procedures for the submission and review of exemption applications to permit alternative transportation methods with an equivalent level of safety, and it also establishes the procedures by which DOT will determine that a state or local regulation is preempted by HMTA;

c. Part 171 contains the definitions used in the regulations and, among other things, includes the regulations for incident reporting; Part 172 contains the Hazardous Materials Table, which lists the hazardous materials and hazard classes subject to regulation, and establishes the detailed hazard communication requirements for labels, shipping papers, marking, and placarding;

e. Part 173 regulates the types of packaging that may be used by shippers of hazardous materials. It also defines specific requirements for certain hazard classes;

Chemical Hazard Communication Guidebook

    f. Part 174 covers rail transportation. It sets forth general handling and loading requirements as well as detailed requirements for certain classes of hazardous materials;

    g. Part 175 provides specific regulations for transportation of hazardous materials by aircraft;

    h. Part 176 addresses the regulation of transportation of hazardous materials by waterborne vessels;

    i. Part 177 contains regulations specific to the transportation of hazardous materials by highway;

    j. Part 178 covers detailed specifications for the manufacture and testing of packagings; and

    k. Part 179 provides detailed specifications for rail tank cars.

Although the requirements of Parts 172 and 173 are of the most immediate concern to shippers, it is necessary to understand the overall structure of DOT's rules because particular requirements applicable to shippers may be included within the modal carrier regulations of Parts 176, 177, and 178. For example, although Subpart C of Part 172 establishes the general rules for shipping papers, Section 177.817 defines particular requirements for shipping papers accompanying hazardous materials transported by highway.

Hazardous Materials Covered

Under HMTA, the Secretary of Transportation may designate as a hazardous material any material, or group or class of materials, that may pose an unreasonable risk to health and safety or property. At present, DOT has defined a number of classes of hazardous materials that are regulated. These are:

    a. Explosives and blasting agents;
    b. Flammable, combustible, and pyrophoric liquids;
    c. Flammable solids, oxidizers, and organic peroxides;
    d. Corrosive materials;
    e. Compressed gases;

f. Poisonous materials, etiologic agents and radioactive materials; and
g. Other regulated material (ORM A, B, C, D, and E).

An important obligation of the shipper is to properly classify the hazardous material being transported. The Hazardous Materials Table indicates the hazard class to which a material belongs.

An explosive is any chemical compound, mixture, or device the primary or common purpose of which is to function by explosion unless such compound, mixture, or device is otherwise classified. Explosives are divided into three subgroups: Class A for detonating explosives, Class B for those which function by rapid combustion rather than detonation, and Class C for manufactured articles such as small arms ammunition and fireworks. An example of a Class A explosive would be a blasting cap or fulminate of mercury. An example of a Class B explosive would be a jet thrust unit (iato).

A flammable liquid is any liquid that has a flashpoint below 100 degrees Fahrenheit (for example, acetone, ethyl alcohol, and gasoline). A combustible liquid is any liquid having a flashpoint at or above 100 degrees Fahrenheit and below 200 degrees Fahrenheit (for example, methyl amyl ketone or fuel oil). A pyrophoric liquid ignites spontaneously in dry or moist air at or below 130 degrees Fahrenheit, such as aluminum alkyl.

In contrast, a flammable solid is any solid material other than an explosive liable to cause fires through friction or retained heat from manufacturing or processing or which can be ignited readily creating a serious transportation hazard because it burns vigorously and persistently. Metallic aluminum powder and phosphorous are included in this class. An oxidizer is defined as a substance such as a chlorate, permanganate, inorganic peroxide, or a nitrate that yields oxygen readily to stimulate combustion. An organic peroxide is defined as an organic compound containing the bivalent -O-O structure and which may be considered a derivative of hydrogen peroxide where one or more of the hydrogen atoms have been replaced by organic radicals; however, certain exceptions are provided for. Benzoyl peroxide is an example of a substance classed in this group.

A corrosive material is defined by its effect on skin or by its effect on steel.

Chemical Hazard Communication Guidebook

Hydrochloric acid is a corrosive material.

Compressed gases include any material having a container absolute pressure exceeding 40 psi at 70°F, or, regardless of the pressure at 70° having an absolute pressure exceeding 104 psi at 130°F, or any liquid flammable material having a vapor pressure exceeding 40 psi absolute at 100°F.

DOT has established three classes of poisonous materials. The class of Poison A materials includes those most dangerous to life, and comprises a list of thirteen that are specifically included (for example, phosgene). Poison B materials are substances, liquids, or solids other than Poison A or irritating materials that are known to be toxic to humans in tests specified for oral, dermal, or inhalation toxicity (for example, strychnine). An irritating material is one that when on contact with fire or exposed to air gives off dangerous or intensely irritating fumes (for example, tear gas).

Etiologic agents are infectious materials that are defined by the Department of Health and Human Services (for example, polio virus and salmonella).

Radioactive materials spontaneously emit ionizing radiation and have a specific activity greater than .0002 microcuries per gram (for example, uranium). Although the shipper may believe it unlikely that he or she will handle radioactive materials, a wide variety of testing devices found in a manufacturing facility or laboratory may contain radioactive materials, requiring a familiarity with the applicable shipping requirements.

DOT has also established five classes for "other regulated materials" (ORM). An ORM is a material that may pose an unreasonable risk to health and safety or property when transported, and does not meet any of the definitions of the other hazard classes, or has been specifically reclassified. The ORM class was created in 1976 when DOT revised its regulations, and ORM A, B, C, and D subsumed materials classifications used by the Federal Aviation Administration and the Coast Guard prior to consolidation. In the case of ORM-E, these materials were initially added in 1980 as a result of the passage of CERCLA.

ORM-A materials have anesthetic, irritating, noxious, or toxic properties that can cause extreme annoyance or discomfort to passengers and crew in the event of a leak. A number of chemicals are listed, and include, for example, chloroform.

ORM-B materials are those capable of causing significant damage if a leak occurs. This class includes listed materials corrosive to aluminum, such as calcium oxide or potassium fluoride.

ORM-C materials have other inherent characteristics not described in ORM-A or ORM-B, that make them unsuitable for shipment unless properly identified and prepared for transportation. The regulations include a list of covered substances and include cotton and inflatable life rafts, for instance.

ORM-D are consumer commodities that although otherwise subject to regulation, are transported in a form, quantity, and packaging that presents a limited hazard.

The definition and regulation of ORM-E materials has been the subject of considerable confusion. These materials incorporate by reference hazardous wastes and hazardous substances not elsewhere specified that are regulated by EPA under the Resource Conservation and Recovery Act or CERCLA/SARA. They are discussed below in greater detail.

General Overview of the Shipper's Responsibilities

In shipping a hazardous material, the shipper should undertake the following principal tasks:

   a. Determine the proper name and hazard classification of the material;
   b. Select the appropriate identification numbers;
   c. Determine the mode(s) to be used and any supplemental requirements that apply;
   d. Determine the package or container that must be used;
   e. Label, mark, and placard to conform with DOT's requirements;
   f. Prepare shipping papers; and
   g. Ensure awareness and understanding of regulatory responsibilities by employees.

Chemical Hazard Communication Guidebook

The discussion below provides an overview of these responsibilities.

The first step in determining the shipper's responsibilities for the transportation of a hazardous material is to determine its proper shipping name and classification. The regulations prohibit the transportation of hazardous materials that have not been properly named and classified. Shipping names as well as hazard classes are listed in the Hazardous Materials Table, Section 172.101.

Where a material is not listed in the Hazardous Materials Table, the shipper must evaluate it against the defined hazard classes, which vary from highly specific and prescriptive tests to general categories. The guidelines for selecting the proper shipping name are found in Section 172.101(c) (See Appendix B).

If a material has more than one hazard class to which it belongs, the regulations establish the following hierarchy of hazards for classification:

- Radioactive material (except a limited quantity);
- Flammable gas;
- Non-flammable gas;
- Flammable liquid;
- Oxidizer;
- Flammable solid;
- Corrosive material (liquid);
- Poison B;
- Corrosive material (solid);
- Irritating materials;
- Combustible liquids (in containers having capacities exceeding 110 gallons);
- ORM-B;
- ORM-A;
- Combustible liquid (in containers having capacities of 110 gallons or less);
- and ORM-E.

The hierarchy does not apply to certain explosives, etiologic agents, and organic peroxides.

It should be noted that where the material has been described by a group designation, such as "n.o.s" (not otherwise specified), the proper shipping name designation may include the technical name of the chemical in parentheses. An example would be the requirements for water transport of an exported material: "Corrosive liquid, n.o.s. (Caprylyl chloride), UN 1760."

In addition, the shipper must determine the proper identification number of the material, which is also found in the Hazardous Materials Table. It must correspond to the shipping name and hazard class appropriate to the material being transported.

The shipper, after selecting the mode of transport, must determine if there are supplemental requirements which apply to that mode. The material may be forbidden aboard aircraft, for example.

The shipper should determine if any exemptions may apply, which will simplify the steps necessary before shipment. For example, DOT has established exceptions for small quantities of hazardous materials being shipped (see 49 CFR §173.4). Exceptions may also exist for consumer commodities being shipped as ORM-D materials (defined in 49 CFR §1200), which, due to their form, quantity, and packaging, present a limited hazard during shipment. The Hazardous Materials Table (see Appendix B) specifies the materials that can be considered as ORM-D, which, if applicable, may relieve the shipper from the packaging, labeling, placarding, and documentation requirements that would otherwise be imposed.

The next major task for the shipper is the selection of the proper packaging. DOT regulations require that the hazardous material be shipped in the appropriate package. This is the shipper's responsibility; however, when a package or container is required to be marked with a DOT specification by the manufacturer, the marking certifies compliance with the DOT requirement and assures the shipper that he has met the packaging standard. If the container is reused, the original manufacturer is not responsible for certifying that the packaging is in compliance.

The selection of the packaging depends on a knowledge of the hazard class of the material, the proper shipping name, and the mode of transport. The Hazardous Materials

Chemical Hazard Communication Guidebook

Table in Appendix B references the packaging restrictions and exceptions that apply to listed hazardous materials. In determining the packaging, the shipper must fulfill any particular requirements that apply to the material, such as a poison, and any packaging requirements that apply to the mode selected.

Most of the packaging specifications are highly detailed. As new materials with hazardous characteristics have been developed, the number of specifications for packages has grown and presently occupies over 400 pages in the Code of Federal Regulations. Indeed, the rigidity of these standards in discouraging innovative packaging, which may now only be used by exemption if it does not meet the terms of the specification, has been one of the arguments of proponents for changing the present U.S. system of regulation.

Each package used for shipping hazardous materials must be designed and limited in such a way that there will be no significant release of the hazardous materials to the environment. DOT regulations prohibit the shipment of leaking containers of hazardous materials.

The next major requirement for the shipper is the selection of an appropriate label. DOT requires that the label contain the requisite wording and numbers, and be of the prescribed dimensions, colors, and shapes. In general, the label indicates the hazard class of the material being shipped. In some cases, additional labeling requirements may apply, for example, when a material is dangerous when wet. Further, labeling requirements may be specific to a particular mode of transportation, such as a "CARGO AIRCRAFT ONLY" label (see Section 172.402).

Another major responsibility of the shipper is marking the shipment in the appropriate manner. Marking serves to identify the contents of a package if it is separated from its shipping papers. DOT has established marking requirements for packages, freight containers, and transport vehicles. (See **Figure 1** on page 87.) Markings include the proper shipping name and identification number, and the consignee's or consignor's name and address. Additional marking requirements may apply, such as orientation markings ("THIS END UP") and marking of ORM.

Hazardous Materials Transportation

The shipper must also determine the necessary placards, which are symbols placed on the ends and sides of motor vehicles, rail cars, and freight containers to indicate the hazards of the cargo. (See **Figure 2** on page 88.) These placarding requirements may be particular to the material; for example, when transporting a substance deemed to be an inhalation hazard, the transport vehicle must be placarded "POISON" on each side and each end in addition to the placards necessary by the general requirements. Certain exceptions exist to the placarding requirements, for example, in the case of limited quantities shipped.

DOT has established requirements for the preparation of shipping papers accompanying hazardous materials being transported. The shipping paper must include the proper shipping name, the hazard class, the identification number, the total quantity, and a description of the material. Importantly, the shipper must certify that the shipment is classified, described, packaged, marked, and labeled, and that the material is in a proper condition for transportation in accordance with DOT requirements.

The documentation requirements are varied, with special requirements for the mode, type, condition, hazards, and packaging of the material. For example, each shipment by water must have as additional shipping paper entries the identification of the type of packages, the number of each type of package, and the gross weight of each type. With regard to types of materials, additional information is required for poisonous materials if the proper shipping name does not include the name of the poisonous constituent. When a hazardous material is dangerous under specified conditions, this information must be included in the shipping paper, such as dangerous when wet. Similar requirements apply to inhalation hazards. Finally, the packaging itself may dictate the necessity of information to be included on the shipping paper. An example of this is the transportation of empty packages containing a residue of the material for which the shipping paper would describe what the package last contained.

The shipper must ensure that employees are properly trained and aware of DOT's regulations. The failure to instruct each officer, agent, and employee regarding the Hazardous Materials regulations is a violation (see 49 CFR 173.1). Although DOT is developing guidelines that may render the requirement more specific, at present it is unlike the companion requirement in OSHA (described in Chapter 2) that spells out the responsibilities of the employer in training employees. In seeking to comply with this requirement, the shipper may wish to contact a trade organization in this area to assist in locating a suitable training program. A list of concerned associations appears at the end of this chapter.

In evaluating the adequacy of the training program, the shipper should consider the breadth of the subjects covered, the experience of the instructors, whether tests of employees are given (followed by spot checks), and the frequency of refresher material as changed circumstances dictate.

Shipment of EPA Hazardous Substances

The confusion that has surrounded the regulation by DOT of hazardous substances defined by EPA warrants a special consideration of the issue. When CERCLA was passed, it provided that each hazardous substance listed or designated by EPA would become a hazardous material under the Hazardous Materials Transportation Act (see Section 306 of CERCLA). This led to the establishment by DOT of ORM-E materials.

In the legislative consideration of SARA, an amendment was adopted to this provision in CERCLA, that required DOT to list "and regulate" within thirty days of passage as hazardous materials those hazardous substances defined under section 101(14) of CERCLA. This prompted DOT to issue a final rule.

Definition of Hazardous Substances

Hazardous substances under CERCLA are defined broadly and include:

a. Any substances designated pursuant to Section 311(b)(2)(A) of the Clean Water Act;
b. Any element, compound, mixture, solution, or substance designated pursuant to section 102 of CERCLA (*i.e.*, those substances additionally designated by the Administrator which, when released into the environment, may present sub-

stantial danger to the public health or welfare, or the environment;
c. Any hazardous waste having the characteristics identified under or listed pursuant to section 3001 of the Resource Conservation and Recovery Act (RCRA);
d. Any toxic pollutant listed under section 307(a) of the Clean Water Act;
e. Any hazardous air pollutant listed under Section 112 of the Clean Air Act; and
f. Any imminently hazardous chemical substance or mixture with respect to which the Administrator has taken action pursuant to Section 7 of the Toxic Substances Control Act.

The term does not include petroleum, including crude oil or any fraction not otherwise designated as a hazardous substance under any provision above. It does not include natural gas, natural gas liquids, liquefied natural gas, or synthetic gas usable for fuel (or mixtures of such). Although DOT had originally removed this exception in its final rule, it amended it on July 1, 1987, to reinstate the exception (52 Fed. Reg. 24474).

Importantly, the definition of hazardous substances encompasses both listed and unlisted substances. In particular, RCRA considers as hazardous wastes that possess one of the four characteristics of ignitibility, corrosivity, reactivity, or EP toxicity. CERCLA incorporates these wastes as hazardous substances.

EPA has established for a number of these substances a reportable quantity which, if released, triggers the emergency notification provisions of CERCLA/SARA. A reportable quantity of one pound applies under the statute in the absence of a designation by EPA.

DOT revised its regulations by establishing an appendix to the Hazardous Materials Table for hazardous substances that are regulated only by virtue of their regulation by EPA under CERCLA. In doing so, DOT removed the special notations in the Hazardous Materials Table of "E" and "RQ." Where a material is both designated as having a DOT hazard class and regulated by EPA as a hazardous substance, it is listed in both the Appendix and the Table, but without special notation.

Chemical Hazard Communication Guidebook

DOT Regulation of Hazardous Substances

In brief, DOT's regulations specify that if a reportable quantity of a hazardous substance is present in a shipment (*i.e.*, one package contains a reportable quantity or more of the designated material) the shipping paper entry must contain the notation "RQ." When the proper shipping name does not contain the name of the constituents that make the material a hazardous substance, that information must be added in parentheses. These requirements for hazardous substances supplement the general obligations of the shipper described below to properly identify, classify, mark, label, package, and placard in a manner consistent with the regulations.

When the hazardous substance is in the form of a mixture or solution, DOT provides a way, in **Table 2** below, to determine whether the mixture is in a concentration by weight that equals or exceeds the concentration corresponding to the RQ of the material.

Table 2

Correspondence of RQ's to Concentrations By Weight
(49 CFR Section 171.8)

Concentration By Weight

RQ (pounds)	Percent	PPM
5000	10	100,000
1000	2	20,000
100	.2	2,000
10	.02	200
1	.002	20

In the case of hazardous wastes, DOT requires the use of the EPA waste number instead of the entire narrative wastestream description, and the waste number must be provided parenthetically with the proper shipping name. Further, if the hazardous substance is covered because it meets the broad criteria of ignitibility, corrosivity, reactivity, or extraction procedure (EP) toxicity defined by EPA for hazardous wastes, DOT requires that the shipping paper include the notation "EPA" followed by the word "ignitibility," "corrosivity," "reactivity," or "EP toxicity."

The procedure for marking non-bulk packagings (those of 110 gallons or less) is to include the notation "RQ" when a hazardous substance is present. If the proper shipping name does not include any constituent(s) which causes the material to be so designated, the name of the constituent must be added. In a similar fashion to the requirement described above for shipping papers, the notation "EPA" followed by the "ICRE" characteristic is required for hazardous substances that are hazardous wastes.

The shipper should be aware of the continual changes likely in the list of hazardous substances and their corresponding reportable quantities by EPA, which will, in turn, necessitate a change in their regulation by DOT.

Incident Reporting for Spills in Transportation

The spill reporting requirements discussed in Chapter 3 were imposed through CERCLA/SARA and apply not only to stationary facilities, but also to spills occurring during transportation. In addition to these requirements, DOT has established regulations under HMTA regarding spills of hazardous materials. Because of the incorporation of hazardous substances as hazardous materials within DOT regulation, spill reporting under the DOT requirements has broad scope.

The reporting system is two-fold, encompassing a telephone report to provide immediate notice and a written report. Under these regulations (49 CFR 171.15-17), any carrier is responsible for reporting by telephone to the National Response Center at the earliest practical moment any incident that occurs during transportation—including loading, unloading, or temporary storage of hazardous materials in which, as a direct result of transportation:

   a. A person is killed;
   b. Estimated carrier or property damage exceeds $50,000;

Chemical Hazard Communication Guidebook

    c. Evacuation of the public lasts for more than one hour;
    d. One or more major transportation facilities or arteries is closed for more than one hour;
    e. The operational flight plan or routine of an aircraft is altered;
    f. Fire, breakage, or spillage of radioactive materials or etiological agents; or,
    g. The carrier judges that the situation should be reported.

The immediate notice is given to the National Response Center (NRC) (phone 800-424-8802) and includes:

    a. The name of the reporter;
    b. Name and address of carrier represented by reporter;
    c. Phone number where the reporter can be contacted;
    d. Date, time, and location of the incident;
    e. The extent of injuries, if any;
    f. Classification, name, and quantity of hazardous materials involved, if such information is available; and
    g. Type of incident and nature of hazardous material involvement and whether a continuing danger to life exists at the scene.

The information required by DOT duplicates much of the information required under CERCLA/SARA for spill reporting, with the only additional element being the information specifying the carrier's identity and address.

A detailed hazardous materials incident report is required in the event of a hazardous or unintentional release of hazardous material. Each carrier who transports hazardous materials must report in writing to DOT on DOT Form F 5800.1 (see **Figure 3**) within thirty days of an incident that occurs during the course of transportation (including loading, unloading, or temporary storage) in which as a direct result of the hazardous materials any of the events necessitating an immediate report occur, or there has been an unintentional release of hazardous materials from a package or tank, or any discharge of hazardous waste. If the report involves a hazardous waste, the report must include a copy of the hazardous waste manifest and an estimate of the quantity of the waste removed from the scene, the name and address of the facility to which it was taken, and the manner of disposition of any unremoved waste.

The incident report is provided by the carrier to the Information Systems Manager, Research and Special Programs Administration, Department of Transportation, Washington, DC 20590.

In sum, the shipper should be aware of the DOT requirements for spill reporting applicable to the carrier that encompass incidents involving the loading or unloading of hazardous materials as well as those occurring in transportation. The information required by DOT's regulations overlaps the information required to be reported under CERCLA/SARA with regard to serious incidents where an immediate report must be given. In addition, DOT provides for a written incident report with certain exceptions. CERCLA/SARA spill reporting has no parallel exceptions equivalent to those established by DOT.

Proposed Changes to the DOT System of Regulation

In 1982, DOT issued an advance notice of proposed rulemaking to establish performance standards for the packaging of hazardous materials in lieu of the specification standards of the present system. The performance standards proposed by DOT are based on the United Nations Recommendations for the Safe Transport of Dangerous Goods, which, in general, divide hazardous materials into three "Packing Groups" depending on their relative hazards. Packing Group I consists of very dangerous materials, such as fuming sulfuric acid. Packing Group II involves materials presenting a moderate degree of danger, such as hydrochloric acid. Packing Group III addresses materials presenting only a minor danger. The UN Dangerous Goods standards also have general requirements for materials, construction, and maximum size. It specifies tests to be met by packages in each group.

One of the reasons prompting the proposed changes in the U.S. system is to conform to international standards. At present, there are two international modal hazardous material regulatory systems in place, one under the auspices of the International Maritime Organization (IMO), and the second under the International Civil Aviation Organization

(ICAO). The IMO system is known as the International Maritime Dangerous Goods Code (or IMDG Code) and the ICAO system is known as the ICAO Technical Instructions for the Safe Transport of Dangerous Goods by Air. Shipments of hazardous materials (both international and within the U.S.) are at present authorized, with certain exceptions, to be shipped under the ICAO system as fully equivalent to DOT hazardous materials regulations where the transportation involves aircraft or motor vehicle only.

In consequence, two systems exist within the U.S., one characterized by very precise specifications for the construction of hazardous materials packagings, as well as the performance-oriented standards used in the international codes.

Another reason behind the proposed changes is the argued dampening that the present system has on innovative packaging. The rigidity and complexity of the present system are evidenced by the fourteen detailed specifications for wooden boxes. Most of these list each acceptable type of wood for construction, the thickness and width of the boards, kinds and dimensions of nails, as well as the spacing of nails used in joining the box.

In 1987, DOT issued a major notice of proposed rulemaking (HM-181) revamping the current system in order to establish performance-oriented packaging standards. A final rule is expected soon.

The shipper faces the necessity of understanding not only the present system, but also the potential impact of the proposed rule and its related dockets on the way in which the transportation of hazardous materials is regulated.

In very brief summary, HM-181 would:

a. Consolidate the hazardous materials tables into one, with conforming changes to eliminate packaging specifications;
b. Numeric hazard classes would replace the present textual description of hazards and be conformed to the UN system;
c. Most hazardous materials would be assigned to packing groups to indicate their relative degree of hazard;

d. Modifications would be made to align the regulations with the ICAO or IMDG provisions, for example, in the case of per-package quantity limitations;
e. Shipping paper requirements would be modified to conform to UN terminology;
f. Marking requirements would be revised;
g. Technical names would be required as part of the package marking for "n.o.s" entries on all non-bulk packages, and on those bulk packages for which proper shipping name markings are required;
h. Requirements applicable to materials presenting an inhalation hazard would be expanded;
i. Labeling requirements would be modified, as would placarding requirements;
j. Definitions of hazard classes would be modified;
k. Packagings for most materials would be authorized by packing group and physical form of a material rather than by hazard class. Arrays of performance-oriented packaging standards would be adopted. Bulk and non-bulk packaging would be affected; and
l. Notice and recordkeeping requirements would be imposed.

The impact of the proposal, when finalized, will be substantial. The greatest impact will be on shipping papers, marking, and placarding used by the shipper. Additional information will be required, such as technical names, and a familiarity with the numeric hazard ranking in lieu of the present systems will become necessary.

Another change will involve packaging and testing requirements. A more flexible approach will be adopted, but with it, a greater responsibility will be demanded in demonstrating that the packaging met the applicable performance-oriented requirements in the event of failure.

Finally, the present policies and procedures used by companies in carrying out their business, and in training their employees, will undergo modification. This may be especially difficult during the transition period, when employees may have to be familiar with the present DOT system, the new DOT system, and perhaps other international standards. The significant changes about to occur in the regulation of hazardous material transpor-

Chemical Hazard Communication Guidebook

tation will have a near-term effect as shippers seek to understand the nature of the proposed changes, an effect in the medium term as a transition occurs and, in the long run, as a new regulatory system is imposed.

Compliance Planning

The following checklist summarizes steps that a shipper of hazardous materials should take:

1. Review the list of hazardous materials being transported;
2. Ensure that employees with responsibilities related to the shipment of hazardous materials are familiar with the regulations that apply. The employer should: identify all personnel who have such responsibilities; determine what additional instruction or training each needs (**Table 3** on page 86 lists associations involved in compliance training); assure that employees receive training; maintain a record of training; and periodically review training needs in order to maintain a thorough knowledge of the regulations that apply to the jobs being carried out by employees;
3. Coordinate the plan for compliance with the OSHA Hazard Communication Standard (see Chapter 2) with procedures taken to comply with DOT's Hazardous Materials Transportation Regulations. In particular, OSHA requires that chemical manufacturers, importers, and distributors must ensure that labels, tags, or markings of hazardous materials leaving the workplace do not conflict with DOT regulations (29 CFR §1200(f)(3), see Appendix B);
4. Develop or review the procedures for hazardous materials transportation to assure that for each material being shipped: the proper shipping name has been used; any applicable hazard classes have been determined; the proper identification numbers have been selected; any supplemental requirements applying to the mode of transport have been determined; the proper package has been selected; the shipment has the proper labeling, marking and placarding; the proper documentation and certification has been followed; and the loading has been carried out properly. The shipper should determine any exceptions that may apply;

Hazardous Materials Transportation

5. Compile or review the list of any extremely hazardous substances and hazardous substances being transported. Establish procedures to ensure coordination of reporting responsibilities in the event of a spill covered by incident reporting under DOT, or under CERCLA/SARA. Review any other federal, state, or local spill reporting requirements that may apply (see Chapter 3 for a discussion of CERCLA/SARA responsibilities and Appendices D, E, F, and G for the lists of extremely hazardous substances and hazardous substances);

6. Review the restrictions that may apply to any chemicals being transported that are hazardous substances. Does any one package in a shipment contain more than the reportable quantity of the material? (See the appendix to the Hazardous Materials Table); and

7. Review HM-181 proposals for their impact on shipments of hazardous materials. Determine whether hazardous materials being transported will be reclassified or if packaging requirements will be different. Review any testing requirements that will be imposed. Determine whether, and to what extent, shipping procedures or recordkeeping will have to be modified to conform to HM-181.

Table 3

Associations Involved in Hazardous Materials Training
Academy of Advanced Traffic, Philadelphia, PA
Air Freight Association, Washington, DC
American Industrial Hygiene Association, Akron, OH
American Trucking Association, Alexandria, VA
Association of American Railroads, Washington, DC
Chemical Manufacturers Association, Washington, DC
Chlorine Institute, Washington, DC
Hazardous Materials Advisory Council, Washington, DC
International Association of Fire Chiefs, Washington, DC
International Association of Fire Service Instructors, Ashland, MA
National Agricultural Chemical Association, Washington, DC
National Fire Protection Association, Quincy, MA

Hazardous Materials Transportation

FIGURE 1

Examples of Labels for Hazardous Materials Packages

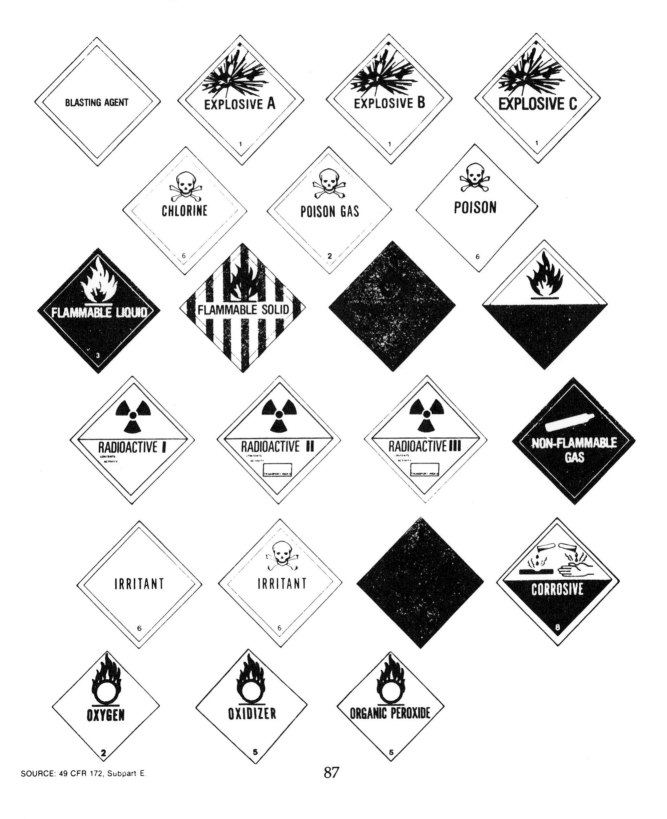

SOURCE: 49 CFR 172, Subpart E.

Chemical Hazard Communication Guidebook

FIGURE 2

Examples of Placards for Hazardous Materials

SOURCE: 49 CFR 172, Subpart F.

Hazardous Materials Transportation

Figure 3

DEPARTMENT OF TRANSPORTATION
HAZARDOUS MATERIALS INCIDENT REPORT
Form Approved OMB No. 2137-0039

INSTRUCTIONS: Submit this report in duplicate to the Information Systems Manager, Office of Hazardous Materials Transportation, DHM-63, Research and Special Programs Administration, U.S. Department of Transportation, Washington, D.C. 20590. If space provided for any item is inadequate, complete that item under Section IX, keying to the entry number being completed. Copies of this form, in limited quantities, may be obtained from the Information Systems Manager, Office of Hazardous Materials Transportation. Additional copies in this prescribed format may be reproduced and used, if on the same size and kind of paper.

I. MODE, DATE, AND LOCATION OF INCIDENT

1. MODE OF TRANSPORATION: ☐ AIR ☐ HIGHWAY ☐ RAIL ☐ WATER ☐ OTHER_____

2. DATE AND TIME OF INCIDENT
   (Use Military Time, e.g. 8:30am = 0830,
   noon = 1200, 6pm = 1800, midnight = 2400).
   Date: __/__/__ TIME: _____

3. LOCATION OF INCIDENT (Include airport name in ROUTE/STREET if incident occurs at an airport.)
   CITY: _____ STATE: _____
   COUNTY: _____ ROUTE/STREET: _____

II. DESCRIPTION OF CARRIER, COMPANY, OR INDIVIDUAL REPORTING

4. FULL NAME
5. ADDRESS (Principal place of business)
6. LIST YOUR OMC MOTOR CARRIER CENSUS NUMBER, REPORTING RAILROAD ALPHABETIC CODE, MERCHANT VESSEL NAME AND ID NUMBER OR OTHER REPORTING CODE OR NUMBER.

III. SHIPMENT INFORMATION (From Shipping Paper or Packaging)

7. SHIPPER NAME AND ADDRESS (Principal place of business)
8. CONSIGNEE NAME AND ADDRESS (Principal place of business)
9. ORIGIN ADDRESS (If different from Shipper address)
10. DESTINATION ADDRESS (If different from Consignee address)
11. SHIPPING PAPER/WAYBILL IDENTIFICATION NO.

IV. HAZARDOUS MATERIAL(S) SPILLED (NOTE: REFERENCE 49 CFR SECTION 172.101.)

12. PROPER SHIPPING NAME
13. CHEMICAL/TRADE NAME
14. HAZARD CLASS
15. IDENTIFICATION NUMBER (e.g. UN 2764, NA 2020)

16. IS MATERIAL A HAZARDOUS SUBSTANCE? ☐ YES ☐ NO
17. WAS THE RQ MET? ☐ YES ☐ NO

V. CONSEQUENCES OF INCIDENT, DUE TO THE HAZARDOUS MATERIAL.

18. ESTIMATED QUANTITY HAZARDOUS MATERIAL RELEASED (Include units of measurement)
19. FATALITIES
20. HOSPITALIZED INJURIES
21. NON-HOSPITALIZED INJURIES
22. NUMBER OF PEOPLE EVACUATED

23. ESTIMATED DOLLAR AMOUNT OF LOSS AND/OR PROPERTY DAMAGE, INCLUDING COST OF DECONTAMINATION OR CLEANUP (Round off in dollars)
    A. PRODUCT LOSS
    B. CARRIER DAMAGE
    C. PUBLIC/PRIVATE PROPERTY DAMAGE
    D. DECONTAMINATION/CLEANUP
    E. OTHER

24. CONSEQUENCES ASSOCIATED WITH THE INCIDENT:
    ☐ SPILLAGE ☐ FIRE ☐ EXPLOSION ☐ VAPOR (GAS) DISPERSION ☐ ENVIRONMENTAL DAMAGE ☐ MATERIAL ENTERED WATERWAY/SEWER ☐ NONE ☐ OTHER: _____

VI. TRANSPORT ENVIRONMENT

25. INDICATE TYPE(S) OF VEHICLE(S) INVOLVED:
    ☐ CARGO TANK ☐ VAN TRUCK/TRAILER ☐ FLAT BED TRUCK/TRAILER
    ☐ TANK CAR ☐ RAIL CAR ☐ TOFC/COFC ☐ AIRCRAFT ☐ BARGE ☐ SHIP ☐ OTHER: _____

26. TRANSPORTATION PHASE DURING WHICH INCIDENT OCCURRED OR WAS DISCOVERED:
    ☐ EN ROUTE BETWEEN ORIGIN/DESTINATION ☐ LOADING ☐ UNLOADING ☐ TEMPORARY STORAGE/TERMINAL

27. LAND USE AT INCIDENT SITE: ☐ INDUSTRIAL ☐ COMMERCIAL ☐ RESIDENTIAL ☐ AGRICULTURAL ☐ UNDEVELOPED

28. COMMUNITY TYPE AT SITE: ☐ URBAN ☐ SUBURBAN ☐ RURAL

29. WAS THE SPILL THE RESULT OF A VEHICLE ACCIDENT/DERAILMENT? ☐ YES ☐ NO
    IF YES AND APPLICABLE, ANSWER PARTS A THRU C.
    A. ESTIMATED SPEED:
    B. HIGHWAY TYPE: ☐ DIVIDED/LIMITED ACCESS ☐ UNDIVIDED
    C. TOTAL NUMBER OF LANES: ☐ ONE ☐ TWO ☐ THREE ☐ FOUR OR MORE
    SPACE FOR DOT USE ONLY

FORM DOT F 5800.1 (Rev. 6/89) Supersedes DOT F 5800.1 (10/70) (9/1/76)   THIS FORM MAY BE REPRODUCED

Chemical Hazard Communication Guidebook

Figure 3, continued

VII. PACKAGING INFORMATION: If the package is overpacked (consists of several packages, e.g. glass jars within a fiberboard box), begin with Column A for information on the innermost package.			
ITEM	A	B	C
30. TYPE OF PACKAGING, INCLUDING INNER RECEPTACLES (e.g. Steel drum, tank car)			
31. CAPACITY OR WEIGHT PER UNIT PACKAGE (e.g. 55 gallons, 65 lbs.)			
32. NUMBER OF PACKAGES OF SAME TYPE WHICH FAILED IN IDENTICAL MANNER			
33. NUMBER OF PACKAGES OF SAME TYPE IN SHIPMENT			
34. PACKAGE SPECIFICATION IDENTIFICATION (e.g. DOT 17E, DOT 105A100, UN 1A1 or none)			
35. ANY OTHER PACKAGING MARKINGS (e.g. STC, 18/16-55-88, Y1.4/150/87)			
36. NAME AND ADDRESS, SYMBOL OR REGISTRATION NUMBER OF PACKAGING MANUFACTURER			
37. SERIAL NUMBER OF CYLINDERS, PORTABLE TANKS, CARGO TANKS, TANK CARS			
38. TYPE OF LABELING OR PLACARDING APPLIED			
39. IF RECONDITIONED OR REQUALIFIED — A. REGISTRATION NUMBER OR SYMBOL			
39. IF RECONDITIONED OR REQUALIFIED — B. DATE OF LAST TEST OR INSPECTION			
40. EXEMPTION/APPROVAL/COMPETENT AUTHORITY NUMBER, IF APPLICABLE (e.g. DOT E1012)			

VIII. DESCRIPTION OF PACKAGING FAILURE: Check all applicable boxes for the package(s) identified above.

41. ACTION CONTRIBUTING TO PACKAGING FAILURE

A B C
- a. ☐ ☐ ☐ TRANSPORT VEHICLE COLLISION
- b. ☐ ☐ ☐ TRANSPORT VEHICLE OVERTURN
- c. ☐ ☐ ☐ OVERLOADING/OVERFILLING
- d. ☐ ☐ ☐ LOOSE FITTINGS, VALVES
- e. ☐ ☐ ☐ DEFECTIVE FITTINGS, VALVES
- f. ☐ ☐ ☐ DROPPED
- g. ☐ ☐ ☐ STRUCK/RAMMED
- h. ☐ ☐ ☐ IMPROPER LOADING
- i. ☐ ☐ ☐ IMPROPER BLOCKING
- j. ☐ ☐ ☐ CORROSION
- k. ☐ ☐ ☐ METAL FATIGUE
- l. ☐ ☐ ☐ FRICTION/RUBBING
- m. ☐ ☐ ☐ FIRE/HEAT
- n. ☐ ☐ ☐ FREEZING
- o. ☐ ☐ ☐ VENTING
- p. ☐ ☐ ☐ VANDALISM
- q. ☐ ☐ ☐ INCOMPATIBLE MATERIALS
- r. ☐ ☐ ☐ OTHER_____

42. OBJECT CAUSING FAILURE

A B C
- a. ☐ ☐ ☐ OTHER FREIGHT
- b. ☐ ☐ ☐ FORKLIFT
- c. ☐ ☐ ☐ NAIL/PROTRUSION
- d. ☐ ☐ ☐ OTHER TRANSPORT VEHICLE
- e. ☐ ☐ ☐ WATER/OTHER LIQUID
- f. ☐ ☐ ☐ GROUND/FLOOR/ROADWAY
- g. ☐ ☐ ☐ ROADSIDE OBSTACLE
- h. ☐ ☐ ☐ NONE
- i. ☐ ☐ ☐ OTHER_____

43. HOW PACKAGE(S) FAILED

A B C
- a. ☐ ☐ ☐ PUNCTURED
- b. ☐ ☐ ☐ CRACKED
- c. ☐ ☐ ☐ BURST/INTERNAL PRESSURE
- d. ☐ ☐ ☐ RIPPED
- e. ☐ ☐ ☐ CRUSHED
- f. ☐ ☐ ☐ RUBBED/ABRADED
- g. ☐ ☐ ☐ RUPTURED
- h. ☐ ☐ ☐ OTHER_____

44. PACKAGE AREA THAT FAILED

A B C
- a. ☐ ☐ ☐ END, FORWARD
- b. ☐ ☐ ☐ END, REAR
- c. ☐ ☐ ☐ SIDE, RIGHT
- d. ☐ ☐ ☐ SIDE, LEFT
- e. ☐ ☐ ☐ TOP
- f. ☐ ☐ ☐ BOTTOM
- g. ☐ ☐ ☐ CENTER
- h. ☐ ☐ ☐ OTHER_____

45. WHAT FAILED ON PACKAGE(S)

A B C
- a. ☐ ☐ ☐ BASIC PACKAGE MATERIAL
- b. ☐ ☐ ☐ FITTING/VALVE
- c. ☐ ☐ ☐ CLOSURE
- d. ☐ ☐ ☐ CHIME
- e. ☐ ☐ ☐ WELD/SEAM
- f. ☐ ☐ ☐ HOSE/PIPING
- g. ☐ ☐ ☐ INNER LINER
- h. ☐ ☐ ☐ OTHER_____

IX. DESCRIPTION OF EVENTS: Describe the sequence of events that led to incident, action taken at time discovered, and action taken to prevent future incidents. Include any recommendations to improve packaging, handling, or transportation of hazardous materials. Photographs and diagrams should be submitted when necessary for clarification. ATTACH A COPY OF THE HAZARDOUS WASTE MANIFEST FOR INCIDENTS INVOLVING HAZARDOUS WASTE. Continue on additional sheets if necessary.

46. NAME OF PERSON RESPONSIBLE FOR PREPARING REPORT	47. SIGNATURE	
48. TITLE OF PERSON RESPONSIBLE FOR PREPARING REPORT	49. TELEPHONE NUMBER (Area Code)	50. DATE REPORT SIGNED

APPENDIX A

MODEL COMPANY PROGRAMS:

HAZARD COMMUNICATION AND EMPLOYEE TRAINING

Introduction

Appendix A includes a sample hazard communication program that may be used as the basis for developing a company program or in comparing the elements of an existing plan to the model. In addition, Appendix A also provides an outline of an employee training course.

The hazard communication program addresses the following elements:

    Policy Statement

    Program Responsibilities

    Employee Notification and Training

    Location and Availability of the Written Hazard Communication Program

    Sample Bulletin Board Notification

    Hazardous Material List

    Chemical Name Index

    Hazard Determinations

    Material Safety Data Sheets (MSDSs)

    Sample MSDS Request Letter

    Guide for Reviewing MSDS Completeness

    Labeling

    Employee Information/Training

    Outside Contractors

    Letter of Notification to Contractor Employers

Chemical Hazard Communication Guidebook

The employee training module outlines:

        Who Should Receive Hazardous Chemical Training

        Who Should Conduct Hazardous Chemical Training

        Training Requirements of the Hazard Communication Standard

Appendix A: Model Company Programs

ABC COMPANY HAZARD COMMUNICATION PROGRAM

1. POLICY STATEMENT

In order to conduct our business, ABC Company must use certain chemicals that require specific precautions to be taken to protect our employees' health. It is the policy of ABC Company to communicate any relevant information regarding hazardous chemicals to potentially exposed employees, as well as to implement appropriate measures to safeguard employee safety and health. The goal of the program shall be to minimize the possibility of employee illness or injury arising from exposure to hazardous chemicals.

It will be the responsibility of management and supervisors to ensure that adequate information is obtained and disseminated to the appropriate employees. It will be the employee's responsibility to follow the recommended practices outlined in product labels, Material Safety Data Sheets, company operating procedures, and/or company-provided training.

This Hazard Communication Program is intended to supplement our existing safety and health program. Current safety and health policies remain in effect.

The effectiveness of this Hazard Communication Program, as with all our programs, depends upon the active support and involvement of all personnel.

Manager, ABC Company

2. PROGRAM RESPONSIBILITIES

A. Facility Manager

1. Ensure compliance by all management personnel with the Hazard Communication Program.
2. Appoint a Program Coordinator.

B. Section/Department Manager

1. Review operations with supervisors to determine which employees require hazard communication training.
2. Follow up to ensure supervisors are carrying out prescribed company policy.
3. Notify the Program Coordinator of any changes in the hazardous chemicals being used.
4. Ensure up-to-date records are maintained of employee training on handling hazardous chemicals.

C. Supervisor

1. Identify all operations requiring the use of hazardous chemicals and provide a list of those chemicals to the Program Coordinator.
2. Ensure MSDSs for each hazardous chemical are available to employees in an easily accessible location.
3. Provide and document training of employees exposed to hazardous chemicals.
4. Periodically inspect engineering controls and personal protective equipment.
5. Make routine surveys of the work area to ensure safe practices are being followed.

Appendix A: Model Company Programs

6. Ensure required labeling practices are being followed.
7. Enforce applicable safety and health rules.

D. Program Coordinator

1. Determine that MSDSs are complete (see MSDS Section).
2. Keep an up-to-date file of all MSDSs or appropriate alternative. Maintain a list of all hazardous chemicals present in the workplace in an up-to-date manner.
3. Ensure employee training is documented.
4. Coordinate emergency procedures and fire department activities related to hazardous chemical leaks and spills.
5. Contract for professional industrial hygiene services as needed.
6. Coordinate compliance with state/community Right-To-Know Requirements.

E. Employee

1. Obey established safety rules.
2. Use personal protective equipment as required by company procedure.
3. Inform your supervisor of:
   a. Any symptoms that may possibly be due to exposure to hazardous chemicals
   b. Missing labels on containers, and
   c. Malfunctioning safety equipment.
4. Review the MSDS for each hazardous chemical with which you work or to which you are exposed. Know the physical and health hazards of the chemical, how to recognize it (color, smell, physical properties), and symptoms of overexposure.

Chemical Hazard Communication Guidebook

       5.   Use hazardous chemicals in accordance with precautions on label, MSDS, and company procedures. Do not remove labels.

       6.   Do not use unapproved containers for hazardous materials.

       7.   Know the location of emergency equipment, first aid supplies, emergency eyewash, etc.

       8.   Know your role in emergency situations.

F.   Shipping/Receiving/Warehousing

       1.   Log in all hazardous chemicals received.

       2.   Ensure MSDSs are received with initial shipment of a hazardous chemical.

       3.   Store hazardous chemicals in designated locations.

       4.   Use prescribed personal protective equipment when handling hazardous chemicals.

       5.   Report damaged containers and leaks or spills to the appropriate supervisor immediately.

Appendix A: Model Company Programs

3. EMPLOYEE NOTIFICATION AND TRAINING

Notification to new employees entering employment with ABC Company of the Company's hazard communication program will be done by the supervisor of the Department in which they are first employed. Supervisors will notify employees of the existence of the Hazard Communication Standard, the company written program, the hazardous materials inventory, and the MSDS file. New employees shall receive training in the hazards present in their workplaces before commencing work in those areas.

Information on the Company's hazard communication program will be posted on all company bulletin boards.

Chemical Hazard Communication Guidebook

4. LOCATION AND AVAILABILITY OF THE WRITTEN HAZARD COMMUNICATION PROGRAM

This program and the full text of the law, 29 CFR 1910.1200, together with Material Safety Data Sheets, will be maintained on file in the Program Coordinator's Office and will be available for review by employees, their designated representatives, and government compliance officers (OSHA) during working hours.

A notice identifying the program's content and its availability will be posted on the Company Bulletin Boards.

Appendix A: Model Company Programs

5. SAMPLE BULLETIN BOARD NOTIFICATION

<u>ABC Company's Hazard Communication Program</u>

The Company's hazard communication program includes:

o   A list of all hazardous chemicals found in the plant, by work area;

o   A Material Safety Data Sheet (MSDS) for each hazardous chemical found in the plant; and

o   A written program explaining:

- How the Company meets the requirements of the OSHA Hazard Communication Standard (29 C.F.R. 1910.1200) for container labeling, MSDSs and employee training,
- How the standard's requirements will be communicated to employees, and
- How emergency and non-routine tasks will be handled.

This information is available for review by employees, their designated representatives, and government compliance officers (OSHA). Requests for the review of this material should be made to:

Hazard Communication Program Coordinator

Chemical Hazard Communication Guidebook

6. HAZARDOUS MATERIAL LIST

Chemicals are listed alphabetically by their commonly known name or trade name. The chemical name and CAS Number (to extent known) are given in a cross-reference list.

Name of Material	Manufacturer/Supplier	Location
_____	_____	_____
_____	_____	_____
_____	_____	_____
_____	_____	_____
_____	_____	_____
_____	_____	_____

Appendix A: Model Company Programs

7. CHEMICAL NAME INDEX

Common or Trade Name	Chemical Name	CAS Number
_____	_____	_____
_____	_____	_____
_____	_____	_____
_____	_____	_____
_____	_____	_____
_____	_____	_____

Chemical Hazard Communication Guidebook

8. HAZARD DETERMINATIONS

ABC Company relies on the chemical manufacturer or importer for the hazard determinations contained in the Material Safety Data Sheets supplied with the product.

Due to the complex and specialized nature of chemical hazard recognition, evaluation, and control, ABC Company will contract for professional industrial hygiene services as needed. Any hazard evaluation information, reports, or data obtained by the company will be maintained by the Program Coordinator.

Requests to review this material should be made to the Program Coordinator.

Appendix A: Model Company Programs

9. MATERIAL SAFETY DATA SHEETS (MSDSs)

MSDSs for every chemical on the hazardous chemical list are available. The MSDS file is maintained by the Program Coordinator and is available for review at any time at the Program Coordinator's office.

MSDSs are obtained by the Purchasing Department for all hazardous chemicals which enter the plant prior to a chemical being placed in use.

The Program Coordinator receives and distributes copies of MSDSs to the department(s) in which the substance is used and they are available from the supervisor of each such department.

<u>Company Procedures for Obtaining an MSDS</u>

<u>Step 1</u>

This letter will be sent if, at any time, an item is received and an MSDS does not accompany it.

<u>Step 2</u>

If, after 30 days, an MSDS has not been received, a follow-up request will be sent.

Chemical Hazard Communication Guidebook

10. SAMPLE MSDS REQUEST LETTER

Dear Supplier:

In an effort to comply with the OSHA Hazard Communication Standard, we are asking your cooperation in providing us with Material Safety Data Sheets for the chemicals(s) listed below:

It is essential that we receive this information no later than _____ . Your cooperation and response directed to me will be greatly appreciated.

Sincerely,

Program Coordinator

Appendix A: Model Company Programs

11. GUIDE FOR REVIEWING MSDS COMPLETENESS

o   Do we have an MSDS for the hazardous chemical used?

o   Is the MSDS in English?

o   Does the MSDS contain at least the following information:

- (a) The identity used on the label?
- (b) The chemical and common name(s) for single substance hazardous chemicals?
- (c) For mixtures tested as a whole:
  The chemical and common name(s) of the ingredients which contribute to hazards?
- (d) For mixtures not tested as a whole:
  The chemical and common name(s) of all ingredients which are health hazards (1 percent or greater), or carcinogens (0.1 percent or greater)?
- (e) The chemical and common name(s) of all ingredients which have been determined to present a physical hazard when present in the mixture?

o   Does the MSDS contain the physical and chemical characteristics of the hazardous chemical (vapor pressure, flash point, etc.)?

o   Does the MSDS contain the physical hazards of the hazardous chemical, including the potential for fire, explosion, and reactivity?

o   Does the MSDS contain the health hazards of the hazardous chemical (including signs and symptoms, medical conditions aggravated)?

o   Does the MSDS contain the primary routes of entry?

Chemical Hazard Communication Guidebook

- o Does the MSDS contain the OSHA PEL, the ACGIH TLV, and other exposure limits (including ceiling and other short-term limits)?

- o Does the MSDS contain information on carcinogen listings (i.e., OSHA regulated carcinogens, those indicated in the National Toxicology Program (NTP) annual report, and those listed by the International Agency for Research on Carcinogens (IARC))?

    NOTE: Negative conclusions regarding carcinogenicity, or the fact that there is no information, do not have to be reported unless there is a specific blank for carcinogenicity on the form.

- o Does the MSDS contain generally applicable procedures and precautions for safe handling and use of the chemical (hygienic practices, maintenance, and spill procedures)?

- o Does the MSDS contain generally applicable control (engineering controls, work practices, or personal protective equipment)?

- o Does the MSDS contain date of preparation or last change?

- o Does the MSDS contain the name, address, and telephone number of a responsible party?

- o Are all sections of the MSDS completed?

Appendix A: Model Company Programs

12. LABELING

Labels are designed to provide information to employees concerning the hazards of various chemicals. Therefore, it is important that hazardous chemicals are not placed in improperly labeled containers. ABC Company does not have its own labeling system. We rely on the labeling systems of our suppliers. The following rules apply:

1. All manufacturers' labels will be left on the containers.
2. All containers will be labeled.
3. At a minimum, each label must contain the following:
    - Identification of the material in the container.
    - Appropriate hazard warnings, such as health, fire, and reactivity hazards and precautions.
    - Name and address of the chemical manufacturer, importer, or other responsible party.

Chemical Hazard Communication Guidebook

13. EMPLOYEE INFORMATION/TRAINING

All employees who are, or may be, exposed to hazardous materials will be trained. The training will include, at a minimum, the following elements.

A. Employee Information:

Employees shall be informed about the hazardous chemicals and suspected hazards of substances to which they are exposed in the normal course of their employment, including:

1. Information on the operations in their work area where hazardous chemicals are present.
2. The location and availability of the Written Hazard Communication Program, including a list of hazardous chemicals and MSDSs.

Employees shall also be informed of the requirements of OSHA Hazard Communication Standard (29 C.F.R. 1910.1200).

B. Employee Training:

All employees exposed to hazardous chemicals, including production employees, line supervisors, repair and maintenance employees, will be trained according to the following training program:

1. Initial training for supervisor and production employees will be conducted by the Program Coordinator.
2. After the initial training, any necessary training required by the introduction of new hazardous chemicals and training for new employees and transferred employees will be conducted by the supervisor of the applicable department.
3. Subjects to be included in the Employee Training Program will include the following:

Appendix A: Model Company Programs

a. Methods and observations that may be used to detect the release of hazardous chemicals, such as employer monitoring, visual sightings, or odors of hazardous chemicals when released.

b. The measures which employees can take to protect themselves from exposure to hazardous chemicals (such as personal protective equipment and emergency procedures).

c. The physical and health hazards of the chemicals in the work area.

d. An explanation of the labeling system, Material Safety Data Sheets, and how employees can obtain and use information on hazardous chemicals.

e. The methods used to inform employees of the hazards of non-routine tasks (e.g., maintenance of enclosed vessels), and the hazards associated with any chemicals contained in unlabeled pipes in the workplace.

All new employees will be trained before being exposed to any hazardous chemicals. Retraining will occur as needed when new hazards become recognized, or when employees become exposed to new hazards as a result of transfer, process changes, or new chemical introductions.

14. OUTSIDE CONTRACTORS

As part of the Hazard Communication Program, contractors will be informed by the Program Coordinator of the chemical hazards to which their employees may be exposed during the period they work in our facility.

Contractors must follow our safety and health program as it applies to their employees while they are in our facility.

In dealing with outside contractors in the area of hazard communication, the following guidelines apply:

1. All contractors will be informed of hazardous chemicals present in the areas in which the contractor's employees will be working.

Chemical Hazard Communication Guidebook

2. Material Safety Data Sheets will be made available to the contractor's management personnel.
3. A signed acknowledgement form that items 1 and 2 were communicated will be obtained.
4. Contractors must supply Material Safety Data Sheets to our firm for hazardous chemicals brought into the workplace by them.

A letter will be used to inform Contractor Employers of ABC Company's Written Hazard Communication Program and the procedure which the Contractor Employer must use to inform his employees of the hazardous substances to which his employees may be exposed.

A copy of this letter will be signed by all Contractor Employers of the ABC Company and will be used as documentation of compliance with the OSHA requirements under 29 C.F.R. 1910.1200.

Appendix A: Model Company Programs

15. LETTER OF NOTIFICATION TO CONTRACTOR EMPLOYERS

ABC Company has a Written Hazard Communication Program as required by the Occupational Safety and Health Administration (OSHA).

OSHA requires that Contractor Employers be informed of the hazardous chemicals to which their employees may be exposed while performing work at this location. The following information is provided to you as a means of fulfilling the OSHA Hazard Communication Standard requirements for notification of Contractor Employers.

1. A copy of the ABC Company Written Hazard Communication Program, including a list of hazardous substances present in the workplace, is available through our Program Coordinator for your review.
2. Each department within this facility has a designated MSDS binder area which contains the MSDSs for substances used in that department. The location of the binder area can be obtained by contacting the supervisor of the department.
3. Medical assistance for the plant is provided by _____ .
4. You have the responsibility for informing your employees of the hazardous substances to which they may be exposed in our facility.

Any questions concerning ABC Company's Written Hazard Communication Program should be directed to the Program Coordinator, whose office is located in _____ .

Chemical Hazard Communication Guidebook

    I have read this letter and understand my responsibility as a Contractor Employer of ABC Company. I also understand that I must inform ABC Company of all hazardous chemicals that I use in the performance of the contracted work and that I must provide the Program Coordinator with an MSDS for each such chemical.

_____ Contractor Employer Company

_____ Contractor Employer Representative

_____   _____

        (Signature)   (Date)

Appendix A: Model Company Programs

HAZARDOUS CHEMICAL EMPLOYEE TRAINING

1. Introduction

Prior to conducting employee hazardous chemical training, course content and organization must be determined. This section outlines the information which must be included in such a course. However, because the chemical hazards present at each facility are different, the following outline will need to be supplemented with material on the specific hazards present at each facility. In addition, it may be too comprehensive for facilities where chemicals are simply stored in closed containers that are not normally opened. Training at such facilities may be restricted to the appropriate response to leaks and spills.

A. Who Should Receive Hazardous Chemical Training

The OSHA Hazard Communication Standard (HCS) applies to all employees who are exposed to hazardous chemicals during routine operations or in any foreseeable emergency. Those employees who should be covered generally include production personnel, supervisors, maintenance, and repair crews. Employees not generally covered include office workers, grounds maintenance crews, security guards, or non-resident management, unless their jobs involve exposure to hazardous materials.

B. Who Should Conduct Hazardous Chemical Training

The person at the facility who coordinates safety and health activities is the logical choice for program instructor. The instructor must have the requisite understanding of safety and health concepts and should be familiar with the specifics of the chemical hazards present at the facility. Two important qualities that the instructor should bring to the training are (1) credibility in dealing with the subject matter, and (2) sensitivity to the group. The instructor should neither alarm employees on the one hand nor appear to minimize hazards on the other. Instructor support, if needed, can usually be obtained from the employer's compensation insurance carrier or from outside consultants.

Chemical Hazard Communication Guidebook

C. Training Requirements of the Hazard Communication Standard

The HCS (29 CFR 1910.1200) provides in subpart (h) as follows:

(h) Employee information and training. Employers shall provide employees with information and training on hazardous chemicals in their work area at the time of their initial assignment, and whenever a new hazard is introduced into their work area.

(1) Information -- Employees shall be informed of:

(i) The requirements of this section;
(ii) Any operations in their work area where hazardous chemicals are present; and
(iii) The location and availability of the written hazard communication program, including the required list(s) of hazardous chemicals, and material safety data sheets required by this section.

(2) Training -- Employee training shall include at least:

(i) Methods and observations that may be used to detect the presence or release of a hazardous chemical in the work area (such as monitoring conducted by the employer, continuous monitoring devices, visual appearance or odor of hazardous chemicals when being released, etc.);
(ii) The physical and health hazards of the chemicals in the work area; and
(iii) The measures employees can take to protect themselves from these hazards, including specific procedures the employer has implemented to protect employees from exposure to hazardous chemicals,

Appendix A: Model Company Programs

such as appropriate work practices, emergency procedures, and personal protective equipment to be used; and,

(iv) The details of the hazard communication program developed by the employer, including an explanation of the labeling system and the material safety data sheet, and how employees can obtain and use the appropriate hazard information.

The HCS relaxes the above training requirements with respect to work operations where employees only handle chemicals in sealed containers which are not opened under normal conditions of use (such as are found in marine cargo handling, warehousing, or retail sales). For such facilities a written hazard communication program is not required so that training concerning the list of hazardous chemicals present at a facility and the location and other content of the program may be deleted. The other training and information required by paragraph (h) must be given to employees but only to the extent necessary to protect them in the event of a leak or spill of a hazardous chemical from a sealed container.

2. OUTLINE OF A MODEL HAZARDOUS CHEMICAL EMPLOYEE TRAINING PROGRAM

The following outline of a model Hazardous Chemical Employee Training Program is provided for use as a guideline in developing a facility's training program. It is presented in two segments: 1) General Information; and 2) Site Specific Information.

Chemical Hazard Communication Guidebook

GENERAL INFORMATION

I. Course Introduction

    A. Chemical Hazard Recognition

    B. Sources of Information on Chemical Hazards

    C. Control of Chemical Hazards

II. Chemical Hazard Recognition

    A. Types of hazards

        1. Physical hazards
- a. Combustible liquids
- b. Compressed gases
- c. Explosives
- d. Flammable substances
- e. Organic peroxides
- f. Oxidizers
- g. Pyrophoric substances
- h. Unstable or water reactive substances

        2. Health Hazards
- a. Acute hazards
    - i. corrosive
    - ii. highly toxic
    - iii. toxic
    - iv. irritant
    - v. sensitizer

Appendix A: Model Company Programs

        vi. other acute effects

    b. Chronic hazards

        i. carcinogens

        ii. mutagens

        iii. teratogens and reproductive toxins

        iv. hepatotoxins

        v. nephrotoxins

        vi. neurotoxins

        vii. other toxic effects

B. Routes of Entry of Chemicals into the Body

1. Inhalation
2. Ingestion
   a. Contamination of eating, drinking, and smoking materials
   b. Ingestion of inhaled substances
3. Skin Absorption

C. Symptoms of Exposure to Hazardous Substances

1. Acute exposure
   a. Short exposure period
   b. Usually high concentrations
   c. Immediate health effect

2. Chronic exposure
   a. Long exposure period
   b. Usually low concentrations
   c. Slowly developing health effect

3. Types of reactions to acute and chronic exposures (individual susceptibility may vary)
   a. Chronic lung diseases - silica, cotton dust
   b. Anesthetics - solvent vapors

Chemical Hazard Communication Guidebook

       c. Irritants - formaldehyde, acids

       d. Chronic liver damage - carbon tetrachloride, nitrosamines

       e. Sensitizers - reactive dyes

       f. Cutaneous hazards - ketones, chlorinated compounds

       g. Eye hazards - methanol, acids

D. Relationship of Dose to Risk

    1. Toxicity of chemical

    2. Concentration of chemical

    3. Mode of exposure and exposure time

    4. The greatest risk is posed by toxic substances that are:

       a. Highly toxic, or

       b. Present in high concentrations, and

       c. To which employees are exposed several hours per day, day-after-day in a manner which permits the chemical to enter the body.

E. Exposure Standards

    1. OSHA Permissible Exposure Limits

       a. 8-hour time weighted averages

       b. 15-minute ceiling

       c. Legally binding

    2. ACGIH Threshold Limit Values

       a. 8-hour time weighted average

       b. Instantaneous Ceiling

       c. 15 Minutes STEL

       d. Not legally binding

    3. Other Relevant Standards or Criteria

       a. NIOSH Criteria Documents

       b. ANSI Standards

Appendix A: Model Company Programs

        c. EPA Health Assessment Documents

        d. Manufacturer recommendations

    4. Common Features of Exposure Limits

        a. Units - Very small amounts

            i. ppm, ppb, ppt

            ii. mg/m3, ug/m3

        b. Not "safe" limits but exposure to concentrations below these levels is generally a low risk

III.   Sources of Information On Chemical Hazards

    A. Summary of HCS

        1. Hazard determinations - performed by manufacturers and importers

        2. MSDS

        3. Labeling

        4. Training requirements

        5. Written Hazard Communication Program

        6. List of hazardous substances in workplace

        7. Trade secrets

    B. Contents of an MSDS

        1. Manufacturer's address and phone number

        2. Hazardous ingredients/identity

        3. OSHA, PEL, ACGIH, TLV, other recommended limits

        4. Physical/chemical characteristics

            a. boiling point

            b. vapor pressure

            c. vapor density

            d. solubility in water

            e. specific gravity

            f. melting point

Chemical Hazard Communication Guidebook

  g. evaporation rate

  h. appearance and odor

5. Fire and explosion hazard data
   a. flash point
   b. flammable limits
   c. explosive levels - upper and lower
   d. extinguishing media
   e. special fire fighting procedures
   f. unusual fire and explosion hazards

6. Reactivity data
   a. stability
   b. conditions to avoid (e.g., heat, impact)

7. Health hazard data
   a. routes of entry
   b. acute and chronic hazards, including carcinogenicity
   c. signs and symptoms of exposure
   d. medical conditions aggravated by exposure
   e. emergency first aid procedures

8. Precautions for safe handling and use
   a. steps to be taken in case material is released or spilled
   b. waste disposal method
   c. precautions to be taken in handling and storing
   d. other precautions

9. Control measures
   a. ventilation
      i. local exhaust, special
      ii. mechanical (general), other
   b. sealed systems
   c. other engineering controls

Appendix A: Model Company Programs

        d. respiratory protection

        e. protective gloves

        f. eye protection

        g. other protective clothing or equipment

        h. Workplace practices, industrial hygiene procedures

  C. Labeling

      1. Labels tell you

        a. What the principal hazards are.

        b. What precautions should be taken.

        c. Emergency first aid procedures.

      2. Labels provide this information by

        a. Words

        b. Symbols

        c. Numbers

        d. Colors

        e. Combinations

  D. Recognizing Hazardous Chemicals are Present

      1. Appearance or odor of hazardous chemicals

      2. Physical or health effects

      3. Monitoring

      4. Inventory control

IV. Control of Chemical Hazards

  A. Chemical Hazards are Controlled by Various Methods

      1. Engineering controls, e.g., ventilation

      2. Workplace practices, e.g., grounding containers of flammable substances

Chemical Hazard Communication Guidebook

      3. Personal protective devices

          a. gloves, shoes

          b. safety glasses

          c. protective clothing

          d. dust masks

          e. respirators

      4. Isolation of chemical

B. Safe Handling of Hazardous Chemicals

      1. Storage practices

      2. Reactivity considerations

      3. Proper containers

      4. Spill prevention

      5. Spill cleanup procedures

      6. Personal protective equipment

Appendix A: Model Company Programs

SITE SPECIFIC INFORMATION

The following site specific training outline will assist in designing a presentation customized to a specific facility. The details of such a presentation can only be prepared by someone familiar with the hazards present at a specific site.

I. The Written Hazard Communication Program

    A. Its relationship to the OSHA Standard

    B. Overview of its Content

    C. Location of Program and How to Obtain Access to It

II. Our Hazard Information Program

    A. The Hazardous Chemicals List: How maintained and where located

    B. The MSDS File

    C. Labeling systems used in the workplace

    D. Other hazard information resources (outside consultants, vendors, air sampling programs, medical monitoring, etc.)

III. The Hazard Control Program

    A. Special Work Practices and Procedures

    B. Engineering Controls/Isolation

    C. Personal Protective Equipment

    D. Employee Training

    E. Outside Contractors

Chemical Hazard Communication Guidebook

IV. Hazards of Specific Chemicals or Classes of Chemicals

    A. We use the following chemicals or types of chemicals:

    B. The hazards of these chemicals are:

    C. You can recognize the presence of these chemicals by:

    D. The precautions you should take in handling and storing these chemicals are:

    E. The control measures available to protect you from over-exposure to these chemicals are:

    F. Emergency and spill information on these chemicals is as follows:

    G. Medical and first aid information on these chemicals is as follows:

APPENDIX B

REFERENCE GUIDE TO THE REGULATIONS

The Reference Guide to the Regulations is intended to assist the reader in obtaining a quick overview of the requirements imposed by the OSHA Hazard Communication Standard, the Emergency Planning and Community Right-To-Know Act, and the requirements of the Department of Transportation applicable to shippers of hazardous materials. The outlines here summarize the principal requirements, briefly describe their application, and provide a reference to the pertinent part of the regulation. The source regulations follow the outlines. However, the regulations of the Department of Transportation are voluminous and are not reprinted.

With regard to the summary of transportation regulations, a checklist of common violations has been included as an aid to the reader. The checklist has been developed from a compilation of actual instances of violations cited by DOT, and, to that extent, represents common errors to be avoided.

The following outlines the topics covered by the Reference Guide:

OSHA HAZARD COMMUNICATION STANDARD

- A. Scope of the OSHA Hazard Communication Standard
- B. Performing a Hazard Determination
- C. Preparing Material Safety Data Sheets
- D. Maintaining Material Safety Data Sheets
- E. Preparing and Maintaining Labels and Other Warnings
- F. Training Employees about Chemical Hazards
- G. Preparing and Providing a Written Hazard Communication Program
- H. Requirements for Those Handling Hazardous Chemical in Sealed Containers
- I. Special Requirements for Laboratories
- J. Requirements for Disclosing and Withholding Trade Secret Information

Chemical Hazard Communication Guidebook

Following the summary is the Standard and OSHA's form for a material safety data sheet. This form is not mandatory, but may be used in complying with the Standard.

EMERGENCY PLANNING AND COMMUNITY RIGHT-TO-KNOW

- A. Emergency Planning Notification
- B. Emergency Release Reporting
- C. MSDS/List Submission
- D. Hazardous Chemical Inventory Reporting
- E. Submission of Toxic Chemical Release Reports
- F. Disclosure of Information

Following the summary are: EPA's emergency planning rule of April 22, 1987; the agency's release reporting regulations under CERCLA; the agency's regulations establishing the Emergency and Hazardous Chemical Inventory form and related reporting requirements; and the agency's toxic chemical release reporting regulation.

DOT REQUIREMENTS APPLICABLE TO SHIPPERS OF HAZARDOUS MATERIALS

- A. Determining the Proper Shipping Name and Classification of the Material
- B. Complying with General Restrictions on Transporting Hazardous Materials
- C. Determining the Proper Packaging
- D. Selecting the Proper Labels(s)
- E. Marking the Packaging in the Manner Required
- F. Determining the Proper Placard(s)
- G. Preparing Shipping Papers

DOT's Hazardous Material Table follows the summary. The appendix also includes DOT's amendment to its final rule that regulates EPA hazardous substances as hazardous materials subject to DOT's regulations when transported. The amendment (52 Fed. Reg. 41300, October 27, 1987) suspends the effective date of DOT's regulation of the hazardous substances for which EPA has proposed to raise the statutory reportable quantity of 1 pound. DOT has suspended its rule until EPA finalizes the proposed reportable quantities, which are listed in the amendment.

Appendix B: Regulations Reference Guide

THE OSHA HAZARD COMMUNICATION STANDARD

A. SCOPE OF THE OSHA HAZARD COMMUNICATION STANDARD

1. OSHA's definition of hazardous chemical is broad, but the Standard includes a number of exemptions.

Under the Standard, a hazardous chemical is broadly defined as any chemical which is a physical or health hazard (52 FR 31878 revising 29 CFR § 1910.1200 (c)).

The standard specifically excludes from its coverage:

1. Any hazardous waste defined by the Solid Waste Disposal Act as amended when subject to regulation by EPA;

2. Tobacco or tobacco products;

3. Wood or wood products;

4. Articles;

5. Foods, drugs, cosmetics, or alcoholic beverages in a retail establishment which are packaged for sale to consumers;

6. Foods, drugs, or cosmetics intended for personal consumption by employees while in the workplace;

7. Any consumer product or hazardous substance covered by the Consumer Product Safety Act and Federal Hazardous Substances Act (see discussion of this definition in Chapter 2); and

8. Any bulk drug (see discussion of this definition in Chapter 2).

Refer To: 52 FR 31878 revising 29 CFR § 1910.1200(b)(6)

Chemical Hazard Communication Guidebook

2. Employers must implement a hazard communication program for each hazardous chemical present at the facility by the deadline applicable to the type of facility.

Chemical manufacturers and importers must assess the hazards of chemicals which they produce or import, and all employers must provide information to their employees about the hazardous chemicals to which they are exposed, by means of:

1. A written hazard communication program;

2. Labels and other forms of warning;

3. Material safety data sheets; and

4. Information and training.

These requirements apply to any chemical which is known to be present in the workplace in such a manner that employees may be exposed under normal conditions of use or in a foreseeable emergency (e.g. vessel or pipe rupture).

Chemical manufacturers, importers, and distributors have been responsible for ensuring that material safety data sheets are provided with shipments of hazardous chemicals to non-manufacturing employers since September 23, 1987. Employers in the non-manufacturing sector must be in compliance with all provisions of the Standard by May 23, 1988. Employers in the manufacturing sector (SIC Codes 20 through 39) are already required to be in compliance with all provisions of the Standard (52 FR 31883, revising 29 CFR § 1910.1200 (j)(1) and (j)(2)).

Refer To: 52 FR 31877 revising 29 CFR § 1910.1200(b)

Appendix B: Regulations Reference Guide

B. PERFORMING A HAZARD DETERMINATION

1. Chemical manufacturers and importers must perform hazard evaluations of the chemicals they make or import.

Chemical manufacturers and importers must evaluate chemicals produced in their workplaces or imported by them to determine if they are hazardous. Other employers are not required to evaluate chemicals unless they choose not to rely on the evaluation performed by the chemical manufacturer or importer for the chemical to satisfy this requirement. Whoever performs the evaluation is responsible for its accuracy.

Refer To: 52 FR 31879 revising 29 CFR § 1910.1200(d)(2)

---

2. Chemical manufacturers, importers, or employers evaluating chemicals must identify and consider the available scientific evidence on hazards.

Chemical manufacturers, importers, or employers evaluating chemicals must identify and consider the available scientific evidence concerning such hazards. For health hazards, evidence which is statistically significant and which is based on at least one positive study conducted in accordance with established scientific principles in considered to be sufficient to establish a hazardous effect if the results of the study meet the definitions of health hazards. See Appendix A of the regulation for the scope of the health hazards covered, and Appendix B for the criteria with respect to completeness of the evaluation and the data to be reviewed.

Refer To: 52 FR 31879 revising 29 CFR § 1910.1200(d)(2)

---

3. Chemical manufacturers, importers, or employers evaluating chemicals must treat specified sources as establishing that the chemicals listed in them are hazardous.

The chemical manufacturer, importer, or employer evaluating chemicals must treat the following sources, at a minimum, as establishing that the chemicals listed in them are hazardous:

1. 29 CFR Part 1920, Subpart Z, Toxic and Hazardous Substances, (OSHA)

Chemical Hazard Communication Guidebook

2. Threshold Limit Values for Chemical Substances and Physical Agents in the Work Environment, American Conference of Governmental Industrial Hygienists (ACGIH) (latest edition)

The chemical manufacturer, importer, or employer is still responsible for evaluating the hazards associated with the chemicals in these source lists in accordance with the requirements of the standard.

Refer To: 52 FR 31879-80 revising 29 CFR § 1910.1200(d)(3)

---

4. Chemical manufacturers and importers performing evaluations must treat specified sources as establishing that a chemical is a carcinogen or potential carcinogen.

Chemical manufacturers, importers, and employers evaluating chemicals must treat the following sources, at a minimum, as establishing that a chemical is a carcinogen or potential carcinogen for hazard communication purposes:

1. National Toxicology Program (NTP) Annual Report on Carcinogens (latest edition);

2. International Agency for Research on Cancer (IARC) Monographs (latest edition); and

3. 29 CFR Part 1920, Subpart Z, Toxic and Hazardous Substances, OSHA.

(The Registry of Toxic Effects of Chemical Substances published by NIOSH indicates whether a chemical has been found by NTP or IARC to be a potential carcinogen.)

Refer To: 52 FR 31880 revising 29 CFR § 1910.1200(d)(4)

---

5. Chemical manufacturers, importers, and employers performing evaluations must determine the hazards of mixtures according to specified criteria.

The chemical manufacturer, importer, or employer must determine the hazards of mixtures of chemicals as follows:

1. If a mixture has been tested as a whole to determine its hazards, the results must be used to determine whether the mixture is hazardous.

Appendix B: Regulations Reference Guide

2. If a mixture has not been tested as a whole to determine whether the mixture is a health hazard, it will be assumed to present the same health hazards as the components which comprise one percent (by weight or volume) or greater of the mixture, except that the mixture will be assumed to present a carcinogenic hazard if it contains a component in concentrations of 0.1 percent or greater which is considered to be a carcinogen.

3. If a mixture has not been tested as a whole to determine whether the mixture is a physical hazard, the chemical manufacturer, importer, or employer may use whatever scientifically valid data is available to evaluate the physical hazard potential or the mixture.

4. If the chemical manufacturer, importer, or employer has evidence to indicate that a component present in the mixture in concentrations of less than one percent (or less than 0.1 percent in the case of carcinogens) could be released in concentrations which would exceed an established OSHA permissible exposure limit or ACGIH Threshold Limit Value, or could present a health hazard to employees in those concentrations, the mixture shall be assumed to present the same hazard.

Refer To: 52 FR 31880 revising 29 CFR § 1910.1200(d)(5)

---

6. Chemical manufacturers, importers, or customers evaluating chemicals must describe the procedures used in making the evaluations.

Chemical manufacturers, importers, or employers evaluating chemicals must describe in writing the procedures they use to determine the hazards of the chemical they evaluate. They must make the written procedures available, upon request, to employees, their designated representatives, the Assistant Secretary and the Director. The written description may be incorporated into the mandatory written hazard communication program.

Refer To: 52 FR 31880 revising 29 CFR § 1910.1200(d)(6)

Chemical Hazard Communication Guidebook

C. PREPARING MATERIAL SAFETY DATA SHEETS

1. Chemical manufacturers and importers must provide distributors and customers with material safety data sheets for all hazardous chemicals.

Chemical manufacturers or importers must ensure that distributors and customers are provided an appropriate MSDS with their initial shipment, and with the first shipment after an MSDS is updated. The chemical manufacturer or importer must either provide material safety data sheets with the shipped containers or send them to the customer prior to or at the time of the shipment. If the MSDS is not provided with a shipment that has been labeled as a hazardous chemical, the recipient must obtain one from the chemical manufacturer, importer, or distributor as soon as possible.

Refer To: 52 FR 31881 revising 29 CFR § 1910.1200(g)(6)

---

2. Chemical manufacturers and importers must develop or obtain a material safety data sheet containing the required information for each hazardous chemical they produce or import.

Each material safety data sheet must be in English and contain at least the following information:

1. The chemical name used on the label and, unless the specific chemical name is a trade secret, the specific name. If the hazardous chemical is:

(a) a single substance, its chemical and common name(s) must be given;

(b) a mixture which has been tested as a whole to determine its hazards, the chemical and common name(s) of the ingredients which contribute to these known hazards, and the common name(s) of the mixture itself must be given;

(c) a mixture which has not been tested as a whole: (1) the chemical and common name(s) of all ingredients which have been determined to be health hazards, and which comprise 1% or greater of the composition must be listed, except that chemicals identified as carcinogens shall be listed if the concentrations are 0.1% or greater; (2) the chemical and common name(s) of all ingredients which have

Appendix B: Regulations Reference Guide

been determined to be health hazards, and which comprise less than 1% (0.1% for carcinogens) of the mixture must be listed, if there is evidence that the ingredient(s) could be released from the mixture in concentrations which would exceed an established OSHA permissible exposure limit or ACGIH Threshold Limit Value, or could present a health hazard to employees; and (3) the chemical and common name(s) of all ingredients which have been determine to present a physical hazard when present in the mixture must be given.

2. Physical and chemical characteristics of the hazardous chemical (such as vapor pressure, flash point);

3. Physical hazards of the hazardous chemical, including the potential for fire, explosion, and reactivity;

4. Health hazards of the hazardous chemical, including signs and symptoms of exposure, and any medical conditions which are generally recognized as being aggravated by exposure to the chemical;

5. Primary route(s) of entry;

6. Where available, OSHA permissible exposure limit, ACGIH Threshold Limit Value, and any other exposure limit used or recommended by the chemical manufacturer, importer, or employer preparing the material safety data sheet;

7. Whether the hazardous chemical is listed in the National Toxicology Program (NTP) "Annual Report on Carcinogens" (latest edition) or has been found to be a potential carcinogen in the International Agency for Research on Cancer (IARC) "Monographs" (latest edition), or by OSHA;

8. Any generally applicable precautions for safe handling and use which are known to the chemical manufacturer, importer or employer preparing the material safety data sheet, including appropriate hygienic practices, protective measures during repair, and maintenance of contaminated equipment, and procedures for clean-up of spills and leaks;

9. Any generally applicable control measures which are known to the chemical manufacturer, importer or employer preparing the material safety data sheet, such as appropriate engineering controls, work practices, or personal protective equipment;

10. Emergency and first aid procedures;

11. Date of preparation of the material safety data sheet or the last change to it; and

12. The name, address, and telephone number of the chemical manufacturer, importer, employer or other responsible party preparing or distributing the material safety data sheet, who can provide additional information on the hazardous chemical and appropriate emergency procedures, if necessary.

In cases where complex mixtures have similar hazards and contents (i.e. the chemical ingredients are essentially the same, but the specific composition varies from mixture to mixture), the chemical manufacturer, importer, or employer may prepare one material safety data sheet to apply to all of these similar mixtures ((g)(4)).

Refer To: 52 FR 31881 revising 29 CFR § 1910.1200(g)(2)

Appendix B: Regulations Reference Guide

2. Chemical manufacturers, importers, or employers preparing the material safety data sheet must ensure that any significant new information that is discovered is used to update the MSDS within three months.

If the chemical manufacturer, importer, or employer preparing the MSDS becomes newly aware of any significant information regarding the hazards of a chemical, or ways to protect against the hazards, this new information must be added to the MSDS within three months. If the chemical is not currently being produced or imported, the chemical manufacturer or importer must add the information to the material safety data sheet before the chemical is introduced into the workplace or shipped again.

Refer To: 52 FR 31881 revising 29 CFR § 1910.1200(g)(5)

---

3. Chemical manufacturers, importers, or employers preparing the material safety data sheet who do not find relevant information for a given category must indicate this by marking that no applicable information was found.

Refer To: 52 FR 31881 revising 29 CFR § 1910.1200(g)(4)

D. MAINTAINING MATERIAL SAFETY DATA SHEETS

1. Employers must maintain a material safety data sheet for each hazardous chemical on site.

Each employer must maintain copies of the required MSDS's for each hazardous chemical in the workplace, and must ensure that they are readily accessible during each work shift to employees when they are in their work areas so that they can immediately obtain the required information in the event of an emergency ((g)(8)).

If the work of employees is carried out at more than one geographical location

Chemical Hazard Communication Guidebook

during a workshift, the MSDS's may be kept at a central location at the primary workplace facility ((g)(9)).

MSDS's may be kept in any form, including operating procedures, and may be designed to cover groups of hazardous chemicals in a work area where it may be more appropriate to address the hazards of a process rather than individual hazardous chemicals ((g)(10)).

Refer To: 52 FR 31881 revising 29 CFR § 1910.1200(g)(1)

---

2. Distributors must ensure that MSDS's and updated information, are provided to other distributors and employers with the first shipment of a hazardous chemical and with the first shipment after an MSDS is updated.

Refer To: 52 FR 31882 revising 29 CFR § 1910.1200(g)(6)

---

3. Employers must make material safety data sheets readily available, upon request, to designated representatives and OSHA.

Refer To: 52 FR 31882 revising 29 CFR § 1910.1200(g)(11)

---

4. Retail distributors which sell hazardous chemical to commercial customers must provide a material safety data sheet to such employers upon request and inform them that an MSDS is available.

Chemical manufacturers, importers, and distributors need not provide MSDS's to a retail distributor which has informed them that it does not sell the product to commercial customers or open the sealed container in its own workplace.

Refer To: 52 FR 31882 revising 29 CFR § 1910.1200(g)(7)

Appendix B: Regulations Reference Guide

E. PREPARING AND MAINTAINING LABELS AND OTHER WARNINGS

1. Employers must label, tag, or mark each container of hazardous chemicals in the workplace with the required data.

Subject to certain limited exceptions, each employer must ensure that each container of hazardous chemicals in the workplace is labeled, tagged, or marked with the following information:

1. Identity of the hazardous chemical(s) contained therein; and

2. Appropriate hazard warnings.

The employer may use signs, placards, process sheets, batch tickets, operating procedures, or other such written materials in lieu of affixing labels to individual stationary process containers, as long as the alternative method identifies the containers to which it is applicable and conveys the information required to be on a label. The written materials must be readily accessible to the employees in their work area throughout each work shift.

The chemical manufacturer, importer, distributor or employer need not affix new labels to comply with this section if existing labels already convey the required information.

Refer To: 52 FR 31881 revising 29 CFR § 1910.1200(f)(5)

---

2. Certain exemptions to the labeling requirement exist.

Labeling requirements do not apply to the following types of chemicals:

1. Any pesticide as defined in the Federal Insecticide, Fungicide, and Rodenticide Act (FIFRA) when subject to the labeling requirements of that Act;

2. Any food, food additive, color additive, drug, cosmetic, or medical or veterinary device, including materials intended for use as ingredients in such

products as defined in the Federal Food, Drug, and Cosmetic Act when they are subject to the labeling requirements of that Act;

3. Any distilled spirits (beverage alcohols), wine, or malt beverage intended for nonindustrial use, as defined in the Federal Alcohol Administration Act when subject to the labeling requirements of that Act;

4. Any consumer product or hazardous substance as defined in the Consumer Product Safety Act and Federal Hazardous Substances Act when subject to the respective labeling requirements of those Acts;

5. In addition, containers used by an employee during one shift do not need to be labeled. Pipes do not need to be labeled. Containers used only "in house" can be labeled in accordance with the HMIS or similar system of providing warning of health and physical hazards by code number or symbol if appropriate training on the system is provided to employees.

Refer To: 52 FR 31877 revising 29 CFR § 1910.1200(b)(5)

---

3. Employers must ensure that labels or other forms of warning are legible, in English, and prominently displayed on the container, or readily available in the work area throughout each work shift.

Refer To: 52 FR 31881 revising 29 CFR § 1910.1200(f)(9)

Appendix B: Regulations Reference Guide

4. Chemical manufacturers, importers, and distributors must ensure that each container of hazardous chemicals leaving the workplace is labeled, tagged, or marked with specified information.	The chemical manufacturer, importer, or distributor must ensure that each container of hazardous chemicals leaving the workplace is labeled, tagged or marked with the following information: 1. Identity of the hazardous chemical(s); 2. Appropriate hazard warnings; and 3. Name and address of the chemical manufacturer, importer, or other responsible party; (For solid metal, such as a steel beam or a metal casting, that is not exempted as an article due to its downstream use, the required label may be transmitted to the customer at the time of the initial shipment, and need not be included with subsequent shipments to the same employer unless the information on the label changes. The label may be transmitted with the initial shipment itself, or with the material safety data sheet that is to be provided prior to or at the time of the first shipment. This exception to requiring labels on every container of hazardous chemicals is only for the solid metal itself and does not apply to hazardous chemicals used in conjunction with, or known to be present with, the metal and to which employees handling the metal may be exposed.)

Refer To: 52 FR 31880 revising 29 CFR § 1910.1200(f)(1)

---

5. Chemical manufacturers, importers, and distributors must ensure that labels, tags, or markings for hazardous materials leaving the workplace do not conflict with DOT regulations.	Chemical manufacturers, importers, or distributors shall ensure that each container of hazardous chemicals leaving the workplace is labeled, tagged, or marked in accordance with the standard in a manner which does not conflict with the requirements of the Hazardous Materials Transportation Act and regulations issued under that Act by the Department of Transportation (See DOT summary below in this appendix)

Refer To: 52 FR 31880 revising 29 CFR § 1910.1200(f)(3)

Chemical Hazard Communication Guidebook

6. Chemical manufacturers, importers, distributors, and employers must label in accordance with applicable requirements if the chemical is regulated by OSHA in a substance-specific standard.

If the hazardous chemical is regulated by OSHA in a substance-specific health standard, the chemical manufacturer, importer, distributor, or employer shall ensure that the labels or other forms of warning used are in accordance with the requirements of that standard.

Refer To: 52 FR 31880 revising 29 CFR § 1910.1200(f)(4)

---

7. Employers shall not remove or deface existing labels on incoming containers of hazardous chemicals unless the container is immediately marked with the required information.

Refer To: 52 FR 31881 revising 29 CFR § 1910.1200(f)(8)

F. TRAINING EMPLOYEES ABOUT CHEMICAL HAZARDS

1. Employers must provide employees with information on hazardous chemicals in their work area at the time of their initial assignment and whenever a new hazard is introduced into their work area.

Employees must be informed of:

1. The requirements of the regulation;

2. Any operations in their work area where hazardous chemicals are present; and

3. The location and availability of the written hazard communication program, including the required list(s) of hazardous chemicals, and material safety data sheets required.

Refer To: 52 FR 31882 revising 29 CFR § 1910.1200(h)(1)

Appendix B: Regulations Reference Guide

2. Employers must train employees regarding hazardous chemicals in their work area at the time of their initial assignment and whenever a new hazard is introduced into their work area.

Employee training must include at least:

1. Methods and observations that may be used to detect the presence or release of a hazardous chemical in the work area (such as monitoring conducted by the employer, continuous monitoring devices, visual appearance or odor of hazardous chemicals when being released, etc.);

2. The physical and health hazards of the chemicals in the work area;

3. The measures employees can take to protect themselves from these hazards, including specific procedures the employer has implemented to protect employees from exposure to hazardous chemicals, such as appropriate work practices, emergency procedures, and personal protective equipment to be used; and

4. The details of the hazard communication program developed by the employer, including an explanation of the labeling system and the material safety data sheet, and how employees can obtain and use the appropriate hazard information

Refer To: 52 FR 31882 revising 29 CFR § 1910.1200(h)(2)

---

3. Employers may use general guidelines provided by OSHA as a model for establishing or evaluating a training program.

OSHA's training guidelines suggest that employers determine whether training is needed, identify training needs, goals, and objectives, develop learning activities, provide training, evaluate the effectiveness of the program, and plan for improvement.

Evaluation of the effectiveness of the program should include consideration of the following questions:

1. If a job analysis was conducted, was it accurate?

2. Was any critical feature of the job overlooked?

3. Were the important gaps in knowledge and skill included?

4. Was material already known by the employees intentionally omitted?

5. Were the instructional objectives presented clearly and concretely?

6. Did the objectives state the level of acceptable performance that was expected of employees?

7. Did the learning activity simulate the actual job?

8. Was the learning activity appropriate for the kinds of knowledge and skills required on the job?

9. When the training was presented, was the organization of the material and its meaning made clear?

10. Were the employees motivated to learn?

11. Were the employees allowed to participate actively in the training process?

12. Was the employer's evaluation of the program thorough?

The general guidelines are not intended, and cannot be used, as evidence of the appropriate level of training in litigation involving either the training requirements of OSHA standards or affirmative defenses based on employer training programs.

Refer To: 49 FR 30293

Appendix B: Regulations Reference Guide

G. PREPARING AND PROVIDING A WRITTEN HAZARD COMMUNICATION PROGRAM

1. Employers must develop, implement, and maintain at the workplace a written hazard communication program.

Employers must develop and maintain at the workplace a written hazard communication program which describes how the requirements for labels and other forms of warning, material safety data sheets, and employee information and training will be met.

The written program must include the following:

1. A list of the hazardous chemicals known to be present using an identity that is referenced on the appropriate material safety data sheet (the list may be compiled for the workplace as a whole or for individual work areas); and

2. The methods the employer will use to inform employees of the hazards of non-routine tasks and the hazards associated with chemicals contained in unlabeled pipes in their work areas.

Refer To: 52 FR 31880 revising 29 CFR § 1910.1200(e)(1)

---

2. Employers must make the written program available, upon request, to employees, their designated representatives, and OSHA.

Refer To: 52 FR 31880 revising 29 CFR § 1910.1200(e)(4)

---

3. Employers whose workplace hazards may expose employees of a different employer must include in their written hazard communication program a procedure for making hazard information available to those employees.

Employers who produce use, or store hazardous chemicals at a workplace in such a way that the employees of other employer(s) may be exposed must additionally ensure that the hazard communication programs developed and implemented include the following:

1. The methods the employer will use to provide the other employer(s) with a

Chemical Hazard Communication Guidebook

copy of the material safety data sheet, or to make it available at a central location in the workplace, for each hazardous chemical the other employer(s)' employees may be exposed to while working;

2. The methods the employer will use to inform the other employer(s) of any precautionary measures that need to be taken to protect employees during the workplace's normal operating conditions and in foreseeable emergencies; and,

3. The methods the employer will use to inform the other employer(s) of the labeling system used in the workplace.

Refer To: 52 FR 31880 revising 29 CFR § 1910.1200(e)(2)

H. REQUIREMENTS FOR THOSE HANDLING HAZARDOUS CHEMICALS IN SEALED CONTAINERS

1. Employers who only handle hazardous chemicals in sealed containers must ensure that employees are provided with information and training on responding to leaks and spills.

Refer To: 52 FR 31877 revising 29 CFR § 1910.1200(b)(4)

---

2. Employers who only handle hazardous chemicals in sealed containers must ensure that labels on incoming containers of hazardous chemicals are not removed or defaced.

Refer To: 52 FR 31877 revising 29 CFR § 1910.1200(b)(4)

Appendix B: Regulations Reference Guide

3. Employers must ensure that MSDS's received are readily accessible during each work shift to employees in their work areas.

Refer To: 52 FR 31877 revising 29 CFR § 1910.1200(b)(4)

---

4. At the request of an employee, employers must obtain an MSDS for sealed containers of hazardous chemicals received without one.

Refer To: 52 FR 31877 revising 29 CFR § 1910.1200(b)(4)

I. SPECIAL REQUIREMENTS FOR LABORATORIES

1. Employers must apprise employees of the hazards of chemicals in their workplaces.

Refer To: 52 FR 31877 revising 29 CFR § 1910.1200(b)(3)

---

2. Employers must ensure that labels on incoming containers of hazardous chemicals are not removed or defaced.

Refer To: 52 FR 31877 revising 29 CFR § 1910.1200(b)(3)

Chemical Hazard Communication Guidebook

3. Employers must maintain material safety data sheets received for laboratory chemicals.

Employers must maintain any material safety data sheets that are received with incoming shipments of hazardous chemicals, and ensure that they are readily accessible to laboratory employees.

Refer To: 52 FR 31877 revising 29 CFR § 1910.1200(b)(3)

J. REQUIREMENTS FOR DISCLOSING AND WITHHOLDING TRADE SECRET INFORMATION

1. Chemical manufacturers, importers, and employers may withhold specific chemical identity information because of trade secret status.

The chemical manufacturer, importer, or employer may withhold the specific chemical identity, including the chemical name and other specific identification of a hazardous chemical from the material safety data sheet, provided that:

1. Information contained in the MSDS concerning the properties and effects of the hazardous chemical is disclosed;

2. The MSDS indicates that the specific chemical identity is being withheld as a trade secret; and

3. The specific chemical identity is made available to health professionals, employees, and designated representatives as required by the standard.

Refer To: 52 FR 31882 revising 29 CFR § 1910.1200(i)(1)

Appendix B: Regulations Reference Guide

2. Chemical manufacturers, importers, and employers withholding chemical identity information because of trade secret status must be able to support their claim of a trade secret.

Refer To: 52 FR 31882 revising 29 CFR § 1910.1200(i)(1)

3. Chemical manufacturers, importers, or employers who withhold chemical identity information based on a trade secret claim must disclose it if deemed necessary by health professionals in a medical emergency.

Where a treating physician or nurse determines that a medical emergency exists and the specific chemical identity of a hazardous chemical is necessary for emergency or first-aid treatment, the chemical manufacturer, importer, or employer must immediately disclose the specific chemical identity of a trade secret chemical to that treating physician or nurse but may require a written statement of need and/or a confidentiality agreement as soon as circumstances permit.

Refer To: 52 FR 31882 revising 29 CFR § 1910.1200(i)(2)

4. Chemical manufacturers, importers, and employers must disclose chemical identity information withheld under a trade secret claim to a health professional in non-emergency situations under certain conditions.

In non-emergency situations, a chemical manufacturer, importer, or employer must, upon request, disclose a specific chemical identity, otherwise permitted to be withheld, to a health professional providing medical or other occupational health services to exposed employees and to employees or designated representatives if the request fulfills certain conditions, including the conditions that it must be in writing and describe with reasonable detail one or more of the following occupational health needs for the information:

Chemical Hazard Communication Guidebook

1. To assess the hazards of the chemicals to which employees will be exposed;

2. To conduct or assess sampling of the workplace atmosphere to determine employee exposure levels;

3. To conduct pre-assignment or periodic medical surveillance of exposed employees;

4. To select or assess appropriate personal protective equipment for exposed employees;

5. To design or assess engineering controls or other protective measures for exposed employees;

6. To conduct studies to determine the health effects of exposure.

Refer To: 52 FR 31882 revising 29 CFR § 1910.1200(i)(3)

---

5. A chemical manufacturer, importer, or employer denying a written request for disclosure of chemical identity information withheld as a trade secret must respond to the requestor within thirty days.

If the chemical manufacturer, importer, or employer denies a written request for disclosure of a specific chemical identity, the denial must:

1. Be in writing;

2. Include evidence to support the claim that the specific chemical identity is a trade secret;

3. State the specific reasons why the request is being denied; and

4. Explain in detail how alternative information may satisfy the specific medical or occupational health need without revealing the specific chemical identity.

OSHA may subject a chemical manufacturer, importer, or employer denying a request to disclose a specific

Appendix B: Regulations Reference Guide

          chemical identity to citation if OSHA deems the trade secret not bona fide or the health need as legitimate, despite the claim for secrecy.

Refer To: 52 FR 31883 revising 29 CFR § 1910.1200(i)(7)

---

6. Chemical manufacturers, importers, or employers must, upon request, disclose to OSHA any information which the regulation requires them to make available, notwithstanding the existence of a trade secret claim.

          Refer To: 52 FR 31883 revising 29 CFR § 1910.1200(i)(12)

OSHA'S HAZARD COMMUNICATION STANDARD

Chemical Hazard Communication Guidebook

29 CFR Ch. XVII (7-1-89 Edition)

§ 1910.1200 Hazard communication.

(a) *Purpose.* (1) The purpose of this section is to ensure that the hazards of all chemicals produced or imported are evaluated, and that information concerning their hazards is transmitted to employers and employees. This transmittal of information is to be accomplished by means of comprehensive hazard communication programs, which are to include container labeling and other forms of warning, material safety data sheets and employee training.

(2) This occupational safety and health standard is intended to address comprehensively the issue of evaluating the potential hazards of chemicals, and communicating information concerning hazards and appropriate protective measures to employees, and to preempt any legal requirements of a state, or political subdivision of a state, pertaining to the subject. Evaluating the potential hazards of chemicals, and communicating information concerning hazards and appropriate protective measures to employees, may include, for example, but is not limited to, provisions for: developing and maintaining a written hazard communication program for the workplace, including lists of hazardous chemicals present; labeling of containers of chemicals in the workplace, as well as of containers of chemicals being shipped to other workplaces; preparation and distribution of material safety data sheets to employees and downstream employers; and development and implementation of employee training programs regarding hazards of chemicals and protective measures. Under section 18 of the Act, no state or political subdivision of a state may adopt or enforce, through any court or agency, any requirement relating to the issue addressed by this Federal standard, except pursuant to a Federally-approved state plan.

(b) *Scope and application.* (1) This section requires chemical manufacturers or importers to assess the hazards of chemicals which they produce or import, and all employers to provide information to their employees about

356

Appendix B: Regulations Reference Guide

Occupational Safety and Health Admin., Labor § 1910.1200

the hazardous chemicals to which they are exposed, by means of a hazard communication program, labels and other forms of warning, material safety data sheets, and information and training. In addition, this section requires distributors to transmit the required information to employers.

(2) This section applies to any chemical which is known to be present in the workplace in such a manner that employees may be exposed under normal conditions of use or in a foreseeable emergency.

(3) This section applies to laboratories only as follows:

(i) Employers shall ensure that labels on incoming containers of hazardous chemicals are not removed or defaced;

(ii) Employers shall maintain any material safety data sheets that are received with incoming shipments of hazardous chemicals, and ensure that they are readily accessible to laboratory employees; and,

(iii) Employers shall ensure that laboratory employees are apprised of the hazards of the chemicals in their workplaces in accordance with paragraph (h) of this section.

(4) In work operations where employees only handle chemicals in sealed containers which are not opened under normal conditions of use (such as are found in marine cargo handling, warehousing, or retail sales), this section applies to these operations only as follows:

(i) Employers shall ensure that labels on incoming containers of hazardous chemicals are not removed or defaced;

(ii) Employers shall maintain copies of any material safety data sheets that are received with incoming shipments of the sealed containers of hazardous chemicals, shall obtain a material safety data sheet for sealed containers of hazardous chemicals received without a material safety data sheet if an employee requests the material safety data sheet, and shall ensure that the material safety data sheets are readily accessible during each work shift to employees when they are in their work area(s); and,

(iii) Employers shall ensure that employees are provided with information and training in accordance with paragraph (h) of this section (except for the location and availability of the written hazard communication program under paragraph (h)(1)(iii)), to the extent necessary to protect them in the event of a spill or leak of a hazardous chemical from a sealed container.

(5) This section does not require labeling of the following chemicals:

(i) Any pesticide as such term is defined in the Federal Insecticide, Fungicide, and Rodenticide Act (7 U.S.C. 136 et seq.), when subject to the labeling requirements of that Act and labeling regulations issued under that Act by the Environmental Protection Agency;

(ii) Any food, food additive, color additive, drug, cosmetic, or medical or veterinary device, including materials intended for use as ingredients in such products (e.g. flavors and fragrances), as such terms are defined in the Federal Food, Drug, and Cosmetic Act (21 U.S.C. 301 *et seq.*) and regulations issued under that Act, when they are subject to the labeling requirements under that Act by the Food and Drug Administration;

(iii) Any distilled spirits (beverage alcohols), wine, or malt beverage intended for nonindustrial use, as such terms are defined in the Federal Alcohol Administration Act (27 U.S.C. 201 *et seq.*) and regulations issued under that Act, when subject to the labeling requirements of that Act and labeling regulations issued under that Act by the Bureau of Alcohol Tobacco, and Firearms; and,

(iv) Any consumer product or hazardous substance as those terms are defined in the Consumer Product Safety Act (15 U.S.C. 2051 et seq.) and Federal Hazardous Substances Act (15 U.S.C. 1261 et seq.) respectively, when subject to a consumer product safety standard or labeling requirement of those Acts, or regulations issued under those Acts by the Consumer Product Safety Commission.

(6) This section does not apply to:

(i) Any hazardous waste as such term is defined by the Solid Waste Disposal Act, as amended by the Resource Conservation and Recovery Act of 1976, as amended (42 U.S.C. 6901 et

357

§ 1910.1200

seq.), when subject to regulations issued under that Act by the Environmental Protection Agency;

(ii) Tobacco or tobacco products;

(iii) Wood or wood products;

(iv) Articles;

(v) Food, drugs, cosmetics, or alcoholic beverages in a retail establishment which are packaged for sale to consumers;

(vi) Foods, drugs, or cosmetics intended for personal consumption by employees while in the workplace;

(vii) Any consumer product or hazardous substance, as those terms are defined in the Consumer Product Safety Act (15 U.S.C. 2051 *et seq.*) and Federal Hazardous Substances Act (15 U.S.C. 1261 *et seq.*) respectively, where the employer can demonstrate it is used in the workplace in the same manner as normal consumer use, and which use results in a duration and frequency of exposure which is not greater than exposures experienced by consumers; and,

(viii) Any drug, as that term is defined in the Federal Food, Drug, and Cosmetic Act (21 U.S.C. 301 *et seq.*), when it is in solid, final form for direct administration to the patient (i.e. tablets or pills).

(c) *Definitions.*

"Article" means a manufactured item: (i) Which is formed to a specific shape or design during manufacture; (ii) which has end use function(s) dependent in whole or in part upon its shape or design during end use; and (iii) which does not release, or otherwise result in exposure to, a hazardous chemical, under normal conditions of use.

"Assistant Secretary" means the Assistant Secretary of Labor for Occupational Safety and Health, U.S. Department of Labor, or designee.

"Chemical" means any element, chemical compound or mixture of elements and/or compounds.

"Chemical manufacturer" means an employer with a workplace where chemical(s) are produced for use or distribution.

"Chemical name" means the scientific designation of a chemical in accordance with the nomenclature system developed by the International Union of Pure and Applied Chemistry (IUPAC) or the Chemical Abstracts Service (CAS) rules of nomenclature, or a name which will clearly identify the chemical for the purpose of conducting a hazard evaluation.

"Combustible liquid" means any liquid having a flashpoint at or above 100 °F (37.8 °C), but below 200 °F (93.3 °C), except any mixture having components with flashpoints of 200 °F (93.3 °C), or higher, the total volume of which make up 99 percent or more of the total volume of the mixture.

"Common name" means any designation or identification such as code name, code number, trade name, brand name or generic name used to identify a chemical other than by its chemical name.

"Compressed gas" means:

(i) A gas or mixture of gases having, in a container, an absolute pressure exceeding 40 psi at 70 °F (21.1 °C); or

(ii) a gas or mixture of gases having, in a container, an absolute pressure exceeding 104 psi at 130 °F (54.4 °C) regardless of the pressure at 70 °F (21.1 °C); or

(iii) A liquid having a vapor pressure exceeding 40 psi at 100 °F (37.8 °C) as determined by ASTM D-323-72.

"Container" means any bag, barrel, bottle, box, can, cylinder, drum, reaction vessel, storage tank, or the like that contains a hazardous chemical. For purposes of this section, pipes or piping systems, and engines, fuel tanks, or other operating systems in a vehicle, are not considered to be containers.

"Designated representative" means any individual or organization to whom an employee gives written authorization to exercise such employee's rights under this section. A recognized or certified collective bargaining agent shall be treated automatically as a designated representative without regard to written employee authorization.

"Director" means the Director, National Institute for Occupational Safety and Health, U.S. Department of Health and Human Services, or designee.

"Distributor" means a business, other than a chemical manufacturer or importer, which supplies hazardous

Appendix B: Regulations Reference Guide

Occupational Safety and Health Admin., Labor § 1910.1200

chemicals to other distributors or to employers.

"Employee" means a worker who may be exposed to hazardous chemicals under normal operating conditions or in foreseeable emergencies. Workers such as office workers or bank tellers who encounter hazardous chemicals only in non-routine, isolated instances are not covered.

"Employer" means a person engaged in a business where chemicals are either used, distributed, or are produced for use or distribution, including a contractor or subcontractor.

"Explosive" means a chemical that causes a sudden, almost instantaneous release of pressure, gas, and heat when subjected to sudden shock, pressure, or high temperature.

"Exposure" or "exposed" means that an employee is subjected to a hazardous chemical in the course of employment through any route of entry (inhalation, ingestion, skin contact or absorption, etc.), and includes potential (e.g. accidental or possible) exposure.

"Flammable" means a chemical that falls into one of the following categories:

(i) "Aerosol, flammable" means an aerosol that, when tested by the method described in 16 CFR 1500.45, yields a flame projection exceeding 18 inches at full valve opening, or a flashback (a flame extending back to the valve) at any degree of valve opening;

(ii) "Gas, flammable" means:

(A) A gas that, at ambient temperature and pressure, forms a flammable mixture with air at a concentration of thirteen (13) percent by volume or less; or

(B) A gas that, at ambient temperature and pressure, forms a range of flammable mixtures with air wider than twelve (12) percent by volume, regardless of the lower limit;

(iii) "Liquid, flammable" means any liquid having a flashpoint below 100 °F (37.8 °C), except any mixture having components with flashpoints of 100 °F (37.8 °C) or higher, the total of which make up 99 percent or more of the total volume of the mixture;

(iv) "Solid, flammable" means a solid, other than a blasting agent or explosive as defined in § 190.109(a), that is liable to cause fire through friction, absorption of moisture, spontaneous chemical change, or retained heat from manufacturing or processing, or which can be ignited readily and when ignited burns so vigorously and persistently as to create a serious hazard. A chemical shall be considered to be a flammable solid if, when tested by the method described in 16 CFR 1500.44, it ignites and burns with a self-sustained flame at a rate greater than one-tenth of an inch per second along its major axis.

"Flashpoint" means the minimum temperature at which a liquid gives off a vapor in sufficient concentration to ignite when tested as follows:

(i) Tagliabue Closed Tester (See American National Standard Method of Test for Flash Point by Tag Closed Tester, Z11.24-1979 (ASTM D 56-79)) for liquids with a viscosity of less than 45 Saybolt University Seconds (SUS) at 100 °F (37.8 °C), that do not contain suspended solids and do not have a tendency to form a surface film under test; or

(ii) Pensky-Martens Closed Tester (See American National Standard Method of Test for Flash Point by Pensky-Martens Closed Tester, Z11.7-1979 (ASTM D 93-79)) for liquids with a viscosity equal to or greater than 45 SUS at 100 °F (37.8 °C), or that contain suspended solids, or that have a tendency to form a surface film under test; or

(iii) Setaflash Closed Tester (see American National Standard Method of Test for Flash Point by Setaflash Closed Tester (ASTMD 3278-78))

Organic peroxides, which undergo autoaccelerating thermal decomposition, are excluded from any of the flashpoint determination methods specified above.

"Foreseeable emergency" means any potential occurrence such as, but not limited to, equipment failure, rupture of containers, or failure of control equipment which could result in an uncontrolled release of a hazardous chemical into the workplace.

"Hazardous chemical" means any chemical which is a physical hazard or a health hazard.

"Hazard warning" means any words, pictures, symbols, or combination

359

Chemical Hazard Communication Guidebook

§ 1910.1200

thereof appearing on a label or other appropriate form of warning which convey the hazard(s) of the chemical(s) in the container(s).

"Health hazard" means a chemical for which there is statistically significant evidence based on at least one study conducted in accordance with established scientific principles that acute or chronic health effects may occur in exposed employees. The term "health hazard" includes chemicals which are carcinogens, toxic or highly toxic agents, reproductive toxins, irritants, corrosives, sensitizers, hepatotoxins, nephrotoxins, neurotoxins, agents which act on the hematopoietic system, and agents which damage the lungs, skin, eyes, or mucous membranes. Appendix A provides further definitions and explanations of the scope of health hazards covered by this section, and Appendix B describes the criteria to be used to determine whether or not a chemical is to be considered hazardous for purposes of this standard.

"Identity" means any chemical or common name which is indicated on the material safety data sheet (MSDS) for the chemical. The identity used shall permit cross-references to be made among the required list of hazardous chemicals, the label and the MSDS.

"Immediate use" means that the hazardous chemical will be under the control of and used only by the person who transfers it from a labeled container and only within the work shift in which it is transferred.

"Importer" means the first business with employees within the Customs Territory of the United States which receives hazardous chemicals produced in other countries for the purpose of supplying them to distributors or employers within the United States.

"Label" means any written, printed, or graphic material, displayed on or affixed to containers of hazardous chemicals.

"Material safety data sheet (MSDS)" means written or printed material concerning a hazardous chemical which is prepared in accordance with paragraph (g) of this section.

"Mixture" means any combination of two or more chemicals if the combination is not, in whole or in part, the result of a chemical reaction.

"Organic peroxide" means an organic compound that contains the bivalent -O-O-structure and which may be considered to be a structural derivative of hydrogen peroxide where one or both of the hydrogen atoms has been replaced by an organic radical.

"Oxidizer" means a chemical other than a blasting agent or explosive as defined in § 1910.109(a), that initiates or promotes combustion in other materials, thereby causing fire either of itself or through the release of oxygen or other gases.

"Physical hazard" means a chemical for which there is scientifically valid evidence that it is a combustible liquid, a compressed gas, explosive, flammable, an organic peroxide, an oxidizer, pyrophoric, unstable (reactive) or water-reactive.

"Produce" means to manufacture, process, formulate, or repackage.

"Pyrophoric" means a chemical that will ignite spontaneously in air at a temperature of 130 °F (54.4 °C) or below.

"Responsible party" means someone who can provide additional information on the hazardous chemical and appropriate emergency procedures, if necessary.

"Specific chemical identity" means the chemical name, Chemical Abstracts Service (CAS) Registry Number, or any other information that reveals the precise chemical designation of the substance.

"Trade secret" means any confidential formula, pattern, process, device, information or compilation of information that is used in an employer's business, and that gives the employer an opportunity to obtain an advantage over competitors who do not know or use it. Appendix D sets out the criteria to be used in evaluating trade secrets.

"Unstable (reactive)" means a chemical which in the pure state, or as produced or transported, will vigorously polymerize, decompose, condense, or will become self-reactive under conditions of shocks, pressure or temperature.

Occupational Safety and Health Admin., Labor § 1910.1200

"Use" means to package, handle, react, or transfer.

"Water-reactive" means a chemical that reacts with water to release a gas that is either flammable or presents a health hazard.

"Work area" means a room or defined space in a workplace where hazardous chemicals are produced or used, and where employees are present.

"Workplace" means an establishment, job site, or project, at one geographical location containing one or more work areas.

(d) *Hazard determination.* (1) Chemical manufacturers and importers shall evaluate chemicals produced in their workplaces or imported by them to determine if they are hazardous. Employers are not required to evaluate chemicals unless they choose not to rely on the evaluation performed by the chemical manufacturer or importer for the chemical to satisfy this requirement.

(2) Chemical manufacturers, importers or employers evaluating chemicals shall identify and consider the available scientific evidence concerning such hazards. For health hazards, evidence which is statistically significant and which is based on at least one positive study conducted in accordance with established scientific principles is considered to be sufficient to establish a hazardous effect if the results of the study meet the definitions of health hazards in this section. Appendix A shall be consulted for the scope of health hazards covered, and Appendix B shall be consulted for the criteria to be followed with respect to the completeness of the evaluation, and the data to be reported.

(3) The chemical manufacturer, importer or employer evaluating chemicals shall treat the following sources as establishing that the chemicals listed in them are hazardous:

(i) 29 CFR Part 1910, Subpart Z, Toxic and Hazardous Substances, Occupational Safety and Health Administration (OSHA); or,

(ii) *Threshold Limit Values for Chemical Substances and Physical Agents in the Work Environment,* American Conference of Governmental Industrial Hygienists (ACGIH) (latest edition).

The chemical manufacturer, importer, or employer is still responsible for evaluating the hazards associated with the chemicals in these source lists in accordance with the requirements of this standard.

(4) Chemical manufacturers, importers and employers evaluating chemicals shall treat the following sources as establishing that a chemical is a carcinogen or potential carcinogen for hazard communication purposes:

(i) National Toxicology Program (NTP), *Annual Report on Carcinogens* (latest edition);

(ii) International Agency for Research on Cancer (IARC) *Monographs* (latest editions); or

(iii) 29 CFR Part 1910, Subpart Z, Toxic and Hazardous Substances, Occupational Safety and Health Administration.

NOTE: The *Registry of Toxic Effects of Chemical Substances* published by the National Institute for Occupational Safety and Health indicates whether a chemical has been found by NTP or IARC to be a potential carcinogen.

(5) The chemical manufacturer, importer or employer shall determine the hazards of mixtures of chemicals as follows:

(i) If a mixture has been tested as a whole to determine its hazards, the results of such testing shall be used to determine whether the mixture is hazardous;

(ii) If a mixture has not been tested as a whole to determine whether the mixture is a health hazard, the mixture shall be assumed to present the same health hazards as do the components which comprise one percent (by weight or volume) or greater of the mixture, except that the mixture shall be assumed to present a carcinogenic hazard if it contains a component in concentrations of 0.1 percent or greater which is considered to be a carcinogen under paragraph (d)(4) of this section;

(iii) If a mixture has not been tested as a whole to determine whether the mixture is a physical hazard, the chemical manufacturer, importer, or employer may use whatever scientifi-

361

cally valid data is available to evaluate the physical hazard potential of the mixture; and,

(iv) If the chemical manufacturer, importer, or employer has evidence to indicate that a component present in the mixture in concentrations of less than one percent (or in the case of carcinogens, less than 0.1 percent) could be released in concentrations which would exceed an established OSHA permissible exposure limit or ACGIH Threshold Limit Value, or could present a health hazard to employees in those concentrations, the mixture shall be assumed to present the same hazard.

(6) Chemical manufacturers, importers, or employers evaluating chemicals shall describe in writing the procedures they use to determine the hazards of the chemical they evaluate. The written procedures are to be made available, upon request, to employees, their designated representatives, the Assistant Secretary and the Director. The written description may be incorporated into the written hazard communication program required under paragraph (e) of this section.

(e) *Written hazard communication program.* (1) Employers shall develop, implement, and maintain at the workplace, a written hazard communication program for their workplaces which at least describes how the criteria specified in paragraphs (f), (g), and (h) of this section for labels and other forms of warning, material safety data sheets, and employee information and training will be met, and which also includes the following:

(i) A list of the hazardous chemicals known to be present using an identity that is referenced on the appropriate material safety data sheet (the list may be compiled for the workplace as a whole or for individual work areas); and,

(ii) The methods the employer will use to inform employees of the hazards of non-routine tasks (for example, the cleaning of reactor vessels), and the hazards associated with chemicals contained in unlabeled pipes in their work areas.

(2) *Multi-employer workplaces.* Employers who produce, use, or store hazardous chemicals at a workplace in such a way that the employees of other employer(s) may be exposed (for example, employees of a construction contractor working on-site) shall additionally ensure that the hazard communication programs developed and implemented under this paragraph (e) include the following:

(i) The methods the employer will use to provide the other employer(s) with a copy of the material safety data sheet, or to make it available at a central location in the workplace, for each hazardous chemical the other employer(s)' employees may be exposed to while working;

(ii) The methods the employer will use to inform the other employer(s) of any precautionary measures that need to be taken to protect employees during the workplace's normal operating conditions and in foreseeable emergencies; and,

(iii) The methods the employer will use to inform the other employer(s) of the labeling system used in the workplace.

(3) The employer may rely on an existing hazard communication program to comply with these requirements, provided that it meets the criteria established in this paragraph (e).

(4) The employer shall make the written hazard communication program available, upon request, to employees, their designated representatives, the Assistant Secretary and the Director, in accordance with the requirements of 29 CFR 1910.20(e).

(f) *Labels and other forms of warning.* (1) The chemical manufacturer, importer, or distributor shall ensure that each container of hazardous chemicals leaving the workplace is labeled, tagged or marked with the following information:

(i) Identity of the hazardous chemical(s);

(ii) Appropriate hazard warnings; and

(iii) Name and address of the chemical manufacturer, importer, or other responsible party.

(2) For solid metal (such as a steel beam or a metal casting) that is not exempted as an article due to its downstream use, the required label may be transmitted to the customer at the time of the intial shipment, and

Occupational Safety and Health Admin., Labor § 1910.1200

need not be included with subsequent shipments to the same employer unless the information on the label changes. The label may be transmitted with the initial shipment itself, or with the material safety data sheet that is to be provided prior to or at the time of the first shipment. This exception to requiring labels on every container of hazardous chemicals is only for the solid metal itself and does not apply to hazardous chemicals used in conjunction with, or known to be present with, the metal and to which employees handling the metal may be exposed (for example, cutting fluids or lubricants).

(3) Chemical manufacturers, importers, or distributors shall ensure that each container of hazardous chemicals leaving the workplace is labeled, tagged, or marked in accordance with this section in a manner which does not conflict with the requirements of the Hazardous Materials Transportation Act (49 U.S.C. 1801 *et seq.*) and regulations issued under that Act by the Department of Transportation.

(4) If the hazardous chemical is regulated by OSHA in a substance-specific health standard, the chemical manufacturer, importer, distributor or employer shall ensure that the labels or other forms of warning used are in accordance with the requirements of that standard.

(5) Except as provided in paragraphs (f)(6) and (f)(7) the employer shall ensure that each container of hazardous chemicals in the workplace is labeled, tagged or marked with the following information:

(i) Identity of the hazardous chemical(s) contained therein; and

(ii) Appropriate hazard warnings.

(6) The employer may use signs, placards, process sheets, batch tickets, operating procedures, or other such written materials in lieu of affixing labels to individual stationary process containers, as long as the alternative method identifies the containers to which it is applicable and conveys the information required by paragraph (f)(5) of this section to be on a label. The written materials shall be readily accessible to the employees in their work area throughout each work shift.

(7) The employer is not required to label portable containers into which hazardous chemicals are transferred from labeled containers, and which are intended only for the immediate use of the employee who performs the transfer.

(8) The employer shall not remove or deface existing labels on incoming containers of hazardous chemicals, unless the container is immediately marked with the required information.

(9) The employer shall ensure that labels or other forms of warning are legible, in English, and prominently displayed on the container, or readily available in the work area throughout each work shift. Employers having employees who speak other languages may add the information in their language to the material presented, as long as the information is presented in English as well.

(10) The chemical manufacturer, importer, distributor or employer need not affix new labels to comply with this section if existing labels already convey the required information.

(g) *Material safety data sheets.* (1) Chemical manufacturers and importers shall obtain or develop a material safety data sheet for each hazardous chemical they produce or import. Employers shall have a material safety data sheet for each hazardous chemical which they use.

(2) Each material safety data sheet shall be in English and shall contain at least the following information:

(i) The identity used on the label, and, except as provided for in paragraph (i) of this section on trade secrets:

(A) If the hazardous chemical is a single substance, its chemical and common name(s);

(B) If the hazardous chemical is a mixture which has been tested as a whole to determine its hazards, the chemical and common name(s) of the ingredients which contribute to these known hazards, and the common name(s) of the mixture itself; or,

(C) If the hazardous chemical is a mixture which has not been tested as a whole:

(*1*) The chemical and common name(s) of all ingredients which have been determined to be health hazards,

363

§ 1910.1200

29 CFR Ch. XVII (7-1-89 Edition)

and which comprise 1% or greater of the composition, except that chemicals identified as carcinogens under paragraph (d)(4) of this section shall be listed if the concentrations are 0.1% or greater; and,

(2) The chemical and common name(s) of all ingredients which have been determined to be health hazards, and which comprise less than 1% (0.1% for carcinogens) of the mixture, if there is evidence that the ingredient(s) could be released from the mixture in concentrations which would exceed an established OSHA permissible exposure limit or ACGIH Threshold Limit Value, or could present a health hazard to employees; and,

(3) The chemical and common name(s) of all ingredients which have been determined to present a physical hazard when present in the mixture;

(ii) Physical and chemical characteristics of the hazardous chemical (such as vapor pressure, flash point);

(iii) The physical hazards of the hazardous chemical, including the potential for fire, explosion, and reactivity;

(iv) The health hazards of the hazardous chemical, including signs and symptoms of exposure, and any medical conditions which are generally recognized as being aggravated by exposure to the chemical;

(v) The primary route(s) of entry;

(vi) The OSHA permissible exposure limit, ACGIH Threshold Limit Value, and any other exposure limit used or recommended by the chemical manufacturer, importer, or employer preparing the material safety data sheet, where available;

(vii) Whether the hazardous chemical is listed in the National Toxicology Program (NTP) *Annual Report on Carcinogens* (latest edition) or has been found to be a potential carcinogen in the International Agency for Research on Cancer (IARC) *Monographs* (latest editions), or by OSHA;

(viii) Any generally applicable precautions for safe handling and use which are known to the chemical manufacturer, importer or employer preparing the material safety data sheet, including appropriate hygienic practices, protective measures during repair and maintenance of contaminated equipment, and procedures for clean-up of spills and leaks;

(ix) Any generally applicable control measures which are known to the chemical manufacturer, importer or employer preparing the material safety data sheet, such as appropriate engineering controls, work practices, or personal protective equipment;

(x) Emergency and first aid procedures;

(xi) The date of preparation of the material safety data sheet or the last change to it; and,

(xii) The name, address and telephone number of the chemical manufacturer, importer, employer or other responsible party preparing or distributing the material safety data sheet, who can provide additional information on the hazardous chemical and appropriate emergency procedures, if necessary.

(3) If no relevant information is found for any given category on the material safety data sheet, the chemical manufacturer, importer or employer preparing the material safety data sheet shall mark it to indicate that no applicable information was found.

(4) Where complex mixtures have similar hazards and contents (i.e. the chemical ingredients are essentially the same, but the specific composition varies from mixture to mixture), the chemical manufacturer, importer or employer may prepare one material safety data sheet to apply to all of these similar mixtures.

(5) The chemical manufacturer, importer or employer preparing the material safety data sheet shall ensure that the information recorded accurately reflects the scientific evidence used in making the hazard determination. If the chemical manufacturer, importer or employer preparing the material safety data sheet becomes newly aware of any significant information regarding the hazards of a chemical, or ways to protect against the hazards, this new information shall be added to the material safety data sheet within three months. If the chemical is not currently being produced or imported the chemical manufacturer or importer shall add the information to the material safety data

Occupational Safety and Health Admin., Labor § 1910.1200

sheet before the chemical is introduced into the workplace again.

(6) Chemical manufacturers or importers shall ensure that distributors and employers are provided an appropriate material safety data sheet with their intitial shipment, and with the first shipment after a material safety data sheet is updated. The chemical manufacturer or importer shall either provide material safety data sheets with the shipped containers or send them to the employer prior to or at the time of the shipment. If the material safety data sheet is not provided with a shipment that has been labeled as a hazardous chemical, the employer shall obtain one from the chemical manufacturer, importer, or distributor as soon as possible.

(7) Distributors shall ensure that material safety data sheets, and updated information, are provided to other distributors and employers. Retail distributors which sell hazardous chemicals to commercial customers shall provide a material safety data sheet to such employers upon request, and shall post a sign or otherwise inform them that a material safety data sheet is available. Chemical manufacturers, importers, and distributors need not provide material safety data sheets to retail distributors which have informed them that the retail distributor does not sell the product to commercial customers or open the sealed container to use it in their own workplaces.

(8) The employer shall maintain copies of the required material safety data sheets for each hazardous chemical in the workplace, and shall ensure that they are readily accessible during each work shift to employees when they are in their work area(s).

(9) Where employees must travel between workplaces during a workshift, *i.e.,* their work is carried out at more than one geographical location, the material safety data sheets may be kept at a central location at the primary workplace facility. In this situation, the employer shall ensure that employees can immediately obtain the required information in an emergency.

(10) Material safety data sheets may be kept in any form, including operating procedures, and may be designed to cover groups of hazardous chemicals in a work area where it may be more appropriate to address the hazards of a process rather than individual hazardous chemicals. However, the employer shall ensure that in all cases the required information is provided for each hazardous chemical, and is readily accessible during each work shift to employees when they are in in their work areas(s).

(11) Material safety data sheets shall also be made readily available, upon request, to designated representatives and to the Assistant Secretary, in accordance with the requirements of 29 CFR 1910.20 (e). The Director shall also be given access to material safety data sheets in the same manner.

(h) *Employee information and training.* Employers shall provide employees with information and training on hazardous chemicals in their work area at the time of their initial assignment, and whenever a new hazard is introduced into their work area.

(1) *Information.* Employees shall be informed of:

(i) The requirements of this section;

(ii) Any operations in their work area where hazardous chemicals are present; and,

(iii) The location and availability of the written hazard communication program, including the required list(s) of hazardous chemicals, and material safety data sheets required by this section.

(2) *Training.* Employee training shall include at least:

(i) Methods and observations that may be used to detect the presence or release of a hazardous chemical in the work area (such as monitoring conducted by the employer, continuous monitoring devices, visual appearance or odor of hazardous chemicals when being released, etc.);

(ii) The physical and health hazards of the chemicals in the work area;

(iii) The measures employees can take to protect themselves from these hazards, including specific procedures the employer has implemented to protect employees from exposure to hazardous chemicals, such as appropriate work practices, emergency procedures, and personal protective equipment to be used; and,

§ 1910.1200

(iv) The details of the hazard communication program developed by the employer, including an explanation of the labeling system and the material safety data sheet, and how employees can obtain and use the appropriate hazard information.

(i) *Trade secrets.* (1) The chemical manufacturer, importer, or employer may withhold the specific chemical identity, including the chemical name and other specific identification of a hazardous chemical, from the material safety data sheet, provided that:

(i) The claim that the information withheld is a trade secret can be supported;

(ii) Information contained in the material safety data sheet concerning the properties and effects of the hazardous chemical is disclosed;

(iii) The material safety data sheet indicates that the specific chemical identity is being withheld as a trade secret; and,

(iv) The specific chemical identity is made available to health professionals, employees, and designated representatives in accordance with the applicable provisions of this paragraph.

(2) Where a treating physician or nurse determines that a medical emergency exists and the specific chemical identity of a hazardous chemical is necessary for emergency or first-aid treatment, the chemical manufacturer, importer, or employer shall immediately disclose the specific chemical identity of a trade secret chemical to that treating physician or nurse, regardless of the existence of a written statement of need of a confidentiality agreement. The chemical manufacturer, importer, or employer may require a written statement of need and confidentiality agreement, in accordance with the provisions of paragraphs (i)(3) and (4) of this section, as soon as circumstances permit.

(3) In non-emergency situations, a chemical manufacturer, importer, or employer shall, upon request, disclose a specific chemical identity, otherwise permitted to be withheld under paragraph (i)(1) of this section, to a health professional (i.e. physician, industrial hygienist, toxicologist, epidemiologist, or occupational health nurse) providing medical or other occupational health services to exposed employee(s), and to employees or designated representatives, if:

(i) The request is in writing;

(ii) The request describes with reasonable detail one or more of the following occupational health needs for the information:

(A) To assess the hazards of the chemicals to which employees will be exposed;

(B) To conduct or assess sampling of the workplace atmosphere to determine employee exposure levels;

(C) To conduct pre-assignment or periodic medical surveillance of exposed employees;

(D) To provide medical treatment to exposed employees;

(E) To select or assess appropriate personal protective equipment for exposed employees;

(F) To design or assess engineering controls or other protective measures for exposed employees; and,

(G) To conduct studies to determine the health effects of exposure.

(iii) The request explains in detail why the disclosure of the specific chemical identity is essential and that, in lieu thereof, the disclosure of the following information to the health professional, employee, or designated representative, would not satisfy the purposes described in paragraph (i)(3)(ii) of this section:

(A) The properties and effects of the chemical;

(B) Measures for controlling workers' exposure to the chemical;

(C) Methods of monitoring and analyzing worker exposure to the chemical; and,

(D) Methods of diagnosing and treating harmful exposures to the chemical;

(iv) The request includes a description of the procedures to be used to maintain the confidentiality of the disclosed information; and,

(v) The health professional, and the employer or contractor of the services of the health professional (i.e. downstream employer, labor organization, or individual employee), employee, or designated representative, agree in a written confidentiality agreement that the health professional, employee, or designated representative, will not use

Appendix B: Regulations Reference Guide

Occupational Safety and Health Admin., Labor § 1910.1200

the trade secret information for any purpose other than the health need(s) asserted and agree not to release the information under any circumstances other than to OSHA, as provided in paragraph (i)(6) of this section, except as authorized by the terms of the agreement or by the chemical manufacturer, importer, or employer.

(4) The confidentiality agreement authorized by paragraph (i)(3)(iv) of this section:

(i) May restrict the use of the information to the health purposes indicated in the written statement of need;

(ii) May provide for appropriate legal remedies in the event of a breach of the agreement, including stipulation of a reasonable pre-estimate of likely damages; and,

(iii) May not include requirements for the posting of a penalty bond.

(5) Nothing in this standard is meant to preclude the parties from pursuing non-contractual remedies to the extent permitted by law.

(6) If the health professional, employee, or designated representative receiving the trade secret information decides that there is a need to disclose it to OSHA, the chemical manufacturer, importer, or employer who provided the information shall be informed by the health professional, employee, or designated representative prior to, or at the same time as, such disclosure.

(7) If the chemical manufacturer, importer, or employer denies a written request for disclosure of a specific chemical identity, the denial must:

(i) Be provided to the health professional, employee, or designated representative, within thirty days of the request;

(ii) Be in writing;

(iii) Include evidence to support the claim that the specific chemical identity is a trade secret;

(iv) State the specific reasons why the request is being denied; and,

(v) Explain in detail how alternative information may satisfy the specific medical or occupational health need without revealing the specific chemical identity.

(8) The health professional, employee, or designated representative whose request for information is denied under paragraph (i)(3) of this section may refer the request and the written denial of the request to OSHA for consideration.

(9) When a health professional, employee, or designated representative refers the denial to OSHA under paragraph (i)(8) of this section, OSHA shall consider the evidence to determine if:

(i) The chemical manufacturer, importer, or employer has supported the claim that the specific chemical identity is a trade secret;

(ii) The health professional, employee, or designated representative has supported the claim that there is a medical or occupational health need for the information; and,

(iii) The health professional, employee, or designated representative has demonstrated adequate means to protect the confidentiality.

(10)(i) If OSHA determines that the specific chemical identity requested under paragraph (i)(3) of this section is not a *bona fide* trade secret, or that it is a trade secret, but the requesting health professional, employee, or designated representative has a legitimate medical or occupational health need for the information, has executed a written confidentiality agreement, and has shown adequate means to protect the confidentiality of the information, the chemical manufacturer, importer, or employer will be subject to citation by OSHA.

(ii) If a chemical manufacturer, importer, or employer demonstrates to OSHA that the execution of a confidentiality agreement would not provide sufficient protection against the potential harm from the unauthorized disclosure of a trade secret specific chemical identity, the Assistant Secretary may issue such orders or impose such additional limitations or conditions upon the disclosure of the requested chemical information as may be appropriate to assure that the occupational health services are provided without an undue risk of harm to the chemical manufacturer, importer, or employer.

(11) If a citation for a failure to release specific chemical identity information is contested by the chemical manufacturer, importer, or employer,

367

§ 1910.1200

the matter will be adjudicated before the Occupational Safety and Health Review Commission in accordance with the Act's enforcement scheme and the applicable Commission rules of procedure. In accordance with the Commission rules, when a chemical manufacturer, importer, or employer continues to withhold the information during the contest, the Administrative Law Judge may review the citation and supporting documentation *in camera* or issue appropriate orders to protect the confidentiality or such matters.

(12) Notwithstanding the existence of a trade secret claim, a chemical manufacturer, importer, or employer shall, upon request, disclose to the Assistant Secretary any information which this section requires the chemical manufacturer, importer, or employer to make available. Where there is a trade secret claim, such claim shall be made no later than at the time the information is provided to the Assistant Secretary so that suitable determinations of trade secret status can be made and the necessary protections can be implemented.

(13) Nothing in this paragraph shall be construed as requiring the disclosure under any circumstances of process or percentage of mixture information which is a trade secret.

(j) *Effective dates.* (1) Chemical manufacturers, importers, and distributors shall ensure that material safety data sheets are provided with the next shipment of hazardous chemicals to employers after September 23, 1987.

(2) Employers in the non-manufacturing sector shall be in compliance with all provisions of this section by May 23, 1988. (Note: Employers in the manufacturing sector (SIC Codes 20 through 39) are already required to be in compliance with this section.)

(Approved by the Office of Management and Budget under control number 1218-0072)

APPENDIX A TO § 1900.1200 HEALTH HAZARD DEFINITIONS (*Mandatory*)

Although safety hazards related to the physical characteristics of a chemical can be objectively defined in terms of testing requirements (e.g. flammability), health hazard definitions are less precise and more subjective. Health hazards may cause measurable changes in the body—such as decreased pulmonary function. These changes are generally indicated by the occurrence of signs and symptoms in the exposed employees—such as shortness of breath, a non-measurable, subjective feeling. Employees exposed to such hazards must be apprised of both the change in body function and the signs and symptoms that may occur to signal that change.

The determination of occupational health hazards is complicated by the fact that many of the effects or signs and symptoms occur commonly in non-occupationally exposed populations, so that effects of exposure are difficult to separate from normally occurring illnesses. Occasionally, a substance causes an effect that is rarely seen in the population at large, such as angiosarcomas caused by vinyl chloride exposure, thus making it easier to ascertain that the occupational exposure was the primary causative factor. More often, however, the effects are common, such as lung cancer. The situation is further complicated by the fact that most chemicals have not been adequately tested to determine their health hazard potential, and data do not exist to substantiate these effects.

There have been many attempts to categorize effects and to define them in various ways. Generally, the terms "acute" and "chronic" are used to delineate between effects on the basis of severity or duration. "Acute" effects usually occur rapidly as a result of short-term exposures, and are of short duration. "Chronic" effects generally occur as a result of long-term exposure, and are of long duration.

The acute effects referred to most frequently are those defined by the American National Standards Institute (ANSI) standard for Precautionary Labeling of Hazardous Industrial Chemicals (Z129.1-1982)—irritation, corrosivity, sensitization and lethal dose. Although these are important health effects, they do not adequately cover the considerable range of acute effects which may occur as a result of occupational exposure, such as, for example, narcosis.

Similarly, the term chronic effect is often used to cover only carcinogenicity, teratogenicity, and mutagenicity. These effects are obviously a concern in the workplace, but again, do not adequately cover the area of chronic effects, excluding, for example, blood dyscrasias (such as anemia), chronic bronchitis and liver atrophy.

The goal of defining precisely, in measurable terms, every possible health effect that may occur in the workplace as a result of chemical exposures cannot realistically be accom-

Occupational Safety and Health Admin., Labor § 1910.1200

plished. This does not negate the need for employees to be informed of such effects and protected from them. Appendix B, which is also mandatory, outlines the principles and procedures of hazardous assessment.

For purposes of this section, any chemicals which meet any of the following definitions, as determined by the criteria set forth in Appendix B are health hazards:

1. *Carcinogen:* A chemical is considered to be a carcinogen if:

(a) It has been evaluated by the International Agency for Research on Cancer (IARC), and found to be a carcinogen or potential carcinogen; or

(b) It is listed as a carcinogen or potential carcinogen in the *Annual Report on Carcinogens* published by the National Toxicology Program (NTP) (latest edition); or,

(c) It is regulated by OSHA as a carcinogen.

2. *Corrosive:* A chemical that causes visible destruction of, or irreversible alterations in, living tissue by chemical action at the site of contact. For example, a chemical is considered to be corrosive if, when tested on the intact skin of albino rabbits by the method described by the U.S. Department of Transportation in Appendix A to 49 CFR Part 173, it destroys or changes irreversibly the structure of the tissue at the site of contact following an exposure period of four hours. This term shall not refer to action on inanimate surfaces.

3. *Highly toxic:* A chemical falling within any of the following categories:

(a) A chemical that has a median lethal dose (LD_{50}) of 50 milligrams or less per kilogram of body weight when administered orally to albino rats weighing between 200 and 300 grams each.

(b) A chemical that has a median lethal does (LD_{50}) of 200 milligrams or less per kilogram of body weight when administered by continuous contact for 24 hours (or less if death occurs within 24 hours) with the bare skin of albino rabbits weighing between two and three kilograms each.

(c) A chemical that has a median lethal concentration (LC_{50}) in air of 200 parts per million by volume or less of gas or vapor, or 2 milligrams per liter or less of mist, fume, or dust, when administered by continuous inhalation for one hour (or less if death occurs within one hour) to albino rats weighing between 200 and 300 grams each.

4. *Irritant:* A chemical, which is not corrosive, but which causes a reversible inflammatory effect on living tissue by chemical action at the site of contact. A chemical is a skin irritant if, when tested on the intact skin of albino rabbits by the methods of 16 CFR 1500.41 for four hours exposure or by other appropriate techniques, it results in an empirical score of five or more. A chemical is an eye irritant if so determined under the procedure listed in 16 CFR 1500.42 or other appropriate techniques.

5. *Sensitizer:* A chemical that causes a substantial proportion of exposed people or animals to develop an allergic reaction in normal tissue after repeated exposure to the chemical.

6. *Toxic.* A chemical falling within any of the following categories:

(a) A chemical that has a median lethal dose (LD_{50}) of more than 50 milligrams per kilogram but not more than 500 milligrams per kilogram of body weight when administered orally to albino rats weighing between 200 and 300 grams each.

(b) A chemical that has a median lethal dose (LD_{50}) of more than 200 milligrams per kilogram but not more than 1,000 milligrams per kilogram of body weight when administered by continuous contact for 24 hours (or less if death occurs within 24 hours) with the bare skin of albino rabbits weighing between two and three kilograms each.

(c) A chemical that has a median lethal concentration (LC_{50}) in air of more than 200 parts per million but not more than 2,000 parts per million by volume of gas or vapor, or more than two milligrams per liter but not more than 20 milligrams per liter of mist, fume, or dust, when administered by continuous inhalation for one hour (or less if death occurs within one hour) to albino rats weighing between 200 and 300 grams each.

7. *Target organ effects.* The following is a target organ categorization of effects which may occur, including examples of signs and symptoms and chemicals which have been found to cause such effects. These examples are presented to illustrate the range and diversity of effects and hazards found in the workplace, and the broad scope employers must consider in this area, but are not intended to be all-inclusive.

a. Hepatotoxins: Chemicals which produce liver damage

Signs & Symptoms: Jaundice; liver enlargement

Chemicals: Carbon tetrachloride; nitrosamines

b. Nephrotoxins: Chemicals which produce kidney damage

Signs & Symptoms: Edema; proteinuria

Chemicals: Halogenated hydrocarbons; uranium

c. Neurotoxins: Chemicals which produce their primary toxic effects on the nervous system

Signs & Symptoms: Narcosis; behavioral changes; decrease in motor functions

Chemicals: Mercury; carbon disulfide

369

§ 1910.1200

d. Agents which act on the blood or hematopoietic system: Decrease hemoglobin function; deprive the body tissues of oxygen

Signs & Symptoms: Cyanosis; loss of consciousness

Chemicals: Carbon monoxide; cyanides

e. Agents which damage the lung: Chemicals which irritate or damage the pulmonary tissue

Signs & Symptoms: Cough; tightness in chest; shortness of breath

Chemicals: Silica; asbestos

f. Reproductive toxins: Chemicals which affect the reproductive capabilities including chromosomal damage (mutations) and effects on fetuses (teratogenesis)

Signs & Symptoms: Birth defects; sterility

Chemicals: Lead; DBCP

g. Cutaneous hazards: Chemicals which affect the dermal layer of the body

Signs & Symptoms: Defatting of the skin; rashes; irritation

Chemicals: Ketones; chlorinated compounds

h. Eye hazards: Chemicals which affect the eye or visual capacity

Signs & Symptoms: Conjunctivitis; corneal damage

Chemicals: Organic solvents; acids

APPENDIX B TO § 1910.1200 HAZARD DETERMINATION (MANDATORY)

The quality of a hazard communication program is largely dependent upon the adequacy and accuracy of the hazard determination. The hazard determination requirement of this standard is performance-oriented. Chemical manufacturers, importers, and employers evaluating chemicals are not required to follow any specific methods for determining hazards, but they must be able to demonstrate that they have adequately ascertained the hazards of the chemicals produced or imported in accordance with the criteria set forth in this Appendix.

Hazard evaluation is a process which relies heavily on the professional judgment of the evaluator, particularly in the area of chronic hazards. The performance-orientation of the hazard determination does not diminish the duty of the chemical manufacturer, importer or employer to conduct a thorough evaluation, examining all relevant data and producing a scientifically defensible evaluation. For purposes of this standard, the following criteria shall be used in making hazard determinations that meet the requirements of this standard.

1. *Carcinogenicity*: As described in paragraph (d)(4) and Appendix A of this section, a determination by the National Toxicology Program, the International Agency for Research on Cancer, or OSHA that a chemical is a carcinogen or potential carcinogen will be considered conclusive evidence for purposes of this section.

2. *Human data*: Where available, epidemiological studies and case reports of adverse health effects shall be considered in the evaluation.

3. *Animal data*: Human evidence of health effects in exposed populations is generally not available for the majority of chemicals produced or used in the workplace. Therefore, the available results of toxicological testing in animal populations shall be used to predict the health effects that may be experienced by exposed workers. In particular, the definitions of certain acute hazards refer to specific animal testing results (see Appendix A).

4. *Adequacy and reporting of data.* The results of any studies which are designed and conducted according to established scientific principles, and which report statistically significant conclusions regarding the health effects of a chemical, shall be a sufficient basis for a hazard determination and reported on any material safety data sheet. The chemical manufacturer, importer, or employer may also report the results of other scientifically valid studies which tend to refute the findings of hazard.

APPENDIX C TO § 1910.1200 INFORMATION SOURCES (ADVISORY)

The following is a list of available data sources which the chemical manufacturer, importer, distributor, or employer may wish to consult to evaluate the hazards of chemicals they produce or import:

—Any information in their own company files, such as toxicity testing results or illness experience of company employees.

—Any information obtained from the supplier of the chemical, such as material safety data sheets or product safety bulletins.

—Any pertinent information obtained from the following source list (latest editions should be used):

Condensed Chemical Dictionary
 Van Nostrand Reinhold Co., 135 West 50th Street, New York, NY 10020.

The Merck Index: An Encyclopedia of Chemicals and Drugs
 Merck and Company, Inc., 126 E. Lincoln Ave., Rahway, NJ 07065.

IARC Monographs on the Evaluation of the Carcinogenic Risk of Chemicals to Man
 Geneva: World Health Organization, International Agency for Research on Cancer, 1972-Present. (Multivolume work). Summaries are available in supplement volumes. 49 Sheridan Street, Albany, NY 12210.

Industrial Hygiene and Toxicology, by F.A. Patty

Appendix B: Regulations Reference Guide

Occupational Safety and Health Admin., Labor § 1910.1200

John Wiley & Sons, Inc., New York, NY (Multivolume work).

Clinical Toxicology of Commercial Products
Gleason, Gosselin, and Hodge

Casarett and Doull's Toxicology; The Basic Science of Poisons
Doull, Klaassen, and Amdur, Macmillan Publishing Co., Inc., New York, NY.

Industrial Toxicology, by Alice Hamilton and Harriet L. Hardy
Publishing Sciences Group, Inc., Acton, MA.

Toxicology of the Eye, by W. Morton Grant
Charles C. Thomas, 301-327 East Lawrence Avenue, Springfield, IL.

Recognition of Health Hazards in Industry
William A. Burgess, John Wiley and Sons, 605 Third Avenue, New York, NY 10158.

Chemical Hazards of the Workplace
Nick H. Proctor and James P. Hughes, J.P. Lipincott Company, 6 Winchester Terrace, New York, NY 10022.

Handbook of Chemistry and Physics
Chemical Rubber Company, 18901 Cranwood Parkway, Cleveland, OH 44128.

Threshold Limit Values for Chemical Substances and Physical Agents in the Work Environment and Biological Exposure Indices with Intended Changes
American Conference of Governmental Industrial Hygienists (ACGIH). 6500 Glenway Avenue, Bldg. D-5, Cincinnati, OH 45211.

Information on the physical hazards of chemicals may be found in publications of the National Fire Protection Association, Boston, MA.

NOTE.—The following documents may be purchased from the Superintendent of Documents, U.S. Government Printing Office, Washington, DC 20402.

Occupational Health Guidelines
NIOSH/OSHA (NIOSH Pub. No. 81-123)

NIOSH Pocket Guide to Chemical Hazards
NIOSH Pub. No. 85-114

Registry of Toxic Effects of Chemical Substances
NIOSH Pub. No. 80-102

Miscellaneous Documents published by the National Institute for Occupational Safety and Health:
Criteria documents.
Special Hazard Reviews.
Occupational Hazard Assessments.
Current Intelligence Bulletins.

OSHA's General Industry Standards (29 CFR Part 1910)

NTP Annual Report on Carcinogens and *Summary of the Annual Report on Carcinogens.*
National Technical Information Service (NTIS), 5285 Port Royal Road, Springfield, VA 22161; (703) 487-4650.

BIBLIOGRAPHIC DATA BASES

Service provider	File name
Bibliographic Retrieval Services (BRS), 1200 Route 7, Latham, NY 12110.	Biosis Previews CA Search Medlars NTIS Hazardline American Chemical Society Journal Excerpta Medica IRCS Medical Science Journal Pre-Med Intl Pharmaceutical Abstracts Paper Chem
Lockheed—DIALOG Information Service, Inc., 3460 Hillview Avenue, Palo Alto, CA 94304.	Biosis Prev. Files CA Search Files CAB Abstracts Chemical Exposure Chemname Chemsis Files Chemzero Embase Files Environmental Bibliographies Enviroline Federal Research in Progress IRL Life Science Collection NTIS Occupational Safety and Health (NIOSH) Paper Chem
SDC—Orbit, SDC Information Service, 2500 Colorado Avenue, Santa Monica, CA 90406.	CAS Files Chemdex, 2, 3 NTIS
National Library of Medicine, Department of Health and Human Services, Public Health Service, National Institutes of Health, Bethesda, MD 20209.	Hazardous Substances Data Bank (NSDB) Medline files Toxline Files Cancerlit RTECS Chemline
Pergamon International Information Corp., 1340 Old Chain Bridge Rd., McLean, VA 22101.	Laboratory Hazard Bulletin
Questel, Inc., 1625 Eye Street, NW., Suite 818, Washington, DC 20006.	CIS/ILO Cancernet
Chemical Information System ICI (ICIS), Bureau of National Affairs, 1133 15th Street, NW., Suite 300, Washington, DC 20005.	Structure and Nomenclature Search System (SANSS) Acute Toxicity (RTECS) Clinical Toxicology of Commercial Products Oil and Hazardous Materials Technical Assistance Data System CCRIS CESARS
Occupational Health Services, 400 Plaza Drive, Secaucus, NJ 07094.	MSDS Hazardline

APPENDIX D TO § 1910.1200 DEFINITION OF "TRADE SECRET" (MANDATORY)

The following is a reprint of the *Restatement of Torts* section 757, comment *b* (1939):

371

§ 1910.1499

b. Definition of trade secret. A trade secret may consist of any formula, pattern, device or compilation of information which is used in one's business, and which gives him an opportunity to obtain an advantage over competitors who do not know or use it. It may be a formula for a chemical compound, a process of manufacturing, treating or preserving materials, a pattern for a machine or other device, or a list of customers. It differs from other secret information in a business (see § 759 of the *Restatement of Torts* which is not included in this Appendix) in that it is not simply information as to single or ephemeral events in the conduct of the business, as, for example, the amount or other terms of a secret bid for a contract or the salary of certain employees, or the security investments made or contemplated, or the date fixed for the announcement of a new policy or for bringing out a new model or the like. A trade secret is a process or device for continuous use in the operations of the business. Generally it relates to the production of goods, as, for example, a machine or formula for the production of an article. It may, however, relate to the sale of goods or to other operations in the business, such as a code for determining discounts, rebates or other concessions in a price list or catalogue, or a list of specialized customers, or a method of bookkeeping or other office management.

Secrecy. The subject matter of a trade secret must be secret. Matters of public knowledge or of general knowledge in an industry cannot be appropriated by one as his secret. Matters which are completely disclosed by the goods which one markets cannot be his secret. Substantially, a trade secret is known only in the particular business in which it is used. It is not requisite that only the proprietor of the business know it. He may, without losing his protection, communicate it to employees involved in its use. He may likewise communicate it to others pledged to secrecy. Others may also know of it independently, as, for example, when they have discovered the process or formula by independent invention and are keeping it secret. Nevertheless, a substantial element of secrecy must exist, so that, except by the use of improper means, there would be difficulty in acquiring the information. An exact definition of a trade secret is not possible. Some factors to be considered in determining whether given information is one's trade secret are: (1) The extent to which the information is known outside of his business; (2) the extent to which it is known by employees and others involved in his business; (3) the extent of measures taken by him to guard the secrecy of the information; (4) the value of the information to him and his competitors; (5) the amount of effort or money expended by him in developing the information; (6) the ease or difficulty with which the information could be properly acquired or duplicated by others.

Novelty and prior art. A trade secret may be a device or process which is patentable; but it need not be that. It may be a device or process which is clearly anticipated in the prior art or one which is merely a mechanical improvement that a good mechanic can make. Novelty and invention are not requisite for a trade secret as they are for patentability. These requirements are essential to patentability because a patent protects against unlicensed use of the patented device or process even by one who discovers it properly through independent research. The patent monopoly is a reward to the inventor. But such is not the case with a trade secret. Its protection is not based on a policy of rewarding or otherwise encouraging the development of secret processes or devices. The protection is merely against breach of faith and reprehensible means of learning another's secret. For this limited protection it is not appropriate to require also the kind of novelty and invention which is a requisite of patentability. The nature of the secret is, however, an important factor in determining the kind of relief that is appropriate against one who is subject to liability under the rule stated in this section. Thus, if the secret consists of a device or process which is a novel invention, one who acquires the secret wrongfully is ordinarily enjoined from further use of it and is required to account for the profits derived from his past use. If, on the other hand, the secret consists of mechanical improvements that a good mechanic can make without resort to the secret, the wrongdoer's liability may be limited to damages, and an injunction against future use of the improvements made with the aid of the secret may be inappropriate.

[52 FR 31877, Aug. 24, 1987, as amended at 52 FR 46080, Dec. 4, 1987; 53 FR 15035, Apr. 27, 1988; 54 FR 24334, June 7, 1989; 54 FR 6888, Feb. 15, 1989]

§ 1910.1499 Source of standards.

Section 1910.1000	41 CFR 50-204.50, except for Table Z-2, the source of which is American National Standards Institute, Z37 series.

[40 FR 23073, May 28, 1975]

§ 1910.1500 Standards organizations.

Specific standards of the following organizations have been referred to in this subpart. Copies of the standards may be obtained from the issuing organization.

Appendix B: Regulations Reference Guide

Wednesday
January 31, 1990

Part II

Department of Labor

Occupational Safety and Health Administration

29 CFR Part 1910
Occupational Exposures to Hazardous Chemicals in Laboratories; Final Rule

Chemical Hazard Communication Guidebook

PART 1910—OCCUPATIONAL SAFETY AND HEALTH STANDARDS

1. The authority citation for part 1910, subpart Z is amended by adding the following citation at the end. (Citation which precedes asterisk indicates general rulemaking authority.)

Authority: Secs. 6 and 8, Occupational Safety and Health Act, 29 U.S.C. 655, 657; Secretary of Labor's Orders Nos. 12-71 (36 FR 8754), 8-76 (41 FR 25059), or 9-83 (48 FR 35736), as applicable; and 29 CFR part 1911. * * * Section 1910.1450 is also issued under sec. 6(b), 8(c) and 8(g)(2), Pub. L. 91-596, 84 Stat. 1593, 1599, 1600; 29 U.S.C. 655, 657.

2. Section 1910.1450 is added to subpart Z, part 1910 to read as follows:

§ 191.1450 Occupational exposure to hazardous chemicals in laboratories.

(a) *Scope and application.* (1) This section shall apply to all employers engaged in the laboratory use of hazardous chemicals as defined below.

(2) Where this section applies, it shall supersede, for laboratories, the requirements of all other OSHA health standards in 29 CFR part 1910, subpart Z, except as follows:

(i) For any OSHA health standard, only the requirement to limit employee exposure to the specific permissible exposure limit shall apply for laboratories, unless that particular standard states otherwise or unless the conditions of paragraph (a)(2)(iii) of this section apply.

(ii) Prohibition of eye and skin contact where specified by any OSHA health standard shall be observed.

(iii) Where the action level (or in the absence of an action level, the permissible exposure limit) is routinely exceeded for an OSHA regulated substance with exposure monitoring and medical surveillance requirements, paragraphs (d) and (g)(1)(ii) of this section shall apply.

(3) This section shall not apply to:

(i) Uses of hazardous chemicals which do not meet the definition of laboratory use, and in such cases, the employer shall comply with the relevant standard in 29 CFR part 1910, subpart 2, even if such use occurs in a laboratory.

(ii) Laboratory uses of hazardous chemicals which provide no potential for employee exposure. Examples of such conditions might include:

(A) Procedures using chemically-impregnated test media such as Dip-and-Read tests where a reagent strip is dipped into the specimen to be tested and the results are interpreted by comparing the color reaction to a color chart supplied by the manufacturer of the test strip; and

(B) Commercially prepared kits such as those used in performing pregnancy tests in which all of the reagents needed to conduct the test are contained in the kit.

(b) *Definitions—*

"*Action level*" means a concentration designated in 29 CFR part 1910 for a specific substance, calculated as an eight (8)-hour time-weighted average, which initiates certain required activities such as exposure monitoring and medical surveillance.

"*Assistant Secretary*" means the Assistant Secretary of Labor for Occupational Safety and Health, U.S. Department of Labor, or designee.

"*Carcinogen*" (see "select carcinogen").

"*Chemical Hygiene Officer*" means an employee who is designated by the employer, and who is qualified by training or experience, to provide technical guidance in the development and implementation of the provisions of the Chemical Hygiene Plan. This definition is not intended to place limitations on the position description or job classification that the designated indvidual shall hold within the employer's organizational structure.

"*Chemical Hygiene Plan*" means a written program developed and implemented by the employer which sets forth procedures, equipment, personal protective equipment and work practices that (i) are capable of protecting employees from the health hazards presented by hazardous chemicals used in that particular workplace and (ii) meets the requirements of paragraph (e) of this section.

"*Combustible liquid*" means any liquid having a flashpoint at or above 100 °F (37.8 °C), but below 200 °F (93.3 °C), except any mixture having components with flashpoints of 200 °F (93.3 °C), or higher, the total volume of which make up 99 percent or more of the total volume of the mixture.

"*Compressed gas*" means:

(i) A gas or mixture of gases having, in a container, an absolute pressure exceeding 40 psi at 70 °F (21.1 °C); or

(ii) A gas or mixture of gases having, in a container, an absolute pressure exceeding 104 psi at 130 °F (54.4 °C) regardless of the pressure at 70 °F (21.1 °C); or

(iii) A liquid having a vapor pressure exceeding 40 psi at 100 °F (37.8 °C) as determined by ASTM D-323-72.

"*Designated area*" means an area which may be used for work with "select carcinogens," reproductive toxins or substances which have a high degree of acute toxicity. A designated area may be the entire laboratory, an area of a laboratory or a device such as a laboratory hood.

"*Emergency*" means any occurrence such as, but not limited to, equipment failure, rupture of containers or failure of control equipment which results in an uncontrolled release of a hazardous chemical into the workplace.

"*Employee*" means an individual employed in a laboratory workplace who may be exposed to hazardous chemicals in the course of his or her assignments.

"*Explosive*" means a chemical that causes a sudden, almost instantaneous release of pressure, gas, and heat when subjected to sudden shock, pressure, or high temperature.

"*Flammable*" means a chemical that falls into one of the following categories:

(i) "*Aerosol, flammable*" means an aerosol that, when tested by the method described in 16 CFR 1500.45, yields a flame protection exceeding 18 inches at full valve opening, or a flashback (a flame extending back to the valve) at any degree of valve opening;

(ii) "*Gas, flammable*" means:

(A) A gas that, at ambient temperature and pressure, forms a flammable mixture with air at a concentration of 13 percent by volume or less; or

(B) A gas that, at ambient temperature and pressure, forms a range of flammable mixtures with air wider than 12 percent by volume, regardless of the lower limit.

(iii) "*Liquid, flammable*" means any liquid having a flashpoint below 100 °F (37.8 °C), except any mixture having components with flashpoints of 100 °F (37.8 °C) or higher, the total of which make up 99 percent or more of the total volume of the mixture.

(iv) "*Solid, flammable*" means a solid, other than a blasting agent or explosive as defined in § 1910.109(a), that is liable to cause fire through friction, absorption of moisture, spontaneous chemical change, or retained heat from manufacturing or processing, or which can be ignited readily and when ignited burns so vigorously and persistently as to create a serious hazard. A chemical shall be considered to be a flammable solid if, when tested by the method described in 16 CFR 1500.44, it ignites and burns with a self-sustained flame at a rate greater than one-tenth of an inch per second along its major axis.

"*Flashpoint*" means the minimum temperature at which a liquid gives off a vapor in sufficient concentration to ignite when tested as follows:

(i) Tagliabue Closed Tester (See American National Standard Method of Test for Flash Point by Tag Closed Tester, Z11.24–1979 (ASTM D 56–79))-for liquids with a viscosity of less than 45 Saybolt Universal Seconds (SUS) at 100 °F (37.8 °C), that do not contain suspended solids and do not have a tendency to form a surface film under test; or

(ii) Pensky-Martens Closed Tester (see American National Standard Method of Test for Flash Point by Pensky-Martens Closed Tester, Z11.7–1979 (ASTM D 93–79))-for liquids with a viscosity equal to or greater than 45 SUS at 100 °F (37.8 °C), or that contain suspended solids, or that have a tendency to form a surface film under test; or

(iii) Setaflash Closed Tester (see American National Standard Method of Test for Flash Point by Setaflash Closed Tester (ASTM D 3278–78)).

Organic peroxides, which undergo autoaccelerating thermal decomposition, are excluded from any of the flashpoint determination methods specified above.

"*Hazardous chemical*" means a chemical for which there is statistically significant evidence based on at least one study conducted in accordance with established scientific principles that acute or chronic health effects may occur in exposed employees. The term "health hazard" includes chemicals which are carcinogens, toxic or highly toxic agents, reproductive toxins, irritants, corrosives, sensitizers, hepatotoxins, nephrotoxins, neurotoxins, agents which act on the hematopoietic systems, and agents which damage the lungs, skin, eyes, or mucous membranes.

Appendices A and B of the Hazard Communication Standard (29 CFR 1910.1200) provide further guidance in defining the scope of health hazards and determining whether or not a chemical is to be considered hazardous for purposes of this standard.

"*Laboratory*" means a facility where the "laboratory use of hazardous chemicals" occurs. It is a workplace where relatively small quantities of hazardous chemicals are used on a non-production basis.

"*Laboratory scale*" means work with substances in which the containers used for reactions, transfers, and other handling of substances are designed to be easily and safely manipulated by one person. "Laboratory scale" excludes those workplaces whose function is to produce commercial quantities of materials.

"*Laboratory-type hood*" means a device located in a laboratory, enclosure on five sides with a moveable sash or fixed partial enclosed on the remaining side; constructed and maintained to draw air from the laboratory and to prevent or minimize the escape of air contaminants into the laboratory; and allows chemical manipulations to be conducted in the enclosure without insertion of any portion of the employee's body other than hands and arms.

Walk-in hoods with adjustable sashes meet the above definition provided that the sashes are adjusted during use so that the airflow and the exhaust of air contaminants are not compromised and employees do not work inside the enclosure during the release of airborne hazardous chemicals.

"*Laboratory use of hazardous chemicals*" means handling or use of such chemicals in which all of the following conditions are met:

(i) Chemical manipulations are carried out on a "laboratory scale;"

(ii) Multiple chemical procedures or chemicals are used;

(iii) The procedures involved are not part of a production process, nor in any way simulate a production process; and

(iv) "Protective laboratory practices and equipment" are available and in common use to minimize the potential for employee exposure to hazardous chemicals.

"*Medical consultation*" means a consultation which takes place between an employee and a licensed physician for the purpose of determining what medical examinations or procedures, if any, are appropriate in cases where a significant exposure to a hazardous chemical may have taken place.

"*Organic peroxide*" means an organic compound that contains the bivalent —O—O—structure and which may be considered to be a structural derivative of hydrogen peroxide where one or both of the hydrogen atoms has been replaced by an organic radical.

"*Oxidizer*" means a chemical other than a blasting agent or explosive as defined in § 1910.109(a), that initiates or promotes combustion in other materials, thereby causing fire either of itself or through the release of oxygen or other gases.

"*Physical hazard*" means a chemical for which there is scientifically valid evidence that it is a combustible liquid, a compressed gas, explosive, flammable, an organic peroxide, an oxidizer, pyrophoric, unstable (reactive) or water-reactive.

"*Protective laboratory practices and equipment*" means those laboratory procedures, practices and equipment accepted by laboratory health and safety experts as effective, or that the employer can show to be effective, in minimizing the potential for employee exposure to hazardous chemicals.

"*Reproductive toxins*" means chemicals which affect the reproductive capabilities including chromosomal damage (mutations) and effects on fetuses (teratogenesis)

"*Select carcinogen*" means any substance which meets one of the following criteria:

(i) It is regulated by OSHA as a carcinogen; or

(ii) It is listed under the category, "known to be carcinogens," in the Annual Report on Carcinogens published by the National Toxicology Program (NTP) (latest edition); or

(iii) It is listed under Group 1 ("carcinogenic to humans") by the International Agency for Research on Cancer Monographs (IARC) (latest editions); or

(iv) It is listed in either Group 2A or 2B by IARC or under the category, "reasonably anticipated to be

Chemical Hazard Communication Guidebook

carcinogens" by NTP, and causes statistically significant tumor incidence in experimental animals in accordance with any of the following criteria:

(A) After inhalation exposure of 6–7 hours per day, 5 days per week, for a significant portion of a lifetime to dosages of less than 10 mg/m³;

(B) After repeated skin application of less than 300 (mg/kg of body weight) per week; or

(C) After oral dosages of less than 50 mg/kg of body weight per day.

"Unstable (reactive)" means a chemical which is the pure state, or as produced or transported, will vigorously polymerize, decompose, condense, or will become self-reactive under conditions of shocks, pressure or temperature.

"Water-reactive" means a chemical that reacts with water to release a gas that is either flammable or presents a health hazard.

(c) *Permissible exposure limits.* For laboratory uses of OSHA regulated substances, the employer shall assure that laboratory employees' exposures to such substances do not exceed the permissible exposure limits specified in 29 CFR part 1910, subpart Z.

(d) *Employee exposure determination*—(1) *Initial monitoring.* The employer shall measure the employee's exposure to any substance regulated by a standard which requires monitoring if there is reason to believe that exposure levels for that substance routinely exceed the action level (or in the absence of an action level, the PEL).

(2) *Periodic monitoring.* If the initial monitoring prescribed by paragraph (d)(1) of this section discloses employee exposure over the action level (or in the absence of an action level, the PEL), the employer shall immediately comply with the exposure monitoring provisions of the relevant standard.

(3) *Termination of monitoring.* Monitoring may be terminated in accordance with the relevant standard.

(4) *Employee notification of monitoring results.* The employer shall, within 15 working days after the receipt of any monitoring results, notify the employee of these results in writing either individually or by posting results in an appropriate location that is accessible to employees.

(e) *Chemical hygiene plan—General.* (Appendix A of this section is non-mandatory but provides guidance to assist employers in the development of the Chemical Hygiene Plan.) (1) Where hazardous chemicals as defined by this standard are used in the workplace, the employer shall develop and carry out the provisions of a written Chemical Hygiene Plan which is:

(i) Capable of protecting employees from health hazards associated with hazardous chemicals in that laboratory and

(ii) Capable of keeping exposures below the limits specified in paragraph (c) of this section.

(2) The Chemical Hygiene Plan shall be readily available to employees, employee representatives and, upon request, to the Assistant Secretary.

(3) The Chemical Hygiene Plan shall include each of the following elements and shall indicate specific measures that the employer will take to ensure laboratory employee protection:

(i) Standard operating procedures relevant to safety and health considerations to be followed when laboratory work involves the use of hazardous chemicals;

(ii) Criteria that the employer will use to determine and implement control measures to reduce employee exposure to hazardous chemicals including engineering controls, the use of personal protective equipment and hygiene practices; particular attention shall be given to the selection of control measures for chemicals that are known to be extremely hazardous;

(iii) A requirement that fume hoods and other protective equipment are functioning properly and specific measures that shall be taken to ensure proper and adequate performance of such equipment;

(iv) Provisions for employee information and training as prescribed in paragraph (f) of this section;

(v) The circumstances under which a particular laboratory operation, procedure or activity shall require prior approval from the employer or the employer's designee before implementation;

(vi) Provisions for medical consultation and medical examinations in accordance with paragraph (g) of this section;

(vii) Designation of personnel responsible for implementation of the Chemical Hygiene Plan including the assignment of a Chemical Hygiene Officer and, if appropriate, establishment of a Chemical Hygiene Committee; and

(viii) Provisions for additional employee protection for work with particularly hazardous substances. These include "select carcinogens," reproductive toxins and substances which have a high degree of acute toxicity. Specific consideration shall be given to the following provisions which shall be included where appropriate:

(A) Establishment of a designated area;

(B) Use of containment devices such as fume hoods or glove boxes;

(C) Procedures for safe removal of contaminated waste; and

(D) Decontamination procedures.

(4) The employer shall review and evaluate the effectiveness of the Chemical Hygiene Plan at least annually and update it as necessary.

(f) *Employee information and training.* (1) The employer shall provide employees with information and training to ensure that they are apprised of the hazards of chemicals present in their work area.

(2) Such information shall be provided at the time of an employee's initial assignment to a work area where hazardous chemicals are present and prior to assignments involving new exposure situations. The frequency of refresher information and training shall be determined by the employer.

(3) *Information.* Employees shall be informed of:

(i) The contents of this standard and its appendices which shall be made available to employees;

(ii) The location and availability of the employer's Chemical Hygiene Plan;

(iii) The permissible exposure limits for OSHA regulated substances or recommended exposure limits for other hazardous chemicals where there is no applicable OSHA standard;

(iv) Signs and symptoms associated with exposures to hazardous chemicals used in the laboratory; and

(v) The location and availability of known reference material on the hazards, safe handling, storage and disposal of hazardous chemicals found in the laboratory including, but not limited to, Material Safety Data Sheets received from the chemical supplier.

(4) *Training.* (i) Employee training shall include:

(A) Methods and observations that may be used to detect the presence or release of a hazardous chemical (such as monitoring conducted by the employer, continuous monitoring devices, visual appearance or odor of hazardous chemicals when being released, etc.);

(B) The physical and health hazards of chemicals in the work area; and

(C) The measures employees can take to protect themselves from these hazards, including specific procedures the employer has implemented to protect employees from exposure to hazardous chemicals, such as appropriate work practices, emergency procedures, and personal protective equipment to be used.

Appendix B: Regulations Reference Guide

(ii) The employee shall be trained on the applicable details of the employer's written Chemical Hygiene Plan.

(g) *Medical consultation and medical examinations*. (1) The employer shall provide all employees who work with hazardous chemicals an opportunity to receive medical attention, including any follow-up examinations which the examining physician determines to be necessary, under the following circumstances:

(i) Whenever an employee develops signs or symptoms associated with a hazardous chemical to which the employee may have been exposed in the laboratory, the employee shall be provided an opportunity to receive an appropriate medical examination.

(ii) Where exposure monitoring reveals an exposure level routinely above the action level (or in the absence of an action level, the PEL) for an OSHA regulated substance for which there are exposure monitoring and medical surveillance requirements, medical surveillance shall be established for the affected employee as prescribed by the particular standard.

(iii) Whenever an event takes place in the work area such as a spill, leak, explosion or other occurrence resulting in the likelihood of a hazardous exposure, the affected employee shall be provided an opportunity for a medical consultation. Such consultation shall be for the purpose of determining the need for a medical examination.

(2) All medical examinations and consultations shall be performed by or under the direct supervision of a licensed physician and shall be provided without cost to the employee, without loss of pay and at a reasonable time and place.

(3) *Information provided to the physician*. The employer shall provide the following information to the physician:

(i) The identity of the hazardous chemical(s) to which the employee may have been exposed;

(ii) A description of the conditions under which the exposure occurred including quantitative exposure data, if available; and

(iii) A description of the signs and symptoms of exposure that the employee is experiencing, if any.

(4) *Physician's written opinion*. (i) For examination or consultation required under this standard, the employer shall obtain a written opinion from the examining physician which shall include the following:

(A) Any recommendation for further medical follow-up;

(B) The results of the medical examination and any associated tests;

(C) Any medical condition which may be revealed in the course of the examination which may place the employee at increased risk as a result of exposure to a hazardous chemical found in the workplace; and

(D) A statement that the employee has been informed by the physician of the results of the consultation or medical examination and any medical condition that may require further examination or treatment.

(ii) The written opinion shall not reveal specific findings of diagnoses unrelated to occupational exposure.

(h) *Hazard identification*. (1) With respect to labels and material safety data sheets:

(i) Employers shall ensure that labels on incoming containers of hazardous chemicals are not removed or defaced.

(ii) Employers shall maintain any material safety data sheets that are received with incoming shipments of hazardous chemicals, and ensure that they are readily accessible to laboratory employees.

(2) The following provisions shall apply to chemical substances developed in the laboratory:

(i) If the composition of the chemical substance which is produced exclusively for the laboratory's use is known, the employer shall determine if it is a hazardous chemical as defined in paragraph (b) of this section. If the chemical is determined to be hazardous, the employer shall provide appropriate training as required under paragraph (f) of this section.

(ii) If the chemical produced is a byproduct whose composition is not known, the employer shall assume that the substance is hazardous and shall implement paragraph (e) of this section.

(iii) If the chemical substance is produced for another user outside of the laboratory, the employer shall comply with the Hazard Communication Standard (29 CFR 1910.1200) including the requirements for preparation of material safety data sheets and labeling.

(i) *Use of respirators*. Where the use of respirators is necessary to maintain exposure below permissible exposure limits, the employer shall provide, at no cost to the employee, the proper respiratory equipment. Respirators shall be selected and used in accordance with the requirements of 29 CFR 1910.134.

(j) *Recordkeeping*. (1) The employer shall establish and maintain for each employee an accurate record of any measurements taken to monitor employee exposures and any medical consultation and examinations including tests or written opinions required by this standard.

(2) The employer shall assure that such records are kept, transferred, and made available in accordance with 29 CFR 1910.20.

(k) *Dates*—(1) *Effective date*. This section shall become effective May 1, 1990.

(2) *Start-up dates*. (i) Employers shall have developed and implemented a written Chemical Hygiene Plan no later than January 31, 1991.

(ii) Paragraph (a)(2) of this section shall not take effect until the employer has developed and implemented a written Chemical Hygiene Plan.

(l) *Appendices*. The information contained in the appendices is not intended, by itself, to create any additional obligations not otherwise imposed or to detract from any existing obligation.

Appendix A to § 1910.1450—National Research Council Recommendations Concerning Chemical Hygiene in Laboratories (Non-Mandatory)

Table of Contents

Foreword

Corresponding Sections of the Standard and This Appendix

A. General Principles
 1. Minimize all Chemical Exposures
 2. Avoid Underestimation of Risk
 3. Provide Adequate Ventilation
 4. Institute a Chemical Hygiene Program
 5. Observe the PELs and TLVs

B. Responsibilities
 1. Chief Executive Officer
 2. Supervisor of Administrative Unit
 3. Chemical Hygiene Officer
 4. Laboratory Supervisor
 5. Project Director
 6. Laboratory Worker

C. The Laboratory Facility
 1. Design
 2. Maintenance
 3. Usage
 4. Ventilation

D. Components of the Chemical Hygiene Plan
 1. Basic Rules and Procedures
 2. Chemical Procurement, Distribution, and Storage
 3. Environmental Monitoring
 4. Housekeeping, Maintenance and Inspections
 5. Medical Program
 6. Personal Protective Apparel and Equipment
 7. Records
 8. Signs and Labels
 9. Spills and Accidents
 10. Training and Information
 11. Waste Disposal

E. General Procedures for Working With Chemicals
 1. General Rules for all Laboratory Work with Chemicals
 2. Allergens and Embryotoxins

Chemical Hazard Communication Guidebook

3. Chemicals of Moderate Chronic or High Acute Toxicity
4. Chemicals of High Chronic Toxicity
5. Animal Work with Chemicals of High Chronic Toxicity

F. Safety Recommendations

G. Material Safety Data Sheets

Foreword

As guidance for each employer's development of an appropriate laboratory Chemical Hygiene Plan, the following non-mandatory recommendations are provided. They were extracted from "Prudent Practices for Handling Hazardous Chemicals in Laboratories" (referred to below as "Prudent Practices"), which was published in 1981 by the National Research Council and is available from the National Academy Press, 2101 Constitution Ave., NW., Washington DC 20418.

"Prudent Practices" is cited because of its wide distribution and acceptance and because of its preparation by members of the laboratory community through the sponsorship of the National Research Council. However, none of the recommendations given here will modify any requirements of the laboratory standard. This Appendix merely presents pertinent recommendations from "Prudent Practices", organized into a form convenient for quick reference during operation of a laboratory facility and during development and application of a Chemical Hygiene Plan. Users of this appendix should consult "Prudent Practices" for a more extended presentation and justification for each recommendation.

"Prudent Practices" deals with both safety and chemical hazards while the laboratory standard is concerned primarily with chemical hazards. Therefore, only those recommendations directed primarily toward control of toxic exposures are cited in this appendix, with the term "chemical hygiene" being substituted for the word "safety". However, since conditions producing or threatening physical injury often pose toxic risks as well, page references concerning major categories of safety hazards in the laboratory are given in section F.

The recommendations from "Prudent Practices" have been paraphrased, combined, or otherwise reorganized, and headings have been added. However, their sense has not been changed.

Corresponding Sections of the Standard and this Appendix

The following table is given for the convenience of those who are developing a Chemical Hygiene Plan which will satisfy the requirements of paragraph (e) of the standard. It indicates those sections of this appendix which are most pertinent to each of the sections of paragraph (e) and related paragraphs.

Paragraph and topic in laboratory standard	Relevant appendix section
(e)(3)(i) Standard operating procedures for handling toxic chemicals.	C, D, E
(e)(3)(ii) Criteria to be used for implementation of measures to reduce exposures.	D
(e)(3)(iii) Fume hood performance.	C4b
(e)(3)(iv) Employee information and training (including emergency procedures).	D10, D9
(e)(3)(v) Requirements for prior approval of laboratory activities.	E2b, E4b
(e)(3)(vi) Medical consultation and medical examinations.	D5, E4f
(e)(3)(vii) Chemical hygiene responsibilities.	B
(e)(3)(viii) Special precautions for work with particularly hazardous substances.	E2, E3, E4

In this appendix, those recommendations directed primarily at administrators and supervisors are given in sections A–D. Those recommendations of primary concern to employees who are actually handling laboratory chemicals are given in section E. (Reference to page numbers in "Prudent Practices" are given in parentheses.)

A. General Principles for Work with Laboratory Chemicals

In addition to the more detailed recommendations listed below in sections B–E, "Prudent Practices" expresses certain general principles, including the following:

1. *It is prudent to minimize all chemical exposures.* Because few laboratory chemicals are without hazards, general precautions for handling all laboratory chemicals should be adopted, rather than specific guidelines for particular chemicals (2, 10). Skin contact with chemicals should be avoided as a cardinal rule (198).

2. *Avoid underestimation of risk.* Even for substances of no known significant hazard, exposure should be minimized; for work with substances which present special hazards, special precautions should be taken (10, 37, 38). One should assume that any mixture will be more toxic than its most toxic component (30, 103) and that all substances of unknown toxicity are toxic (3, 34).

3. *Provide adequate ventilation.* The best way to prevent exposure to airborne substances is to prevent their escape into the working atmosphere by use of hoods and other ventilation devices (32, 198).

4. *Institute a chemical hygiene program.* A mandatory chemical hygiene program designed to minimize exposures is needed; it should be a regular, continuing effort, not merely a standby or short-term activity (6, 11). Its recommendations should be followed in academic teaching laboratories as well as by full-time laboratory workers (13).

5. *Observe the PELs, TLVs.* The Permissible Exposure Limits of OSHA and the Threshold Limit Values of the American Conference of Governmental Industrial Hygienists should not be exceeded (13).

B. Chemical Hygiene Responsibilities

Responsibility for chemical hygiene rests at all levels (6, 11, 21) including the:

1. *Chief executive officer,* who has ultimate responsibility for chemical hygiene within the institution and must, with other administrators, provide continuing support for institutional chemical hygiene (7, 11).

2. *Supervisor of the department or other administrative unit,* who is responsible for chemical hygiene in that unit (7).

3. *Chemical hygiene officer(s),* whose appointment is essential (7) and who must:

(a) Work with administrators and other employees to develop and implement appropriate chemical hygiene policies and practices (7);

(b) Monitor procurement, use, and disposal of chemicals used in the lab (8);

(c) See that appropriate audits are maintained (8);

(d) Help project directors develop precautions and adequate facilities (10);

(e) Know the current legal requirements concerning regulated substances (50); and

(f) Seek ways to improve the chemical hygiene program (8, 11).

4. *Laboratory supervisor,* who has overall responsibility for chemical hygiene in the laboratory (21) including responsibility to:

(a) Ensure that workers know and follow the chemical hygiene rules, that protective equipment is available and in working order, and that appropriate training has been provided (21, 22);

(b) Provide regular, formal chemical hygiene and housekeeping inspections including routine inspections of emergency equipment (21, 171);

(c) Know the current legal requirements concerning regulated substances (50, 231);

(d) Determine the required levels of protective apparel and equipment (156, 160, 162); and

(e) Ensure that facilities and training for use of any material being ordered are adequate (215).

5. *Project director or director of other specific operation,* who has primary responsibility for chemical hygiene procedures for that operation (7).

6. *Laboratory worker,* who is responsible for:

(a) Planning and conducting each operation in accordance with the institutional chemical hygiene procedures (7, 21, 22, 230); and

(b) Developing good personal chemical hygiene habits (22).

C. The Laboratory Facility

1. *Design.* The laboratory facility should have:

(a) An appropriate general ventilation system (see C4 below) with air intakes and exhausts located so as to avoid intake of contaminated air (194);

(b) Adequate, well-ventilated stockrooms/storerooms (218, 219);

(c) Laboratory hoods and sinks (12, 162);

(d) Other safety equipment including eyewash fountains and drench showers (162, 169); and

(e) Arrangements for waste disposal (12, 240).

Appendix B: Regulations Reference Guide

2. *Maintenance.* Chemical-hygiene-related equipment (hoods, incinerator, etc.) should undergo continuing appraisal and be modified if inadequate (11, 12).

3. *Usage.* The work conducted (10) and its scale (12) must be appropriate to the physical facilities available and, especially, to the quality of ventilation (13).

4. *Ventilation*—(a) *General laboratory ventilation.* This system should: Provide a source of air for breathing and for input to local ventilation devices (199); it should not be relied on for protection from toxic substances released into the laboratory (198); ensure that laboratory air is continually replaced, preventing increase of air concentrations of toxic substances during the working day (194); direct air flow into the laboratory from non-laboratory areas and out to the exterior of the building (194).

(b) *Hoods.* A laboratory hood with 2.5 linear feet of hood space per person should be provided for every 2 workers if they spend most of their time working with chemicals (199); each hood should have a continuous monitoring device to allow convenient confirmation of adequate hood performance before use (200, 209). If this is not possible, work with substances of unknown toxicity should be avoided (13) or other types of local ventilation devices should be provided (199). See pp. 201–206 for a discussion of hood design, construction, and evaluation.

(c) *Other local ventilation devices.* Ventilated storage cabinets, canopy hoods, snorkels, etc. should be provided as needed (199). Each canopy hood and snorkel should have a separate exhaust duct (207).

(d) *Special ventilation areas.* Exhaust air from glove boxes and isolation rooms should be passed through scrubbers or other treatment before release into the regular exhaust system (208). Cold rooms and warm rooms should have provisions for rapid escape and for escape in the event of electrical failure (209).

(e) *Modifications.* Any alteration of the ventilation system should be made only if thorough testing indicates that worker protection from airborne toxic substances will continue to be adequate (12, 193, 204).

(f) *Performance.* Rate: 4–12 room air changes/hour is normally adequate general ventilation if local exhaust systems such as hoods are used as the primary method of control (194).

(g) *Quality.* General air flow should not be turbulent and should be relatively uniform throughout the laboratory, with no high velocity or static areas (194, 195); airflow into and within the hood should not be excessively turbulent (200); hood face velocity should be adequate (typically 60–100 lfm) (200, 204).

(h) *Evaluation.* Quality and quantity of ventilation should be evaluated on installation (202), regularly monitored (at least every 3 months) (6, 12, 14, 195), and reevaluated whenever a change in local ventilation devices is made (12, 195, 207). See pp. 195–198 for methods of evaluation and for calculation of estimated airborne contaminant concentrations.

D. *Components of the Chemical Hygiene Plan*

1. Basic Rules and Procedures (Recommendations for these are given in section E, below)

2. Chemical Procurement, Distribution, and Storage

(a) *Procurement.* Before a substance is received, information on proper handling, storage, and disposal should be known to those who will be involved (215, 216). No container should be accepted without an adequate identifying label (216). Preferably, all substances should be received in a central location (216).

(b) *Stockrooms/storerooms.* Toxic substances should be segregated in a well-identified area with local exhaust ventilation (221). Chemicals which are highly toxic (227) or other chemicals whose containers have been opened should be in unbreakable secondary containers (219). Stored chemicals should be examined periodically (at least annually) for replacement, deterioration, and container integrity (218–19).

Stockrooms/storerooms should not be used as preparation or repackaging areas, should be open during normal working hours, and should be controlled by one person (219).

(c) *Distribution.* When chemicals are hand carried, the container should be placed in an outside container or bucket. Freight-only elevators should be used if possible (223).

(d) *Laboratory storage.* Amounts permitted should be as small as practical. Storage on bench tops and in hoods is inadvisable. Exposure to heat or direct sunlight should be avoided. Periodic inventories should be conducted, with unneeded items being discarded or returned to the storeroom/stockroom (225–6, 229).

3. Environmental Monitoring

Regular instrumental monitoring of airborne concentrations is not usually justified or practical in laboratories but may be appropriate when testing or redesigning hoods or other ventilation devices (12) or when a highly toxic substance is stored or used regularly (e.g., 3 times/week) (13).

4. Housekeeping, Maintenance, and Inspections

(a) *Cleaning.* Floors should be cleaned regularly (24).

(b) *Inspections.* Formal housekeeping and chemical hygiene inspections should be held at least quarterly (6, 21) for units which have frequent personnel changes and semiannually for others; informal inspections should be continual (21).

(c) *Maintenance.* Eye wash fountains should be inspected at intervals of not less than 3 months (6). Respirators for routine use should be inspected periodically by the laboratory supervisor (169). Safety showers should be tested routinely (169). Other safety equipment should be inspected regularly. (e.g., every 3–6 months) (6, 24, 171). Procedures to prevent restarting of out-of-service equipment should be established (25).

(d) *Passageways.* Stairways and hallways should not be used as storage areas (24). Access to exits, emergency equipment, and utility controls should never be blocked (24).

5. Medical Program

(a) *Compliance with regulations.* Regular medical surveillance should be established to the extent required by regulations (12).

(b) *Routine surveillance.* Anyone whose work involves regular and frequent handling of toxicologically significant quantities of a chemical should consult a qualified physician to determine on an individual basis whether a regular schedule of medical surveillance is desirable (12, 50).

(c) *First aid.* Personnel trained in first aid should be available during working hours and an emergency room with medical personnel should be nearby (173). See pp. 176–178 for description of some emergency first aid procedures.

6. Protective Apparel and Equipment

These should include for each laboratory:

(a) Protective apparel compatible with the required degree of protection for substances being handled (158–161);

(b) An easily accessible drench-type safety shower (162, 169);

(c) An eyewash fountain (162);

(d) A fire extinguisher (162–164);

(e) Respiratory protection (164–9), fire alarm and telephone for emergency use (162) should be available nearby; and

(f) Other items designated by the laboratory supervisor (156, 160).

7. Records

(a) Accident records should be written and retained (174).

(b) Chemical Hygiene Plan records should document that the facilities and precautions were compatible with current knowledge and regulations (7).

(c) Inventory and usage records for high-risk substances should be kept as specified in sections E3e below.

(d) Medical records should be retained by the institution in accordance with the requirements of state and federal regulations (12).

8. Signs and Labels

Prominent signs and labels of the following types should be posted:

(a) Emergency telephone numbers of emergency personnel/facilities, supervisors, and laboratory workers (28);

(b) Identity labels, showing contents of containers (including waste receptacles) and associated hazards (27, 48);

(c) Location signs for safety showers, eyewash stations, other safety and first aid equipment, exits (27) and areas where food and beverage consumption and storage are permitted (24); and

(d) Warnings at areas or equipment where special or unusual hazards exist (27).

9. Spills and Accidents

(a) A written emergency plan should be established and communicated to all personnel; it should include procedures for ventilation failure (200), evacuation, medical care, reporting, and drills (172).

(b) There should be an alarm system to alert people in all parts of the facility including isolation areas such as cold rooms (172).

(c) A spill control policy should be developed and should include consideration of prevention, containment, cleanup, and reporting (175).

(d) All accidents or near accidents should be carefully analyzed with the results distributed to all who might benefit (8, 28).

10. Information and Training Program

(a) Aim: To assure that all individuals at risk are adequately informed about the work in the laboratory, its risks, and what to do if an accident occurs (5, 15).

(b) Emergency and Personal Protection Training: Every laboratory worker should know the location and proper use of available protective apparel and equipment (154, 169).

Some of the full-time personnel of the laboratory should be trained in the proper use of emergency equipment and procedures (6).

Such training as well as first aid instruction should be available to (154) and encouraged for (176) everyone who might need it.

(c) Receiving and stockroom/storeroom personnel should know about hazards, handling equipment, protective apparel, and relevant regulations (217).

(d) Frequency of Training: The training and education program should be a regular, continuing activity—not simply an annual presentation (15).

(e) Literature/Consultation: Literature and consulting advice concerning chemical hygiene should be readily available to laboratory personnel, who should be encouraged to use these information resources (14).

11. *Waste Disposal Program.*

(a) Aim: To assure that minimal harm to people, other organisms, and the environment will result from the disposal of waste laboratory chemicals (5).

(b) Content (14, 232, 233, 240): The waste disposal program should specify how waste is to be collected, segregated, stored, and transported and include consideration of what materials can be incinerated. Transport from the institution must be in accordance with DOT regulations (244).

(c) Discarding Chemical Stocks: Unlabeled containers of chemicals and solutions should undergo prompt disposal; if partially used, they should not be opened (24, 27).

Before a worker's employment in the laboratory ends, chemicals for which that person was responsible should be discarded or returned to storage (226).

(d) Frequency of Disposal: Waste should be removed from laboratories to a central waste storage area at least once per week and from the central waste storage area at regular intervals (14).

(e) Method of Disposal: Incineration in an environmentally acceptable manner is the most practical disposal method for combustible laboratory waste (14, 238, 241).

Indiscriminate disposal by pouring waste chemicals down the drain (14, 231, 242) or adding them to mixed refuse for landfill burial is unacceptable (14).

Hoods should not be used as a means of disposal for volatile chemicals (40, 200).

Disposal by recycling (233, 243) or chemical decontamination (40, 230) should be used when possible.

E. Basic Rules and Procedures for Working with Chemicals

The Chemical Hygiene Plan should require that laboratory workers know and follow its rules and procedures. In addition to the procedures of the sub programs mentioned above, these should include the rules listed below.

1. General Rules

The following should be used for essentially all laboratory work with chemicals:

(a) *Accidents and spills*—Eye Contact: Promptly flush eyes with water for a prolonged period (15 minutes) and seek medical attention (33, 172).

Ingestion: Encourage the victim to drink large amounts of water (178).

Skin Contact: Promptly flush the affected area with water (33, 172, 178) and remove any contaminated clothing (172, 178). If symptoms persist after washing, seek medical attention (33).

Clean-up. Promptly clean up spills, using appropriate protective apparel and equipment and proper disposal (24 33). See pp. 233–237 for specific clean-up recommendations.

(b) *Avoidance of "routine" exposure:* Develop and encourage safe habits (23); avoid unnecessary exposure to chemicals by any route (23);

Do not smell or taste chemicals (32). Vent apparatus which may discharge toxic chemicals (vacuum pumps, distillation columns, etc.) into local exhaust devices (199).

Inspect gloves (157) and test glove boxes (208) before use.

Do not allow release of toxic substances in cold rooms and warm rooms, since these have contained recirculated atmospheres (209).

(c) *Choice of chemicals:* Use only those chemicals for which the quality of the available ventilation system is appropriate (13).

(d) *Eating, smoking, etc.:* Avoid eating, drinking, smoking, gum chewing, or application of cosmetics in areas where laboratory chemicals are present (22, 24, 32, 40); wash hands before conducting these activities (23, 24).

Avoid storage, handling or consumption of food or beverages in storage areas, refrigerators, glassware or utensils which are also used for laboratory operations (23, 24, 226).

(e) *Equipment and glassware:* Handle and store laboratory glassware with care to avoid damage; do not use damaged glassware (25). Use extra care with Dewar flasks and other evacuated glass apparatus; shield or wrap them to contain chemicals and fragments should implosion occur (25). Use equipment only for its designed purpose (23, 26).

(f) *Exiting:* Wash areas of exposed skin well before leaving the laboratory (23).

(g) *Horseplay:* Avoid practical jokes or other behavior which might confuse, startle or distract another worker (23).

(h) *Mouth suction:* Do not use mouth suction for pipeting or starting a siphon (23, 32).

(i) *Personal apparel:* Confine long hair and loose clothing (23, 158). Wear shoes at all times in the laboratory but do not wear sandals, perforated shoes, or sneakers (158).

(j) *Personal housekeeping:* Keep the work area clean and uncluttered, with chemicals and equipment being properly labeled and stored; clean up the work area on completion of an operation or at the end of each day (24).

(k) *Personal protection:* Assure that appropriate eye protection (154–156) is worn by all persons, including visitors, where chemicals are stored or handled (22, 23, 33, 154).

Wear appropriate gloves when the potential for contact with toxic materials exists (157); inspect the gloves before each use, wash them before removal, and replace them periodically (157). (A table of resistance to chemicals of common glove materials is given p. 159).

Use appropriate (164–168) respiratory equipment when air contaminant concentrations are not sufficiently restricted by engineering controls (164–5), inspecting the respirator before use (169).

Use any other protective and emergency apparel and equipment as appropriate (22, 157–162).

Avoid use of contact lenses in the laboratory unless necessary; if they are used, inform supervisor so special precautions can be taken (155).

Remove laboratory coats immediately on significant contamination (161).

(l) *Planning:* Seek information and advice about hazards (7), plan appropriate protective procedures, and plan positioning of equipment before beginning any new operation (22, 23).

(m) *Unattended operations:* Leave lights on, place an appropriate sign on the door, and provide for containment of toxic substances in the event of failure of a utility service (such as cooling water) to an unattended operation (27, 128).

(n) *Use of hood:* Use the hood for operations which might result in release of toxic chemical vapors or dust (198–9).

As a rule of thumb, use a hood or other local ventilation device when working with any appreciably volatile substance with a TLV of less than 50 ppm (13).

Confirm adequate hood performance before use; keep hood closed at all times except when adjustments within the hood are being made (200); keep materials stored in hoods to a minimum and do not allow them to block vents or air flow (200).

Leave the hood "on" when it is not in active use if toxic substances are stored in it or if it is uncertain whether adequate general laboratory ventilation will be maintained when it is "off" (200).

(o) *Vigilance:* Be alert to unsafe conditions and see that they are corrected when detected (22).

(p) *Waste disposal:* Assure that the plan for each laboratory operation includes plans and training for waste disposal (230).

Deposit chemical waste in appropriately labeled receptacles and follow all other waste disposal procedures of the Chemical Hygiene Plan (22, 24).

Do not discharge to the sewer concentrated acids or bases (231); highly toxic, malodorous, or lachrymatory substances

(231); or any substances which might interfere with the biological activity of waste water treatment plants, create fire or explosion hazards, cause structural damage or obstruct flow (242).

(q) *Working alone:* Avoid working alone in a building; do not work alone in a laboratory if the procedures being conducted are hazardous (28).

2. Working with Allergens and Embryotoxins

(a) *Allergens* (examples: diazomethane, isocyanates, bichromates): Wear suitable gloves to prevent hand contact with allergens or substances of unknown allergenic activity (35).

(b) *Embryotoxins* (34–5) (examples: organomercurials, lead compounds, formamide): If you are a woman of childbearing age, handle these substances only in a hood whose satisfactory performance has been confirmed, using appropriate protective apparel (especially gloves) to prevent skin contact.

Review each use of these materials with the research supervisor and review continuing uses annually or whenever a procedural change is made.

Store these substances, properly labeled, in an adequately ventilated area in an unbreakable secondary container.

Notify supervisors of all incidents of exposure or spills; consult a qualified physician when appropriate.

3. Work with Chemicals of Moderate Chronic or High Acute Toxicity

Examples: diisopropylflurophosphate (41), hydrofluoric acid (43), hydrogen cyanide (45).

Supplemental rules to be followed in addition to those mentioned above (Procedure B of "Prudent Practices", pp. 39–41):

(a) *Aim:* To minimize exposure to these toxic substances by any route using all reasonable precautions (39).

(b) *Applicability:* These precautions are appropriate for substances with moderate chronic or high acute toxicity used in significant quantities (39).

(c) *Location:* Use and store these substances only in areas of restricted access with special warning signs (40, 229).

Always use a hood (previously evaluated to confirm adequate performance with a face velocity of at least 60 linear feet per minute) (40) or other containment device for procedures which may result in the generation of aerosols or vapors containing the substance (39); trap released vapors to prevent their discharge with the hood exhaust (40).

(d) *Personal protection:* Always avoid skin contact by use of gloves and long sleeves (and other protective apparel as appropriate) (39). Always wash hands and arms immediately after working with these materials (40).

(e) *Records:* Maintain records of the amounts of these materials on hand, amounts used, and the names of the workers involved (40, 229).

(f) *Prevention of spills and accidents:* Be prepared for accidents and spills (41).

Assure that at least 2 people are present at all times if a compound in use is highly toxic or of unknown toxicity (39).

Store breakable containers of these substances in chemically resistant trays; also work and mount apparatus above such trays or cover work and storage surfaces with removable, absorbent, plastic backed paper (40).

If a major spill occurs outside the hood, evacuate the area; assure that cleanup personnel wear suitable protective apparel and equipment (41).

(g) *Waste:* Thoroughly decontaminate or incinerate contaminated clothing or shoes (41). If possible, chemically decontaminate by chemical conversion (40).

Store contaminated waste in closed, suitably labeled, impervious containers (for liquids, in glass or plastic bottles half-filled with vermiculite) (40).

4. Work with Chemicals of High Chronic Toxicity

(Examples: dimethylmercury and nickel carbonyl (48), benzo-a-pyrene (51), N-nitrosodiethylamine (54), other human carcinogens or substances with high carcinogenic potency in animals (38).)

Further supplemental rules to be followed, in addition to all these mentioned above, for work with substances of known high chronic toxicity (in quantities above a few milligrams to a few grams, depending on the substance) (47). (Procedure A of "Prudent Practices" pp. 47–50).

(a) *Access:* Conduct all transfers and work with these substances in a "controlled area": a restricted access hood, glove box, or portion of a lab, designated for use of highly toxic substances, for which all people with access are aware of the substances being used and necessary precautions (48).

(b) *Approvals:* Prepare a plan for use and disposal of these materials and obtain the approval of the laboratory supervisor (48).

(c) *Non-contamination/Decontamination:* Protect vacuum pumps against contamination by scrubbers or HEPA filters and vent them into the hood (49). Decontaminate vacuum pumps or other contaminated equipment, including glassware, in the hood before removing them from the controlled area (49, 50).

Decontaminate the controlled area before normal work is resumed there (50).

(d) *Exiting:* On leaving a controlled area, remove any protective apparel (placing it in an appropriate, labeled container) and thoroughly wash hands, forearms, face, and neck (49).

(e) *Housekeeping:* Use a wet mop or a vacuum cleaner equipped with a HEPA filter instead of dry sweeping if the toxic substance was a dry powder (50).

(f) *Medical surveillance:* If using toxicologically significant quantities of such a substance on a regular basis (*e.g.*, 3 times per week), consult a qualified physician concerning desirability of regular medical surveillance (50).

(g) *Records:* Keep accurate records of the amounts of these substances stored (229) and used, the dates of use, and names of users (48).

(h) *Signs and labels:* Assure that the controlled area is conspicuously marked with warning and restricted access signs (49) and that all containers of these substances are appropriately labeled with identity and warning labels (48).

(i) *Spills:* Assure that contingency plans, equipment, and materials to minimize exposures of people and property in case of accident are available (233–4).

(j) *Storage:* Store containers of these chemicals only in a ventilated, limited access (48, 227, 229) area in appropriately labeled, unbreakable, chemically resistant, secondary containers (48, 229).

(k) *Glove boxes:* For a negative pressure glove box, ventilation rate must be at least 2 volume changes/hour and pressure at least 0.5 inches of water (48). For a positive pressure glove box, thoroughly check for leaks before each use (49). In either case, trap the exit gases or filter them through a HEPA filter and then release them into the hood (49).

(l) *Waste:* Use chemical decontamination whenever possible; ensure that containers of contaminated waste (including washings from contaminated flasks) are transferred from the controlled area in a secondary container under the supervision of authorized personnel (49, 50, 233).

5. Animal Work with Chemicals of High Chronic Toxicity

(a) *Access:* For large scale studies, special facilities with restricted access are preferable (56).

(b) *Administration of the toxic substance:* When possible, administer the substance by injection or gavage instead of in the diet. If administration is in the diet, use a caging system under negative pressure or under laminar air flow directed toward HEPA filters (56).

(c) *Aerosol suppression:* Devise procedures which minimize formation and dispersal of contaminated aerosols, including those from food, urine, and feces (e.g., use HEPA filtered vacuum equipment for cleaning, moisten contaminated bedding before removal from the cage, mix diets in closed containers in a hood) (55, 56).

(d) *Personal protection:* When working in the animal room, wear plastic or rubber gloves, fully buttoned laboratory coat or jumpsuit and, if needed because of incomplete suppression of aerosols, other apparel and equipment (shoe and head coverings, respirator) (56).

(e) *Waste disposal:* Dispose of contaminated animal tissues and excreta by incineration if the available incinerator can convert the contaminant to non-toxic products (238); otherwise, package the waste appropriately for burial in an EPA-approved site (239).

F. Safety Recommendations

The above recommendations from "Prudent Practices" do not include those which are directed primarily toward prevention of physical injury rather than toxic exposure. However, failure of precautions against injury will often have the secondary effect of causing toxic exposures. Therefore, we list below page references for recommendations concerning some of the major categories of safety hazards which also have implications for chemical hygiene:

1. Corrosive agents: (35–6)

Chemical Hazard Communication Guidebook

Federal Register / Vol. 55, No. 21 / Wednesday, January 31, 1990 / Rules and Regulations   3335

2. Electrically powered laboratory apparatus: (179–92)
3. Fires, explosions: (26, 57–74, 162–4, 174–5, 219–20, 226–7)
4. Low temperature procedures: (26, 88)
5. Pressurized and vacuum operations (including use of compressed gas cylinders): (27, 75–101)

G. Material Safety Data Sheets

Material safety data sheets are presented in "Prudent Practices" for the chemicals listed below. (Asterisks denote that comprehensive material safety data sheets are provided).

*Acetyl peroxide (105)
*Acrolein (106)
*Acrylonilrile (107)
Ammonia (anhydrous) (91)
*Aniline (109)
*Benzene (110)
*Benzo[a]pyrene (112)
*Bis(chloromethyl) ether (113)
Boron trichloride (91)
Boron trifluoride (92)
Bromine (114)
*Tert-butyl hydroperoxide (148)
*Carbon disulfide (116)
Carbon monoxide (92)
*Carbon tetrachloride (118)
*Chlorine (119)
Chlorine trifluoride (94)
*Chloroform (121)
Chloromethane (93)
*Diethyl ether (122)
Diisopropyl fluorophosphate (41)
*Dimethylformamide (123)
*Dimethyl sulfate (125)
*Dioxane (126)
*Ethylene dibromide (128)
*Fluorine (95)
*Formaldehyde (130)
*Hydrazine and salts (132)
Hydrofluoric acid (43)
Hydrogen bromide (98)
Hydrogen chloride (98)
*Hydrogen cyanide (133)
*Hydrogen sulfide (135)
Mercury and compounds (52)
*Methanol (137)
*Morpholine (138)
*Nickel carbonyl (99)
*Nitrobenzene (139)
Nitrogen dioxide (100)
N-nitrosodiethylamine (54)
*Peracetic acid (141)
*Phenol (142)
*Phosgene (143)
*Pyridine (144)
*Sodium azide (145)
*Sodium cyanide (147)
Sulfur dioxide (101)
*Trichloroethylene (149)
*Vinyl chloride (150)

Appendix B to § 1910.1450—References (Non-Mandatory)

The following references are provided to assist the employer in the development of a Chemical Hygiene Plan. The materials listed below are offered as non-mandatory guidance. References listed here do not imply specific endorsement of a book, opinion, technique, policy or a specific solution for a safety or health problem. Other references not listed here may better meet the needs of a specific laboratory. (a) Materials for the development of the Chemical Hygiene Plan:

1. American Chemical Society, Safety in Academic Chemistry Laboratories. 4th edition, 1985.
2. Fawcett, H.H. and W. S. Wood, Safety and Accident Prevention in Chemical Operations, 2nd edition, Wiley-Interscience, New York, 1982.
3. Flury, Patricia A., Environmental Health and Safety in the Hospital Laboratory, Charles C. Thomas Publisher, Springfield IL, 1978.
3. Green, Michael E. and Turk, Amos, Safety in Working with Chemicals, Macmillan Publishing Co., NY, 1978.
5. Kaufman, James A., Laboratory Safety Guidelines, Dow Chemical Co., Box 1713, Midland, MI 48640, 1977.
6. National Institutes of Health, NIH Guidelines for the Laboratory use of Chemical Carcinogens, NIH Pub. No. 81–2385, GPO, Washington, DC 20402, 1981.
7. National Research Council, Prudent Practices for Disposal of Chemicals from Laboratories, National Academy Press, Washington, DC, 1983.
8. National Research Council, Prudent Practices for Handling Hazardous Chemicals in Laboratories, National Academy Press, Washington, DC, 1981.
9. Renfrew, Malcolm, Ed., Safety in the Chemical Laboratory, Vol. IV, *J. Chem. Ed.*, American Chemical Society, Easlon, PA, 1981.
10. Steere, Norman V., Ed., Safety in the Chemical Laboratory, *J. Chem. Ed.* American Chemical Society, Easlon, PA, 18042, Vol. I, 1967, Vol. II, 1971, Vol. III 1974.
11. Steere, Norman V., Handbook of Laboratory Safety, the Chemical Rubber Company Cleveland, OH, 1971.
12. Young, Jay A., Ed., Improving Safety in the Chemical Laboratory, John Wiley & Sons, Inc. New York, 1987.

(b) Hazardous Substances Information:

1. American Conference of Governmental Industrial Hygienists, Threshold Limit Values for Chemical Substances and Physical Agents in the Workroom Environment with Intended Changes, P.O. Box 1937 Cincinnati, OH 45201 (latest edition).
2. Annual Report on Carcinogens, National Toxicology Program U.S. Department of Health and Human Services, Public Health Service, U.S. Government Printing Office, Washington, DC, (latest edition).
3. Best Company, Best Safety Directory, Vols. I and II, Oldwick, N.J., 1981.
4. Bretherick, L., Handbook of Reactive Chemical Hazards, 2nd edition, Butterworths, London, 1979.
5. Bretherick, L., Hazards in the Chemical Laboratory, 3rd edition, Royal Society of Chemistry, London, 1986.
6. Code of Federal Regulations, 29 CFR part 1910 subpart Z. U.S. Govt. Printing Office, Washington, DC 20402 (latest edition).
7. IARC Monographs on the Evaluation of the Carcinogenic Risk of Chemicals to Man, World Health Organization Publications Center, 49 Sheridan Avenue, Albany, New York 12210 (latest editions).
8. NIOSH/OSHA Pocket Guide to Chemical Hazards. NIOSH Pub. No. 85–114, U.S. Government Printing Office, Washington, DC, 1985 (or latest edition).
9. Occupational Health Guidelines, NIOSH/OSHA NIOSH Pub. No. 81–123 U.S. Government Printing Office, Washington, DC, 1981.
10. Patty, F.A., Industrial Hygiene and Toxicology. John Wiley & Sons, Inc., New York, NY (Five Volumes).
11. Registry of Toxic Effects of Chemical Substances, U.S. Department of Health and Human Services, Public Health Service, Centers for Disease Control, National Institute for Occupational Safety and Health, Revised Annually, for sale from Superintendent of Documents U.S. Govt. Printing Office, Washington, DC 20402.
12. The Merck Index: An Encyclopedia of Chemicals and Drugs. Merck and Company Inc. Rahway, N.J., 1976 (or latest edition).
13. Sax, N.I. Dangerous Properties of Industrial Materials, 5th edition, Van Nostrand Reinhold, NY., 1979.
14. Sittig, Marshall, Handbook of Toxic and Hazardous Chemicals, Noyes Publications, Park Ridge, NJ, 1981.

(c) Information on Ventilation:

1. American Conference of Governmental Industrial Hygienists Industrial Ventilation, 16th edition Lansing, MI, 1980.
2. American National Standards Institute, Inc. American National Standards Fundamentals Governing the Design and Operation of Local Exhaust Systems ANSI Z 9.2–1979 American National Standards Institute, N.Y. 1979.
3. Imad, A.P. and Watson, C.L. Ventilation Index: An Easy Way to Decide about Hazardous Liquids, Professional Safety pp 15–18, April 1980.
4. National Fire Protection Association, Fire Protection for Laboratories Using Chemicals NFPA-45, 1982.
Safety Standard for Laboratories in Health Related Institutions, NFPA, 56c, 1980.
Fire Protection Guide on Hazardous Materials, 7th edition, 1978.
National Fire Protection Association, Batterymarch Park, Quincy, MA 02269.
5. Scientific Apparatus Makers Association (SAMA), Standard for Laboratory Fume Hoods, SAMA LF7–1980, 1101 16th Street, NW., Washington, DC 20036.

(d) Information on Availability of Referenced Material:

1. American National Standards Institute (ANSI), 1430 Broadway, New York, NY 10018.
2. American Society for Testing and Materials (ASTM), 1916 Race Street, Philadelphia, PA 19103.

(Approved by the Office of Management and Budget under control number 1218–0131)

[FR Doc. 90–1717 Filed 1–30–90; 8:45 am]
BILLING CODE 4510-26-M

Appendix B: Regulations Reference Guide

Material Safety Data Sheet
May be used to comply with
OSHA's Hazard Communication Standard,
29 CFR 1910.1200. Standard must be
consulted for specific requirements.

U.S. Department of Labor
Occupational Safety and Health Administration
(Non-Mandatory Form)
Form Approved
OMB No. 1218-0072

IDENTITY (As Used on Label and List)	Note: Blank spaces are not permitted. If any item is not applicable, or no information is available, the space must be marked to indicate that.

Section I

Manufacturer's Name	Emergency Telephone Number
Address (Number, Street, City, State, and ZIP Code)	Telephone Number for Information
	Date Prepared
	Signature of Preparer (optional)

Section II — Hazardous Ingredients/Identity Information

Hazardous Components (Specific Chemical Identity; Common Name(s))	OSHA PEL	ACGIH TLV	Other Limits Recommended	% (optional)

Section III — Physical/Chemical Characteristics

Boiling Point		Specific Gravity (H_2O = 1)	
Vapor Pressure (mm Hg.)		Melting Point	
Vapor Density (AIR = 1)		Evaporation Rate (Butyl Acetate = 1)	

Solubility in Water

Appearance and Odor

Section IV — Fire and Explosion Hazard Data

Flash Point (Method Used)	Flammable Limits	LEL	UEL

Extinguishing Media

Special Fire Fighting Procedures

Unusual Fire and Explosion Hazards

(Reproduce locally) OSHA 174, Sept. 1985

Chemical Hazard Communication Guidebook

Section V — Reactivity Data

Stability	Unstable		Conditions to Avoid
	Stable		

Incompatibility (*Materials to Avoid*)

Hazardous Decomposition or Byproducts

Hazardous Polymerization	May Occur		Conditions to Avoid
	Will Not Occur		

Section VI — Health Hazard Data

Route(s) of Entry: Inhalation?   Skin?   Ingestion?

Health Hazards (*Acute and Chronic*)

Carcinogenicity: NTP?   IARC Monographs?   OSHA Regulated?

Signs and Symptoms of Exposure

Medical Conditions Generally Aggravated by Exposure

Emergency and First Aid Procedures

Section VII — Precautions for Safe Handling and Use

Steps to Be Taken in Case Material Is Released or Spilled

Waste Disposal Method

Precautions to Be Taken in Handling and Storing

Other Precautions

Section VIII — Control Measures

Respiratory Protection (*Specify Type*)

Ventilation	Local Exhaust	Special
	Mechanical (*General*)	Other
Protective Gloves		Eye Protection

Other Protective Clothing or Equipment

Work/Hygienic Practices

Appendix B: Regulations Reference Guide

EMERGENCY PLANNING AND COMMUNITY RIGHT-TO-KNOW

A. EMERGENCY PLANNING NOTICE

1. A Facility with Extremely Hazardous Substances in Amounts Exceeding the TPQ Must Notify the State Commission.

The owner or operator of a covered facility must notify the state emergency response commission that the facility is subject to the emergency planning provisions of the Act. A facility is covered if there is present at the facility an amount of the extremely hazardous substance above the threshold planning quantity established by EPA.

1. The original notice was required to be submitted by May 17, 1987;

2. If the facility subsequently becomes subject to the requirement the notification must be made within 60 days.

Statutory Reference: Section 302(c)
Refer To: 40 CFR § 355.30 (52 FR 13396)

---

2. A Covered Facility Must Appoint a Facility Representative and Notify the Local Committee.

The owner or operator of the facility is to notify the emergency planning committee (or the Governor if there is no committee) of a representative of the facility who will participate as a "facility emergency coordinator."

This notification was originally due by September 17, 1987 or as soon thereafter as a facility becomes subject to the notification requirement and files the required notification.

Statutory Reference: Section 303(d)
Refer To: 40 CFR § 355.30 (52 FR 13396)

---

3. A covered Facility Must Notify the Local Committee of Changes That Affect Emergency Planning.

The owner or operator of a facility subject to emergency planning must promptly inform the local committee of any relevant changes occurring at the facility as such changes occur or are expected to occur.

Statutory Reference: Section 303(d)
Refer To: 40 CFR § 355.30 (52 FR 13396)

Chemical Hazard Communication Guidebook

4. A Facility Must Submit An Emergency Planning Notice After a Covered Substance is First Present or After A Revision of the List.	If a substance on the list of extremely hazardous substances first becomes present at the facility in excess of the threshold, or if EPA revises the list of extremely hazardous substances and the substance is present at the facility in excess of the established threshold, notice must be provided to the state commission. The facility must notify the state emergency response commission and the local committee within 60 days.

Statutory Reference: Section 302(c)
Refer To: 40 CFR § 355.30 (52 FR 13396)

---

5. A Facility May Be Specially Designated to Be Required to Provide Emergency Planning Information.	A Governor or state commission may designate additional facilities which will be subject to emergency response planning and notification requirements. Notice and comment must be provided in advance. The facility concerned must be notified of such designation.

Statutory Reference: Section 302(b)

---

6. A Facility Must Provide Plan Development Information To Committee Upon Request.	On request by the local committee, the facility is to promptly provide information necessary for developing and implementing the emergency plan.

Statutory Reference: Section 303(d)
Refer To: 40 CFR § 355.30 (52 FR 13396)

Appendix B: Regulations Reference Guide

B. EMERGENCY RELEASE REPORTING

1. A Facility Must Notify the Community Emergency Coordinator and State Commission if a Covered Substance is Released.

Emergency notification requirements under SARA/CERCLA apply to any facility at which a hazardous chemical is produced, used or stored, and at which there is a release of a reportable quantity of an extremely hazardous substance or hazardous substance defined under CERCLA. Other federal and state reporting requirements may apply to a spill (See discussion in Chapters 3-4).

Certain exemptions apply to:

1. Releases which only result in exposure within a facility (for purposes of CERCLA) or only to persons within the facility boundaries (for purposes of SARA).

2. Defined releases which are "federally permitted".

3. Any release which is "continuous" as defined in section 103(f) of CERCLA.

4. Any release below an applicable RQ or, in the absence of an RQ, one pound.

5. Releases of a pesticide consistent with its registration and labelling under FIFRA.

Importantly, releases incident to transportation are covered.

The emergency notice required in the event of a reportable release must be given immediately to the community emergency coordinator for the local emergency planning committee for any area likely to be affected by the release and to the state emergency planning commission of any state likely to be affected by the release.

Statutory Reference: Section 304(b)
Refer To: 40 CFR § 355.40 (52 FR 13396)

Chemical Hazard Communication Guidebook

2. A Facility Must Provide Follow-up Emergency Notice.	In the event of a reportable release which triggers the emergency notification requirements, a follow-up notice in writing is required setting forth information on actions taken, any known or anticipated health risks and, where appropriate, advice regarding medical attention.

Statutory Reference: Section 304(c)
Refer To: 40 CFR § 355.40 (52 FR 13396)

---

3. A Transporter Must Provide Immediate Emergency Notice By Dialing 911 or Operator for Releases in Transportation.	In the event of a reportable release which triggers SARA's emergency notification requirements for a transportation incident, notice is required immediately by dialing 911 or the operator. It should be emphasized that other reporting requirements may apply, such as spill reporting to the Department of Transportation (see discussion of DOT requirements in Chapter 4).

Statutory Reference: Section 304(b)
Refer To: 40 CFR § 355.40 (52 FR 13396)

C. MSDS/LIST SUBMISSION

1. A Facility Required to Prepare or Have Available MSDS's Must Submit Copies or Lists of Hazardous Chemicals.	The owner or operator of any facility which is required to prepare or have available an MSDS for a hazardous chemical (defined according to the OSHA Hazard Communication rule) must submit a copy of the MSDS or an alternative list of hazardous chemicals to : i) the local emergency planning committee; ii) the state commission; and iii) the fire department with jurisdiction over the facility.

The MSDS or list must be supplied 3 months after the facility is required to have available an MSDS under OSHA regulations.

Manufacturers were originally required to submit MSDS's by October 17, 1987. The expansion of the OSHA Standard

Appendix B: Regulations Reference Guide

requires submission of MSDS's by the non-manufacturing sector by August 23, 1988.

EPA has set reporting thresholds for hazardous substances at 10,000 pounds for the first two years and will finalize a threshold for the third and subsequent years for hazardous chemicals. Facilities must submit MSDS's for extremely hazardous substances in excess of 500 pounds (or 55 gallons) or the threshold planning quantity, whichever is lower.

Statutory Reference: Section 311
Refer To: 40 CFR § 370.21 (52 FR 38365)

---

2. A Facility Can Submit a List of Hazardous Chemicals In Lieu of Actual Copies of MSDS's.

A list of the hazardous chemicals may be submitted in lieu of the MSDS's. In this case, the chemicals must be grouped in categories of five health and physical hazards. The chemical name and hazardous components must be specified. Mixtures may be reported as a whole or by their components but consistency must be maintained. EPA has set thresholds at 10,000 pounds for the first two years and will finalize a threshold for third and subsequent years for hazardous chemicals; for extremely hazardous substances, the threshold is 500 pounds (or 55 gallons) or the threshold planning quantity, whichever is less. Certain exemptions apply.

Statutory Reference: Section 311(a)
Refer To: 40 CFR § 370.21 (52 FR 38365)

---

3. A Facility Must Submit Revised MSDS's Within Three Months Of Discovery of Significant New Information.

Within 3 months following the discovery of significant new information concerning a hazardous chemical for which an MSDS was previously submitted, the owner or operator must provide a revised MSDS. If a facility has submitted a list instead of MSDS's, it must provide the MSDS upon request.

Statutory Reference: Section 311(d)(2)
Refer To: 40 CFR § 370.21 (52 FR 38365)

Chemical Hazard Communication Guidebook

4. A Facility Must Provide an MSDS Upon Request If a List of Hazardous Chemicals Was Originally Submitted In Lieu of the MSDS.

If the facility chooses to submit a list of chemicals in lieu of the MSDS, the MSDS must be submitted if the local committee subsequently requests it.

Statutory Reference: Section 311(c)
Refer To: 40 CFR 370.21 (52 FR 38365)

D. HAZARDOUS CHEMICAL INVENTORY REPORTING

1. Covered Facility Must Submit an Annual Hazardous Chemical Inventory Form.

The owner or operator of any facility which is required to submit an MSDS or list must prepare and annually submit an Emergency and Hazardous Chemical Inventory Form. It must be submitted to:
i) the local committee; ii) the state commission; and iii) the fire department with jurisdiction. The basic inventory form, referred to as the "tier one" form, contains an estimate of the maximum amount, average daily amount (both in ranges), and general location of hazardous chemicals aggregated by categories of physical and health hazards. The facility has the option to submit chemical specific information on the "tier two" form.

The report is due March 1, 1988 and annually thereafter

Those non-manufacturing facilities newly subject to the OSHA Hazard Communication Standard are subject to this inventory requirement beginning, March 1, 1989.

Statutory Reference: Section 312 (a)
Refer To: 40 CFR §370.25 (52 FR 38365)

Appendix B: Regulations Reference Guide

2. A Facility Must Provide A Tier Two Inventory for Specific Chemicals Upon Request by Authorized Officials.

On request by a state commission, local committee, or fire department, the facility must provide chemical specific, "tier two" information to the person making the request. The request must be with regard to a specific facility.

In addition, a member of the public may ask the state commission or local committee for tier two information regarding a facility. Any information in their possession is to be made available. If the tier two information is not available, the commission or committee is to request it of the facility. If the hazardous chemical was present at the facility in excess of 10,000 pounds during the preceding calendar year, the request is compulsory. Below this threshold, the decision to request the tier two information from the facility rests with the commission or committee.

Statutory Reference: Section 312 (e)
Refer To: 40 CFR § 370.30 (52 FR 38366)

---

3. A Facility Must Permit Access to Fire Department Officials.

Upon receiving a request, the facility must permit the fire department to enter to conduct an on-site inspection and obtain information.

Statutory Reference: Section 312(f)
Refer To: 40 CFR § 370.25(d) (52 FR 38365)

Chemical Hazard Communication Guidebook

E. SUBMISSION OF TOXIC CHEMICAL RELEASE REPORTS

1. Submission of the Toxic Chemical Release Inventory (See Chapter 3 For Discussion of EPA Final Rule).

The owner or operator of a covered facility must submit the emissions inventory on or before July 1, 1988 to the Administrator of EPA and to state officials designated by the Governor.

A facility is covered if it manufactures, processes, or otherwise uses specified toxic chemicals above a specified threshold, has 10 or more full-time employees and is in Standard Industrial Classification Codes 20-39 (as of July 1, 1985), which includes the manufacturing sector.

The established thresholds are:

1987 Reporting-- manufacture or process above 75,000 pounds of a covered chemical

1988 Reporting-- manufacture or process above 50,000 pounds of a covered chemical

1989 Reporting and Thereafter-- manufacture or process above 25,000 pounds of a covered chemical,

Use -- The threshold for use is set at 10,000 pounds.

A facility must file the emissions report for all listed toxic chemicals present at the facility above the specified thresholds.

Statutory Reference: Section 313(a)
Refer To: 40 CFR § 372.22, .25

Appendix B: Regulations Reference Guide

3. A Facility May Become Responsible for Release Reporting if EPA Adds or Deletes SIC Codes Defining Covered Facilities or Operations.

The Administrator of EPA may add or delete Standard Industrial Classification codes that are included within the requirement to submit a toxic chemical release form, i.e. the emissions report, but only to the extent necessary to ensure that the covered SIC's are relevant to the legislative purpose.

Statutory Reference: Section 313(b)
Refer To: 40 CFR § 372.65

---

4. A Facility's Responsibilities to Report May Be Modified if EPA Adds or Deletes Chemicals from the List of Toxic Chemicals.

The Administrator may by rule add or delete a chemical from the statutory list at any time. States and the public may petition EPA to have chemicals added to the list. With regard to additions, a chemical may be added if the Administrator determines, in his judgment, that there is sufficient evidence to establish: i) significant adverse acute human health effects beyond facility site boundaries; ii) specified chronic human health effects; or iii) serious environmental effects. With regard to deletions, a chemical may be deleted if the EPA determines that there is not sufficient evidence to establish any of the criteria above.

EPA has published a statement of policy, regarding petitions under section 313 (52 FR 3479 (February 4, 1987).

Statutory Reference: Section 313(d)

189

Chemical Hazard Communication Guidebook

5. A Facility May be Required to Submit Emissions Report at the Motion of EPA.

The Administrator, on his own motion, or at the request of a Governor, may apply emissions reporting to the owners or operators of any particular facility that manufactures, processes, or otherwise uses a toxic chemical included on the statutory list. The Administrator must consider certain criteria. The request of the Governor is limited to facilities in that state.

Statutory Reference: Section 313(b)

---

6. A Facility May Be Required to Submit Information For Mass Balance Study.

The statute calls upon the Administrator to carry out a mass balance study to be submitted no later than 5 years after enactment. Information is to be collected from states which currently conduct (or establish within 5 years) a mass balance survey. If EPA determines that the information is inadequate, it may require the submission of information by a state or facility.

Statutory Reference: Section 313(l)

A Facility Must Estimate Releases Subject to Reporting for the Toxic Chemical Inventory.

The reporting period for the emissions inventory, submitted on the Toxic Chemical Release Forms, is each calendar year. The first deadline is July 1, 1988, and covers releases from January 1, 1987. Facilities must keep submissions and supporting documentation for five years.

Readily available data may be used as the basis for the estimation or where such data are not available, reasonable estimates

The statute does not require monitoring or measurement of the releases, but such data must be used if available.

Statutory Reference: Section 313(a)
Refer To: 40 CFR § 372.30

Appendix B: Regulations Reference Guide

F. DISCLOSURE OF INFORMATION

1. A Facility May Withhold a Trade Secret Chemical Identity From Reports.

Any person required to submit an MSDS, chemical inventory, or emissions report may withhold the specific chemical identity if it is a trade secret and meets specified criteria.

A detailed procedure is established to substantiate trade secret claims.

Statutory Reference: Section 322
Refer To: 40 CFR § 350.1, et seq.

---

2. A Trade Secret Claim May Be Challenged.

Any person may petition the Administrator for the disclosure of the specific chemical identity of a hazardous chemical (subject to OSHA standard), an extremely hazardous chemical (subject to emergency planning), or toxic chemical (subject to emissions reporting). The Administrator may initiate his own action.

The statute establishes a complex procedure by which a claim of a trade secret chemical identity may be challenged. On October 15, 1987 EPA proposed its regulations regarding trade secret claims (5 2 Fed. Reg. 38312).

Statutory Reference: Section 322(d)
Refer To: 40 CFR § 350.15

---

3. A Facility May Be Required to Disclose Complete Chemical Information to a Physician/Nurse In a Medical Emergency.

In the event of a medical emergency the facility is required to supply to a treating physician or nurse specific chemical identity information.

Statutory Reference: Section 323(b)
Refer To: 40 CFR § 350.40

Chemical Hazard Communication Guidebook

4. A Facility May Be Required to Disclose the Specific Chemical Identity of a Trade Secret Chemical to a Health Professional.

An owner or operator of a facility subject to the MSDS, chemical inventory, or emissions inventory requirements of the Act must provide specific chemical identity information to a health professional if it is requested in connection with diagnosis or treatment. The statute and the regulations spell out the written statement of need and confidentiality agreement that may apply.

Statutory Reference: Section 323(a)
Refer To: 40 CFR § 350.40

---

5. A Facility May Be Required to Disclose the Chemical Identity of a Trade Secret Chemical to a Local Government Health Professional.

The owner or operator of a facility subject to MSDS, chemical inventory, or emissions inventory reporting requirements must provide specific chemical identity for any covered substance to a health professional who is a local government employee (or contractor) and who requests the information in writing.

Statutory Reference: Section 323(c)
Refer To: 40 CFR § 350.40

EPA'S EMERGENCY PLANNING RULE

Chemical Hazard Communication Guidebook

PART 355—EMERGENCY PLANNING AND NOTIFICATION

Sec.
355.10 Purpose
355.20 Definitions
355.30 Emergency planning
355.40 Emergency release notification
355.50 Penalties

APPENDIX A—THE LIST OF EXTREMELY HAZARDOUS SUBSTANCES, AND THEIR THRESHOLD PLANNING QUANTITIES (ALPHABETICAL ORDER)

APPENDIX B—THE LIST OF EXTREMELY HAZARDOUS SUBSTANCES AND THEIR THRESHOLD PLANNING QUANTITIES (CAS NUMBER ORDER)

AUTHORITY: 42 U.S.C. 11002 and 11048.

SOURCE: 52 FR 13395, Apr. 22, 1987, unless otherwise noted.

§ 355.10 Purpose.

This regulation establishes the list of extremely hazardous substances, threshold planning quantities, and facility notification responsibilities necessary for the development and implementation of State and local emergency response plans.

§ 355.20 Definitions.

Act means the Superfund Amendments and Reauthorization Act of 1986.

CERCLA means the Comprehensive Emergency Response, Compensation and Liability Act of 1980, as amended.

CERCLA Hazardous Substance means a substance on the list defined in section 101(14) of CERCLA.

NOTE: Listed CERCLA hazardous substances appear in Table 302.4 of 40 CFR Part 302.

Commission means the emergency response commission, or the Governor if there is no commission, for the State in which the facility is located.

Environment includes water, air, and land and the interrelationship which exists among and between water, air, and land and all living things.

Extremely hazardous substance means a substance listed in Appendices A and B of this part.

Facility means all buildings, equipment, structures, and other stationary items which are located on a single site or on contiguous or adjacent sites and which are owned or operated by the same person (or by any person which controls, is controlled by, or under common control with, such person). For purposes of emergency release notification, the term includes motor vehicles, rolling stock, and aircraft.

Hazardous chemical means any hazardous chemical as defined under § 1910.1200(c) of Title 29 of the Code of Federal Regulations, except that such term does not include the following substances:

(1) Any food, food additive, color additive, drug, or cosmetic regulated by the Food and Drug Administration.

(2) Any substance present as a solid in any manufactured item to the extent exposure to the substance does not occur under normal conditions of use.

(3) Any substance to the extent it is used for personal, family, or household purposes, or is present in the same form and concentration as a product packaged for distribution and use by the general public.

(4) Any substance to the extent it is used in a research laboratory or a hospital or other medical facility under the direct supervision of a technically qualified individual.

Environmental Protection Agency

§ 355.30

(5) Any substance to the extent it is used in routine agricultural operations or is a fertilizer held for sale by a retailer to the ultimate customer.

Mixture means a heterogenous association of substances where the various individual substances retain their identities and can usually be separated by mechanical means. Includes solutions or compounds but does not include alloys or amalgams.

Person means any individual, trust, firm, joint stock company, corporation (including a government corporation), partnership, association, State, municipality, commission, political subdivision of a State, or interstate body.

Release means any spilling, leaking, pumping, pouring, emitting, emptying, discharging, injecting, escaping, leaching, dumping, or disposing into the environment (including the abandonment or discarding of barrels, containers, and other closed receptacles) of any hazardous chemical, extremely hazardous substance, or CERCLA hazardous substance.

Reportable quantity means, for any CERCLA hazardous substance, the reportable quantity established in Table 302.4 of 40 CFR Part 302, for such substance, for any other substance, the reportable quantity is one pound.

Threshold planning quantity means, for a substance listed in Appendices A and B, the quantity listed in the column "threshold planning quantity" for that substance.

§ 355.30 **Emergency planning.**

(a) *Applicability.* The requirements of this section apply to any facility at which there is present an amount of any extremely hazardous substance equal to or in excess of its threshold planning quantity, or designated, after public notice and opportunity for comment, by the Commission or the Governor for the State in which the facility is located. For purposes of this section, an "amount of any extremely hazardous substance" means the total amount of an extremely hazardous substance present at any one time at a facility at concentrations greater than one percent by weight, regardless of location, number of containers, or method of storage.

(b) *Emergency planning notification.* The owner or operator of a facility subject to this section shall provide notification to the Commission that it is a facility subject to the emergency planning requirements of this Part. Such notification shall be provided: on or before May 17, 1987 or within sixty days after a facility first becomes subject to the requirements of this section, whichever is later.

(c) *Facility emergency coordinator.* The owner or operator of a facility subject to this section shall designate a facility representative who will participate in the local emergency planning process as a facility emergency response coordinator. The owner or operator shall notify the local emergency planning committee (or the Governor if there is no committee) of the facility representative on or before September 17, 1987 or 30 days after establishment of a local emergency planning committee, whichever is earlier.

(d) *Provision of information.* (1) The owner or operator of a facility subject to this section shall inform the local emergency planning committee of any changes occurring at the facility which may be relevant to emergency planning.

(2) Upon request of the local emergency planning committee, the owner or operator of a facility subject to this section shall promptly provide to the committee any information necessary for development or implementation of the local emergency plan.

(e) *Calculation of TPQs for solids and mixtures.* (1) If a container or storage vessel holds a mixture or solution of an extremely hazardous substance, then the concentration of extremely hazardous substance, in weight percent (greater than 1%), shall be multiplied by the mass (in pounds) in the vessel to determine the actual quantity of extremely hazardous substance therein.

(2)(i) Extremely hazardous substances that are solids are subject to either of two threshold planning quantities as shown on Appendices A and B (i.e., 500/10,000 pounds). The lower quantity applies only if the solid exists in powdered form and has a particle size less than 100 microns; or is handled in solution or in molten form; or

225

§ 355.40

meets the criteria for a National Fire Protection Association (NFPA) rating of 2, 3 or 4 for reactivity. If the solid does not meet any of these criteria, it is subject to the upper (10,000 pound) threshold planning quantity as shown in Appendices A and B.

(ii) The 100 micron level may be determined by multiplying the weight percent of solid with a particle size less than 100 microns in a particular container by the quantity of solid in the container.

(iii) The amount of solid in solution may be determined by multiplying the weight percent of solid in the solution in a particular container by the quantity of solution in the container.

(iv) The amount of solid in molten form must be multipled by 0.3 to determine whether the lower threshold planning quantity is met.

(Approved by the Office of Management and Budget under control number 2050-0046)

§ 355.40 Emergency release notification.

(a) *Applicability.* (1) The requirements of this section apply to any facility: (i) at which a hazardous chemical is produced, used or stored and (ii) at which there is release of a reportable quantity of any extremely hazardous substance or CERCLA hazardous substance.

(2) This section does not apply to:

(i) Any release which results in exposure to persons solely within the boundaries of the facility;

(ii) Any release which is a "federally permitted release" as defined in section 101 (10) of CERCLA;

(iii) Any release which is "continuous," as defined under section 103 (f) of CERCLA (except for "statistically significant increases" as defined under section 103(e) of CERCLA; and

(iv) Any release of a pesticide product exempt from CERCLA section 103(a) reporting under section 103(e) of CERCLA;

(v) Any release not meeting the definition of release under Section 101(22) of CERCLA, and therefore exempt from Section 103(a) reporting; and

(vi) Any radionuclide release which occurs (A) naturally in soil from land holdings such as parks, golf courses, or other large tracts of land; (B) naturally from the disturbance of land for purposes other than mining, such as for agricultural or construction activities; (C) from the dumping of coal and coal ash at utility and industrial facilities with coal-fired boilers; and (D) from coal and coal ash piles at utility and industrial facilities with coal-fired boilers.

Note to paragraph (a): Releases of CERCLA hazardous substances are subject to the release reporting requirements of CERCLA section 103, codified at 40 CFR Part 302, in addition to the requirements of this part.

(b) *Notice requirements.* (1) The owner or operator of a facility subject to this section shall immediately notify the community emergency coordinator for the local emergency planning committee of any area likely to be affected by the release and the State emergency response commission of any State likely to be affected by the release. If there is no local emergency planning committee, notification shall be provided under this section to relevant local emergency response personnel.

(2) The notice required under this section shall include the following to the extent known at the time of notice and so long as no delay in notice or emergency response results:

(i) The chemical name or identity of any substance involved in the release.

(ii) An indication of whether the substance is an extremely hazardous substance.

(iii) An estimate of the quantity of any such substance that was released into the environment.

(iv) The time and duration of the release.

(v) The medium or media into which the release occurred.

(vi) Any known or anticipated acute or chronic health risks associated with the emergency and, where appropriate, advice regarding medical attention necessary for exposed individuals.

(vii) Proper precautions to take as a result of the release, including evacuation (unless such information is readily available to the community emergency coordination pursuant to the emergency plan).

Appendix B: Regulations Reference Guide

Environmental Protection Agency

(viii) The names and telephone number of the person or persons to be contacted for further information.

(3) As soon as practicable after a release which requires notice under (b)(1) of this section, such owner or operator shall provide a written follow-up emergency notice (or notices, as more information becomes available) setting forth and updating the information required under paragraph (b)(2) of this section, and including additional information with respect to:

(i) Actions taken to respond to and contain the release,

(ii) Any known or anticipated acute or chronic health risks associated with the release, and,

(iii) Where appropriate, advice regarding medical attention necessary for exposed individuals.

(4) *Exceptions.* (i) Until April 30, 1988, in lieu of the notice specified in paragraph (b)(2) of this section, any owner or operator of a facility subject to this section from which there is a release of a CERCLA hazardous substance which is not an extremely hazardous substance and has a statutory reportable quantity may provide the same notice required under CERCLA section 103(a) to the local emergency planning committee.

(ii) An owner or operator of a facility from which there is a transportation-related release may meet the requirements of this section by providing the information indicated in paragraph (b)(2) to the 911 operator, or in the absence of a 911 emergency telephone number, to the operator. For purposes of this paragraph, a "transportation-related release" means a release during transportation, or storage incident to transportation if the stored substance is moving under active shipping papers and has not reached the ultimate consignee.

(Approved by the Office of Management and Budget under control number 2050-0046)

§ 355.50

[52 FR 13395, Apr. 22, 1987, as amended at 54 FR 22543, May 24, 1989]

EFFECTIVE DATE NOTE: At 54 FR 22543, May 24, 1989, § 355.40 was amended by adding paragraphs (a)(2) (v) and (vi) revising paragraph (a)(2)(iv) and republishing (a)(2) introductory text, effective July 24, 1989. For the convenience of the user, the superseded text follows.

§ 355.40 Emergency release notification.

(a) * * *

(2) * * *

(iv) Any release exempt from CERCLA section 103(a) reporting under section 101(22) of CERCLA.

* * * * *

§ 355.50 Penalties.

(a) *Civil penalties.* Any person who fails to comply with the requirements of § 355.40 shall be subject to civil penalties of up to $25,000 for each violation in accordance with section 325(b)(1) of the Act.

(b) *Civil penalties for continuing violations.* Any person who fails to comply with the requirements of § 355.40 shall be subject to civil penalties of up to $25,000 for each day during which the violation continues, in accordance with section 325(b)(2) of the Act. In the case of a second or subsequent violation, any such person may be subject to civil penalties of up to $75,000 for each day the violation continues, in accordance with section 325(b)(2) of the Act.

(c) *Criminal penalties.* Any person who knowingly and willfully fails to provide notice in accordance with § 355.40 shall, upon conviction, be fined not more than $25,000 or imprisoned for not more than two (2) years, or both (or, in the case of a second or subsequent conviction, shall be fined not more than $50,000 or imprisoned for not more than five (5) years, or both) in accordance with section 325(b)(4) of the Act.

EPA'S RELEASE REPORTING REGULATIONS UNDER CERCLA

Chemical Hazard Communication Guidebook

PART 302—DESIGNATION, REPORTABLE QUANTITIES, AND NOTIFICATION

40 CFR Ch. I (7-1-89 Edition)

Sec.
302.1 Applicability.
302.2 Abbreviations.
302.3 Definitions.
302.4 Designation of hazardous substances.
302.5 Determination of reportable quantities.
302.6 Notification requirements.
302.7 Penalties.

AUTHORITY: Sec. 102 of the Comprehensive Environmental Response, Compensation, and Liability Act of 1980, 42 U.S.C. 9602; secs. 311 and 501(a) of the Federal Water Pollution Control Act, 33 U.S.C. 1321 and 1361.

SOURCE: 50 FR 13474, Apr. 4, 1985, unless otherwise noted.

§ 302.1 Applicability.

This regulation designates under section 102(a) of the Comprehensive Environmental Response, Compensation, and Liability Act of 1980 ("the Act") those substances in the statutes referred to in section 101(14) of the Act, identifies reportable quantities for these substances, and sets forth the notification requirements for releases of these substances. This regulation also sets forth reportable quantities for hazardous substances designated under section 311(b)(2)(A) of the Clean Water Act.

§ 302.2 Abbreviations.

CASRN = Chemical Abstracts Service Registry Number
RCRA = Resource Conservation and Recovery Act of 1976, as amended
lb = pound
kg = kilogram
RQ = reportable quantity

§ 302.3 Definitions.

As used in this part, all terms shall have the meaning set forth below:

"The Act", *"CERCLA"*, or *"Superfund"* means the Comprehensive Environmental Response, Compensation, and Liability Act of 1980 (Pub. L. 96-510);

"Administrator" means the Administrator of the United States Environmental Protection Agency ("EPA");

"Consumer product" shall have the meaning stated in 15 U.S.C. 2052;

"Environment" means (1) the navigable waters, the waters of the contiguous zone, and the ocean waters of which the natural resources are under the exclusive management authority of the United States under the Fishery Conservation and Management Act of 1976, and (2) any other surface water, ground water, drinking water supply, land surface or subsurface strata, or ambient air within the United States or under the jurisdiction of the United States;

"Facility" means (1) any building, structure, installation, equipment, pipe or pipeline (including any pipe into a sewer or publicly owned treatment works), well, pit, pond, lagoon, impoundment, ditch, landfill, storage container, motor vehicle, rolling stock, or aircraft, or (2) any site or area where a hazardous substance has been deposited, stored, disposed of, or placed, or otherwise come to be located; but does not include any consumer product in consumer use or any vessel;

"Hazardous substance" means any substance designated pursuant to 40 CFR Part 302;

"Hazardous waste" shall have the meaning provided in 40 CFR 261.3;

"Navigable waters" or "navigable waters of the United States means waters of the United States, including the territorial seas;

"Offshore facility" means any facility of any kind located in, on, or under, any of the navigable waters of the United States, and any facility of any kind which is subject to the jurisdiction of the United States and is located in, on, or under any other waters, other than a vessel or a public vessel;

"Onshore facility" means any facility (including, but not limited to, motor vehicles and rolling stock) of any kind located in, on, or under, any land or non-navigable waters within the United States;

"Person" means an individual, firm, corporation, association, partnership, consortium, joint venture, commercial entity, United States Government, State, municipality, commission, political subdivision of a State, or any interstate body;

"Release" means any spilling, leaking, pumping, pouring, emitting, emptying, discharging, injecting, escaping, leaching, dumping, or disposing into the environment, but excludes (1) any release which results in exposure to persons solely within a workplace, with respect to a claim which such

200

Environmental Protection Agency § 302.4

persons may assert against the employer of such persons, (2) emissions from the engine exhaust of a motor vehicle, rolling stock, aircraft, vessel, or pipeline pumping station engine, (3) release of source, byproduct, or special nuclear material from a nuclear incident, as those terms are defined in the Atomic Energy Act of 1954, if such release is subject to requirements with respect to financial protection established by the Nuclear Regulatory Commission under section 170 of such Act, or for the purposes of section 104 of the Comprehensive Environmental Response, Compensation, and Liability Act or any other response action, any release of source, byproduct, or special nuclear material from any processing site designated under section 102(a)(1) or 302(a) of the Uranium Mill Tailings Radiation Control Act of 1978, and (4) the normal application of fertilizer;

"Reportable quantity" means that quantity, as set forth in this part, the release of which requires notification pursuant to this part;

"United States" include the several States of the United States, the District of Columbia, the Commonwealth of Puerto Rico, Guam, American Samoa, the United States Virgin Islands, the Commonwealth of the Northern Marianas, and any other territory or possession over which the United States has jurisdiction; and

"Vessel" means every description of watercraft or other artificial contrivance used, or capable of being used, as a means of transportation on water.

§ 302.4 Designation of hazardous substances.

(a) *Listed hazardous substances.* The elements and compounds and hazardous wastes appearing in Table 302.4 are designated as hazardous substances under section 102(a) of the Act.

(b) *Unlisted hazardous substances.* A solid waste, as defined in 40 CFR 261.2, which is not excluded from regulation as a hazardous waste under 40 CFR 261.4(b), is a hazardous substance under section 101(14) of the Act if it exhibits any of the characteristics identified in 40 CFR 261.20 through 261.24.

Appendix B: Regulations Reference Guide

**Tuesday
July 24, 1990**

Part VII

Environmental Protection Agency

40 CFR Parts 302 and 355
Reporting Continuous Releases of Hazardous Substances; Final Rule

PART 355—EMERGENCY PLANNING AND NOTIFICATION

3. The authority citation for part 355 is revised to read as follows:

Authority: 42 U.S.C. 11002, 11004, and 11048.

4. Section 355.40 is amended by revising paragraph (a)(2)(iii) to read as follows:

§ 355.40 Emergency release notification.

(a) * * *

(2) * * *

(iii) Any release that is continuous and stable in quantity and rate under the definitions in 40 CFR 302.8(b). Exemption from notification under this subsection does not include exemption from:

(A) Initial notifications as defined in 40 CFR 302.8(d) and (e);

(B) Notification of a "statistically significant increase," defined in 40 CFR 302.8(b) as any increase above the upper bound of the reported normal range, which is to be submitted to the community emergency coordinator for the local emergency planning committee for any area likely to be affected by the release and to the State emergency response commission of any State likely to be affected by the release;

(C) Notification of a "new release" as defined in 40 CFR 302.8(g)(1); or

(D) Notification of a change in the normal range of the release as required under 40 CFR 302.8(g)(2).

(Approved by the Office of Management and Budget under the control number 2050-0092)

* * * * *

[FR Doc. 90-15260 Filed 7-23-90; 8:45 am]

BILLING CODE 6560-50-M

EPA'S REGULATION ESTABLISHING EMERGENCY AND HAZARDOUS CHEMICAL INVENTORY FORMS AND COMMUNITY RIGHT-TO-KNOW REPORTING REQUIREMENTS

PART 370—HAZARDOUS CHEMICAL REPORTING: COMMUNITY RIGHT-TO-KNOW

Subpart A—General Provisions

Sec.
370.1 Purpose.
370.2 Definitions.
370.5 Penalties.

Subpart B—Reporting Requirements

370.20 Applicability.
370.21 MSDS reporting.
370.25 Inventory reporting.
370.28 Mixtures.

Subpart C—Public Access and Availability of Information

370.30 Requests for information.
370.31 Provision of information.

Subpart D—Inventory Forms

370.40 Tier I emergency and hazardous chemical inventory form.
370.41 Tier II emergency and hazardous chemical inventory form.

AUTHORITY: Secs. 311, 312, 324, 325, 328, 329 of Pub. L. 99-499, 100 Stat. 1613, 42 U.S.C. 11011, 11012, 11024, 11025, 11028, 11029.

SOURCE: 52 FR 38364, Oct. 15, 1987, unless otherwise noted.

Subpart A—General Provisions

§ 370.1 Purpose.

These regulations establish reporting requirements which provide the public with important information on the hazardous chemicals in their communities for the purpose of enhancing community awareness of chemical hazards and facilitating development of State and local emergency response plans.

§ 370.2 Definitions.

"*Commission*" means the State emergency response commission, or the Governor if there is no commission, for the State in which the facility is located.

"*Committee*" means the local emergency planning committee for the emergency planning district in which the facility is located.

"*Environment*" includes water, air, and land and the interrelationship that exists among and between water, air, and land and all living things.

"*Extremely hazardous substance*" means a substance listed in the Appendices to 40 CFR Part 355, Emergency Planning and Notification.

"*Facility*" means all buildings, equipment, structures, and other stationary items that are located on a single site or on contiguous or adjacent sites and that are owned or operated by the same person (or by any person which controls, is controlled by, or under common control with, such person). For purposes of emergency release notification, the term includes motor vehicles, rolling stock, and aircraft.

"*Hazard category*" means any of the following:

(1) "Immediate (acute) health hazard," including "highly toxic," "toxic," "irritant," "sensitizer," "corrosive," (as defined under § 1910.1200 of Title 29 of the Code of Federal Regulations) and other hazardous chemicals that cause an adverse effect to a target organ and which effect usually occurs rapidly as a result of short term exposure and is of short duration;

(2) "Delayed (chronic) health hazard," including "carcinogens" (as defined under § 1910.1200 of Title 29 of the Code of Federal Regulations) and other hazardous chemicals that cause an adverse effect to a target organ and which effect generally occurs as a result of long term exposure and is of long duration;

(3) "Fire hazard," including "flammable," "combustible liquid," "pyrophoric," and "oxidizer" (as defined under § 1910.1200 of Title 29 of the Code of Federal Regulations);

(4) "Sudden release of pressure," including "explosive" and "compressed gas" (as defined under § 1910.1200 of Title 29 of the Code of Federal Regulations); and

(5) "Reactive," including "unstable reactive," "organic peroxide," and "water reactive" (as defined under § 1910.1200 of Title 29 of the Code of Federal Regulations).

"*Hazardous chemical*" means any hazardous chemical as defined under § 1910.1200(c) of Title 29 of the Code of Federal Regulations, except that

240

Environmental Protection Agency § 370.20

such term does not include the following substances:

(1) Any food, food additive, color additive, drug, or cosmetic regulated by the Food and Drug Administration.

(2) Any substance present as a solid in any manufactured item to the extent exposure to the substance does not occur under normal conditions of use.

(3) Any substance to the extent it is used for personal, family, or household purposes, or is present in the same form and concentration as a product packaged for distribution and use by the general public.

(4) Any substance to the extent it is used in a research laboratory or a hospital or other medical facility under the direct supervision of a technically qualified individual.

(5) Any substance to the extent it is used in routine agricultural operations or is a fertilizer held for sale by a retailer to the ultimate customer.

"Inventory form" means the Tier I and Tier II emergency and hazardous chemical inventory forms set forth in Subpart D of this Part

"Material Safety Data Sheet" or *"MSDS"* means the sheet required to be developed under § 1910.1200(g) of Title 29 of the Code of Federal Regulations.

"Person" means any individual, trust, firm, joint stock company, corporation (including a government corporation), partnership, association, State, municipality, commission, political subdivision of State, or interstate body.

"Present in the same form and concentration as a product packaged for distribution and use by the general public" means a substance packaged in a similar manner and present in the same concentration as the substance when packaged for use by the general public, whether or not it is intended for distribution to the general public or used for the same purpose as when it is packaged for use by the general public.

"State" means any State of the United States, the District of Columbia, the Commonwealth of Puerto Rico, Guam, American Samoa, the United States Virgin Islands, the Northern Mariana Islands, and any other territory or possession over which the United States has jurisdiction.

"TPQ" means the threshold planning quantity for an extremely hazardous substance as defined in 40 CFR Part 355.

§ 370.5 **Penalties.**

(a) *MSDA reporting.* Any person other than a governmental entity who violates any requirement of § 370.21 shall be liable for civil and administrative penalties of not more than $10,000 for each violation.

(b) *Inventory reporting.* Any person other than a governmental entity who violates any requirement of § 370.25 shall be liable for civil and administrative penalties of not more than $25,000 for each violation.

(c) *Continuing violations.* Each day a violation described in paragraph (a) or (b) of this section continues shall constitute a separate violation.

Subpart B—Reporting Requirements

§ 370.20 Applicability.

(a) *General.* The requirements of this subpart apply to any facility that is required to prepare or have available a material safety data sheet (or MSDS) for a hazardous chemical under the Occupational Safety and Health Act of 1970 and regulations promulgated under that Act.

(b) *Minimum threshold levels.* Except as provided in paragraph (b)(3) of this section, the minimum threshold level for reporting under this subpart shall be according to the following schedule.

(1) The owner or operator of a facility subject to this Subpart shall submit an MSDS:

(i) On or before October 17, 1987 (or 3 months after the facility first becomes subject to this subpart), for all hazardous chemicals present at the facility in amounts equal to or greater than 10,000 pounds, or that are extremely hazardous substances present at the facility in an amount greater than or equal to 500 pounds (or 55 gallons) or the TPQ, whichever is less, and

241

§ 370.21

(ii) On or before October 17, 1989 (or 2 years and 3 months after the facility first becomes subject to this Subpart), for all hazardous chemicals present at the facility between 10,000 and zero pounds for which an MSDS has not yet been submitted.

(2) The owner or operator of a facility subject to this Subpart shall submit the Tier I form:

(i) On or before March 1, 1988 (or March 1 of the first year after the facility first becomes subject to this Subpart), covering all hazardous chemicals present at the facility during the preceding calendar year in amounts equal to or greater than 10,000 pounds, or that are extremely hazardous substances present at the facility in an amount greater than or equal to 500 pounds (or 55 gallons) or the TPQ, whichever is less, and

(ii) On or before March 1, 1989 (or March 1 of the second year after the facility first becomes subject to this Subpart), covering all hazardous chemicals present at the facility during the preceding calendar year in amounts equal to or greater than 10,000 pounds, or that are extremely hazardous substances present at the facility in an amount greater than or equal to 500 pounds (or 55 gallons) or the TPQ, whichever is less, and

(iii) On or before March 1990 (or March 1 of the third year after the facility first becomes subject to this Subpart), and annually thereafter, covering all hazardous chemicals present at the facility during the preceding calendar year in amounts equal to or greater than zero pounds or that are extremely hazardous substances present at the facility in an amount equal to or greater than 500 pounds (or 55 gallons) or the TPQ, whichever is less.

(3) The minimum threshold for reporting in response to requests for submission of an MSDS or a Tier II form pursuant to §§ 370.21(d) and 370.25(c) of this part shall be zero.

§ 370.21 MSDS reporting.

(a) *Basic requirement.* The owner or operator of a facility subject to this Subpart shall submit an MSDS for each hazardous chemical present at the facility according to the minimum threshold schedule provided in paragraph (b) of § 370.20 to the committee, the commission, and the fire department with jurisdiction over the facility.

(b) *Alternative reporting.* In lieu of the submission of an MSDS for each hazardous chemical under paragraph (a) of this section, the owner or operator may submit the following:

(1) A list of the hazardous chemicals for which the MSDS is required, grouped by hazard category as defined under § 370.2 of this Part;

(2) The chemical or common name of each hazardous chemical as provided on the MSDS; and

(3) Except for reporting of mixtures under § 370.28(a)(2), any hazardous component of each hazardous chemical as provided on the MSDS.

(c) *Supplemental reporting.* (1) The owner or operator of a facility that has submitted an MSDS under this section shall provide a revised MSDS to the committee, the commission, and the fire department with jurisdiction over the facility within three months after discovery of significant new information concerning the hazardous chemical for which the MSDS was submitted.

(2) After October 17, 1987, the owner or operator of a facility subject to this section shall submit an MSDS for a hazardous chemical pursuant to paragraph (a) of this section or a list pursuant to paragraph (b) of this section within three months after the owner or operator is first required to prepare or have available the MSDS or after a hazardous chemical requiring an MSDS becomes present in an amount exceeding the threshold established in § 370.20(b).

(d) *Submission of MSDS upon request.* The owner or operator of a facility that has not submitted the MSDS for a hazardous chemical present at the facility shall submit the MSDS for any such hazardous chemical to the committee upon its request. The MSDS shall be submitted within 30 days of the receipt of such request.

§ 370.25 Inventory reporting.

(a) *Basic requirement.* The owner or operator of a facility subject to this

242

Environmental Protection Agency §370.31

Subpart shall submit an inventory form to the commission, the committee, and the fire department with jurisdiction over the facility. The inventory form containing Tier I information on hazardous chemicals present at the facility during the preceding calendar year above the threshold levels established in §370.20(b) shall be submitted on or before March 1 of each year, beginning in 1988.

(b) *Alternative reporting.* With respect to any specific hazardous chemical at the facility, the owner or operator may submit a Tier II form in lieu of the Tier I information.

(c) *Submission of Tier II information.* The owner or operator of a facility subject to this Section shall submit the Tier II form to the commission, committee, or the fire department having jurisdiction over the facility upon request of such persons. The Tier II form shall be submitted within 30 days of the receipt of each request.

(d) *Fire department inspection.* The owner or operator of a facility that has submitted an inventory form under this section shall allow on-site inspection by the fire department having jurisdiction over the facility upon request of the department, and shall provide to the department specific location information on hazardous chemicals at the facility.

§370.28 Mixtures.

(a) *Basic reporting.* The owner or operator of a facility may meet the reporting requirements of §§370.21 (MSDS reporting) and 370.25 (inventory form reporting) of this Subpart for a hazardous chemical that is a mixture of hazardous chemicals by:

(1) Providing the required information on each component in the mixture which is a hazardous chemical, or

(2) Providing the required information on the mixture itself, so long as the reporting of mixtures by a facility under §370.21 is in the same manner as under §370.25, where practicable.

(b) *Calculation of the quantity.* (1) if the reporting is on each component of the mixture which is a hazardous chemical, then the concentration of the hazardous chemical, in weight percent (greater than 1% or 0.1% if carcinogenic) shall be multiplied by the mass (in pounds) of the mixture to determine the quantity of the hazardous chemical in the mixture.

(2) If the reporting is on the mixture itself, the total quantity of the mixture shall be reported.

Subpart C—Public Access and Availability of Information

§370.30 Requests for information.

(a) *Request for MSDS information.* (1) Any person may obtain an MSDS with respect to a specific facility by submitting a written request to the committee.

(2) If the committee does not have in its possession the MSDS requested in paragraph (a)(1) of this section, it shall request a submission of the MSDS from the owner or operator of the facility that is the subject of the request.

(b) *Requests for Tier II information.* (1) Any person may request Tier II information with respect to a specific facility by submitting a written request to the commission or committee in accordance with the requirements of this section.

(2) If the committee or commission does not have in its possession the Tier II information requested in paragraph (b)(1) of this section, it shall request a submission of the Tier II form from the owner or operator of the facility that is the subject of the request, provided that the request is from a State or local official acting in his or her official capacity or the request is limited to hazardous chemicals stored at the facility in an amount in excess of 10,000 pounds.

(3) If the request under paragraph (b)(1) of this section does not meet the requirements of paragraph (b)(2) of this section, the committee or commission may request submission of the Tier II form from the owner or operator of the facility that is the subject of the request if the request under paragraph (b)(1) of this section includes a general statement of need.

§370.31 Provision of information.

All information obtained from an owner or operator in response to a request under this subpart and any re-

243

§ 370.40

quested Tier II form or MSDS otherwise in possession of the commission or the committee shall be made available to the person submitting the request under this Subpart; provided upon request of the owner or operator, the commission or committee shall withhold from disclosure the location of any specific chemical identified in the Tier II form.

40 CFR Ch. I (7-1-89 Edition)

Subpart D—Inventory Forms

§ 370.40 Tier I emergency and hazardous chemical inventory form.

(a) The form set out in paragraph (b) of this section shall be completed and submitted as required in § 370.25(a). In lieu of the form set out in paragraph (b) of this section, the facility owner or operator may submit a State or local form that contains identical content.

(b) Tier I Emergency and Hazardous Chemical Inventory Form.

Appendix B: Regulations Reference Guide

Environmental Protection Agency § 370.40

Page ____ of ____ pages
Form Approved OMB No. 2050-0072

Tier One — EMERGENCY AND HAZARDOUS CHEMICAL INVENTORY
Aggregate Information by Hazard Type

FOR OFFICIAL USE ONLY: ID #; Date Received

Important: Read instructions before completing form

Reporting Period From January 1 to December 31, 19____

Facility Identification
- Name
- Street Address
- City / State / Zip
- SIC Code
- Dun & Brad Number

Owner/Operator
- Name
- Mail Address
- Phone ( )

Emergency Contacts
- Name / Title / Phone ( ) / 24 Hour Phone ( )
- Name / Title / Phone ( ) / 24 Hour Phone ( )

☐ Check if site plan is attached

	Hazard Type	Max Amount*	Average Daily Amount*	Number of Days On-Site	General Location
Physical Hazards	Fire				
	Sudden Release of Pressure				
	Reactivity				
Health Hazards	Immediate (acute)				
	Delayed (Chronic)				

Certification (Read and sign after completing all sections)

I certify under penalty of law that I have personally examined and am familiar with the information submitted in this and all attached documents, and that based on my inquiry of those individuals responsible for obtaining the information, I believe that the submitted information is true, accurate and complete.

Name and official title of owner/operator OR owner/operator's authorized representative

Signature / Date signed

* Reporting Ranges

Range Value	Weight Range in Pounds From...	To...
00	0	99
01	100	999
02	1000	9,999
03	10,000	99,999
04	100,000	999,999
05	1,000,000	9,999,999
06	10,000,000	49,999,999
07	50,000,000	99,999,999
08	100,000,000	499,999,999
09	500,000,000	999,999,999
10	1 billion	higher than 1 billion

245

Chemical Hazard Communication Guidebook

§ 370.40   40 CFR Ch. I (7-1-89 Edition)

TIER ONE INSTRUCTIONS

GENERAL INFORMATION

Submission of this form is required by Title III of the Superfund Amendments and Reauthorization Act of 1986, Section 312, Public Law 99-499.

The purpose of this form is to provide State and local officials and the public with information on the general types and locations of hazardous chemicals present at your facility during the past year.

YOU MUST PROVIDE ALL INFORMATION REQUESTED ON THIS FORM.

You may substitute the Tier Two form for this Tier One form. (The Tier Two form provides detailed information and must be submitted in response to a specific request from State or local officials.)

WHO MUST SUBMIT THIS FORM
Section 312 of Title III requires that the owner or operator of a facility submit this form if, under regulations implementing the Occupational Safety and Health Act of 1970, the owner or operator is required to prepare or have available Material Safety Data Sheets (MSDS) for hazardous chemicals present at the facility. MSDS requirements are specified in the Occupational Safety and Health Administration (OSHA) Hazard Communication Standard, found in Title 29 of the Code of Federal Regulations at §1910.1200.

WHAT CHEMICALS ARE INCLUDED
You must report the information required on this form for every hazardous chemical for which you are required to prepare or have available an MSDS under the Hazard Communication Standard. However, OSHA regulations and Title III exempt some chemicals from reporting.

Section 1910.1200(b) of the OSHA regulations currently provides the following exemptions:

(i) Any hazardous waste as such term is defined by the Solid Waste Disposal Act, as amended (42 U.S.C. 6901 et seq.) when subject to regulations issued under that Act;

(ii) Tobacco or tobacco products;

(iii) Wood or wood products;

(iv) "Articles" - defined under §1910.1200 (b) as a manufactured item:

- Which is formed to a specific shape or design during manufacture;
- Which has end use function(s) dependent in whole or in part upon the shape or design during end use; and
- Which does not release, or otherwise result in exposure to a hazardous chemical under normal conditions of use.

(v) Food, drugs, cosmetics or alcoholic beverages in a retail establishment which are packaged for sale to consumers;

(vi) Foods, drugs, or cosmetics intended for personal consumption by employees while in the workplace;

(vii) Any consumer product or hazardous substance, as those terms are defined in the Consumer Product Safety Act (15 U.S.C. 1251 et seq.) respectively, where the employer can demonstrate it is used in the workplace in the same manner as normal consumer use, and which use results in a duration and frequency of exposure which is not greater than exposures experienced by consumers; and

(viii) Any drug, as that term is defined in the Federal Food, Drug, and Cosmetic Act (21 U.S.C. 301 et seq.), when it is in solid, final form for direct administration to the patient (i.e., tablets or pills).

In addition, Section 311(e) of Title III excludes the following substances:

(i) Any food, food additive, color additive, drug, or cosmetic regulated by the Food and Drug Administration;

(ii) Any substance present as a solid in any manufactured item to the extent exposure to the substance does not occur under normal conditions of use;

(iii) Any substance to the extent it is used for personal, family, or household purposes, or is present in the same form and concentration as a product packaged for distribution and use by the general public;

(iv) Any substance to the extent it is used in a research laboratory or a hospital or other medical facility under the direct supervision of a technically qualified individual;

(v) Any substance to the extent it is used in routine agricultural operations or is a fertilizer held for sale by a retailer to the ultimate customer.

Also, minimum reporting thresholds have been established under Title III, Section 312. You need to report only those hazardous chemicals that were present at your facility at any time during the preceding calendar year at or above the levels listed below:

- January to December 1987 (or first year of reporting) ...10,000 lbs.
- January to December 1988 (or second year of reporting) ...10,000 lbs.
- January to December 1989 (or third year of reporting) ...zero lbs.*
    * EPA will publish the final threshold, effective in the third year, after additional analysis.
- For extremely hazardous substances...500 lbs. or the threshold planning quantity, whichever is less, from the first year of reporting and thereafter.

WHEN TO SUBMIT THIS FORM
Beginning March 1, 1988, owners or operators must submit the Tier One form (or substitute the Tier Two form) on or before March 1 of every year.

Appendix B: Regulations Reference Guide

Environmental Protection Agency § 370.40

INSTRUCTIONS

Please read these instructions carefully. Print or type all responses.

WHERE TO SUBMIT THIS FORM
Send one completed inventory form to each of the following organizations:

1. Your State emergency planning commission
2. Your local emergency planning committee
3. The fire department with jurisdiction over your facility.

PENALTIES
Any owner or operator of a facility who fails to submit or supplies false Tier One information shall be liable to the United States for a civil penalty of up to $25,000 for each such violation. Each day a violation continues shall constitute a separate violation. In addition, any citizen may commence a civil action on his or her own behalf against any owner or operator who fails to submit Tier One information.

You may use the Tier Two form as a worksheet for completing Tier One. Filling in the Tier Two chemical information section should help you assemble your Tier One responses.

If your responses require more than one page, fill in the page number at the top of the form.

REPORTING PERIOD
Enter the appropriate calendar year, beginning January 1 and ending December 31.

FACILITY IDENTIFICATION
Enter the complete name of your facility (and company identifier where appropriate).

Enter the full street address or state road. If a street address is not available, enter other appropriate identifiers that describe the physical location of your facility (e.g., longitude and latitude). Include city, state, and zip code.

Enter the primary Standard Industrial Classification (SIC) code and the Dun & Bradstreet number for your facility. The financial officer of your facility should be able to provide the Dun & Bradstreet number. If your firm does not have this information, contact the state or regional office of Dun & Bradstreet to obtain your facility number or have one assigned.

OWNER/OPERATOR
Enter the owner's or operator's full name, mailing address, and phone number.

EMERGENCY CONTACT
Enter the name, title, and work phone number of at least one local person or office that can act as a referral if emergency responders need assistance in responding to a chemical accident at the facility.

Provide an emergency phone number where such emergency information will be available 24 hours a day, every day.

PHYSICAL AND HEALTH HAZARDS
Descriptions, Amounts, and Locations
This section requires aggregate information on chemicals by hazard categories as defined in 40 CFR 370.3. The two health hazard categories and three physical hazard categories are a consolidation of the 23 hazard categories defined in the OSHA Hazard Communication Standard, 29 CFR 1910.1200. For each hazard type, indicate the total amounts and general locations of all applicable chemicals present at your facility during the past year.

- **What units should I use?**

  Calculate all amounts as weight in pounds. To convert gas or liquid volume to weight in pounds, multiply by an appropriate density factor.

- **What about mixtures?**

  If a chemical is part of a mixture, you have the option of reporting either the weight of the entire mixture or only the portion of the mixture that is a particular hazardous chemical (e.g., if a hazardous solution weighs 100 lbs. but is composed of only 5% of a particular hazardous chemical, you can indicate either 100 lbs. of the mixture or 5 lbs. of the chemical).

  Select the option consistent with your Section 311 reporting of the chemical on the MSDS or list of MSDS chemicals.

- **Where do I count a chemical that is a fire reactivity physical hazard and an immediate (acute) health hazard?**

  Add the chemical's weight to your totals for all three hazard categories and include its location in all three categories. Many chemicals fall into more than one hazard category, which results in double-counting.

MAXIMUM AMOUNT
The amounts of chemicals you have on hand may vary throughout the year. The peak weights — greatest single-day weights during the year — are added together in this column to determine the maximum weight for each hazard type. Since the peaks for different chemicals often occur on different days, this maximum amount will seem artificially high.

To complete this and the following sections, you may choose to use the Tier Two form as a worksheet.

To determine the Maximum Amount:

1. List all of your hazardous chemicals individually.
2. For each chemical...
   a. Indicate all physical and health hazards that the chemical presents. Include all chemicals, even if they are present for only a short period of time during the year.

2

247

§ 370.40   40 CFR Ch. I (7-1-89 Edition)

b. Estimate the maximum weight in pounds that was present at your facility on any single day of the reporting period.

3. For each hazard type -- beginning with Fire and repeating for all physical and health hazard types...
   a. Add the maximum weights of all chemicals you indicated as the particular hazard type.
   b. Look at the Reporting Ranges at the bottom of the Tier One form. Find the appropriate range value code.
   c. Enter this range value as the Maximum Amount.

> **EXAMPLE:**
> You are using the Tier Two form as a worksheet and have listed raw weights in pounds for each of your hazardous chemicals. You have marked an X in the immediate (acute) hazard column for phenol and sulfuric acid. The maximum amount raw weight you listed were 10,000 lbs. and 50 lbs. respectively. You add these together to reach a total of 10,050 lbs. Then you look at the Reporting Range at the bottom of your Tier One form and find that the value of 03 corresponds to 10,050 lbs. Enter 03 as your Maximum Amount for immediate (acute) hazards materials.
>
> You also marked an X in the Fire hazard box for phenol. When you calculate your Maximum Amount totals for fire hazards, add the 10,000 lb. weight again.

AVERAGE DAILY AMOUNT

This column should represent the average daily amount of chemicals *of each hazard type* that were present at your facility at any point during the year.

To determine this amount:

1. List all of your hazardous chemicals individually (same as for Maximum Amount).

2. For each chemical...
   a. Indicate all physical and health hazards that the chemical presents (same as for Maximum Amount).
   b. Estimate the average weight in pounds that was present at your facility throughout the year. To do this, total all daily weights and divide by the number of days the chemical was present on the site.

3. For each hazard type -- beginning with Fire and repeating for all physical and health hazards...
   a. Add the average weights of all chemicals you indicated for the particular hazard type.
   b. Look at the Reporting Ranges at the bottom of the Tier One form. Find the appropriate range value code.
   c. Enter this range value as the Average Daily Amount.

> **EXAMPLE:**
> You are using the Tier Two form, and have marked an X in the immediate (acute) hazard column for nicotine and phenol. Nicotine is present at your facility 100 days during the year, and the sum of the daily weights is 100,000 lbs. By dividing 100,000 lbs. by 100 days on-site, you calculate an Average Daily Amount of 1,000 lbs. for nicotine. Phenol is present at your facility 50 days during the year, and the sum of the daily weights is 10,000 lbs. By dividing 10,000 lbs. by 50 days on-site, you calculate an Average Daily Amount of 200 lbs. for phenol. You then add the two average daily amounts together to reach a total of 1,200 lbs. Then you look at the Reporting Range on your Tier One form and find that the value 02 corresponds to 1,200 lbs. Enter 02 as your Average Daily Amount for Immediate (acute) Hazard.
>
> You also marked an X in the Fire hazard column for phenol. When you calculate your Average Daily Amount for fire hazards, use the 200 lb. weight again.

NUMBER OF DAYS ON-SITE

Enter the greatest number of days that a single chemical within that hazard category was present on-site.

> **EXAMPLE:**
> At your facility, nicotine is present for 100 days and phosgene is present for 150 days. Enter 150 in the space provided.

GENERAL LOCATION

Enter the general location within your facility where each hazard may be found. General locations should include the names or identifications of buildings, tank fields, lots, sheds, or other such areas.

For each hazard type, list the locations of all applicable chemicals. As an alternative you may also attach a site plan and list the site coordinates related to the appropriate locations. If you do so, check the Site Plan box.

> **EXAMPLE:**
> On your worksheet you have marked an X in the Fire hazard column for acetone and butane. You noted that these are kept in steel drums in Room C of the Main Building, and in pressurized cylinders in Storage Shed 13, respectively. You could enter Main Building and Storage Shed 13 as the General Locations of your fire hazards. However, you choose to attach a site plan and list coordinates. Check the Site Plan box at the top of the column and enter site coordinates for the Main Building and Storage Shed 13 under General Locations.

If you need more space to list locations, attach an additional Tier One form and continue your list on the proper line. Number all pages.

CERTIFICATION

This must be completed by the owner or operator or the officially designated representative of the owner or operator. Enter your full name and official title. Sign your name and enter the current date.

Appendix B: Regulations Reference Guide

Federal Register / Vol. 55, No. 144 / Thursday, July 26, 1990 / Rules and Regulations   30651

Tier Two — EMERGENCY AND HAZARDOUS CHEMICAL INVENTORY

Specific Information by Chemical

Revised June 1990

Page ___ of ___ pages
Form Approved OMB No. 2050-0072

Facility Identification
- Name
- Street
- City
- County
- State
- Zip
- SIC Code
- Dun & Brad Number

FOR OFFICIAL USE ONLY: ID / Date Received

Owner/Operator Name
- Name
- Mail Address
- Phone ( )

Emergency Contact
- Name / Title / Phone ( ) / 24 Hr. Phone ( )
- Name / Title / Phone ( ) / 24 Hr. Phone ( )

☐ Check if information below is identical to the information submitted last year.

Reporting Period From January 1 to December 31, 19___

Important: Read all instructions before completing form

Chemical Description	Physical and Health Hazards (check all that apply)	Inventory	Storage Codes and Locations (Non-Confidential)
CAS ☐☐☐☐☐☐☐☐☐☐ ☐ Trade Secret Chem. Name _____ Check all that apply: ☐ Pure ☐ Mix ☐ Solid ☐ Liquid ☐ Gas ☐ EHS EHS Name _____	☐ Fire ☐ Sudden Release of Pressure ☐ Reactivity ☐ Immediate (acute) ☐ Delayed (chronic)	Max. Daily Amount (code) ☐☐ Avg. Daily Amount (code) ☐☐ No. of Days On-site (days) ☐☐☐	Container Type / Pressure / Temperature *Storage Locations* ☐ Optional
CAS ☐☐☐☐☐☐☐☐☐☐ ☐ Trade Secret Chem. Name _____ Check all that apply: ☐ Pure ☐ Mix ☐ Solid ☐ Liquid ☐ Gas ☐ EHS EHS Name _____	☐ Fire ☐ Sudden Release of Pressure ☐ Reactivity ☐ Immediate (acute) ☐ Delayed (chronic)	Max. Daily Amount (code) ☐☐ Avg. Daily Amount (code) ☐☐ No. of Days On-site (days) ☐☐☐	☐ Optional
CAS ☐☐☐☐☐☐☐☐☐☐ ☐ Trade Secret Chem. Name _____ Check all that apply: ☐ Pure ☐ Mix ☐ Solid ☐ Liquid ☐ Gas ☐ EHS EHS Name _____	☐ Fire ☐ Sudden Release of Pressure ☐ Reactivity ☐ Immediate (acute) ☐ Delayed (chronic)	Max. Daily Amount (code) ☐☐ Avg. Daily Amount (code) ☐☐ No. of Days On-site (days) ☐☐☐	☐ Optional

Certification (Read and sign after completing all sections)

I certify under penalty of law that I have personally examined and am familiar with the information submitted in pages one through ___ and that based on my inquiry of those individuals responsible for obtaining the information, I believe that the submitted information is true, accurate, and complete.

Name and official title of owner/operator OR owner/operator's authorized representative _____ Signature _____ Date signed _____

Optional Attachments
☐ I have attached a site plan
☐ I have attached a list of site coordinate abbreviations
☐ I have attached a description of dikes and other safeguard measures

Chemical Hazard Communication Guidebook

Environmental Protection Agency § 370.41

Appendix B: Regulations Reference Guide

§ 370.41 **40 CFR Ch. I (7-1-89 Edition)**

TIER TWO INSTRUCTIONS

GENERAL INFORMATION

Submission of this Tier Two form (when requested) is required by Title III of the Superfund Amendments and Reauthorization Act of 1986, Section 312, Public Law 99-499. The purpose of this Tier Two form is to provide State and local officials and the public with specific information on hazardous chemicals present at your facility during the past year.

YOU MUST PROVIDE ALL INFORMATION REQUESTED ON THIS FORM TO FULFILL TIER TWO REPORTING REQUIREMENTS.

This form may also be used as a worksheet for completing the Tier One form or may be submitted in place of the Tier One form.

WHO MUST SUBMIT THIS FORM

Section 312 of Title III requires that the owner or operator of a facility submit this Tier Two form if so requested by a State emergency planning commission, a local emergency planning committee, or a fire department with jurisdiction over the facility.

This request may apply to the owner or operator of any facility that is required, under regulations implementing the Occupational Safety and Health Act of 1970, to prepare or have available a Material Safety Data Sheet (MSDS) for a hazardous chemical present at the facility. MSDS requirements are specified in the Occupational Safety and Health Administration (OSHA) Hazard Communications Standard, found in Title 29 of the Code of Federal Regulations at § 1910.1200.

WHAT CHEMICALS ARE INCLUDED

You must report the information required on this form for each hazardous chemical for which Tier Two information is requested. However, OSHA regulations and Title III exempt some chemicals from reporting.

Section 1910.1200(b) of the OSHA regulations currently provides the following exemptions:

(i) Any hazardous waste as such term is defined by the Solid Waste Disposal Act as amended (42 U.S.C. 6901 et seq.) when subject to regulations issued under that Act;

(ii) Tobacco or tobacco products;

(iii) Wood or wood products;

(iv) "Articles" - defined under § 1910.1200(b) as a manufactured item:

- Which is formed to a specific shape or design during manufacture;
- Which has end use function(s) dependent in whole or in part upon the shape or design during end use; and
- Which does not release, or otherwise result in exposure to a hazardous chemical under normal conditions of use.

(v) Food, drugs, cosmetics or alcoholic beverages in a retail establishment which are packaged for sale to consumers;

(vi) Foods, drugs, or cosmetics intended for personal consumption by employees while in the workplace.

(vii) Any consumer product or hazardous substance, as those terms are defined in the Consumer Product Safety Act (15 U.S.C. 1251 et seq.) respectively, where the employer can demonstrate it is used in the workplace in the same manner as normal consumer use, and which use results in a duration and frequency of exposure which is not greater than exposures experienced by consumers.

(viii) Any drug, as that term is defined in the Federal Food, Drug, and Cosmetic Act (21 U.S.C. 301 et seq.), when it is in solid, final form for direct administration to the patient (i.e., tablets or pills).

In addition, Section 311(e) of Title III excludes the following substances:

(i) Any food, food additive, color additive, drug, or cosmetic regulated by the Food and Drug Administration;

(ii) Any substance present as a solid in any manufactured item to the extent exposure to the substance does not occur under normal conditions of use;

(iii) Any substance to the extent it is used for personal, family, or household purposes, or is present in the same form and concentration as a product packaged for distribution and use by the general public;

(iv) Any substance to the extent it is used in a research laboratory or a hospital or other medical facility under the direct supervision of a technically qualified individual;

(v) Any substance to the extent it is used in routine agricultural operations or is a fertilizer held for sale by a retailer to the ultimate customer.

Also, minimum reporting thresholds have been established for Tier One under Title III, Section 312. You need to report only those hazardous chemicals that were present at your facility at any time during the preceding calendar year at or above the levels listed below:

- January to December 1987 (or first year of reporting) ...10,000 lbs.
- January to December 1988 (or second year of reporting) ...10,000 lbs.
- January to December 1989 (or third year of reporting) ...zero lbs.*

  * EPA will publish the final threshold, effective in the third year, after additional analysis.

- For extremely hazardous substances...500 lbs. or the threshold planning quantity, whichever is less, from the first year of reporting and thereafter.

A requesting official may limit the responses required under Tier Two by specifying particular chemicals or groups of chemicals. Such requests apply to hazardous chemicals regardless of established thresholds.

252

Chemical Hazard Communication Guidebook

Environmental Protection Agency § 370.41

INSTRUCTIONS

Please read these instructions carefully. Print or type all responses.

WHEN TO SUBMIT THIS FORM
Owners or operators must submit the Tier Two form to the requesting agency within 30 days of receipt of a written request from an authorized official.

WHERE TO SUBMIT THIS FORM
Send the completed Tier Two form to the requesting agency.

PENALTIES
Any owner or operator who violates any Tier Two reporting requirements shall be liable to the United States for a civil penalty of up to $25,000 for each such violation. Each day a violation continues shall constitute a separate violation.

You may use the Tier Two form as a worksheet for completing the Tier One form. Filling in the Tier Two Chemical Information section should help you assemble your Tier One responses.

If your responses require more than one page, fill in the page number at the top of the form.

REPORTING PERIOD
Enter the appropriate calendar year, beginning January 1 and ending December 31.

FACILITY IDENTIFICATION
Enter the full name of your facility (and company identifier where appropriate).

Enter the full street address or state road. If a street address is not available, enter other appropriate identifiers that describe the physical location of your facility (e.g., longitude and latitude). Include city, state, and zip code.

Enter the primary Standard Industrial Classification (SIC) code and the Dun & Bradstreet number for your facility. The financial officer of your facility should be able to provide the Dun & Bradstreet number. If your firm does not have this information, contact the state or regional office of Dun & Bradstreet to obtain your facility number or have one assigned.

OWNER/OPERATOR
Enter the owner's or operator's full name, mailing address, and phone number.

EMERGENCY CONTACT
Enter the name, title, and work phone number of at least one local person or office who can act as a referral if emergency responders need assistance in responding to a chemical accident at the facility.

Provide an emergency phone number where such emergency chemical information will be available 24 hours a day, every day.

CHEMICAL INFORMATION: Description, Hazards, Amounts, and Locations
The main section of the Tier Two form requires specific information on amounts and locations of hazardous chemicals, as defined in the OSHA Hazard Communication Standard.

- What units should I use?

    Calculate all amounts as *weight in pounds*. To convert gas or liquid volume to weight in pounds, multiply by an appropriate density factor.

- What about mixtures?

    If a chemical is part of a mixture, *you have the option* of reporting either the weight of the entire mixture or only the portion of the mixture that is a particular hazardous chemical (e.g., if a hazardous solution weighs 100 lbs. but is composed of only 5% of a particular hazardous chemical, you can indicate either 100 lbs. of the mixture *or* 5 lbs. of the chemical.

    Select the option consistent with your Section 311 reporting of the chemical on the MSDS or list of MSDS chemicals.

CHEMICAL DESCRIPTION

1. Enter the Chemical Abstract Service number (CAS#).

    For mixtures, enter the CAS number of the mixture as a whole if it has been assigned a number distinct from its components. For a mixture that has no CAS number, leave this item blank or report the CAS numbers of as many constituent chemicals as possible.

    If you are withholding the name of a chemical in accordance with criteria specified in Title III, Section 322, enter the generic chemical class (e.g., list toluene diisocyanate as organic isocyanate) and check the box marked Trade Secret. Trade secret information should be submitted to EPA and must include a substantiation. Please refer to Section 322 of Title III for detailed information on how to comply with trade secret requests.

2. Enter the chemical name or common name of each hazardous chemical.

3. Circle *ALL* applicable descriptors: pure or mixture, and solid, liquid, or gas.

> **EXAMPLE:**
> You have pure chlorine gas on hand, as well as two mixtures that contain liquid chlorine. You write "chlorine" and enter the CAS#. Then you circle "pure" and "mix" -- as well as "liq" and "gas".

2

Appendix B: Regulations Reference Guide

§ 370.41

PHYSICAL AND HEALTH HAZARDS

For each chemical you have listed, check all the physical and health hazard boxes that apply. These hazard categories are defined in 40 CFR 370.3. The two health hazard categories and three physical hazard categories are a consolidation of the 23 hazard categories defined in the OSHA Hazard Communication Standard, 29 CFR 1910.1200.

MAXIMUM AMOUNT

1. For each hazardous chemical, estimate the greatest amount present at your facility on any single day during the reporting period.
2. Find the appropriate range value code in Table I.
3. Enter this range value as the Maximum Amount.

Table I REPORTING RANGES

Range Value	Weight Range in Pounds From...	To...
00	0	99
01	100	999
02	1,000	9,999
03	10,000	99,999
04	100,000	999,999
05	1,000,000	9,999,999
06	10,000,000	49,999,999
07	50,000,000	99,999,999
08	100,000,000	499,999,999
09	500,000,000	999,999,999
10	1 billion	higher than 1 billion

If you are using this form as a worksheet for completing Tier One, enter the actual weight in pounds in the shaded space below the response blocks. Do this for both Maximum Amount and Average Daily Amount.

EXAMPLE:

You received one large shipment of a solvent mixture last year. The shipment filled your 5,000-gallon storage tank. You know that the solvent contains 10% benzene, which is a hazardous chemical.

You figure that 10% of 5,000 gallons is 500 gallons. You also know that the density of benzene is 7.29 pounds per gallon, so you multiply 500 by 7.29 to get a weight of 3,645 pounds.

Then you look at Table I and find that the range value 02 corresponds to 3,645. You enter 02 as the Maximum Amount.

(If you are using the form as a worksheet for completing a Tier One form, you should write 3,645 in the shaded area.)

AVERAGE DAILY AMOUNT

40 CFR Ch. I (7-1-89 Edition)

1. For each hazardous chemical, estimate the average weight in pounds that was present at your facility during the year.

   To do this, total all daily weights and divide by the number of days the chemical was present on the site.

2. Find the appropriate range value in Table I.
3. Enter this range value as the Average Daily Amount.

EXAMPLE:

The 5,000-gallon shipment of solvent you received last year was gradually used up and completely gone in 315 days. The sum of the daily volume levels in the tank is 929,250 gallons. By dividing 929,250 gallons by 315 days on-site, you calculate an average daily amount of 2,950 gallons.

You already know that the solvent contains 10% benzene, which is a hazardous chemical. Since 10% of 2,950 is 295, you figure that you had an average of 295 gallons of benzene. You also know that the density of benzene is 7.29 pounds per gallon, so you multiply 295 by 7.29 to get a weight of 2,150 pounds.

Then you look at Table I and find that the range value 02 corresponds to 2,150. You enter 02 as the Average Daily Amount.

(If you are using the form as a worksheet for completing a Tier One form, you should write 2,150 in the shaded area.)

NUMBER OF DAYS ON-SITE

Enter the number of days that the hazardous chemical was found on-site.

EXAMPLE:

The solvent composed of 10% benzene was present for 315 days at your facility. Enter 315 in the space provided.

STORAGE CODES AND STORAGE LOCATIONS

List all non-confidential chemical locations in this column, along with storage types/conditions associated with each location.

Storage Codes: Indicate the types and conditions of storage present.

a. *Look at Table II.* For each location, find the appropriate storage type(s). Enter the corresponding code(s) in front of the parentheses.

b. *Look at Table III.* For each storage type, find the temperature and pressure conditions. Enter the applicable pressure code in the first space within the parentheses. Enter the applicable temperature code in the last space within the parentheses.

Environmental Protection Agency

§ 370.41

Table II – STORAGE TYPES

CODES	Types of Storage
A	Above ground tank
B	Below ground tank
C	Tank inside building
D	Steel drum
E	Plastic or non-metallic drum
F	Can
G	Carboy
H	Silo
I	Fiber drum
J	Bag
K	Box
L	Cylinder
M	Glass bottles or jugs
N	Plastic bottles or jugs
O	Tote bin
P	Tank wagon
Q	Rail car
R	Other

Table III – TEMPERATURE AND PRESSURE CONDITIONS

CODES	Storage Conditions
	(PRESSURE)
1	Ambient pressure
2	Greater than ambient pressure
3	Less than ambient pressure
	(TEMPERATURE)
4	Ambient temperature
5	Greater than ambient temperature
6	Less than ambient temperature but not cryogenic
7	Cryogenic conditions

EXAMPLE:

The benzene in the main building is kept in a tank inside the building, at ambient pressure and less than ambient temperature.

Table II shows you that the code for a tank inside a building is C. Table III shows you that code for ambient pressure is 1, and the code for less than ambient temperature is 6.

You enter: C(1,6)

Storage Locations:

Provide a brief description of the precise location of the chemical, so that emergency responders can locate the area easily. You may find it advantageous to provide the optional site plan or site coordinates as explained below.

For each chemical, indicate at a minimum the building or lot. Additionally, where practical, the room or area may be indicated. You may respond in narrative form with appropriate site coordinates or abbreviations.

If the chemical is present in more than one building, lot, or area location, continue your responses down the page as needed. If the chemical exists everywhere at the plant site simultaneously, you may report that the chemical is ubiquitous at the site.

<u>Optional attachments</u>: If you choose to attach one of the following, check the appropriate Attachments box at the bottom of the Tier Two form.

a. *A site plan* with site coordinates indicated for buildings, lots, areas, etc. throughout your facility.

b. *A list of site coordinate abbreviations* that correspond to buildings, lots, areas, etc. throughout your facility.

EXAMPLE:

You have benzene in the main room of the main building, and in tank 2 in tank field 10. You attach a site plan with coordinates as follows: main building = G-2, tank field 10 = B-6. Fill in the Storage Location as follows:

B-6 [Tank 2]   G-2 [Main Room]

Under Title III, Section 324, you may elect to withhold location information on a specific chemical from disclosure to the public. If you choose to do so:

- Enter the word "confidential" in the Non-Confidential Location section of the Tier Two form.

- On a separate Tier Two Confidential Location Information Sheet, enter the name and CAS# of each chemical for which you are keeping the location confidential.

- Enter the appropriate location and storage information, as described above for non-confidential locations.

- Attach the Tier Two Confidential Location Information Sheet to the Tier Two form. This separates confidential locations from other information that will be disclosed to the public.

CERTIFICATION.

This must be completed by the owner or operator or the officially designated representative of the owner or operator. Enter your full name and official title. Sign your name and enter the current date.

EPA'S REGULATION ESTABLISHING TOXIC CHEMICAL RELEASE REPORTING

§ 372.1

PART 372—TOXIC CHEMICAL RELEASE REPORTING: COMMUNITY RIGHT-TO-KNOW

Subpart A—General Provisions

Sec.
372.1 Scope and purpose.
372.3 Definitions.
372.5 Persons subject to this part.
372.10 Recordkeeping.
372.18 Compliance and enforcement.

Subpart B—Reporting Requirements

372.22 Covered facilities for toxic chemical release reporting.
372.25 Thresholds for reporting.
372.30 Reporting requirements and schedule for reporting.
372.38 Exemptions.

Subpart C—Supplier Notification Requirements

372.45 Notification about toxic chemicals.

Subpart D—Specific Toxic Chemical Listings

372.65 Chemicals and chemical categories to which this part applies.

Subpart E—Forms and Instructions

372.85 Toxic chemical release reporting form and instructions.

AUTHORITY: 42 U.S.C. 11013, 11028.

SOURCE: 53 FR 4525, Feb. 16, 1988, unless otherwise noted.

Subpart A—General Provisions

§ 372.1 Scope and purpose.

This part sets forth requirements for the submission of information relating to the release of toxic chemicals under section 313 of Title III of the Superfund Amendments and Reauthorization Act of 1986. The information collected under this part is intended to inform the general public and the communities surrounding covered facilities about releases of toxic chemicals, to assist research, to aid in the development of regulations, guidelines, and standards, and for other purposes. This part also sets forth requirements for suppliers to notify persons to whom they distribute mixtures or trade name products containing toxic chemicals that they contain such chemicals.

40 CFR Ch. I (7-1-89 Edition)

§ 372.3 Definitions.

Terms defined in sections 313(b)(1)(c) and 329 of Title III and not explicitly defined herein are used with the meaning given in Title III. For the purpose of this part:

"*Acts*" means Title III.

"*Article*" means a manufactured item: (1) Which is formed to a specific shape or design during manufacture; (2) which has end use functions dependent in whole or in part upon its shape or design during end use; and (3) which does not release a toxic chemical under normal conditions of processing or use of that item at the facility or establishments.

"*Customs territory of the United States*" means the 50 States, the District of Columbia, and Puerto Rico.

"*EPA*" means the United States Environmental Protection Agency.

"*Establishment*" means an economic unit, generally at a single physical location, where business is conducted or where services or industrial operations are performed.

"*Facility*" means all buildings, equipment, structures, and other stationary items which are located on a single site or on contiguous or adjacent sites and which are owned or operated by the same person (or by any person which controls, is controlled by, or under common control with such person). A facility may contain more than one establishment.

"*Full-time employee*" means 2,000 hours per year of full-time equivalent employment. A facility would calculate the number of full-time employees by totaling the hours worked during the calendar year by all employees, including contract employees, and dividing that total by 2,000 hours.

"*Import*" means to cause a chemical to be imported into the customs territory of the United States. For purposes of this definition, "to cause" means to intend that the chemical be imported and to control the identity of the imported chemical and the amount to be imported.

"*Manufacture*" means to produce, prepare, import, or compound a toxic chemical. Manufacture also applies to a toxic chemical that is produced coincidentally during the manufacture,

Environmental Protection Agency § 372.10

processing, use, or disposal of another chemical or mixture of chemicals, including a toxic chemical that is separated from that other chemical or mixture of chemicals as a byproduct, and a toxic chemical that remains in that other chemical or mixture of chemicals as an impurity.

"*Mixture*" means any combination of two or more chemicals, if the combination is not, in whole or in part, the result of a chemical reaction. However, if the combination was produced by a chemical reaction but could have been produced without a chemical reaction, it is also treated as a mixture. A mixture also includes any combination which consists of a chemical and associated impurities.

"*Otherwise use*" or "*use*" means any use of a toxic chemical that is not covered by the terms "manufacture" or "process" and includes use of a toxic chemical contained in a mixture or trade name product. Relabeling or redistributing a container of a toxic chemical where no repackaging of the toxic chemical occurs does not constitute use or processing of the toxic chemical.

"*Process*" means the preparation of a toxic chemical, after its manufacture, for distribution in commerce:

(1) In the same form or physical state as, or in a different form or physical state from, that in which it was received by the person so preparing such substance, or

(2) As part of an article containing the toxic chemical. Process also applies to the processing of a toxic chemical contained in a mixture or trade name product.

"*Release*" means any spilling, leaking, pumping, pouring, emitting, emptying, discharging, injecting, escaping, leaching, dumping, or disposing into the environment (including the abandonment or discarding of barrels, containers, and other closed receptacles) of any toxic chemical.

"*Senior management official*" means an official with management responsibility for the person or persons completing the report, or the manager of environmental programs for the facility or establishments, or for the corporation owning or operating the facility or establishments responsible for certifying similar reports under other environmental regulatory requirements.

"*Title III*" means Title III of the Superfund Amendments and Reauthorization Act of 1986, also titled the Emergency Planning and Community Right-To-Know Act of 1986.

"*Toxic chemical*" means a chemical or chemical category listed in § 372.65.

"*Trade name product*" means a chemical or mixture of chemicals that is distributed to other persons and that incorporates a toxic chemical component that is not identified by the applicable chemical name or Chemical Abstracts Service Registry number listed in § 372.65.

§ 372.5 **Persons subject to this part.**

Owners and operators of facilities described in §§ 372.22 and 372.45 are subject to the requirements of this part. If the owner and operator of a facility are different persons, only one need report under § 372.17 or provide a notice under § 372.45 for each toxic chemical in a mixture or trade name product distributed from the facility. However, if no report is submitted or notice provided, EPA will hold both the owner and the operator liable under section 325(c) of Title III, except as provided in §§ 372.38(e) and 372.45(g).

§ 372.10 **Recordkeeping.**

(a) Each person subject to the reporting requirements of this part must retain the following records for a period of 3 years from the date of the submission of a report under § 372.30:

(1) A copy of each report submitted by the person under § 372.30.

(2) All supporting materials and documentation used by the person to make the compliance determination that the facility or establishments is a covered facility under § 372.22 or § 372.45.

(3) Documentation supporting the report submitted under § 372.30 including:

(i) Documentation supporting any determination that a claimed allowable exemption under § 372.38 applies.

(ii) Data supporting the determination of whether a threshold under

§ 372.18

§ 372.25 applies for each toxic chemical.

(iii) Documentation supporting the calculations of the quantity of each toxic chemical released to the environment or transferred to an off-site location.

(iv) Documentation supporting the use indications and quantity on site reporting for each toxic chemical, including dates of manufacturing, processing, or use.

(v) Documentation supporting the basis of estimate used in developing any release or off-site transfer estimates for each toxic chemical.

(vi) Receipts or manifests associated with the transfer of each toxic chemical in waste to off-site locations.

(vii) Documentation supporting reported waste treatment methods, estimates of treatment efficiencies, ranges of influent concentration to such treatment, the sequential nature of treatment steps, if applicable, and the actual operating data, if applicable, to support the waste treatment efficiency estimate for each toxic chemical.

(b) Each person subject to the notification requirements of this part must retain the following records for a period of 3 years from the date of the submission of a notification under § 372.45.

(1) All supporting materials and documentation used by the person to determine whether a notice is required under § 372.45.

(2) All supporting materials and documentation used in developing each required notice under § 372.45 and a copy of each notice.

(c) Records retained under this section must be maintained at the facility to which the report applies or from which a notification was provided. Such records must be readily available for purposes of inspection by EPA.

§ 372.18 Compliance and enforcement.

Violators of the requirements of this part shall be liable for a civil penalty in an amount not to exceed $25,000 each day for each violation as provided in section 325(c) of Title III.

Subpart B—Reporting Requirements

§ 372.22 Covered facilities for toxic chemical release reporting.

A facility that meets all of the following criteria for a calendar year is a covered facility for that calendar year and must report under § 372.30.

(a) The facility has 10 or more full-time employees.

(b) The facility is in Standard Industrial Classification Codes 20 through 39 (as in effect on January 1, 1987) by virtue of the fact that it meets one of the following criteria:

(1) The facility is an establishment with a primary SIC code of 20 through 39.

(2) The facility is a multi-establishment complex where all establishments have a primary SIC code of 20 through 39.

(3) The facility is a multi-establishment complex in which one of the following is true:

(i) The sum of the value of products shipped and/or produced from those establishments that have a primary SIC code of 20 through 39 is greater than 50 percent of the total value of all products shipped and/or produced from all establishments at the facility.

(ii) One establishment having a primary SIC code of 20 through 39 contributes more in terms of value of products shipped and/or produced than any other establishment within the facility.

(c) The facility manufactured (including imported), processed, or otherwise used a toxic chemical in excess of an applicable threshold quantity of that chemical set forth in § 372.25.

§ 372.25 Thresholds for reporting.

The threshold amounts for purposes of reporting under § 372.30 for toxic chemicals are as follows:

(a) With respect to a toxic chemical manufactured (including imported) or processed at a facility during the following calendar years:

1987—75,000 pounds of the chemical manufactured or processed for the year.
1988—50,000 pounds of the chemical manufactured or processed for the year.

Environmental Protection Agency § 372.30

1989 and thereafter—25,000 pounds of the chemical manufactured or processed for the year.

(b) With respect to a chemical otherwise used at a facility, 10,000 pounds of the chemical used for the applicable calendar year.

(c) With respect to activities involving a toxic chemical at a facility, when more than one threshold applies to the activities, the owner or operator of the facility must report if it exceeds any applicable threshold and must report on all activities at the facility involving the chemical, except as provided in § 372.38.

(d) When a facility manufactures, processes, or otherwise uses more than one member of a chemical category listed in § 372.65(c), the owner or operator of the facility must report if it exceeds any applicable threshold for the total volume of all the members of the category involved in the applicable activity. Any such report must cover all activities at the facility involving members of the category.

(e) A facility may process or otherwise use a toxic chemical in a recycle/reuse operation. To determine whether the facility has processed or used more than an applicable threshold of the chemical, the owner or operator of the facility shall count the amount of the chemical added to the recycle/reuse operation during the calendar year. In particular, if the facility starts up such an operation during a calendar year, or in the event that the contents of the whole recycle/reuse operation are replaced in a calendar year, the owner or operator of the facility shall also count the amount of the chemical placed into the system at these times.

(f) A toxic chemical may be listed in § 372.65 with the notation that only persons who manufacture the chemical, or manufacture it by a certain method, are required to report. In that case, only owners or operators of facilities that manufacture that chemical as described in § 372.65 in excess of the threshold applicable to such manufacture in § 372.25 are required to report. In completing the reporting form, the owner or operator is only required to account for the quantity of the chemical so manufactured and releases associated with such manufacturing, but not releases associated with subsequent processing or use of the chemical at that facility. Owners and operators of facilities that solely process or use such a chemical are not required to report for that chemical.

(g) A toxic chemical may be listed in § 372.65 with the notation that it is in a specific form (e.g., fume or dust, solution, or friable) or of a specific color (e.g., yellow or white). In that case, only owners or operators of facilities that manufacture, process, or use that chemical in the form or of the color, specified in § 372.65 in excess of the threshold applicable to such activity in § 372.25 are required to report. In completing the reporting form, the owner or operator is only required to account for the quantity of the chemical manufactured, processed, or used in the form or color specified in § 372.65 and for releases associated with the chemical in that form or color. Owners or operators of facilities that solely manufacture, process, or use such a chemical in a form or color other than those specified by § 372.65 are not required to report for that chemical.

(h) Metal compound categories are listed in § 372.65(c). For purposes of determining whether any of the thresholds specified in § 372.25 are met for metal compound category, the owner or operator of a facility must make the threshold determination based on the total amount of all members of the metal compound category manufactured, processed, or used at the facility. In completing the release portion of the reporting form for releases of the metal compounds, the owner or operator is only required to account for the weight of the parent metal released. Any contribution to the mass of the release attributable to other portions of each compound in the category is excluded.

§ 372.30 Reporting requirements and schedule for reporting.

(a) For each toxic chemical known by the owner or operator to be manufactured (including imported), processed, or otherwise used in excess of an applicable threshold quantity in

259

§ 372.25 at its covered facility described in § 372.22 for a calendar year, the owner or operator must submit to EPA and to the State in which the facility is located a completed EPA Form R (EPA Form 9350-1) in accordance with the instructions in Subpart E.

(b)(1) The owner or operator of a covered facility is required to report as described in paragraph (a) of this section on a toxic chemical that the owner or operator knows is present as a component of a mixture or trade name product which the owner or operator receives from another person, if that chemical is imported, processed, or otherwise used by the owner or operator in excess of an applicable threshold quantity in § 372.25 at the facility as part of that mixture or trade name product.

(2) The owner or operator knows that a toxic chemical is present as a component of a mixture or trade name product (i) if the owner or operator knows or has been told the chemical identity or Chemical Abstracts Service Registry Number of the chemical and the identity or Number corresponds to an identity or Number in § 372.65, or (ii) if the owner or operator has been told by the supplier of the mixture or trade name product that the mixture or trade name product contains a toxic chemical subject to section 313 of the Act or this part.

(3) To determine whether a toxic chemical which is a component of a mixture or trade name product has been imported, processed, or otherwise used in excess of an applicable threshold in § 372.25 at the facility, the owner or operator shall consider only the portion of the mixture or trade name product that consists of the toxic chemical and that is imported, processed, or otherwise used at the facility, together with any other amounts of the same toxic chemical that the owner or operator manufactures, imports, processes, or otherwise uses at the facility as follows:

(i) If the owner or operator knows the specific chemical identity of the toxic chemical and the specific concentration at which it is present in the mixture or trade name product, the owner or operator shall determine the weight of the chemical imported, processed, or otherwise used as part of the mixture or trade name product at the facility and shall combine that with the weight of the toxic chemical manufactured (including imported) processed, or otherwise used at the facility other than as part of the mixture or trade name product. After combining these amounts, if the owner or operator determines that the toxic chemical was manufactured, processed, or otherwise used in excess of an applicable threshold in § 372.25, the owner or operator shall report the specific chemical identity and all releases of the toxic chemical on EPA Form R in accordance with the instructions in Subpart E.

(ii) If the owner or operator knows the specific chemical identity of the toxic chemical and does not know the specific concentration at which the chemical is present in the mixture or trade name product, but has been told the upper bound concentration of the chemical in the mixture or trade name product, the owner or operator shall assume that the toxic chemical is present in the mixture or trade name product at the upper bound concentration, shall determine whether the chemical has been manufactured, processed, or otherwise used at the facility in excess of an applicable threshold as provided in paragraph (b)(3)(i) of this section, and shall report as provided in paragraph (b)(3)(i) of this section.

(iii) If the owner or operator knows the specific chemical identity of the toxic chemical, does not know the specific concentration at which the chemical is present in the mixture or trade name product, has not been told the upper bound concentration of the chemical in the mixture or trade name product, and has not otherwise developed information on the composition of the chemical in the mixture or trade name product, then the owner or operator is not required to factor that chemical in that mixture or trade name product into threshold and release calculations for that chemical.

(iv) If the owner or operator has been told that a mixture or trade name product contains a toxic chemical, does not know the specific chemical identity of the chemical and knows

260

Appendix B: Regulations Reference Guide

Environmental Protection Agency § 372.38

the specific concentration at which it is present in the mixture or trade name product, the owner or operator shall determine the weight of the chemical imported, processed, or otherwise used as part of the mixture or trade name product at the facility. Since the owner or operator does not know the specific identity of the toxic chemical, the owner or operator shall make the threshold determination only for the weight of the toxic chemical in the mixture or trade name product. If the owner or operator determines that the toxic chemical was imported, processed, or otherwise used as part of the mixture or trade name product in excess of an applicable threshold in § 372.25, the owner or operator shall report the generic chemical name of the toxic chemical, or a trade name if the generic chemical name is not known, and all releases of the toxic chemical on EPA Form R in accordance with the instructions in Subpart E.

(v) If the owner or operator has been told that a mixture or trade name product contains a toxic chemical, does not know the specific chemical identity of the chemical, and does not know the specific concentration at which the chemical is present in the mixture or trade name product, but has been told the upper bound concentration of the chemical in the mixture or trade name product, the owner or operator shall assume that the toxic chemical is present in the mixture or trade name product at the upper bound concentration, shall determine whether the chemical has been imported, processed, or otherwise used at the facility in excess of an applicable threshold as provided in paragraph (b)(3)(iv) of this section, and shall report as provided in paragraph (b)(3)(iv) of this section.

(vi) If the owner or operator has been told that a mixture or trade name product contains a toxic chemical, does not know the specific chemical identity of the chemical, does not know the specific concentration at which the chemical is present in the mixture or trade name product, including information they have themselves developed, and has not been told the upper bound concentration of the chemical in the mixture or trade name product, the owner or operator is not required to report with respect to that toxic chemical.

(c) A covered facility may consist of more than one establishment. The owner or operator of such a facility at which a toxic chemical was manufactured (including imported), processed, or otherwise used in excess of an applicable threshold may submit a separate Form R for each establishment or for each group of establishments within the facility to report the activities involving the toxic chemical at each establishment or group of establishments, provided that activities involving that toxic chemical at all the establishments within the covered facility are reported. If each establishment or group of establishments files separate reports then for all other chemicals subject to reporting at that facility they must also submit separate reports. However, an establishment or group of establishments does not have to submit a report for a chemical that is not manufactured (including imported), processed, otherwise used, or released at that establishment or group of establishments.

(d) Each report under this section for activities involving a toxic chemical that occurred during a calendar year at a covered facility must be submitted on or before July 1 of the next year. The first such report for calendar year 1987 activities must be submitted on or before July 1, 1988.

(e) For reports applicable to activities for calendar years 1987, 1988, and 1989 only, the owner or operator of a covered facility may report releases of a specific toxic chemical to an environmental medium, or transfers of wastes containing a specific toxic chemical to an off-site location, of less than 1,000 pounds using the ranges provided in the form and instructions in Subpart E. For reports applicable to activities in calendar year 1990 and beyond, these ranges may not be used.

[53 FR 4525, Feb. 16, 1988; 53 FR 12748, Apr. 18, 1988]

§ 372.38 Exemptions.

(a) *De minimis concentrations of a toxic chemical in a mixture.* If a toxic

261

§ 372.38

chemical is present in a mixture of chemicals at a covered facility and the toxic chemical is in a concentration in the mixture which is below 1 percent of the mixture, or 0.1 percent of the mixture in the case of a toxic chemical which is a carcinogen as defined in 29 CFR 1910.1200(d)(4), a person is not required to consider the quantity of the toxic chemical present in such mixture when determining whether an applicable threshold has been met under § 372.25 or determining the amount of release to be reported under § 372.30. This exemption applies whether the person received the mixture from another person or the person produced the mixture, either by mixing the chemicals involved or by causing a chemical reaction which resulted in the creation of the toxic chemical in the mixture. However, this exemption applies only to the quantity of the toxic chemical present in the mixture. If the toxic chemical is also manufactured (including imported), processed, or otherwise used at the covered facility other than as part of the mixture or in a mixture at higher concentrations, in excess of an applicable threshold quantity set forth in § 372.25, the person is required to report under § 372.30.

(b) *Articles.* If a toxic chemical is present in an article at a covered facility, a person is not required to consider the quantity of the toxic chemical present in such article when determining whether an applicable threshold has been met under § 372.25 or determining the amount of release to be reported under § 372.30. This exemption applies whether the person received the article from another person or the person produced the article. However, this exemption applies only to the quantity of the toxic chemical present in the article. If the toxic chemical is manufactured (including imported), processed, or otherwise used at the covered facility other than as part of the article, in excess of an applicable threshold quantity set forth in § 372.25, the person is required to report under § 372.30. Persons potentially subject to this exemption should carefully review the definitions of "article" and "release" in § 372.3. If a release of a toxic chemical occurs as a result of the processing or use of an item at the facility, that item does not meet the definition of "article."

(c) *Uses.* If a toxic chemical is used at a covered facility for a purpose described in this paragraph (c), a person is not required to consider the quantity of the toxic chemical used for such purpose when determining whether an applicable threshold has been met under § 372.25 or determining the amount of releases to be reported under § 372.30. However, this exemption only applies to the quantity of the toxic chemical used for the purpose described in this paragraph (c). If the toxic chemical is also manufactured (including imported), processed, or otherwise used at the covered facility other than as described in this paragraph (c), in excess of an applicable threshold quantity set forth in § 372.25, the person is required to report under § 372.30.

(1) Use as a structural component of the facility.

(2) Use of products for routine janitorial or facility grounds maintenance. Examples include use of janitorial cleaning supplies, fertilizers, and pesticides similar in type or concentration to consumer products.

(3) Personal use by employees or other persons at the facility of foods, drugs, cosmetics, or other personal items containing toxic chemicals, including supplies of such products within the facility such as in a facility operated cafeteria, store, or infirmary.

(4) Use of products containing toxic chemicals for the purpose of maintaining motor vehicles operated by the facility.

(5) Use of toxic chemicals present in process water and non-contact cooling water as drawn from the environment or from municipal sources, or toxic chemicals present in air used either as compressed air or as part of combustion.

(d) *Activities in laboratories.* If a toxic chemical is manufactured, processed, or used in a laboratory at a covered facility under the supervision of a technically qualified individual as defined in § 720.3(ee) of this title, a person is not required to consider the quantity so manufactured, processed, or used when determining whether an

Appendix B: Regulations Reference Guide

Environmental Protection Agency

applicable threshold has been met under § 372.25 or determining the amount of release to be reported under § 372.30. This exemption does not apply in the following cases:

(1) Specialty chemical production.

(2) Manufacture, processing, or use of toxic chemicals in pilot plant scale operations.

(3) Activities conducted outside the laboratory.

(e) *Certain owners of leased property.* The owner of a covered facility is not subject to reporting under § 372.30 if such owner's only interest in the facility is ownership of the real estate upon which the facility is operated. This exemption applies to owners of facilities such as industrial parks, all or part of which are leased to persons who operate establishments within SIC code 20 through 39 where the owner has no other business interest in the operation of the covered facility.

(f) *Reporting by certain operators of establishments on leased property such as industrial parks.* If two or more persons, who do not have any common corporate or business interest (including common ownership or control), operate separate establishments within a single facility, each such person shall treat the establishments it operates as a facility for purposes of this part. The determinations in § 372.22 and § 372.25 shall be made for those establishments. If any such operator determines that its establishment is a covered facility under § 372.22 and that a toxic chemical has been manufactured (including imported), processed, or otherwise used at the establishment in excess of an applicable threshold in § 372.25 for a calendar year, the operator shall submit a report in accordance with § 372.30 for the establishment. For purposes of this paragraph (f), a common corporate or business interest includes ownership, partnership, joint ventures, ownership of a controlling interest in one person by the other, or ownership of a controlling interest in both persons by a third person.

§ 372.45

Subpart C—Supplier Notification Requirement

§ 372.45 Notification about toxic chemicals.

(a) Except as provided in paragraphs (c), (d), and (e) of this section and § 372.65, a person who owns or operates a facility or establishment which:

(1) Is in Standard Industrial Classification codes 20 through 39 as set forth in paragraph (b) of § 372.22,

(2) Manufactures (including imports) or processes a toxic chemical, and

(3) Sells or otherwise distributes a mixture or trade name product containing the toxic chemical, to (i) a facility described in § 372.22, or (ii) to a person who in turn may sell or otherwise distributes such mixture or trade name product to a facility described in § 372.22(b), must notify each person to whom the mixture or trade name product is sold or otherwise distributed from the facility or establishment in accordance with paragraph (b) of this section.

(b) The notification required in paragraph (a) of this section shall be in writing and shall include:

(1) A statement that the mixture or trade name product contains a toxic chemical or chemicals subject to the reporting requirements of section 313 of Title III of the Superfund Amendments and Reauthorization Act of 1986 and 40 CFR Part 372.

(2) The name of each toxic chemical, and the associated Chemical Abstracts Service registry number of each chemical if applicable, as set forth in § 372.65.

(3) The percent by weight of each toxic chemical in the mixture or trade name product.

(c) Notification under this section shall be provided as follows:

(1) For a mixture or trade name product containing a toxic chemical listed in § 373.65 with an effective date of January 1, 1987, the person shall provide the written notice described in paragraph (b) of this section to each recipient of the mixture or trade name product with at least the first shipment of each mixture or trade name

263

§ 372.45

product to each recipient in each calendar year beginning January 1, 1989.

(2) For a mixture or trade name product containing a toxic chemical listed in § 372.65 with an effective date of January 1, 1989 or later, the person shall provide the written notice described in paragraph (b) of this section to each recipient of the mixture or trade name product with at least the first shipment of the mixture or trade name product to each recipient in each calendar year beginning with the applicable effective date.

(3) If a person changes a mixture or trade name product for which notification was previously provided under paragraph (b) of this section by adding a toxic chemical, removing a toxic chemical, or changing the percent by weight of a toxic chemical in the mixture or trade name product, the person shall provide each recipient of the changed mixture or trade name product a revised notification reflecting the change with the first shipment of the changed mixture or trade name product to the recipient.

(4) If a person discovers (i) that a mixture or trade name product previously sold or otherwise distributed to another person during the calendar year of the discovery contains one or more toxic chemicals and (ii), that any notification providied to such other persons in that calendar year for the mixture or trade name product either did not properly identify any of the toxic chemicals or did not accurately present the percent by weight of any of the toxic chemicals in the mixture or trade name product, the person shall provide a new notification to the recipient within 30 days of the discovery which contains the information described in paragraph (b) of this section and identifies the prior shipments of the mixture or product in that calendar year to which the new notification applies.

(5) If a Material Safety Data Sheet (MSDS) is required to be prepared and distributed for the mixture or trade name product in accordance with 29 CFR 1910.1200, the notification must be attached to or otherwise incorporated into such MSDS. When the notification is attached to the MSDS, the notice must contain clear instructions

40 CFR Ch. I (7-1-89 Edition)

that the notifications must not be detached from the MSDS and that any copying and redistribution of the MSDS shall include copying and redistribution of the notice attached to copies of the MSDS subsequently redistributed.

(d) Notifications are not required in the following instances:

(1) If a mixture or trade name product contains no toxic chemical in excess of the applicable de minimis concentration as specified in § 372.38(a).

(2) If a mixture or trade name product is one of the following:

(i) An "article" as defined in § 372.3

(ii) Foods, drugs, cosmetics, alcoholic beverages, tobacco, or tobacco products packaged for distribution to the general public.

(iii) Any consumer product as the term is defined in the Consumer Product Safety Act (15 U.S.C. 1251 et seq.) packaged for distribution to the general public.

(e) If the person considers the specific identity of a toxic chemical in a mixture or trade name product to be a trade secret under provisions of 29 CFR 1910.1200, the notice shall contain a generic chemical name that is descriptive of that toxic chemical.

(f) If the person considers the specific percent by weight composition of a toxic chemical in the mixture or trade name product to be a trade secret under applicable State law or under the Restatement of Torts section 757, comment b, the notice must contain a statement that the chemical is present at a concentration that does not exceed a specified upper bound concentration value. For example, a mixture contains 12 percent of a toxic chemical. However, the supplier considers the specific concentration of the toxic chemical in the product to be a trade secret. The notice would indicate that the toxic chemical is present in the mixture in a concentration of no more than 15 percent by weight. The upper bound value chosen must be no larger than necessary to adequately protect the trade secret.

(g) A person is not subject to the requirements of this section to the extent the person does not know that the facility or establishment(s) is sell-

264

Environmental Protection Agency

ing or otherwise distributing a toxic chemical to another person in a mixture or trade name product. However, for purposes of this section, a person has such knowledge if the person receives a notice under this section from a supplier of a mixture or trade name product and the person in turn sells or otherwise distributes that mixture or trade name product to another person.

(h) If two or more persons, who do not have any common corporate or business interest (including common ownership or control), as described in § 372.38(f), operate separate establishments within a single facility, each such persons shall treat the establishment(s) it operates as a facility for purposes of this section. The determination under paragraph (a) of this section shall be made for those establishments.

[53 FR 4525, Feb. 16, 1988; 53 FR 12748, Apr. 18, 1988]

§ 372.65

Subpart D—Specific Toxic Chemical Listings

§ 372.65 Chemicals and chemical categories to which this part applies.

The requirements of this part apply to the following chemicals and chemical categories. This section contains three listings. Paragraph (a) of this section is an alphabetical order listing of those chemicals that have an associated Chemical Abstracts Service (CAS) Registry number. Paragraph (b) of this section contains a CAS number order list of the same chemicals listed in paragraph (a) of this section. Paragraph (c) of this section contains the chemical categories for which reporting is required. These chemical categories are listed in alphabetical order and do not have CAS numbers. Each listing identifies the effective date for reporting under § 372.30.

(a) *Alphabetical listing.*

Chemical name	CAS No.	Effective date
Acetaldehyde	75-07-0	01/01/87
Acetamide	60-35-5	01/01/87
Acetone	67-64-1	01/01/87
Acetonitrile	75-05-8	01/01/87
2-Acetylaminofluorene	53-96-3	01/01/87
Acrolein	107-02-8	01/01/87
Acrylamide	79-06-1	01/01/87
Acrylic acid	79-10-7	01/01/87
Acrylonitrile	107-13-1	01/01/87
Aldrin[1,4:5,8-Dimethanonaphthalene,1,2,3,4,10,10-hexachloro-1,4,4a,5,8,8a-hexahydro-(1.alpha.,4.alpha.,4a.beta.,5.alpha.,8.alpha., 8a.beta.)-]	309-00-2	01/01/87
Allyl chloride	107-05-1	01/01/87
Aluminum (fume or dust)	7429-90-5	01/01/87
Aluminum oxide	1344-28-1	01/01/87
2-Aminoanthraquinone	117-79-3	01/01/87
4-Aminoazobenzene	60-09-3	01/01/87
4-Aminobiphenyl	92-67-1	01/01/87
1-Amino-2-methylanthraquinone	82-28-0	01/01/87
Ammonia	7664-41-7	01/01/87
Ammonium nitrate (solution)	6484-52-2	01/01/87
Ammonium sulfate (solution)	7783-20-2	01/01/87
Aniline	62-53-3	01/01/87
o-Anisidine	90-04-0	01/01/87
p-Anisidine	104-94-9	01/01/87
o-Anisidine hydrochloride	134-29-2	01/01/87
Anthracene	120-12-7	01/01/87
Antimony	7440-36-0	01/01/87
Arsenic	7440-38-2	01/01/87
Asbestos (friable)	1332-21-4	01/01/87
Barium	7440-39-3	01/01/87
Benzal chloride	98-87-3	01/01/87
Benzamide	55-21-0	01/01/87
Benzene	71-43-2	01/01/87
Benzidine	92-87-5	01/01/87
Benzoic trichloride (Benzotrichloride)	98-07-7	01/01/87
Benzoyl chloride	98-88-4	01/01/87
Benzoyl peroxide	94-36-0	01/01/87
Benzyl chloride	100-44-7	01/01/87
Beryllium	7440-41-7	01/01/87

265

Chemical Hazard Communication Guidebook

§ 372.65
40 CFR Ch. I (7-1-89 Edition)

Chemical name	CAS No.	Effective date
Biphenyl	92-52-4	01/01/87
Bis(2-chloroethyl) ether	111-44-4	01/01/87
Bis(chloromethyl) ether	542-88-1	01/01/87
Bis(2-chloro-1-methylethyl) ether	108-60-1	01/01/87
Bis(2-ethylhexyl) adipate	103-23-1	01/01/87
Bromoform (Tribromomethane)	75-25-2	01/01/87
Bromomethane (Methyl bromide)	74-83-9	01/01/87
1,3-Butadiene	106-99-0	01/01/87
Butyl acrylate	141-32-2	01/01/87
n-Butyl alcohol	71-36-3	01/01/87
sec-Butyl alcohol	78-92-2	01/01/87
tert-Butyl alcohol	75-65-0	01/01/87
Butyl benzyl phthalate	85-68-7	01/01/87
1,2-Butylene oxide	106-88-7	01/01/87
Butyraldehyde	123-72-8	01/01/87
C.I. Acid Green 3	4680-78-8	01/01/87
C.I. Basic Green 4	569-64-2	01/01/87
C.I. Basic Red 1	989-38-8	01/01/87
C.I. Direct Black 38	1937-37-7	01/01/87
C.I. Direct Blue 6	2602-46-2	01/01/87
C.I. Direct Brown 95	16071-86-6	01/01/87
C.I. Disperse Yellow 3	2832-40-8	01/01/87
C.I. Food Red 5	3761-53-3	01/01/87
C.I. Food Red 15	81-88-9	01/01/87
C.I. Solvent Orange 7	3118-97-6	01/01/87
C.I. Solvent Yellow 3	97-56-3	01/01/87
C.I. Solvent Yellow 14	842-07-9	01/01/87
C.I. Solvent Yellow 34 (Aurimine)	492-80-8	01/01/87
C.I. Vat Yellow 4	128-66-5	01/01/87
Cadmium	7440-43-9	01/01/87
Calcium cyanamide	156-62-7	01/01/87
Captan [1H-Isoindole-1,3(2H)-dione,3a,4,7,7a-tetrahydro-2-[(trichloromethyl)thio]-]	133-06-2	01/01/87
Carbaryl [1-Naphthalenol, methylcarbamate]	63-25-2	01/01/87
Carbon disulfide	75-15-0	01/01/87
Carbon tetrachloride	56-23-5	01/01/87
Carbonyl sulfide	463-58-1	01/01/87
Catechol	120-80-9	01/01/87
Chloramben [Benzoic acid,3-amino-2,5-dichloro-]	133-90-4	01/01/87
Chlordane [4,7-Methanoindan,1,2,4,5,6,7,8,8-octachloro-2,3,3a,4,7,7a-hexahydro-]	57-74-9	01/01/87
Chlorine	7782-50-5	01/01/87
Chlorine dioxide	10049-04-4	01/01/87
Chloroacetic acid	79-11-8	01/01/87
2-Chloroacetophenone	532-27-4	01/01/87
Chlorobenzene	108-90-7	01/01/87
Chlorobenzilate [Benzeneacetic acid, 4-chloro-.alpha.-(4-.chlorophenyl)-.alpha.-hydroxy-, ethyl ester]	510-15-6	01/01/87
Chloroethane (Ethyl chloride)	75-00-3	01/01/87
Chloroform	67-66-3	01/01/87
Chloromethane (Methyl chloride)	74-87-3	01/01/87
Chloromethyl methyl ether	107-30-2	01/01/87
Chloroprene	126-99-8	01/01/87
Chlorothalonil [1,3-Benzenedicarbonitrile,2,4,5,6-tetrachloro-]	1897-45-6	01/01/87
Chromium	7440-47-3	01/01/87
Cobalt	7440-48-4	01/01/87
Copper	7440-50-8	01/01/87
p-Cresidine	120-71-8	01/01/87
Cresol (mixed isomers)	1319-77-3	01/01/87
m-Cresol	108-39-4	01/01/87
o-Cresol	95-48-7	01/01/87
p-Cresol	106-44-5	01/01/87
Cumene	98-82-8	01/01/87
Cumene hydroperoxide	80-15-9	01/01/87
Cupferron [Benzeneamine, N-hydroxy-N-nitroso, ammonium salt]	135-20-6	01/01/87
Cyclohexane	110-82-7	01/01/87
2,4-D [Acetic acid, (2,4-dichlorophenoxy)-]	94-75-7	01/01/87
Decabromodiphenyl oxide	1163-19-5	01/01/87
Diallate [Carbamothioic acid, bis(1-methylethyl)-, S-(2,3-dichloro-2-propenyl) ester]	2303-16-4	01/01/87
2,4-Diaminoanisole	615-05-4	01/01/87
2,4-Diaminoanisole sulfate	39156-41-7	01/01/87
4,4'-Diaminodiphenyl ether	101-80-4	01/01/87
Diaminotoluene (mixed isomers)	25376-45-8	01/01/87
2,4-Diaminotoluene	95-80-7	01/01/87
Diazomethane	334-88-3	01/01/87

266

Appendix B: Regulations Reference Guide

Environmental Protection Agency §**372.65**

Chemical name	CAS No.	Effective date
Dibenzofuran	132-64-9	01/01/87
1,2-Dibromo-3-chloropropane (DBCP)	96-12-8	01/01/87
1,2-Dibromoethane (Ethylene dibromide)	106-93-4	01/01/87
Dibutyl phthalate	84-74-2	01/01/87
Dichlorobenzene (mixed isomers)	25321-22-6	01/01/87
1,2-Dichlorobenzene	95-50-1	01/01/87
1,3-Dichlorobenzene	541-73-1	01/01/87
1,4-Dichlorobenzene	106-46-7	01/01/87
3,3'-Dichlorobenzidine	91-94-1	01/01/87
Dichlorobromomethane	75-27-4	01/01/87
1,2-Dichloroethane (Ethylene dichloride)	107-06-2	01/01/87
1,2-Dichlorethylene	540-59-0	01/01/87
Dichloromethane (Methylene chloride)	75-09-2	01/01/87
2,4-Dichlorophenol	120-83-2	01/01/87
1,2-Dichloropropane	78-87-5	01/01/87
1,3-Dichloropropylene	542-75-6	01/01/87
Dichlorvos [Phosphoric acid, 2,2-dichloroethenyl dimethyl ester]	62-73-7	01/01/87
Dicofol [Benzenemethanol,4-chloro-.alpha.-(4-chlorophenyl)-.alpha.-(trichloromethyl)-]	115-32-2	01/01/87
Diepoxybutane	1464-53-5	01/01/87
Diethanolamine	111-42-2	01/01/87
Di-(2-ethylhexyl) phthalate (DEHP)	177-81-7	01/01/87
Diethyl phthalate	84-66-2	01/01/87
Diethyl sulfate	64-67-5	01/01/87
3,3'-Dimethoxybenzidine	119-90-4	01/01/87
4-Dimethylaminoazobenzene	60-11-7	01/01/87
3,3'-Dimethylbenzidine (o-Tolidine)	119-93-7	01/01/87
Dimethylcarbamyl chloride	79-44-7	01/01/87
1,1-Dimethyl hydrazine	57-14-7	01/01/87
2,4-Dimethylphenol	105-67-9	01/01/87
Dimethyl phthalate	131-11-3	01/01/87
Dimethyl sulfate	77-78-1	01/01/87
4,6-Dinitro-o-cresol	534-52-1	01/01/87
2,4-Dinitrophenol	51-28-5	01/01/87
2,4-Dinitrotoluene	121-14-2	01/01/87
2,6-Dinitrotoluene	606-20-2	01/01/87
n-Dioctyl phthalate	117-84-0	01/01/87
1,4-Dioxane	123-91-1	01/01/87
1,2-Diphenylhydrazine (Hydrazobenzene)	122-66-7	01/01/87
Epichlorohydrin	106-89-8	01/01/87
2-Ethoxyethanol	110-80-5	01/01/87
Ethyl acrylate	140-88-5	01/01/87
Ethylbenzene	100-41-4	01/01/87
Ethyl chloroformate	541-41-3	01/01/87
Ethylene	74-85-1	01/01/87
Ethylene glycol	107-21-1	01/01/87
Ethyleneimine (Aziridine)	151-56-4	01/01/87
Ethylene oxide	75-21-8	01/01/87
Ethylene thiourea	96-45-7	01/01/87
Fluometuron [Urea, N,N-dimethyl-N-[3-(trifluoromethyl)phenyl]-]	2164-17-2	01/01/87
Formaldehyde	50-00-0	01/01/87
Freon 113 [Ethane, 1,1,2-trichloro-1,2,2-trifluoro-]	76-13-1	01/01/87
Heptachlor [1,4,5,6,7,8,8-Heptachloro-3a,4,7,7a-tetrahydro-4,7-methano-1H-indene]	76-44-8	01/01/87
Hexachlorobenzene	118-74-1	01/01/87
Hexachloro-1,3-butadiene	87-68-3	01/01/87
Hexachlorocyclopentadiene	77-47-4	01/01/87
Hexachloroethane	67-72-1	01/01/87
Hexachloronaphthalene	1335-87-1	01/01/87
Hexamethylphosphoramide	680-31-9	01/01/87
Hydrazine	302-01-2	01/01/87
Hydrazine sulfate	10034-93-2	01/01/87
Hydrochloric acid	7647-01-0	01/01/87
Hydrogen cyanide	74-90-8	01/01/87
Hydrogen fluoride	7664-39-3	01/01/87
Hydroquinone	123-31-9	01/01/87
Isobutyraldehyde	78-84-2	01/01/87
Isopropyl alcohol (Only persons who manufacture by the strong acid process are subject, no supplier notifiction.)	67-63-0	01/01/87
4,4'-Isopropylidenediphenol	80-05-7	01/01/87
Lead	7439-92-1	01/01/87
Lindane [Cyclohexane, 1,2,3,4,5,6-hexachloro-(1.alpha.,2.alpha.,3.beta.,4.alpha.,5.alpha.,6.beta.)-]	58-89-9	01/01/87
Maleic anhydride	108-31-6	01/01/87
Maneb [Carbamodithioic acid, 1,2-ethanediylbis-, manganese complex]	12427-38-2	01/01/87

267

Chemical Hazard Communication Guidebook

§ 372.65
40 CFR Ch. I (7-1-89 Edition)

Chemical name	CAS No.	Effective date
Manganese	7439-96-5	01/01/87
Mercury	7439-97-6	01/01/87
Methanol	67-56-1	01/01/87
Methoxychlor [Benzene, 1,1'-(2,2,2-trichloroethylidene)bis[4-methoxy-]	72-43-5	01/01/87
2-Methoxyethanol	109-86-4	01/01/87
Methyl acrylate	96-33-3	01/01/87
Methyl *tert*-butyl ether	1634-04-4	01/01/87
4,4'-Methylenebis(2-chloroaniline) (MBOCA)	101-14-4	01/01/87
4,4'-Methylenebis(*N,N*-dimethyl) benzenamine	101-61-1	01/01/87
Methylenebis(phenylisocyanate) (MBI)	101-68-8	01/01/87
Methylene bromide	74-95-3	01/01/87
4,4'-Methylenedianiline	101-77-9	01/01/87
Methyl ethyl ketone	78-93-3	01/01/87
Methyl hydrazine	60-34-4	01/01/87
Methyl iodide	74-88-4	01/01/87
Methyl isobutyl ketone	108-10-1	01/01/87
Methyl isocyanate	624-83-9	01/01/87
Methyl methacrylate	80-62-6	01/01/87
Michler's ketone	90-94-8	01/01/87
Molybdenum trioxide	1313-27-5	01/01/87
Mustard gas [Ethane, 1,1'-thiobis[2-chloro-]	505-60-2	01/01/87
Naphthalene	91-20-3	01/01/87
alpha-Naphthylamine	134-32-7	01/01/87
beta-Naphthylamine	91-59-8	01/01/87
Nickel	7440-02-0	01/01/87
Nitric acid	7697-37-2	01/01/87
Nitrilotriacetic acid	139-13-9	01/01/87
5-Nitro-*o*-anisidine	99-59-2	01/01/87
Nitrobenzene	98-95-3	01/01/87
4-Nitrobiphenyl	92-93-3	01/01/87
Nitrofen [Benzene, 2,4-dichloro-1-(4-nitrophenoxy)-]	1836-75-5	01/01/87
Nitrogen mustard [2-Chloro-N-(2-chloroethyl)-N-methylethanamine]	51-75-2	01/01/87
Nitroglycerin	55-63-0	01/01/87
2-Nitrophenol	88-75-5	01/01/87
4-Nitrophenol	100-02-7	01/01/87
2-Nitropropane	79-46-9	01/01/87
p-Nitrosodiphenylamine	156-10-5	01/01/87
N,N-Dimethylaniline	121-69-7	01/01/87
N-Nitrosodi-*n*-butylamine	924-16-3	01/01/87
N-Nitrosodiethylamine	55-18-5	01/01/87
N-Nitrosodimethylamine	62-75-9	01/01/87
N-Nitrosodiphenylamine	86-30-6	01/01/87
N-Nitrosodi-*n*-propylamine	621-64-7	01/01/87
N-Nitrosomethylvinylamine	4549-40-0	01/01/87
N-Nitrosomorpholine	59-89-2	01/01/87
N-Nitroso-*N*-ethylurea	759-73-9	01/01/87
N-Nitroso-*N*-methylurea	684-93-5	01/01/87
N-Nitrosonornicotine	16543-55-8	01/01/87
N-Nitrosopiperidine	100-75-4	01/01/87
Octachloronaphthalene	2234-13-1	01/01/87
Osmium tetroxide	20816-12-0	01/01/87
Parathion [Phosphorothioic acid, O,O-diethyl-O-(4-nitrophenyl) ester]	56-38-2	01/01/87
Pentachlorophenol (PCP)	87-86-5	01/01/87
Peracetic acid	79-21-0	01/01/87
Phenol	108-95-2	01/01/87
p-Phenylenediamine	106-50-3	01/01/87
2-Phenylphenol	90-43-7	01/01/87
Phosgene	75-44-5	01/01/87
Phosphoric acid	7664-38-2	01/01/87
Phosphorus (yellow or white)	7723-14-0	01/01/87
Phthalic anhydride	85-44-9	01/01/87
Picric acid	88-89-1	01/01/87
Polychlorinated biphenyls (PCBs)	1336-36-3	01/01/87
Propane sultone	1120-71-4	01/01/87
beta-Propiolactone	57-57-8	01/01/87
Propionaldehyde	123-38-6	01/01/87
Propoxur [Phenol, 2-(1-methylethoxy)-, methylcarbamate]	114-26-1	01/01/87
Propylene (Propene)	115-07-1	01/01/87
Propyleneimine	75-55-8	01/01/87
Propylene oxide	75-56-9	01/01/87
Pyridine	110-86-1	01/01/87
Quinoline	91-22-5	01/01/87
Quinone	106-51-4	01/01/87

268

Appendix B: Regulations Reference Guide

Environmental Protection Agency § 372.65

Chemical name	CAS No.	Effective date
Quintozene [Pentachloronitrobenzene]	82-68-8	01/01/87
Saccharin (only persons who manufacture are subject, no supplier notification) [1,2-Benzisothiazol-3(2H)-one,1,1-dioxide]	81-07-2	01/01/87
Safrole	94-59-7	01/01/87
Selenium	7782-49-2	01/01/87
Silver	7440-22-4	01/01/87
Sodium hydroxide (solution)	1310-73-2	01/01/87
Styrene	100-42-5	01/01/87
Styrene oxide	96-09-3	01/01/87
Sulfuric acid	7664-93-9	01/01/87
Terephthalic acid	100-21-0	01/01/87
1,1,2,2-Tetrachloroethane	79-34-5	01/01/87
Tetrachloroethylene (Perchloroethylene)	127-18-4	01/01/87
Tetrachlorvinphos [Phosphoric acid, 2-chloro-1-(2,4,5-trichlorophenyl)ethenyl dimethyl ester]	961-11-5	01/01/87
Thallium	7440-28-0	01/01/87
Thioacetamide	62-55-5	01/01/87
4,4'-Thiodianiline	139-65-1	01/01/87
Thiourea	62-56-6	01/01/87
Thorium dioxide	1314-20-1	01/01/87
Titanium tetrachloride	7550-45-0	01/01/87
Toluene	108-88-3	01/01/87
Toluene-2,4-diisocyanate	584-84-9	01/01/87
Toluene-2,6-diisocyanate	91-08-7	01/01/87
o-Toluidine	95-53-4	01/01/87
o-Toluidine hydrochloride	636-21-5	01/01/87
Toxaphene	8001-35-2	01/01/87
Triaziquone [2,5-Cyclohexadiene-1,4-dione,2,3,5-tris(1-aziridinyl)-]	68-76-8	01/01/87
Trichlorfon [Phosphonic acid, (2,2,2-trichloro-1-hydroxyethyl)-, dimethyl ester]	52-68-6	01/01/87
1,2,4-Trichlorobenzene	120-82-1	01/01/87
1,1,1-Trichloroethane (Methyl chloroform)	71-55-6	01/01/87
1,1,2-Trichloroethane	79-00-5	01/01/87
Trichloroethylene	79-01-6	01/01/87
2,4,5-Trichlorophenol	95-95-4	01/01/87
2,4,6-Trichlorophenol	88-06-2	01/01/87
Trifluralin [Benzeneamine, 2,6-dinitro-N,N-dipropyl-4-(trifluoromethyl)-]	1582-09-8	01/01/87
1,2,4-Trimethylbenzene	95-63-6	01/01/87
Tris(2,3-dibromopropyl) phosphate	126-72-7	01/01/87
Urethane (Ethyl carbamate)	51-79-6	01/01/87
Vanadium (fume or dust)	7440-62-2	01/01/87
Vinyl acetate	108-05-4	01/01/87
Vinyl bromide	593-60-2	01/01/87
Vinyl chloride	75-01-4	01/01/87
Vinylidene chloride	75-35-4	01/01/87
Xylene (mixed isomers)	1330-20-7	01/01/87
m-Xylene	108-38-3	01/01/87
o-Xylene	95-47-6	01/01/87
p-Xylene	106-42-3	01/01/87
2,6-Xylidine	87-62-7	01/01/87
Zinc (fume or dust)	7440-66-6	01/01/87
Zineb [Carbamodithioic acid, 1,2-ethanediylbis-, zinc complex]	12122-67-7	01/01/87

(b) *CAS Number listing.*

CAS No.	Chemical name	Effective date
50-00-0	Formaldehyde	01/01/87
51-28-5	2,4-Dinitrophenol	01/01/87
51-75-2	Nitrogen mustard [2-Chloro-N-(2-chloroethyl)-N-methylethanamine]	01/01/87
51-79-6	Urethane (Ethyl carbamate)	01/01/87
52-68-6	Trichlorfon [Phosphonic acid, (2,2,2-trichloro-1-hydroxyethyl)-dimethyl ester]	01/01/87
53-96-3	2-Acetylaminofluorene	01/01/87
55-18-5	N-Nitrosodiethylamine	01/01/87
55-21-0	Benzamide	01/01/87
55-63-0	Nitroglycerin	01/01/87
56-23-5	Carbon tetrachloride	01/01/87
56-38-2	Parathion [Phosphorothioic acid, O,O-diethyl-O-(4-nitrophenyl)ester]	01/01/87
57-14-7	1,1-Dimethyl hydrazine	01/01/87
57-57-8	*beta*-Propiolactone	01/01/87

269

Chemical Hazard Communication Guidebook

§ 372.65   40 CFR Ch. I (7-1-89 Edition)

CAS No.	Chemical name	Effective date
57-74-9	Chlordane [4,7-Methanoindan, 1,2,4,5,6,7,8,8-octachloro-2,3,3a,4,7,7a-hexahydro-]	01/01/87
58-89-9	Lindane [Cyclohexane, 1,2,3,4,5,6-hexachloro-(1.alpha.,2.alpha.,3.beta.,4.alpha.,5.alpha.,6.beta.)-]	01/01/87
59-89-2	N-Nitrosomorpholine	01/01/87
60-09-3	4-Aminoazobenzene	01/01/87
60-11-7	4-Dimethylaminoazobenzene	01/01/87
60-34-4	Methyl hydrazine	01/01/87
60-35-5	Acetamide	01/01/87
62-53-3	Aniline	01/01/87
62-55-5	Thioacetamide	01/01/87
62-56-6	Thiourea	01/01/87
62-73-7	Dichlorvos [Phosphoric acid, 2,2-dichloroethenyl dimethyl ester]	01/01/87
62-75-9	N-Nitrosodimethylamine	01/01/87
63-25-2	Carbaryl [1-Naphthalenol, methylcarbamate]	01/01/87
64-67-5	Diethyl sulfate	01/01/87
67-56-1	Methanol	01/01/87
67-63-0	Isopropyl alcohol (only persons who manufacture by the strong acid process are subject, supplier notification not required.)	01/01/87
67-64-1	Acetone	01/01/87
67-66-3	Chloroform	01/01/87
67-72-1	Hexachloroethane	01/01/87
68-76-8	Triaziquone [2,5-Cyclohexadiene-1,4-dione,2,3,5-tris(1-aziridinyl)-]	01/01/87
71-36-3	n-Butyl alcohol	01/01/87
71-43-2	Benzene	01/01/87
71-55-6	1,1,1-Trichloroethane (Methyl chloroform)	01/01/87
72-43-5	Methoxychlor [Benzene, 1,1'-(2,2,2,-trichloroethylidene)bis [4-methoxy-]	01/01/87
74-83-9	Bromomethane (Methyl bromide)	01/01/87
74-85-1	Ethylene	01/01/87
74-87-3	Chloromethane (Methyl chloride)	01/01/87
74-88-4	Methyl iodide	01/01/87
74-90-8	Hydrogen cyanide	01/01/87
74-95-3	Methylene bromide	01/01/87
75-00-3	Chloroethane (Ethyl chloride)	01/01/87
75-01-4	Vinyl chloride	01/01/87
75-05-8	Acetonitrile	01/01/87
75-07-0	Acetaldehyde	01/01/87
75-09-2	Dichloromethane (Methylene chloride)	01/01/87
75-15-0	Carbon disulfide	01/01/87
75-21-8	Ethylene oxide	01/01/87
75-25-2	Bromoform (Tribromomethane)	01/01/87
75-27-4	Dichlorobromomethane	01/01/87
75-35-4	Vinylidene chloride	01/01/87
75-44-5	Phosgene	01/01/87
75-55-8	Propyleneimine	01/01/87
75-56-9	Propylene oxide	01/01/87
75-65-0	tert-Butyl alcohol	01/01/87
77-13-1	Freon 113 [Ethane, 1,1,2-trichloro-1,2,2-trifluoro-]	01/01/87
76-44-8	Heptachlor [1,4,5,6,7,8,8-Heptachloro-3a,4,7,7a-tetrahydro-4,7-methano-1H-indene]	01/01/87
77-47-4	Hexachlorocyclopentadiene	01/01/87
77-78-1	Dimethyl sulfate	01/01/87
78-84-2	Isobutyraldehyde	01/01/87
78-87-5	1,2-Dichloropropane	01/01/87
78-92-2	sec-Butyl alcohol	01/01/87
78-93-3	Methyl ethyl ketone	01/01/87
79-00-5	1,1,2-Trichloroethane	01/01/87
79-01-6	Trichloroethylene	01/01/87
79-06-1	Acrylamide	01/01/87
79-10-7	Acrylic acid	01/01/87
79-11-8	Chloroacetic acid	01/01/87
79-21-0	Peracetic acid	01/01/87
79-34-5	1,1,2,2-Tetrachloroethane	01/01/87
79-44-7	Dimethylcarbamyl chloride	01/01/87
79-46-9	2-Nitropropane	01/01/87
80-05-7	4,4'-Isopropylidenediphenol	01/01/87
80-15-9	Cumene hydroperoxide	01/01/87
80-62-6	Methyl methacrylate	01/01/87
81-07-2	Saccharin (only persons who manufacture are subject, no supplier notification) [1,2-Benzisothiazol-3(2H)-one,1,1-dioxide]	01/01/87
81-88-9	C.I. Food Red 15	01/01/87
82-28-0	1-Amino-2-methylanthraquinone	01/01/87
82-68-8	Quintozene [Pentachloronitrobenzene]	C12
84-66-2	Diethyl phthalate	01/01/87
84-74-2	Dibutyl phthalate	01/01/87

Appendix B: Regulations Reference Guide

Environmental Protection Agency § 372.65

CAS No.	Chemical name	Effective date
85-44-9	Phthalic anhydride	01/01/87
85-68-7	Butyl benzyl phthalate	01/01/87
86-30-6	N-Nitrosodiphenylamine	01/01/87
87-62-7	2,6-Xylidine	01/01/87
87-68-3	Hexachloro-1,3-butadiene	01/01/87
87-86-5	Pentachlorophenol (PCP)	01/01/87
88-06-2	2,4,6-Trichlorophenol	01/01/87
88-75-5	2-Nitrophenol	01/01/87
88-89-1	Picric acid	01/01/87
90-04-0	o-Anisidine	01/01/87
90-43-7	2-Phenylphenol	01/01/87
90-94-8	Michler's ketone	01/01/87
91-08-7	Toluene-2,6-diisocyanate	01/01/87
91-20-3	Naphthalene	01/01/87
91-22-5	Quinoline	01/01/87
91-59-8	beta-Naphthylamine	01/01/87
91-94-1	3,3'-Dichlorobenzidine	01/01/87
92-52-4	Biphenyl	01/01/87
92-67-1	4-Aminobiphenyl	01/01/87
92-87-5	Benzidine	01/01/87
92-93-3	4-Nitrobiphenyl	01/01/87
94-36-0	Benzoyl peroxide	01/01/87
94-59-7	Safrole	01/01/87
94-75-7	2,4-D [Acetic acid, (2,4-dichlorophenoxy)-]	01/01/87
95-47-6	o-Xylene	01/01/87
95-48-7	o-Cresol	01/01/87
95-50-1	1,2-Dichlorobenzene	01/01/87
95-53-4	o-Toluidine	01/01/87
95-63-6	1,2,4-Trimethylbenzene	01/01/87
95-80-7	2,4-Diaminotoluene	01/01/87
95-95-4	2,4,5-Trichlorophenol	01/01/87
96-09-3	Styrene oxide	01/01/87
96-12-8	1,2-Dibromo-3-chloropropane (DBCP)	01/01/87
96-33-3	Methyl acrylate	01/01/87
96-45-7	Ethylene thiourea	01/01/87
97-56-3	C.I. Solvent Yellow 3	01/01/87
98-07-7	Benzoic trichloride (Benzotrichloride)	01/01/87
98-82-8	Cumene	01/01/87
98-87-3	Benzal chloride	01/01/87
98-88-4	Benzoyl chloride	01/01/87
98-95-3	Nitrobenzene	01/01/87
99-59-2	5-Nitro-o-anisidine	01/01/87
100-02-7	4-Nitrophenol	01/01/87
100-21-0	Terephthalic acid	01/01/87
100-41-4	Ethylbenzene	01/01/87
100-42-5	Styrene	01/01/87
100-44-7	Benzyl chloride	01/01/87
100-75-4	N-Nitrosopiperidine	01/01/87
101-14-4	4,4'-Methylenebis(2-chloroaniline) (MBOCA)	01/01/87
101-61-1	4,4'-Methylenebis(N,N-dimethyl)benzenamine	01/01/87
101-68-8	Methylenebis(phenylisocyanate) (MBI)	01/01/87
101-77-9	4,4'-Methylenedianiline	01/01/87
101-80-4	4,4'-Diaminodiphenyl ether	01/01/87
103-23-1	Bis(2-ethylhexyl) adipate	01/01/87
104-94-9	p-Anisidine	01/01/87
105-67-9	2,4-Dimethylphenol	01/01/87
106-42-3	p-Xylene	01/01/87
106-44-5	p-Cresol	01/01/87
106-46-7	1,4-Dichlorobenzene	01/01/87
106-50-3	p-Phenylenediamine	01/01/87
106-51-4	Quinone	01/01/87
106-88-7	1,2-Butylene oxide	01/01/87
106-89-8	Epichlorohydrin	01/01/87
106-93-4	1,2-Dibromoethane (Ethylene dibromide)	01/01/87
106-99-0	1,3-Butadiene	01/01/87
107-02-8	Acrolein	01/01/87
107-05-1	Allyl chloride	01/01/87
107-06-2	1,2-Dichloroethane (Ethylene dichloride)	01/01/87
107-13-1	Acrylonitrile	01/01/87
107-21-1	Ethylene glycol	01/01/87
107-30-2	Chloromethyl methyl ether	01/01/87
108-05-4	Vinyl acetate	01/01/87
108-10-1	Methyl isobutyl ketone	01/01/87

271

§ 372.65

40 CFR Ch. I (7-1-89 Edition)

CAS No.	Chemical name	Effective date
108-31-6	Maleic anhydride	01/01/87
108-38-3	m-Xylene	01/01/87
108-39-4	m-Cresol	01/01/87
108-60-1	Bis(2-chloro-1-methylethyl)ether	01/01/87
108-88-3	Toluene	01/01/87
108-90-7	Chlorobenzene	01/01/87
108-95-2	Phenol	01/01/87
109-86-4	2-Methoxyethanol	01/01/87
110-80-5	2-Ethoxyethanol	01/01/87
110-82-7	Cyclohexane	01/01/87
110-86-1	Pyridine	01/01/87
111-42-2	Diethanolamine	01/01/87
111-44-4	Bis(2-chloroethyl) ether	01/01/87
114-26-1	Propoxur [Phenol, 2-(1-methylethoxy)-, methylcarbamate]	01/01/87
115-07-1	Propylene (Propene)	01/01/87
115-32-2	Dicofol [Benzenemethanol, 4-chloro-.alpha.-(4-chlorophenyl)-.alpha.-(trichloromethyl)-]	01/01/87
117-79-3	2-Aminoanthraquinone	01/01/87
117-81-7	Di(2-ethylhexyl) phthalate (DEHP)	01/01/87
117-84-0	n-Dioctyl phthalate	01/01/87
118-74-1	Hexachlorobenzene	01/01/87
119-90-4	3,3'-Dimethoxybenzidine	01/01/87
119-93-7	3,3'-Dimethylbenzidine (o-Tolidine)	01/01/87
120-12-7	Anthracene	01/01/87
120-71-8	p-Cresidine	01/01/87
120-80-9	Catechol	01/01/87
120-82-1	1,2,4-Trichlorobenzene	01/01/87
120-83-2	2,4-Dichlorophenol	01/01/87
121-14-2	2,4-Dinitrotoluene	01/01/87
121-69-7	N,N-Dimethylaniline	01/01/87
122-66-7	1,2-Diphenylhydrazine (Hydrazobenzene)	01/01/87
123-31-9	Hydroquinone	01/01/87
123-38-6	Propionaldehyde	01/01/87
123-72-8	Butyraldehyde	01/01/87
123-91-1	1,4-Dioxane	01/01/87
126-72-7	Tris-2,3-dibromopropyl) phosphate	01/01/87
126-99-8	Chloroprene	01/01/87
127-18-4	Tetrachloroethylene (Perchloroethylene)	01/01/87
128-66-5	C.I. Vat Yellow 4	01/01/87
131-11-3	Dimethyl phthalate	01/01/87
132-64-9	Dibenzofuran	01/01/87
133-06-2	Captan [1H-Isoindole-1,3(2H)-dione,3a,4,7,7a-tetrahydro-2-[(trichloromethyl)thio]-]	01/01/87
133-90-4	Chloramben [Benzoic acid, 3-amino-2,5-dichloro-]	01/01/87
134-29-2	o-Anisidine hydrochloride	01/01/87
134-32-7	alpha-Naphthylamine	01/01/87
135-20-6	Cupferron [Benzeneamine, N-hydroxy-N-nitroso, ammonium salt]	01/01/87
139-13-9	Nitrilotriacetic acid	01/01/87
139-65-1	4,4'-Thiodianiline	01/01/87
140-88-5	Ethyl acrylate	01/01/87
141-32-2	Butyl acrylate	01/01/87
151-56-4	Ethyleneimine (Aziridine)	01/01/87
156-10-5	p-Nitrosodiphenylamine	01/01/87
156-62-7	Calcium cyanamide	01/01/87
302-01-2	Hydrazine	01/01/87
309-00-2	Aldrin[1,4:5,8-Dimethanonaphthalene,1,2,3,4,10,10-hexachloro-1,4,4a,5,8,8a-hexahydro-(1.alpha.,4.alpha.,4a.beta.,5.alpha., 8.alpha.,8a.beta.)-]	01/01/87
334-88-3	Diazomethane	01/01/87
463-58-1	Carbonyl sulfide	01/01/87
492-80-8	C.I. Solvent Yellow 34 (Auramine)	01/01/87
505-60-2	Mustard gas [Ethane, 1,1'-thiobis[2-chloro-]	01/01/87
510-15-6	Chlorobenzilate[Benzeneacetic acid, 4-chloro-.alpha.-(4-chlorophenyl)-.alpha.-hydroxy-, ethyl ester]	01/01/87
532-27-4	2-Chloroacetophenone	01/01/87
534-52-1	4,6-Dinitro-o-cresol	01/01/87
540-59-0	1,2-Dichloroethylene	01/01/87
541-41-3	Ethyl chloroformate	01/01/87
541-73-1	1,3-Dichlorobenzene	01/01/87
542-75-6	1,3-Dichloropropylene	01/01/87
542-88-1	Bis(chloromethyl) ether	01/01/87
569-64-2	C.I. Basic Green 4	01/01/87
606-20-2	2,6-Dinitrotoluene	01/01/87
615-05-4	2,4-Diaminoanisole	01/01/87
621-64-7	N-Nitrosodi-n-propylamine	01/01/87
624-83-9	Methyl isocyanate	01/01/87

272

Appendix B: Regulations Reference Guide

Environmental Protection Agency §372.65

CAS No.	Chemical name	Effective date
636-21-5	o-Toluidine hydrochloride	01/01/87
680-31-9	Hexamethylphosphoramide	01/01/87
684-93-5	N-Nitroso-N-methylurea	01/01/87
759-73-9	N-Nitroso-N-ethylurea	01/01/87
842-07-9	C.I. Solvent Yellow 14	01/01/87
924-16-3	N-Nitrosodi-n-butylamine	01/01/87
961-11-5	Tetrachlorvinphos [Phosphoric acid, 2-chloro-1-(2,4,5-trichlorophenyl)ethenyl dimethyl ester]	01/01/87
989-38-8	C.I. Basic Red I	01/01/87
1120-71-4	Propane sultone	01/01/87
1163-19-5	Decabromodiphenyl oxide	01/01/87
1310-73-2	Sodium hydroxide (solution)	01/01/87
1313-27-5	Molybdenum trioxide	01/01/87
1314-20-1	Thorium dioxide	01/01/87
1319-77-3	Cresol (mixed isomers)	01/01/87
1330-20-7	Xylene (mixed isomers)	01/01/87
1332-21-4	Asbestos (friable)	01/01/87
1335-87-1	Hexachloronaphthalene	01/01/87
1336-36-3	Polychlorinated biphenyls (PCBs)	01/01/87
1344-28-1	Aluminum oxide	01/01/87
1464-53-5	Diepoxybutane	01/01/87
1582-09-8	Trifluralin [Benzeneamine, 2,6-dinitro-N,N-dipropyl-4-(trifluoromethyl)-]	01/01/87
1634-04-4	Methyl tert-butyl ether	01/01/87
1836-75-5	Nitrofen [Benzene, 2,4-dichloro-1-(4-nitrophenoxy)-]	01/01/87
1897-45-6	Chlorothalonil [1-3-Benzenedicarbonitrile,2,4,5,6-tetrachloro-]	01/01/87
1937-37-7	C.I. Direct Black 38	01/01/87
2164-17-2	Fluometuron [Urea, N,N-dimethyl-N'-[3-(trifluoromethyl)phenyl]-]	01/01/87
2234-13-1	Octachloronaphthalene	01/01/87
2303-16-4	Diallate [Carbamothioic acid, bis(1-methylethyl)-, S-(2,3-dichloro-2-propenyl)ester]	01/01/87
2602-46-2	C.I. Direct Blue 6	01/01/87
2832-40-8	C.I. Disperse Yellow 3	01/01/87
3118-97-6	C.I. Solvent Orange 7	01/01/87
3761-53-3	C.I. Food Red 5	01/01/87
4549-40-0	N-Nitrosomethylvinylamine	01/01/87
4680-78-8	C.I. Acid Green 3	01/01/87
6484-52-2	Ammonium nitrate (solution)	01/01/87
7429-90-5	Aluminum (fume or dust)	01/01/87
7439-92-1	Lead	01/01/87
7439-96-5	Manganese	01/01/87
7439-97-6	Mercury	01/01/87
7440-02-0	Nickel	01/01/87
7440-22-4	Silver	01/01/87
7440-28-0	Thallium	01/01/87
7440-36-0	Antimony	01/01/87
7440-38-2	Arsenic	01/01/87
7440-39-3	Barium	01/01/87
7440-41-7	Beryllium	01/01/87
7440-43-9	Cadmium	01/01/87
7440-47-3	Chromium	01/01/87
7440-48-4	Cobalt	01/01/87
7440-50-8	Copper	01/01/87
7440-62-2	Vanadium (fume or dust)	01/01/87
7440-66-6	Zinc (fume or dust)	01/01/87
7550-45-0	Titanium tetrachloride	01/01/87
7647-01-0	Hydrochloric acid	01/01/87
7664-38-2	Phosphoric acid	01/01/87
7664-39-3	Hydrogen fluoride	01/01/87
7664-41-7	Ammonia	01/01/87
7664-93-9	Sulfuric acid	01/01/87
7697-37-2	Nitric acid	01/01/87
7723-14-0	Phosphorus (yellow or white)	01/01/87
7782-49-2	Selenium	01/01/87
7782-50-5	Chlorine	01/01/87
7783-20-2	Ammonium sulfate (solution)	01/01/87
8001-35-2	Toxaphene	01/01/87
10034-93-2	Hydrazine sulfate	01/01/87
10049-04-4	Chlorine dioxide	01/01/87
12122-67-7	Zineb [Carbamodithioic acid, 1,2-ethanediylbis-, zinc complex]	01/01/87
12427-38-2	Maneb [Carbamodithioic acid, 1,2-ethanediylbis-, manganese complex]	01/01/87
16071-86-6	C.I. Direct Brown 95	01/01/87
16543-55-8	N-Nitrosonornicotine	01/01/87
20816-12-0	Osmium tetroxide	01/01/87
25321-22-6	Dichlorobenzene (mixed isomers)	01/01/87

273

Chemical Hazard Communication Guidebook

§ 372.65　　　　　　　　　　　　　　　　　　　　　40 CFR Ch. I (7-1-89 Edition)

CAS No.	Chemical name	Effective date
25376-45-8	Diaminotoluene (mixed isomers)	01/01/87
39156-41-7	2,4-Diaminoanisole sulfate	01/01/87

(c) *Chemical categories in alphabetical order.*

Category name	Effective date.
Antimony Compounds: Includes any unique chemical substance that contains antimony as part of that chemical's infrastructure	01/01/87
Arsenic Compounds: Includes any unique chemical substance that contains arsenic as part of that chemical's infrastructure	01/01/87
Barium Compounds: Includes any unique chemical substance that contains barium as part of that chemical's infrastructure	01/01/87
Beryllium Compounds: Includes any unique chemical substance that contains beryllium as part of that chemical's infrastructure	01/01/87
Cadmium Compounds: Includes any unique chemical substance that contains cadmium as part of that chemical's infrastructure	01/01/87
Chlorophenols	01/01/87

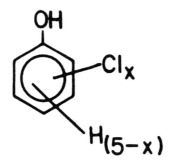

Where x = 1 to 5

Category name	Effective date.
Chromium Compounds: Includes any unique chemical substance that contains chromium as part of that chemical's infrastructure	01/01/87
Cobalt Compounds: Includes any unique chemical substance that contains cobalt as part of that chemical's infrastructure	01/01/87
Copper Compounds: Includes any unique chemical substance that contains copper as part of that chemical's infrastructure	01/01/87
Cyanide Compounds: $X^+ CN^-$ where $X = H^-$ or any other group where a formal dissociation can be made. For example KCN, or $Ca(CN)_2$	01/01/87
Glycol Ethers: Includes mono- and di- ethers of ethylene glycol, diethylene glycol, and triethylene glycol	01/01/87

$$R-(OCH_2CH_2)n^{-OR'}$$

Where:
　n = 1, 2, or 3
　R = alkyl or aryl groups.
　R = R' H, or groups which, when removed, yield glycol ethers with the structure:

274

Appendix B: Regulations Reference Guide

Environmental Protection Agency § 372.65

Category name	Effective date.
R—(OCH$_2$CH)$_n$—OH Polymers are excluded from this category.	
Lead Compounds: Includes any unique chemical substance that contains lead as part of that chemical's infrastructure...	01/01/87
Manganese Compounds: Includes any unique chemical substance that contains manganese as part of that chemical's infrastructure...	01/01/87
Mercury Compounds: Includes any unique chemical substance that contains mercury as part of that chemical's infrastructure...	01/01/87
Nickel Compounds: Includes any unique chemical substance that contains nickel as part of that chemical's infrastructure...	01/01/87
Polybrominated Biphenyls (PBBs)	01/01/87

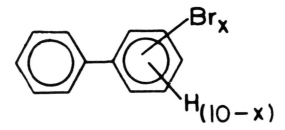

Where x = 1 to 10

Category name	Effective date.
Selenium Compounds: Includes any unique chemical substance that contains selenium as part of that chemical's infrastructure...	01/01/87
Silver Compounds: Includes any unique chemical substance that contains silver as part of that chemical's infrastructure...	01/01/87
Thallium Compounds: Includes any unique chemical substance that contains thallium as part of that chemical's infrastructure...	01/01/87
Zinc Compounds: Includes any unique chemical substance that contains zinc as part of that chemical's infrastructure...	01/01/87

[53 FR 4525, Feb. 16, 1988; 53 FR 12748, Apr. 18, 1988, as amended at 53 FR 23112, June 20, 1988; 53 FR 39475, Oct. 8, 1988; 54 FR 12913, Mar. 29, 1989; 54 FR 25851, June 20, 1989]

275

United States
Environmental Protection
Agency

Office of Toxic
Substances
Washington, D.C. 20460

January 1990
EPA 560/4-90-007

&EPA # Toxic Chemical Release Inventory Reporting Form R and Instructions

Revised 1989 Version

Section 313
of the Emergency Planning and Community Right-to-Know Act
(Title III of the Superfund Amendments and Reauthorization Act of 1986)

Chemical Hazard Communication Guidebook

Form Approved OMB No.: 2070-0093
Approval Expires: 01/91

(Important: Type or print; read instructions before completing form.)

Page 1 of 5

♦EPA U.S. Environmental Protection Agency

TOXIC CHEMICAL RELEASE INVENTORY REPORTING FORM
Section 313 of the Emergency Planning and Community Right-to-Know Act of 1986, also known as Title III of the Superfund Amendments and Reauthorization Act

EPA FORM R

PART I. FACILITY IDENTIFICATION INFORMATION

(This space for your optional use.)

Public reporting burden for this collection of information is estimated to vary from 30 to 34 hours per response, with an average of 32 hours per response, including time for reviewing instructions, searching existing data sources, gathering and maintaining the data needed, and completing and reviewing the collection of information. Send comments regarding this burden estimate or any other aspect of this collection of information, including suggestions for reducing this burden, to Chief, Information Policy Branch (PM-2231), US EPA, 401 M St., SW, Washington, D.C. 20460 Attn: TRI Burden and to the Office of Information and Regulatory Affairs, Office of Management and Budget Paperwork Reduction Project (2070-0093), Washington, D.C. 20603.

1.
- 1.1 Are you claiming the chemical identity on page 3 trade secret?
  - [ ] Yes (Answer question 1.2; Attach substantiation forms.)
  - [ ] No (Do not answer 1.2; Go to question 1.3.)
- 1.2 If "Yes" in 1.1, is this copy:
  - [ ] Sanitized [ ] Unsanitized
- 1.3 Reporting Year 19____

2. CERTIFICATION (Read and sign after completing all sections.)
I hereby certify that I have reviewed the attached documents and that, to the best of my knowledge and belief, the submitted information is true and complete and that the amounts and values in this report are accurate based on reasonable estimates using data available to the preparers of this report.

Name and official title of owner/operator or senior management official

Signature   Date signed

3. FACILITY IDENTIFICATION

3.1
- Facility or Establishment Name
- Street Address
- City | County
- State | Zip Code
- TRI Facility Identification Number

WHERE TO SEND COMPLETED FORMS:

1. EPCRA REPORTING CENTER
   P.O. BOX 23779
   WASHINGTON, DC 20026-3779
   ATTN: TOXIC CHEMICAL RELEASE INVENTORY

2. APPROPRIATE STATE OFFICE (See instructions in Appendix G)

3.2 This report contains information for (Check only one):
   a. [ ] An entire facility   b. [ ] Part of a facility.

3.3 Technical Contact | Telephone Number (include area code)

3.4 Public Contact | Telephone Number (include area code)

3.5 SIC Code (4 digit)
   a.   b.   c.   d.   e.   f.

3.6 Latitude: Degrees | Minutes | Seconds   Longitude: Degrees | Minutes | Seconds

3.7 Dun & Bradstreet Number(s)
   a.   b.

3.8 EPA Identification Number(s) (RCRA I.D. No.)
   a.   b.

3.9 NPDES Permit Number(s)
   a.   b.

3.10 Receiving Streams or Water Bodies (enter one name per box)
   a.   b.
   c.   d.
   e.   f.

3.11 Underground Injection Well Code (UIC) Identification Number(s)
   a.   b.

4. PARENT COMPANY INFORMATION

4.1 Name of Parent Company
4.2 Parent Company's Dun & Bradstreet Number

EPA Form 9350-1 (1-90) Revised – Do not use previous versions.

Appendix B: Regulations Reference Guide

(Important: Type or print; read instructions before completing form.)

Page 2 of 5

⊕EPA   **EPA FORM R**
PART II. OFF-SITE LOCATIONS TO WHICH TOXIC CHEMICALS ARE TRANSFERRED IN WASTES

(This space for your optional use.)

1. PUBLICLY OWNED TREATMENT WORKS (POTWs)

1.1 POTW name	1.2 POTW name
Street Address	Street Address
City / County	City / County
State / Zip	State / Zip

2. OTHER OFF-SITE LOCATIONS (DO NOT REPORT LOCATIONS TO WHICH WASTES ARE SENT ONLY FOR RECYCLING OR REUSE).

2.1 Off-site location name	2.2 Off-site location name
EPA Identification Number (RCRA ID. No.)	EPA Identification Number (RCRA ID. No.)
Street Address	Street Address
City / County	City / County
State / Zip	State / Zip
Is location under control of reporting facility or parent company? [ ] Yes [ ] No	Is location under control of reporting facility or parent company? [ ] Yes [ ] No

2.3 Off-site location name	2.4 Off-site location name
EPA Identification Number (RCRA ID. No.)	EPA Identification Number (RCRA ID. No.)
Street Address	Street Address
City / County	City / County
State / Zip	State / Zip
Is location under control of reporting facility or parent company? [ ] Yes [ ] No	Is location under control of reporting facility or parent company? [ ] Yes [ ] No

2.5 Off-site location name	2.6 Off-site location name
EPA Identification Number (RCRA ID. No.)	EPA Identification Number (RCRA ID. No.)
Street Address	Street Address
City / County	City / County
State / Zip	State / Zip
Is location under control of reporting facility or parent company? [ ] Yes [ ] No	Is location under control of reporting facility or parent company? [ ] Yes [ ] No

[ ] Check if additional pages of Part II are attached. How many? _____

EPA Form 9350-1 (1-90) Revised - Do not use previous versions.

Chemical Hazard Communication Guidebook

(Important: Type or print; read instructions before completing form.) Page 3 of 5

⊕EPA
EPA FORM R
PART III. CHEMICAL-SPECIFIC INFORMATION

(This space for your optional use.)

1. CHEMICAL IDENTITY (Do not complete this section if you complete Section 2.)

1.1	[Reserved]
1.2	CAS Number (Enter only one number exactly as it appears on the 313 list. Enter NA if reporting a chemical category.)
1.3	Chemical or Chemical Category Name (Enter only one name exactly as it appears on the 313 list.)
1.4	Generic Chemical Name (Complete only if Part I, Section 1.1 is checked "Yes." Generic name must be structurally descriptive.)

2. MIXTURE COMPONENT IDENTITY (Do not complete this section if you complete Section 1.)

Generic Chemical Name Provided by Supplier (Limit the name to a maximum of 70 characters (e.g., numbers, letters, spaces, punctuation).)

3. ACTIVITIES AND USES OF THE CHEMICAL AT THE FACILITY (Check all that apply.)

3.1	Manufacture the chemical: a.[ ] Produce b.[ ] Import	If produce or import: c.[ ] For on-site use/processing e.[ ] As a byproduct	d.[ ] For sale/distribution f.[ ] As an impurity	
3.2	Process the chemical: a.[ ] As a reactant d.[ ] Repackaging only	b.[ ] As a formulation component	c.[ ] As an article component	
3.3	Otherwise use the chemical: a.[ ] As a chemical processing aid	b.[ ] As a manufacturing aid	c.[ ] Ancillary or other use	

4. MAXIMUM AMOUNT OF THE CHEMICAL ON-SITE AT ANY TIME DURING THE CALENDAR YEAR

[ ] (enter code)

5. RELEASES OF THE CHEMICAL TO THE ENVIRONMENT ON-SITE

You may report releases of less than 1,000 pounds by checking ranges under A.1. (Do not use both A.1 and A.2)

	A. Total Release (pounds/year)		B. Basis of Estimate	C. % From Stormwater
	A.1 Reporting Ranges 0 1-499 500-999	A.2 Enter Estimate	(enter code)	
5.1 Fugitive or non-point air emissions	5.1a [ ] [ ] [ ]		5.1b [ ]	
5.2 Stack or point air emissions	5.2a [ ] [ ] [ ]		5.2b [ ]	
5.3 Discharges to receiving streams or water bodies 5.3.1 [ ]	5.3.1a [ ] [ ] [ ]		5.3.1b [ ]	5.3.1c %
(Enter letter code for stream from Part I Section 3.10 in the box provided.) 5.3.2 [ ]	5.3.2a [ ] [ ] [ ]		5.3.2b [ ]	5.3.2c %
5.3.3 [ ]	5.3.3a [ ] [ ] [ ]		5.3.3b [ ]	5.3.3c %
5.4 Underground Injection on-site	5.4a [ ] [ ] [ ]		5.4b [ ]	
5.5 Releases to land on-site 5.5.1 Landfill	5.5.1a [ ] [ ] [ ]		5.5.1b [ ]	
5.5.2 Land treatment/application farming	5.5.2a [ ] [ ] [ ]		5.5.2b [ ]	
5.5.3 Surface impoundment	5.5.3a [ ] [ ] [ ]		5.5.3b [ ]	
5.5.4 Other disposal	5.5.4a [ ] [ ] [ ]		5.5.4b [ ]	

[ ] (Check if additional information is provided on Part IV-Supplemental Information.)

EPA Form 9350-1 (1-90) Revised - Do not use previous versions.

Appendix B: Regulations Reference Guide

(Important: Type or print; read instructions before completing form.)

Page 4 of 5

♻ EPA — EPA FORM **R**
PART III. CHEMICAL-SPECIFIC INFORMATION
(continued)

(This space for your optional use.)

6. TRANSFERS OF THE CHEMICAL IN WASTE TO OFF-SITE LOCATIONS

You may report transfers of less than 1,000 pounds by checking ranges under A.1. (Do not use both A.1 and A.2)

	A. Total Transfers (pounds/year)		B. Basis of Estimate	C. Type of Treatment/ Disposal
	A.1 Reporting Ranges (0, 1–499, 500–999)	A.2 Enter Estimate	(enter code)	(enter code)
6.1.1 Discharge to POTW (enter location number from Part II, Section 1.) [1] [ ] [ ] [ ]			6.1.1b [ ]	■■■
6.2.1 Other off-site location (enter location number from Part II, Section 2.) [2] [ ] [ ] [ ]			6.2.1b [ ]	6.2.1c [M][ ][ ]
6.2.2 Other off-site location (enter location number from Part II, Section 2.) [2] [ ] [ ] [ ]			6.2.2b [ ]	6.2.2c [M][ ][ ]
6.2.3 Other off-site location (enter location number from Part II, Section 2.) [2] [ ] [ ] [ ]			6.2.3b [ ]	6.2.3c [M][ ][ ]

[ ] (Check if additional information is provided on Part IV-Supplemental Information.)

7. WASTE TREATMENT METHODS AND EFFICIENCY

[ ] Not Applicable (NA) – Check if no on-site treatment is applied to any wastestream containing the chemical or chemical category.

A. General Wastestream (enter code)	B. Treatment Method (enter code)	C. Range of Influent Concentration (enter code)	D. Sequential Treatment? (check if applicable)	E. Treatment Efficiency Estimate	F. Based on Operating Data? Yes No
7.1a [ ]	7.1b [ ][ ][ ]	7.1c [ ]	7.1d [ ]	7.1e ___%	7.1f [ ] [ ]
7.2a [ ]	7.2b [ ][ ][ ]	7.2c [ ]	7.2d [ ]	7.2e ___%	7.2f [ ] [ ]
7.3a [ ]	7.3b [ ][ ][ ]	7.3c [ ]	7.3d [ ]	7.3e ___%	7.3f [ ] [ ]
7.4a [ ]	7.4b [ ][ ][ ]	7.4c [ ]	7.4d [ ]	7.4e ___%	7.4f [ ] [ ]
7.5a [ ]	7.5b [ ][ ][ ]	7.5c [ ]	7.5d [ ]	7.5e ___%	7.5f [ ] [ ]
7.6a [ ]	7.6b [ ][ ][ ]	7.6c [ ]	7.6d [ ]	7.6e ___%	7.6f [ ] [ ]
7.7a [ ]	7.7b [ ][ ][ ]	7.7c [ ]	7.7d [ ]	7.7e ___%	7.7f [ ] [ ]
7.8a [ ]	7.8b [ ][ ][ ]	7.8c [ ]	7.8d [ ]	7.8e ___%	7.8f [ ] [ ]
7.9a [ ]	7.9b [ ][ ][ ]	7.9c [ ]	7.9d [ ]	7.9e ___%	7.9f [ ] [ ]
7.10a [ ]	7.10b [ ][ ][ ]	7.10c [ ]	7.10d [ ]	7.10e ___%	7.10f [ ] [ ]

[ ] (Check if additional information is provided on Part IV-Supplemental Information.)

8. POLLUTION PREVENTION: OPTIONAL INFORMATION ON WASTE MINIMIZATION

(Indicate actions taken to reduce the amount of the chemical being released from the facility. See the instructions for coded items and an explanation of what information to include.)

A. Type of Modification (enter code)	B. Quantity of the Chemical in Wastes Prior to Treatment or Disposal			C. Index	D. Reason for Action (enter code)
	Current reporting year (pounds/year)	Prior year (pounds/year)	Or percent change (Check (+) or (–)) [ ] + [ ] – ___%		
[M]	_____	_____		[ ].[ ]	[R]

EPA Form 9350-1 (1-90) Revised – Do not use previous versions.

Chemical Hazard Communication Guidebook

(Important: Type or print; read instructions before completing form.)   Page 5 of 5

♻ EPA   EPA FORM **R**   PART IV. SUPPLEMENTAL INFORMATION
Use this section if you need additional space for answers to questions in Part III.
Number the lines used sequentially from lines in prior sections (e.g., 5.3.4, 6.1.2, 7.11).

(This space for your optional use.)

ADDITIONAL INFORMATION ON RELEASES OF THE CHEMICAL TO THE ENVIRONMENT ON-SITE (Part III, Section 5.3)

You may report releases of less than 1,000 pounds by checking ranges under A.1. (Do not use both A.1 and A.2)	A. Total Release (pounds/year)		B. Basis of Estimate (enter code in box provided)	C. % From Stormwater
	A.1 Reporting Ranges 0 1–499 500–999	A.2 Enter Estimate		
5.3 Discharges to receiving streams or water bodies 5.3.___ ☐	5.3.___a [ ] [ ] [ ]		5.3.___b ☐	5.3.___c %
(Enter letter code for stream from Part I Section 3.10 in the box provided.) 5.3.___ ☐	5.3.___a [ ] [ ] [ ]		5.3.___b ☐	5.3.___c %
5.3.___ ☐	5.3.___a [ ] [ ] [ ]		5.3.___b ☐	5.3.___c %

ADDITIONAL INFORMATION ON TRANSFERS OF THE CHEMICAL IN WASTE TO OFF-SITE LOCATIONS (Part III, Section 6)

You may report transfers of less than 1,000 pounds by checking ranges under A.1. (Do not use both A.1 and A.2)	A. Total Transfers (pounds/year)		B. Basis of Estimate (enter code in box provided)	C. Type of Treatment/ Disposal (enter code in box provided)
	A.1 Reporting Ranges 0 1–499 500–999	A.2 Enter Estimate		
6.1.___ Discharge to POTW (enter location number from Part II, Section 1.) **1**.☐	[ ] [ ] [ ]		6.1.___b ☐	▓▓▓
6.2.___ Other off-site location (enter location number from Part II, Section 2.) **2**.☐	[ ] [ ] [ ]		6.2.___b ☐	6.2.___c M☐
6.2.___ Other off-site location (enter location number from Part II, Section 2.) **2**.☐	[ ] [ ] [ ]		6.2.___b ☐	6.2.___c M☐
6.2.___ Other off-site location (enter location number from Part II, Section 2.) **2**.☐	[ ] [ ] [ ]		6.2.___b ☐	6.2.___c M☐

ADDITIONAL INFORMATION ON WASTE TREATMENT METHODS AND EFFICIENCY (Part III, Section 7)

A. General Wastestream (enter code in box provided)	B. Treatment Method (enter code in box provided)	C. Range of Influent Concentration (enter code)	D. Sequential Treatment? (check if applicable)	E. Treatment Efficiency Estimate	F. Based on Operating Data? Yes No
7.___a ☐	7.___b ☐☐☐	7.___c ☐	7.___d [ ]	7.___e %	7.___f [ ] [ ]
7.___a ☐	7.___b ☐☐☐	7.___c ☐	7.___d [ ]	7.___e %	7.___f [ ] [ ]
7.___a ☐	7.___b ☐☐☐	7.___c ☐	7.___d [ ]	7.___e %	7.___f [ ] [ ]
7.___a ☐	7.___b ☐☐☐	7.___c ☐	7.___d [ ]	7.___e %	7.___f [ ] [ ]
7.___a ☐	7.___b ☐☐☐	7.___c ☐	7.___d [ ]	7.___e %	7.___f [ ] [ ]
7.___a ☐	7.___b ☐☐☐	7.___c ☐	7.___d [ ]	7.___e %	7.___f [ ] [ ]
7.___a ☐	7.___b ☐☐☐	7.___c ☐	7.___d [ ]	7.___e %	7.___f [ ] [ ]
7.___a ☐	7.___b ☐☐☐	7.___c ☐	7.___d [ ]	7.___e %	7.___f [ ] [ ]
7.___a ☐	7.___b ☐☐☐	7.___c ☐	7.___d [ ]	7.___e %	7.___f [ ] [ ]

EPA Form 9350-1 (1-90) Revised - Do not use previous versions.

Appendix B: Regulations Reference Guide

APPENDIX B

REPORTING CODES FOR EPA FORM R

Part III, Section 4 - Maximum Amount of the Chemical On-Site at Any Time During the Calendar Year

Weight Range in Pounds

Range Code	From...	To....
01	0	99
02	100	999
03	1,000	9,999
04	10,000	99,999
05	100,000	999,999
06	1,000,000	9,999,999
07	10,000,000	49,999,999
08	50,000,000	99,999,999
09	100,000,000	499,999,999
10	500,000,000	999,999,999
11	1 billion	more than 1 billion

Part III, Section 5 - Releases of the Chemical to the Environment On-Site and Section 6 - Transfers of the Chemical in Waste to Off-Site Locations

M - Estimate is based on monitoring data or measurements for the toxic chemical as released to the environment and/or off-site facility.

C - Estimate is based on mass balance calculations, such as calculation of the amount of the toxic chemical in streams entering and leaving process equipment.

E - Estimate is based on published emission factors, such as those relating release quantity to through-put or equipment type (e.g., air emission factors).

O - Estimate is based on other approaches such as engineering calculations (e.g., estimating volatilization using published mathematical formulas) or best engineering judgment. This would include applying an estimated removal efficiency to a wastestream, even if the composition of the stream before treatment was fully characterized by monitoring data.

Part III, Section 6 - Transfers of the Chemical in Waste to Off-Site Locations

Type of Treatment/Disposal

M10	Storage Only
M40	Solidification/Stabilization
M50	Incineration/Thermal Treatment
M61	Wastewater Treatment (Excluding POTW)
M69	Other Treatment
M71	Underground Injection
M72	Landfill/Disposal Surface Impoundment
M73	Land Treatment
M79	Other Land Disposal
M90	Other Off-Site Management
M91	Transfer to Waste Broker
M99	Unknown

Part III, Section 7 - Waste Treatment Methods and Efficiency

General Wastestream

A = Gaseous (gases, vapors, airborne particulates)
W = Wastewater (aqueous waste)
L = Liquid waste (non-aqueous waste)
S = Solid waste (including sludges and slurries)

Part III, Section 7 - Waste Treatment Methods and Efficiency

Air Emissions Treatment

A01	Flare
A02	Condenser
A03	Scrubber
A04	Absorber
A05	Electrostatic Precipitator
A06	Mechanical Separation
A07	Other Air Emission Treatment

Biological Treatment

B11	Biological Treatment -- Aerobic
B21	Biological Treatment -- Anaerobic
B31	Biological Treatment -- Facultative
B99	Biological Treatment -- Other

Chemical Hazard Communication Guidebook

Chemical Treatment

- C01 Chemical Precipitation -- Lime or Sodium Hydroxide
- C02 Chemical Precipitation -- Sulfide
- C09 Chemical Precipitation -- Other
- C11 Neutralization
- C21 Chromium Reduction
- C31 Complexed Metals Treatment (other than pH Adjustment)
- C41 Cyanide Oxidation -- Alkaline Chlorination
- C42 Cyanide Oxidation -- Electrochemical
- C43 Cyanide Oxidation -- Other
- C44 General Oxidation (including Disinfection) -- Chlorination
- C45 General Oxidation (including Disinfection) -- Ozonation
- C46 General Oxidation (including Disinfection) -- Other
- C99 Other Chemical Treatment

Incineration/Thermal Treatment

- F01 Liquid Injection
- F11 Rotary Kiln with Liquid Injection Unit
- F19 Other Rotary Kiln
- F31 Two Stage
- F41 Fixed Hearth
- F42 Multiple Hearth
- F51 Fluidized Bed
- F61 Infra-Red
- F71 Fume/Vapor
- F81 Pyrolytic Destructor
- F82 Wet Air Oxidation
- F83 Thermal Drying/Dewatering
- F99 Other Incineration/Thermal Treatment

Physical Treatment

- P01 Equalization
- P09 Other Blending
- P11 Settling/Clarification
- P12 Filtration
- P13 Sludge Dewatering (non-thermal)
- P14 Air Flotation
- P15 Oil Skimming
- P16 Emulsion Breaking -- Thermal
- P17 Emulsion Breaking -- Chemical
- P18 Emulsion Breaking -- Other
- P19 Other Liquid Phase Separation
- P21 Adsorption -- Carbon
- P22 Adsorption -- Ion Exchange (other than for recovery/reuse)
- P23 Adsorption -- Resin
- P29 Adsorption -- Other
- P31 Reverse Osmosis (other than for recovery/reuse)
- P41 Stripping -- Air
- P42 Stripping -- Steam
- P49 Stripping -- Other
- P51 Acid Leaching (other than for recovery/reuse)
- P61 Solvent Extraction (other than recovery/reuse)
- P99 Other Physical Treatment

Recovery/Reuse

- R01 Reuse as Fuel -- Industrial Kiln
- R02 Reuse as Fuel -- Industrial Furnace
- R03 Reuse as Fuel -- Boiler
- R04 Reuse as Fuel -- Fuel Blending
- R09 Reuse as Fuel -- Other
- R11 Solvents/Organics Recovery -- Batch Still Distillation
- R12 Solvents/Organics Recovery -- Thin-Film Evaporation
- R13 Solvents/Organics Recovery -- Fractionation
- R14 Solvents/Organics Recovery -- Solvent Extraction
- R19 Solvents/Organics Recovery -- Other
- R21 Metals Recovery -- Electrolytic
- R22 Metals Recovery -- Ion Exchange
- R23 Metals Recovery -- Acid Leaching
- R24 Metals Recovery -- Reverse Osmosis
- R26 Metals Recovery -- Solvent Extraction
- R29 Metals Recovery -- Other
- R99 Other Reuse or Recovery

Solidification/Stabilization

- G01 Cement Processes (including Silicates)
- G09 Other Pozzolonic Processes (including Silicates)
- G11 Asphaltic Processes
- G21 Thermoplastic Techniques
- G99 Other Solidification Processes

Part III, Section 7 - Waste Treatment Methods and Efficiency

Range of Influent Concentration

- 1 = Greater than 1 percent
- 2 = 100 parts per million (0.01 percent) to 1 percent (10,000 parts per million)
- 3 = 1 part per million to 100 parts per million
- 4 = 1 part per billion to 1 part per million
- 5 = Less than 1 part per billion

[Note: Parts per million (ppm) is milligrams/kilogram (mass/mass) for solids and liquids; cubic centimeters/cubic meter (volume/volume) for gases; milligrams/liter for solutions or dispersions of the chemical in water; and milligrams of chemical/kilogram of air for particulates in air. If you have particulate concentrations (at standard temperature and pressure) as grains/cubic foot of air, multiply by 1766.6 to convert to parts per million; if in milligrams/cubic meters, multiply by 0.773 to obtain parts per million. Factors are for standard conditions of 0°C (32°F) and 760 mmHg atmospheric pressure.]

Appendix B: Regulations Reference Guide

Part III, Section 8 - Optional Information on Waste Minimization

<u>Type of Modification</u>

- M1 - Recycling/Reuse On-Site
- M2 - Recycling/Reuse Off-Site
- M3 - Equipment/Technology Modifications
- M4 - Process Procedure Modifications
- M5 - Reformulation/Redesign of Product
- M6 - Substitution of Raw Materials
- M7 - Improved Housekeeping, Training, Inventory Control
- M8 - Other Waste Minimization Technique

<u>Reason for Action</u>

- R1 - Regulatory Requirement for the Waste
- R2 - Reduction of Treatment/Disposal Costs
- R3 - Other Process Cost Reduction
- R4 - Self-Initiated Review
- R5 - Other (e.g., discontinuation of product, occupational safety, etc.)

Chemical Hazard Communication Guidebook

C. INSTRUCTIONS FOR COMPLETING EPA FORM R

The following are specific instructions for completing each part of EPA Form R. The number designations of the parts and sections of these instructions correspond to those in Form R unless otherwise indicated.

A sample of a completed Form R for a hypothetical facility reporting under Title III, section 313, is included as Appendix C. You may want to refer to this sample as you read through these instructions.

Instructions for Completing All Parts of Form R:

1. Type or print information on the form in the units and format requested.

2. All information on Form R is required except Part III, Section 8.

3. Do not leave items on Form R blank unless specifically directed to do so; if an item does not apply to you, enter "NA," not applicable, in the space provided. If your information does not fill all the spaces provided for a type of information, enter NA, in the next blank space in the sequence.

4. Report releases and off-site transfers to the nearest pound. Do not report fractions of pounds.

5. Do not submit an incomplete form. The certification statement (Part I) specifies that the report is complete as submitted. See page 1 of these instructions for the definition of a complete submission.

6. When completing Part IV, supplemental information, or additional pages for Part II of the form, number the additional information sequentially from the prior sections of the form.

7. The box labelled "This space for your optional use" on each page may be used to differentiate one chemical-specific submission from another. This box is used to identify a voluntary revision of a previous submission (see page 2).

PART I. FACILITY IDENTIFICATION INFORMATION

1.1 Are you claiming the chemical identity on page 3 trade secret?

Answer this question only after you have completed the rest of the report. The specific identity of the toxic chemical being reported in Part III, Sections 1.2 and 1.3, may be designated as trade secret. If you are making a trade secret claim, mark "yes" and proceed to Section 1.2. Only check "Yes" if it is your manufacturing, processing, or use of the chemical that is a trade secret. (See page 1 of these instructions for specific information on trade secrecy claims.) If you checked "no," proceed to Section 1.3; do not answer Section 1.2.

1.2 If "yes" in 1.1, is this copy sanitized or unsanitized?

Answer this question only after you have completed the rest of the report. Check "sanitized" if this copy of the report is the public version and you have claimed the chemical identity trade secret in Part I, Section 1.1. Otherwise, check "unsanitized."

1.3 Reporting Year

Enter the last two digits of the calendar year to which the reported information applies, not the year in which you are submitting the report. Information for the 1989 reporting year must be submitted on or before July 1, 1990.

2. Certification

The certification statement must be signed by the owner or operator or a senior official with management responsibility for the person (or persons) completing the form. The owner, operator, or official must certify the accuracy and completeness of the information reported on the form by signing and dating the certification statement. Each report must contain an original signature. Print or type in the space provided the name and title of the person who signs the statement. This certification statement applies to all the information supplied on the form and should be signed only after the form has been completed.

Appendix B: Regulations Reference Guide

Form R - Part I   Page 15

3.1 Facility Name and Location

Enter the name of your facility (plant site name or appropriate facility designation), street address, city, county, state, zip code, and TRI Facility Identification number (if appropriate), in the space provided. Do not use a post office box number as the address. The address provided should be the location where the chemicals are manufactured, processed, or otherwise used.

If you have submitted a Form R for previous reporting years, a TRI Facility Identification Number has been assigned to your facility. The TRI Facility Identification Number appears on the peel-off mailing label on the cover the Toxic Chemical Release Inventory Reporting Package for 1989 (EPA 560/4-90-001) you should have received directly from EPA. Remove this mailing label from the back cover of the reporting package and apply it to Part I, Section 3.1 of the blank Form R in Appendix A. Then photocopy that page for use as the master copy of page 1 for all the reports you are submitting.

If you do not have a mailing label or cannot locate your TRI Facility Identification Number, please contact the Emergency Planning and Community Right-to-Know Information Hotline. Enter your TRI Facility Identification number to each Form R that your facility submits.

Enter NA to the space for the TRI Facility Identification number, if this is your first submission of a Form R.

3.2 Full or Partial Facility Indication

A covered facility must report all releases of a listed chemical if it meets a reporting threshold for that chemical. However, if the facility is composed of several distinct establishments, EPA allows these establishments to submit separate reports for the chemical as long as all releases of the chemical from the entire facility are accounted for. Indicate in Section 3.2 whether your report is for the entire covered facility as a whole or for part of a covered facility. Check box a. if the chemical information applies to the entire covered facility. Check box b. if the chemical information applies only to part of a covered facility.

Section 313 requires reports by "facilities," which are defined as "all buildings, equipment, structures, and other stationary items which are located on a single site or on contiguous or adjacent sites and which are owned or operated by the same person."

The SIC code system defines business "establishments" as "distinct and separate economic activities [that] are performed at a single physical location." Under section 372.30(c) of the reporting rule, you may submit a separate Form R for each establishment, or for groups of establishments, in your covered facility, provided that all releases of the toxic chemicals from the entire covered facility are reported. This allows you the option of reporting separately on the activities involving a toxic chemical at each establishment, or group of establishments (e.g., part of a covered facility), rather than submitting a single Form R for that chemical for the entire facility. However, if an establishment or group of establishments does not manufacture, process, or otherwise use or release a toxic chemical, you do not have to submit a report for that establishment or group of establishments. (See also Section B.2.a on page 5.)

3.3 Technical Contact

Enter the name and telephone number (including area code) of a technical representative whom EPA or State officials may contact for clarification of the information reported on Form R. This contact person does not have to be the same person who prepares the report or signs the certification statement and does not necessarily need to be someone at the location of the reporting facility; however, this person must be familiar with the details of the report so that he or she can answer questions about the information provided.

3.4 Public Contact

Enter the name and telephone number (including area code) of a person who can respond to questions from the public about the report. If you choose to designate the same person as both the technical and the public contact, you may enter "Same as Section 3.3" in this space. This contact person does not have to be the same person who prepares the report or signs the certification statement and does not necessarily need to be someone at the location of the reporting facility.

3.5 Standard Industrial Classification (SIC) Code

Enter the appropriate 4-digit primary Standard Industrial Classification (SIC) code for your facility (Table I, pages 34-39, lists the SIC codes within the 20-39 range). If the report covers more than one establishment, enter the primary 4-digit SIC code for each establishment. You are required to enter SIC codes only for those establishments within the facility that fall within SIC codes 20 to 39. If you do not know your SIC code, check with your financial office or contact your local Chamber of Commerce or State Department of Labor.

3.6 Latitude and Longitude

Enter the latitudinal and longitudinal coordinates of your facility. Sources of these data include EPA permits (e.g., NPDES permits), county property records, facility blueprints, and site plans. Instructions on how to develop these coordinates can be found in Appendix F. Enter only numerical data. <u>Do not</u> preface numbers with letters such as N or W to denote the hemisphere.

Chemical Hazard Communication Guidebook

3.7 Facility Dun and Bradstreet Number

Enter the 9-digit number assigned by Dun and Bradstreet (D&B) for your facility or each establishment within your facility. These numbers code the facility for financial purposes. This number may be available from your facility's treasurer or financial officer. You can also obtain the numbers from your local Dun and Bradstreet office (check the telephone book White Pages). If none of your establishments has been assigned a D & B number, enter not applicable, NA, in box a. If only some of your establishments have been assigned Dun and Bradstreet numbers, enter those numbers in Section 3.7.

3.8 EPA Identification Number

The EPA I.D. Number is a 12-digit number assigned to facilities covered by hazardous waste regulations under the Resource Conservation and Recovery Act (RCRA). Facilities not covered by RCRA are not likely to have an assigned I.D. Number. If your facility is not required to have an I.D. Number, enter not applicable, NA, in box a. If your facility has been assigned EPA Identification Numbers, you must enter those numbers in the spaces provided in Section 3.8.

3.9 NPDES Permit Number

Enter the numbers of any permits your facility holds under the National Pollutant Discharge Elimination System (NPDES) even if the permit(s) do not pertain to the toxic chemical being reported. This 9-digit permit number is assigned to your facility by EPA or the State under the authority of the Clean Water Act. If your facility does not have a permit, enter not applicable, NA, in box a.

3.10 Receiving Streams or Water Bodies

In Section 3.10 you are to enter the name(s) of the stream(s) or water body(ies) to which your facility directly discharges the chemicals you are reporting. A total of six spaces are provided, lettered a through f. The information you provide relates directly to the discharge quantity information required in Part III, Section 5.3. You can complete Section 3.10 in one of two ways. You can enter only those stream names that relate to the specific chemical that is the subject of the report or, you can enter all stream names that relate to all covered chemicals being reported by the facility. Enter the name of each receiving stream or surface water body to which the chemical being reported is directly discharged. Report the name of the receiving stream or water body as it appears on the NPDES permit for the facility. If the stream is not covered by a permit, enter the name of the off-site stream or water body by which it is publicly known. Also do not list a series of streams through which the chemical flows. Be sure to include the receiving stream(s) or water body(ies) that receive stormwater runoff from your facility. Do not enter names of streams to which off-site treatment plants discharge. Enter not applicable, NA, in Section 3.10a. if you do not discharge any listed toxic chemicals to surface water bodies.

3.11 Underground Injection Well Code (UIC) Identification Number

If your facility has a permit to inject a waste containing the toxic chemical into Class 1 deep wells, enter the 12-digit Underground Injection Well Code (UIC) identification number assigned by EPA or by the State under the authority of the Safe Drinking Water Act. If your facility does not hold such a permit(s), enter not applicable, NA, in Section 3.11a. You are only required to provide the UIC number for wells that receive the toxic chemical being reported.

4. Parent Company Information

You must provide information on your parent company. For purposes of Form R, a parent company is defined as the highest level company, located in the United States, that directly owns at least 50 percent of the voting stock of your company. If your facility is owned by a foreign entity, enter not applicable, NA, in this space. Corporate names should be treated as parent company names for companies with multiple facility sites. For example, the Bestchem Corporation is not owned or controlled by any other corporation but has sites throughout the country whose names begin with Bestchem. In this case, Bestchem Corporation would be listed as the "parent" company.

4.1 Name of Parent Company

Enter the name of the corporation or other business entity that is your ultimate US parent company. If your facility has no parent company, enter not applicable, NA.

4.2 Parent Company's Dun & Bradstreet Number

Enter the Dun and Bradstreet Number for your ultimate US parent company, if applicable. The number may be obtained from the treasurer or financial officer of the company. If your parent company does not have a Dun and Bradstreet number, enter not applicable, NA.

Appendix B: Regulations Reference Guide

Form R - Part III   Page 17

PART II. OFF-SITE LOCATIONS TO WHICH TOXIC CHEMICALS ARE TRANSFERRED IN WASTES

In this part of the form, you are required to list all off-site locations to which you transfer wastes containing toxic chemicals. Do not list locations to which products containing toxic chemicals are shipped for sale or distribution in commerce or for further use. Also, do not list locations to which wastes containing chemicals are sold or sent for recovery, recycling, or reuse of the toxic chemicals. The information that you enter in this section relates to data you will report in Part III, Section 6.

You may complete Part II for only the off-site locations that apply to the specific chemical cited in a particular report or you can list all off-site locations that apply to all chemicals being reported and include a photostatic copy of Part II with each individual report. List only publicly owned treatment works (POTWs) and off-site treatment or disposal facilities.

1. Publicly Owned Treatment Works (POTWs)

Enter the name and address of each POTW to which your facility discharges wastewater containing toxic chemicals for which you are reporting. If you do not discharge wastewater containing the reported toxic chemicals to a POTW, enter not applicable, NA, in the POTW name line in Part II, Section 1.1.

If you discharge such wastewater to more than two POTWs, use additional copies of Part II. Cross through the printed numbers and write in numbers for these locations in ascending order (e.g., 1.3, 1.4). Check the box at the bottom of the page and indicate the number of additional pages of Part II that are attached.

2. Other Off-Site Locations

Enter in the spaces provided, the name and address of each location (other than POTWs) to which you ship or transfer wastes containing toxic chemicals. Do not include locations to which you ship the toxic chemical for recycle or reuse. If you do not ship or transfer wastes containing toxic chemicals to off-site locations, enter not applicable, NA in the off-site location name line of 2.1. Also enter the EPA Identification Number (RCRA I.D. Number) for each such location if known to you. This number may be found on the Uniform Hazardous Waste Manifest, which is required by RCRA regulations. Also indicate in the space provided whether the location is owned or controlled by your facility or your parent company. If the facility does not have a RCRA I.D. number, enter not applicable, NA, in this space.

If your facility transfers toxic chemicals to more than six off-site locations, use additional copies of Part II. Cross through the printed numbers and write in numbers for these locations in ascending order (i.e., 2.7, 2.8). Check the box at the bottom of the page and indicate the number of additional pages of Part II that are attached.

EXAMPLE 5: Off-Site Locations

Your facility is involved in chrome plating of metal parts, which produces an aqueous plating waste that is treated on-site to recover chromium sludge. The effluent from the on-site treatment plant, which contains chromium compounds (a listed toxic chemical), is piped to a POTW. The chromium sludge is transferred to an off-site, privately owned firm for the recovery of the chromium.

You must report the location of the POTW in Section 1 in Part II of Form R. Do not report any information about the on-site treatment plant in this section. You are not required to report the location of the off-site, privately owned recovery firm or provide any information concerning off-site recovery because recycling or reuse of toxic chemicals is exempt from reporting.

PART III. CHEMICAL-SPECIFIC INFORMATION

In Part III, you are to identify the toxic chemical being reported. You must indicate the general uses and activities involving the chemical at your facility. In Part III, you will also enter quantitative data relating to releases of the chemical from the facility to air, water, and land. Quantities of the chemical transferred to off-site locations, identified in Part II, are also reported in this part. Any waste treatment information for on-site treatment of wastestreams containing the toxic chemical are also required to be reported on Part III. An optional section is included in this part that allows you to report waste minimization information associated with the chemical.

1.1 [Reserved]

1.2 CAS Number

Enter the Chemical Abstracts Service (CAS) registry number in Section 1.2 exactly as it appears in Table II, pages 40-48, for the chemical being reported. CAS numbers are cross-referenced with an alphabetical list of chemical names in Table II of these instructions. If you are reporting one of the chemical categories in Table II (e.g., copper compounds), enter not applicable, NA, in the CAS number space.

If you are making a trade secret claim, you must report the CAS number on your unsanitized Form R and unsanitized substantiation form. Do not include the CAS number on your sanitized Form R and sanitized substantiation form (see page 1 for more information).

1.3 Chemical or Chemical Category Name

Enter the name of the chemical or chemical category exactly as it appears in Table II. If the chemical name is followed by a synonym in parentheses, report the chemical by the name that directly follows the CAS number (i.e., not the synonym).

Chemical Hazard Communication Guidebook

If the listed chemical identity is actually a product trade name (e.g., dicofol), the 9th Collective Index name is listed below it in brackets. You may report either name in this case.

Do not list the name of a chemical that does not appear in Table II, such as individual members of a reportable category. For example, if you use silver nitrate, do not report silver nitrate with its CAS number. Report this chemical as "silver compounds" which has no CAS number.

If you are making a trade secret claim, you must report the specific chemical identity on your unsanitized Form R and unsanitized substantiation form. Do not report the chemical name on your sanitized Form R and sanitized substantiation form. Include a generic name in Part III, Section 1.4 of your sanitized Form R report.

1.4 Generic Chemical Name

Complete Section 1.4 only if you are claiming the specific chemical identity of the toxic chemical as a trade secret and have marked the trade secret block in Part I, Section 1.1 on page 1 of Form R. Enter a generic chemical name that is descriptive of the chemical structure. You must limit the generic name to seventy characters (e.g., numbers, letters, spaces, punctuation) or less. Do not enter mixture names in Section 1.4; see Section 2 below.

In-house plant codes and other substitute names that are not structurally descriptive of the chemical identity being withheld as a trade secret are not acceptable as a generic name. The generic name must appear on both sanitized and unsanitized Form R's, and the name must be the same as that used on your substantiation forms. The Emergency Planning and Community Right-to-Know Information Hotline can provide you with assistance in selecting an appropriate generic name.

2. Mixture Component Identity

Do not complete this section if you have completed Section 1 of Part III. Report the generic name provided to you by your supplier in the section if your supplier is claiming the chemical identity proprietary or trade secret. Do not answer "yes" in Part I, Section 1.1 on page 1 of the form if you complete this section. You do not need to supply trade secret substantiation forms since it is your supplier who is claiming the material a trade secret.

Enter the generic chemical name in this section only if the following three conditions apply:

1. You determine that the mixture contains a listed toxic chemical but the only identity you have for that chemical is a generic name;

2. You know either the specific concentration of that toxic chemical component or a maximum concentration level; and

3. You multiply the concentration level by the total annual amount of the whole mixture used (or processed) and determine that you meet the use or process threshold for that single, generically identified mixture component.

EXAMPLE 6: Mixture Containing Unidentified Toxic Chemical

Your facility uses 20,000 pounds of a solvent that your supplier has told you contains 80 percent "chlorinated aromatic," their generic name for a chemical subject to reporting under section 313. You therefore know that you have used 16,000 pounds of some listed toxic chemical which exceeds the "otherwise use" threshold. You would file a Form R and enter the name "chlorinated aromatic" in the space provided in Part III, Section 2.

---

3. Activities and Uses of the Chemical at the Facility

Indicate whether the chemical is manufactured (including imported), processed, or otherwise used at the facility and the general nature of such activities and uses at the facility during the calendar year. Report activities that take place only at your facility, not activities that take place at other facilities involving your products. You must check all the blocks in this section that apply. If you are a manufacturer of the chemical, you must check a and/or b, and at least one of c, d, e, or f. Refer to the definitions of "manufacture," "process," and "otherwise use" in the general information section of these instructions or section 372.3 of the rule for additional explanations.

3.1 Manufacture the Chemical

Persons who manufacture (including import) the toxic chemical must check at least one:

a. *Produce* - the chemical is produced at the facility.

b. *Import* - the chemical is imported by the facility into the Customs Territory of the United States. (See page 6 of these instructions for further clarification of import.)

And check at least one:

c. *For on-site use/processing* - the chemical is produced or imported and then further processed or otherwise used at the same facility. If you check this block, you must also check at least one item in Part III, Section 3.2 or 3.3.

Appendix B: Regulations Reference Guide

Form R - Part III   Page 19

d. *For sale/distribution* - the chemical is produced or imported specifically for sale or distribution outside the manufacturing facility.

e. *As a byproduct* - the chemical is produced coincidentally during the production, processing, otherwise use, or disposal of another chemical substance or mixture and, following its production, is separated from that other chemical substance or mixture. Chemicals produced and released as a result of waste treatment or disposal are also considered byproducts.

f. *As an impurity* - the chemical is produced coincidentally as a result of the manufacture, processing, or otherwise use of another chemical but is not separated and remains primarily in the mixture or product with that other chemical.

3.2 Process the Chemical (incorporative-type activities)

a. *As a reactant* - A natural or synthetic chemical used in chemical reactions for the manufacture of another chemical substance or of a product. Includes, but is not limited to, feedstocks, raw materials, intermediates, and initiators.

b. *As a formulation component* - A chemical added to a product (or product mixture) prior to further distribution of the product that acts as a performance enhancer during use of the product. Examples of chemicals used in this capacity include, but are not limited to, additives, dyes, reaction diluents, initiators, solvents, inhibitors, emulsifiers, surfactants, lubricants, flame retardants, and rheological modifiers.

c. *As an article component* - A chemical substance that becomes an integral component of an article distributed for industrial, trade, or consumer use. One example is the pigment components of paint applied to a chair that is sold.

d. *Repackaging only* - Processing or preparation of a chemical (or product mixture) for distribution in commerce in a different form, state, or quantity. This includes, but is not limited to, the transfer of material from a bulk container, such as a tank truck to smaller cans or bottles.

3.3 Otherwise Use the Chemical (non-incorporative-type activities)

a. *As a chemical processing aid* - A chemical that is added to a reaction mixture to aid in the manufacture or synthesis of another chemical substance but is not intended to remain in or become part of the product or product mixture. Examples of such chemicals include, but are not limited to, process solvents, catalysts, inhibitors, initiators, reaction terminators, and solution buffers.

b. *As a manufacturing aid* - A chemical that aids the manufacturing process but does not become part of the resulting product and is not added to the reaction mixture during the manufacture or synthesis of another chemical substance. Examples include, but are not limited to, process lubricants, metalworking fluids, coolants, refrigerants, and hydraulic fluids.

c. *Ancillary or other use* - A chemical in this category that is used at a facility for purposes other than as a chemical processing aid or manufacturing aid as described above. Includes, but is not limited to, cleaners, degreasers, lubricants, fuels, and chemicals used for treating wastes.

EXAMPLE 7: Activities and Uses of Toxic Chemicals

In the example below, it is assumed that the threshold quantities for manufacture, process, or otherwise use (25,000 pounds, 25,000 pounds, and 10,000 pounds, respectively, for 1989) have been exceeded and the reporting of listed chemicals is therefore required.

Your facility receives toluene and naphthalene (both listed toxic chemicals) from an off-site location. You react the toluene with air to form benzoic acid and react the naphthalene with sulfuric acid, which forms phthalic acid and also produces sulfur dioxide fumes. Your facility processes toluene and naphthalene. Both are used as reactants to produce benzoic acid and phthalic acid, chemicals not on the section 313 list.

The phthalic acid and benzoic acid are reacted to form a reaction intermediate. The reaction intermediate is dissolved in sulfuric acid, which precipitates terephthalic acid (TPA). Fifty percent of the TPA is sold as a product and 50 percent is further processed at your facility into polyester fiber. The TPA is treated with ethylene glycol to form an intermediate product, which is condensed to polyester.

Your company manufactures terephthalic acid, a listed chemical, both for sale/distribution as a commercial product and for on-site use/processing as a feedstock in the polyester process. Because it is a reactant, it is also processed. See Figure D for how this information would be reported in Part III, Section 3 of Form R.

Your facility also uses, as well as processes, sulfuric acid, a listed substance, as it serves as a process solvent to precipitate terephthalic acid.

Chemical Hazard Communication Guidebook

Page 20   Form R - Part III

Figure D

(For more information, see Example 7 on page 19)

(Important: Type or print; read instructions before completing form.)   Page 3 of 5

⊕EPA   **EPA FORM R**   **PART III. CHEMICAL-SPECIFIC INFORMATION**   (This space for your optional use.)

1. CHEMICAL IDENTITY (Do not complete this section if you complete Section 2.)

1.1	[Reserved]
1.2	CAS Number (Enter only one number exactly as it appears on the 313 list. Enter NA if reporting a chemical category.) 100-21-0
1.3	Chemical or Chemical Category Name (Enter only one name exactly as it appears on the 313 list.) Terephthalic Acid
1.4	Generic Chemical Name (Complete only if Part I, Section 1.1 is checked "Yes." Generic name must be structurally descriptive.)

2. MIXTURE COMPONENT IDENTITY (Do not complete this section if you complete Section 1.)

Generic Chemical Name Provided by Supplier (Limit the name to a maximum of 70 characters (e.g., numbers, letters, spaces, punctuation).)

3. ACTIVITIES AND USES OF THE CHEMICAL AT THE FACILITY (Check all that apply.)

3.1	Manufacture the chemical:	a. [X] Produce b. [ ] Import	If produce or import: c. [X] For on-site use/processing e. [ ] As a byproduct	d. [X] For sale/distribution f. [ ] As an impurity
3.2	Process the chemical:	a. [X] As a reactant d. [ ] Repackaging only	b. [ ] As a formulation component	c. [ ] As an article component
3.3	Otherwise use the chemical:	a. [ ] As a chemical processing aid	b. [ ] As a manufacturing aid	c. [ ] Ancillary or other use

Appendix B: Regulations Reference Guide

Form R - Part IIIPage 21

4. Maximum Amount of the Chemical On-Site at Any Time During the Calendar Year

Insert the appropriate code (see below) that indicates the maximum quantity of the chemical (e.g., in storage tanks, process vessels, on-site shipping containers) at your facility at any time during the calendar year. If the chemical was present at several locations within your facility, use the maximum total amount present at the entire facility at any one time.

Weight Range in Pounds

Range Code	From...	To....
01	0	99
02	100	999
03	1,000	9,999
04	10,000	99,999
05	100,000	999,999
06	1,000,000	9,999,999
07	10,000,000	49,999,999
08	50,000,000	99,999,999
09	100,000,000	499,999,999
10	500,000,000	999,999,999
11	1 billion	more than 1 billion

If the toxic chemical present at your facility was part of a mixture or trade name product, determine the maximum quantity of the chemical present at the facility by calculating the weight of the toxic chemical only. Do not include the weight of the entire mixture or trade name product. See section 372.30(b) of the reporting rule for further information on how to calculate the weight of the chemical in the mixture or trade name product. For chemical categories (e.g., copper compounds), include all chemicals in the category when calculating the weight of the toxic chemical.

5. Releases of the Chemical to the Environment On-Site

In Section 5, you must account for the total aggregate releases of the toxic chemical to the environment from your facility for the calendar year. Releases to the environment include emissions to the air, discharges to surface waters, and on-site releases to land and underground injection wells. If you have no releases to a particular media (e.g., stack air), enter not applicable, NA; do not leave any part of Section 5 blank. Check the box on the last line of this section if you use Part IV, the supplemental information sheet.

You are not required to count, as a release, quantities of a toxic chemical that are lost due to natural weathering or corrosion, normal/natural degradation of a product, or normal migration of a chemical from a product. For example, amounts of a covered toxic chemical that migrate from plastic products in storage do not have to be counted in estimates of releases of that chemical from the facility. Also, amounts of listed metal compounds (e.g., copper compounds) that are lost due to normal corrosion of process equipment do not have to be considered as releases of copper compounds from the facility.

All air releases of the chemical from the facility must be accounted for. Do not enter information on individual emission points or releases. Enter only the total release. If there is doubt about whether an air release is a point or non-point release, you must identify the release as one or the other rather than leave items 5.1 and 5.2 blank. Instructions for columns A, B, and C follow the discussions of Sections 5.1 through 5.5.

5.1 Fugitive or Non-Point Air Emissions

Report the total of all releases to the air that are not released through stacks, vents, ducts, pipes, or any other confined air stream. You must include (1) fugitive equipment leaks from valves, pump seals, flanges, compressors, sampling connections, open-ended lines, etc.; (2) evaporative losses from surface impoundments and spills; (3) releases from building ventilation systems; and (4) any other fugitive or non-point air emissions.

5.2 Stack or Point Air Emissions

Report the total of all releases to the air that occur through stacks, vents, ducts, pipes, or other confined air streams. You must include storage tank emissions. Air releases from air pollution control equipment would generally fall in this category.

5.3 Discharges to Receiving Streams or Water Bodies

Enter the applicable letter code for the receiving stream or water body from Section 3.10 of Part I of the form. Also, enter the total annual amount of the chemical released from all discharge points at the facility to each receiving stream or water body. Include process outfalls such as pipes and open trenches, releases from on-site wastewater treatment systems, and the contribution from stormwater runoff, if applicable (see instructions for column C below). Do not include discharges to a POTW or other off-site wastewater treatment facilities in this section. These off-site transfers must be reported in Part III, Section 6 of the form.

Discharges of listed acids (e.g., hydrogen flouride; hydrogen chloride; nitric acid; phosphoric acid; and sulfuric acid) may be reported as zero if the discharges have been neutralized to pH 6 or above. For discharges of listed bases, a zero release may be reported if the discharge has been neutralized to pH 9 or below.

Chemical Hazard Communication Guidebook

5.4 Underground Injection On-Site

Enter the total annual amount of the chemical that was injected into all wells, including Class I wells, at the facility.

5.5 Releases to Land On-Site

Four predefined subcategories for reporting quantities released to land within the boundaries of the facility are provided. Do not report land disposal at off-site locations in this section.

5.5.1 Landfill -- Typically, the ultimate disposal method for solid wastes is landfilling. Leaks from landfills need not be reported as a release because the amount of the toxic chemical in the landfill as already been reported as a release.

5.5.2 Land treatment/application farming -- Another disposal method is land treatment in which a waste containing a listed chemical is applied onto or incorporated into soil. While this disposal method is considered a release to land, any volatilization of listed chemicals into the air occurring during the disposal operation must be reported as a fugitive air release in Section 5.1 of Form R.

5.5.3 Surface Impoundment -- A natural topographic depression, man-made excavation, or diked area formed primarily of earthen materials (although some may be lined with man-made materials), which is designed to hold an accumulation of liquid wastes or wastes containing free liquids. Examples of surface impoundments are holding, settling, storage, and elevation pits; ponds; and lagoons. If the pit, pond, or lagoon is intended for storage or holding without discharge, it would be considered to be a surface impoundment used as a final disposal method.

Quantities of the chemical released to surface impoundments that are used merely as part of a wastewater treatment process generally must not be reported in this section. However, if the impoundment accumulates sludges containing the chemical, you must include an estimate in this section unless the sludges are removed and otherwise disposed of (in which case they should be reported under the appropriate section of the form). For the purposes of this reporting, storage tanks are not considered to be a type of disposal and are not to be reported in this section of the form.

5.5.4 Other disposal -- Includes any amount of a listed toxic chemical released to land that does not fit the categories of landfills, land treatment, or surface impoundment. This other disposal would include any spills or leaks of listed toxic chemicals to land. For example, 2,000 pounds benzene leaks from a underground pipeline into the land at a facility. Because the pipe was only a few feet from the surface at the erupt point, 30 percent of the benzene evaporates into the air. The 600 pounds released to the air would be reported as a fugitive air release (Section 5.1) and the remaining 1400 pounds would be reported as a release to land, other disposal (Section 5.5.4).

5.A Total Release

Only on-site releases of the toxic chemical to the environment for the calendar year are to be reported in this section of the form. The total releases from your facility do not include transfers or shipments of the chemical from your facility for sale or distribution in commerce, or of wastes to other facilities for treatment or disposal (see Part III, Section 6). Both routine releases, such as fugitive air emissions, and accidental or non-routine releases, such as chemical spills, must be included in your estimate of the quantity released. EPA requires no more than two significant digits when reporting releases (e.g., 7521 pounds would be reported as 7500 pounds).

Releases of Less Than One Pound. Total annual releases or off-site transfers of a toxic chemical from the facility of less than 1 pound may be reported in one of several ways. You should round the value to the nearest pound. If the estimate is 0.5 pounds or greater, you should either check the range bracket of "1-499" in column A.1 or enter "1" in column A.2. Do not use both columns A.1 and A.2. If the release is less than 0.5 pounds, you may round to zero and check the "0" bracket in A.1.

Note that total annual releases of less than 0.5 pounds from the processing or otherwise use of an article maintains the article status of that item. Thus, if the only releases you have are from processing an article, and such releases are less than 0.5 pounds per year, you are not required to submit a report for that chemical. The 0.5 pound release determination does not apply to just a single article. It applies to the cumulative releases from the processing or use of that same type of article (e.g., sheet metal or plastic film) that occurs over the course of the calendar year.

Zero Releases. If you have no releases of a toxic chemical to a particular medium, report either NA, not applicable, or 0, as appropriate. Report NA only when there is no possibility a release could occur to a specific media or off-site location. If a release to a specific media or off-site location could occur, but either no release occurred or the annual aggregate release was less than 0.5 pounds, report zero. However, if you report zero releases, a basis of estimate must be provided in column B. For example, if hydrochloric acid is involved in the facility processing activities but the facility neutralizes the wastestreams to a pH of 6-9, then the facility reports a 0 release for the chemical. If the facility has no underground injection well, it enters NA for that item on the form. If the facility does not landfill the acidic waste, it enters NA for landfills.

5.A.1 Reporting Ranges

For reports submitted for calendar years 1987, 1988, and 1989 only, you may take advantage of range reporting for releases to an environmental medium that are less than 1,000 pounds for the year. If you choose this option, mark one of the three boxes, 0, 1-499, or 500-999, that corresponds to releases of the chemical to the appropriate environmental medium (i.e., any line item). You are not required, however, to use these range check boxes; you have the option of providing a specific value in column A.2, as described below. However, do not mark a range and also enter a specific estimate in A.2.

5.A.2 Enter Estimate

For releases to any medium that amount to 1,000 pounds or more for the year, you must provide an estimate in pounds per year in column A.2. Any estimate provided in column A.2 should be reported to no more than two significant figures. This estimate should be in whole numbers. Do not use decimal points.

If you do not use the range reporting option, provide your estimates of total annual releases (in pounds) in column A.2.

Calculating Releases - To provide the release information required in columns A.1 and A.2 in this section, you must use all readily available data (including relevant monitoring data and emissions measurements) collected at your facility to meet other regulatory requirements or as part of routine plant operations, to the extent you have such data for the toxic chemical.

When relevant monitoring data or emission measurements are not readily available, reasonable estimates of the amounts released must be made using published emission factors, material balance calculations, or engineering calculations. You may not use emission factors or calculations to estimate releases if more accurate data are available.

No additional monitoring or measurement of the quantities or concentrations of any toxic chemical released into the environment, or of the frequency of such releases, is required for the purpose of completing this form, beyond that which is required under other provisions of law or regulation or as part of routine plant operations.

You must estimate, as accurately as possible, the quantity (in pounds) of the chemical or chemical category that is released annually to each environmental medium. Include only the quantity of the toxic chemical contained in the wastestream in this estimate. If the toxic chemical present at your facility was part of a mixture or trade name product, calculate only the releases of the chemical, not the other components of the mixture or trade name product. If you are only able to estimate the releases of the mixture or trade name product as a whole, you must assume that the release of the toxic chemical is proportional to its concentration in the mixture or trade name product. See section 372.30(b) of the reporting rule for further information on how to calculate the concentration and weight of the toxic chemical in the mixture or trade name product.

If you are reporting a chemical category listed in Table II of these instructions, rather than a specific chemical, you must combine the release data for all chemicals in the listed chemical category (e.g., all glycol ethers or all chlorophenols) and report the aggregate amount for that chemical category. Do not report releases of each individual chemical in that category separately. For example, if your facility releases 3,000 pounds per year of 2-chlorophenol, 4,000 pounds per year of 3-chlorophenol, and 4,000 pounds per year of 4-chlorophenol, you should report that your facility releases 11,000 pounds per year of chlorophenols.

For listed chemicals with the qualifier "solution," such as ammonium nitrate, at concentrations of 1 percent (or 0.1 percent in the case of a carcinogen) or greater, the chemical concentrations must be factored into threshold and release calculations because threshold and release amounts relate to the amount of chemical in solution, not the amount of solution.

For metal compound categories (e.g., chromium compounds), report releases of only the parent metal. For example, a user of various inorganic chromium salts would report the total chromium released in each waste type regardless of the chemical form (e.g., as the original salts, chromium ion, oxide) and exclude any contribution to mass made by other species in the molecule.

EXAMPLE 8:   Calculating Releases

Your facility disposes of 14,000 pounds of lead chromate ($PbCrO_4 \cdot PbO$) in an on-site landfill and transfers 16,000 pounds of lead selenate ($PbSeO_4$) to an off-site land disposal facility. You would therefore be submitting three separate reports on the following: lead compounds, selenium compounds, and chromium compounds. However, the quantities you would be reporting would be the pounds of "parent" metal being released or transferred off-site. All quantities are based on mass balance calculations (See Section 5.B for information on Basis of Estimate and Section 6.C for treatment/disposal codes and information on transfers of chemical wastes). You would calculate releases of lead, chromium, and selenium by first determining the percentage by weight of these metals in the materials you use as follows:

Lead Chromate ($PbCrO_4 \cdot PbO$) -
    Molecular weight   =   546.37
Lead 2 Pb -
    Molecular weight   =   207.2 x 2 = 414.4
Chromate 1 Cr -
    Molecular weight   =   51.996

Lead chromate is therefore (% by weight)
(414.4/546.37) = 75.85% lead and (51.996/546.37) = 9.52% chromium

You can then calculate the total amount of the metals that you must report, based on your knowledge that 14,000 pounds of lead chromate contains:

14,000 x 0.7585 = 10,619 pounds of lead
14,000 x 0.0952 = 1,334 pounds of chromium

Similarly, lead selenate is (207.2/350.17) = 59.17% lead and (78.96/350.17) = 22.55% selenium.

The total pounds of lead, chromium, and selenium released or transferred from your facility are as follows:

<u>Lead</u>

Release:
0.7585 x 14,000 = 10,619 pounds from lead chromate
(round to 11,000 pounds)

Transfer:
0.5917 x 16,000 = 9,467 pounds from lead selenate
(round to 9,500 pounds)

(As an example, the releases and transfers of <u>lead</u> should be reported as illustrated in Figure E on the pages 24-25.)

<u>Chromium</u>

Release:
0.0952 x 14,000 = 1,333 pounds from lead chromate
(round to 1,300 pounds)

<u>Selenium</u>

Transfer:
0.2255 x 16,000 = 3,608 pounds of selenium from lead selenate (round to 3,600 pounds)

5.B Basis of Estimate

For each release estimate, you are required to indicate the principal method used to determine the amount of release reported. You will enter a letter code that identifies the method that applies to the largest portion of the total estimated release quantity.

For example, if 40 percent of stack emissions of the reported substance were derived using monitoring data, 30 percent by mass balance, and 30 percent by emission factors, you would enter the code letter "M" for monitoring.

The codes are as follows:

M - Estimate is based on monitoring data or measurements for the toxic chemical as released to the environment and/or off-site facility.

C - Estimate is based on mass balance calculations, such as calculation of the amount of the toxic chemical in streams entering and leaving process equipment.

E - Estimate is based on published emission factors, such as those relating release quantity to through-put or equipment type (e.g., air emission factors).

O - Estimate is based on other approaches such as engineering calculations (e.g., estimating volatilization using published mathematical formulas) or best engineering judgment. This would include applying an estimated removal efficiency to a waste stream, even if the composition of the stream before treatment was fully identified through monitoring data.

If the monitoring data, mass balance, or emission factor used to estimate the release is not specific to the toxic chemical being reported, the form should identify the estimate as based on engineering calculations or best engineering judgment.

If a mass balance calculation yields the flow rate of a wastestream, but the quantity of reported chemical in the waste-

Figure E

1.	CHEMICAL IDENTITY (Do not complete this section if you complete Section 2.)
1.1	[Reserved]
1.2	CAS Number (Enter only one number exactly as it appears on the 313 list. Enter NA if reporting a chemical category.) NA
1.3	Chemical or Chemical Category Name (Enter only one name exactly as it appears on the 313 list.) Lead Compounds
1.4	Generic Chemical Name (Complete only if Part I, Section 1.1 is checked "Yes." Generic name must be structurally descriptive.)
	MIXTURE COMPONENT IDENTITY (Do not complete this section if you complete Section 1.)
2.	Generic Chemical Name Provided by Supplier (Limit the name to a maximum of 70 characters (e.g., numbers, letters, spaces, punctuation).)

Appendix B: Regulations Reference Guide

Form R - Part III   Page 25

Figure E

(Continued)

5. RELEASES OF THE CHEMICAL TO THE ENVIRONMENT ON-SITE		A. Total Release (pounds/year)		B. Basis of Estimate	C. % From Stormwater
You may report releases of less than 1,000 pounds by checking ranges under A.1. (Do not use both A.1 and A.2)		A.1 Reporting Ranges 0 1-499 500-999	A.2 Enter Estimate	(enter code)	
5.1 Fugitive or non-point air emissions	5.1a	[ ] [ ] [ ]	NA	5.1b [ ]	
5.2 Stack or point air emissions	5.2a	[ ] [ ] [ ]	NA	5.2b [ ]	
5.3 Discharges to receiving streams or water bodies 5.3.1 [ ]	5.3.1a	[ ] [ ] [ ]	NA	5.3.1b [ ]	5.3.1c NA %
(Enter letter code for stream from Part I Section 3.10 in the box provided.) 5.3.2 [ ]	5.3.2a	[ ] [ ] [ ]		5.3.2b [ ]	5.3.2c %
5.3.3 [ ]	5.3.3a	[ ] [ ] [ ]		5.3.3b [ ]	5.3.3c %
5.4 Underground injection on-site	5.4a	[ ] [ ] [ ]	NA	5.4b [ ]	
5.5 Releases to land on-site 5.5.1 Landfill	5.5.1a	[ ] [ ] [ ]	11,000	5.5.1b [C]	
5.5.2 Land treatment/application farming	5.5.2a	[ ] [ ] [ ]	NA	5.5.2b [ ]	
5.5.3 Surface impoundment	5.5.3a	[ ] [ ] [ ]	NA	5.5.3b [ ]	
5.5.4 Other disposal	5.5.4a	[ ] [ ] [ ]	NA	5.5.4b [ ]	

[ ] (Check if additional information is provided on Part IV-Supplemental Information.)

6. TRANSFERS OF THE CHEMICAL IN WASTE TO OFF-SITE LOCATIONS		A. Total Transfers (pounds/year)		B. Basis of Estimate	C. Type of Treatment/ Disposal
You may report transfers of less than 1,000 pounds by checking ranges under A.1. (Do not use both A.1 and A.2)		A.1 Reporting Ranges 0 1-499 500-999	A.2 Enter Estimate	(enter code)	(enter code)
6.1.1 Discharge to POTW (enter location number from Part II, Section 1.) [1]		[ ] [ ] [ ]	NA	6.1.1b [ ]	
6.2.1 Other off-site location (enter location number from Part II, Section 2.) [2][1]		[ ] [ ] [ ]	9,500	6.2.1b [C]	6.2.1c [M][7][2]
6.2.2 Other off-site location (enter location number from Part II, Section 2.) [2][ ]		[ ] [ ] [ ]	NA	6.2.2b [ ]	6.2.2c [M][ ][ ]
6.2.3 Other off-site location (enter location number from Part II, Section 2.) [2][ ]		[ ] [ ] [ ]		6.2.3b [ ]	6.2.3c [M][ ][ ]

[ ] (Check if additional information is provided on Part IV-Supplemental Information.)

Chemical Hazard Communication Guidebook

stream is based on solubility data, report "O" because "engineering calculations" were used as the basis of estimate of the quantity of the chemical in the wastestream.

If the concentration of the chemical in the wastestream was measured by monitoring equipment and the flow rate of the wastestream was determined by mass balance, then the primary basis of estimate is "monitoring" (M). Even though a mass balance calculation also contributed to the estimate, "Monitoring" should be indicated because monitoring data was used to estimate the concentration of the waste stream.

Mass balance (C) should only be indicated if it is _directly_ used to calculate the mass (weight) of chemical released. Monitoring data should be indicated as the basis of estimate _only_ if the chemical concentration is measured in the wastestream being released into the environment. Monitoring data should _not_ be indicated, for example, if the monitoring data relates to a concentration of the toxic chemical in other process streams within the facility.

5.C Percent From Stormwater

This column relates only to Section 5.3 -- Discharges to receiving streams or water bodies. If your facility has monitoring data on the amount of the chemical in stormwater runoff (including unchanneled runoff), you must include that quantity of the chemical in your water release in column A _and_ indicate the percentage of the total quantity (by weight) of the chemical contributed by stormwater in column C (Section 5.3c).

If your facility has monitoring data on the chemical and an estimate of flow rate, you must use this data to determine the percent stormwater.

If you have monitored stormwater but did not detect the chemical, enter zero (0) in column C. If your facility has no stormwater monitoring data for the chemical, enter not applicable, NA, in this space on the form.

EXAMPLE 9: Releases from Stormwater

Bi-monthly stormwater monitoring data shows that the average concentration of zinc in the stormwater runoff from your facility from a biocide containing a zinc compound is 1.4 milligrams per liter, and the total annual stormwater discharge from the facility is 7.527 million gallons. The total amount of zinc discharged to surface water through the plant wastewater discharge (non-stormwater) is 250 pounds per year. The total amount of zinc discharged with stormwater is:

(7,527,000 gallons stormwater) x (3.785 liters/gallon)
= 28,489,695 liters stormwater

(28,489,695 liters stormwater) x (1.4 mg. zinc/liter)
= 39,885.6 grams zinc
= 88 pounds zinc

The total amount of zinc discharged from all sources of your facility is:

250	pounds zinc from wastewater discharge
+ 88	pounds zinc from stormwater runoff
338	pounds zinc total water discharge

Round to 340 pounds of zinc for report.

The _percentage_ of zinc discharged through stormwater is:

88/338 x 100 = 26%

If your facility does not have periodic measurements of stormwater releases of the chemical, but has submitted chemical-specific monitoring data in permit applications, then these data must be used to calculate the percent contribution from stormwater. Rates of flow can be estimated by multiplying the annual amount of rainfall by the land area of the facility and then multiplying that figure by the runoff coefficient. The runoff coefficient represents the fraction of rainfall that does not infiltrate into the ground but runs off as stormwater. The runoff coefficient is directly related to how the land in the drainage area is used. (See table below.)

Description of Land Area	Runoff Coefficient
Business	
Downtown areas	0.70-0.95
Neighborhood areas	0.50-0.70
Industrial	
Light areas	0.50-0.80
Heavy areas	0.60-0.90
Railroad yard areas	0.20-0.40
Unimproved areas	0.10-0.30
Streets	
Asphaltic	0.70-0.95
Concrete	0.80-0.95
Brick	0.70-0.85
Drives and walks	0.70-0.85
Roofs	0.75-0.95
Lawns: Sandy Soil	
Flat, 2%	0.05-0.10
Average, 2-7%	0.10-0.15
Steep, 7%	0.15-0.20
Lawns: Heavy Soil	
Flat, 2%	0.13-0.17
Average, 2-7%	0.18-0.22
Steep, 7%	0.25-0.35

Choose the most appropriate runoff coefficient for your site or calculate a weighted-average coefficient, which takes into account different types of land use at your facility:

$$\text{Weighted-average runoff coefficient} = \frac{(Area_1 C_1 + Area_2 C_2 + \ldots\ldots A_i C_i)}{\text{Total Site Area}}$$

where C_i = runoff coefficient for a specific land use of $Area_i$.

Appendix B: Regulations Reference Guide

Form R - Part III   Page 27

EXAMPLE 10: Stormwater Runoff

Your facility is located in a semi-arid region of the United States which has an annual precipitation (including snowfall) of 12 inches of rain. (Snowfall should be converted to the equivalent inches of rain; assume one foot of snow is equivalent to one inch of rain.) The area covered by your facility is 42 acres (about 170,000 square meters or 1,829,520 square feet). The area of your facility is 50 percent unimproved area, 10 percent asphaltic streets, and 40 percent concrete pavement.

The total stormwater runoff from your facility is therefore calculated as follows:

Land Use	% Area	Runoff Coefficient
Unimproved area	50	0.20
Asphaltic streets	10	0.85
Concrete pavement	40	0.90

Weighted-average runoff coefficient =
$$\frac{(50\%) \times (0.20) + (10\%) \times (0.85) + (40\%) \times (0.90)}{100\% \text{ Area}} = 0.545$$

(Rainfall) x (land area) x (conversion factor) x (runoff coefficient) = stormwater runoff

(1 foot) x (1,829,520 ft^2) x (7.48 gal/ft^3) x (0.545)
= 7,458,221 gallons/year

Total stormwater runoff = 7.45 million gallons/year

---

6. Transfers of the Chemical in Waste to Off-Site Locations

You must report in this section the total annual quantity of the chemical sent to any of the off-site disposal, treatment, or storage facilities for which you have provided an address in Part II. You are not required to report quantities of the chemical sent off-site for purposes of recycle or reuse. Report the amount of the toxic chemical transferred off-site after any on-site treatment or removal is completed. Report zero for releases of listed acids and bases if they have been neutralized to pH 6-9 prior to discharge to a POTW. See the discussion under Section 5.3, Discharges to Receiving Streams or Water Bodies (see page 21).

On line 6.1.1, report the amount of the listed chemical transferred to a POTW listed in Part II, Section 1. In the block provided, enter the number from Part II, Section 1 corresponding to the POTW to which the discharge is sent. For example, if the discharge is sent to the location listed in Part II, Section 1.1, then enter "1" in the block provided (the first digit of this section number has been precoded). If you transfer waste containing the toxic chemical to more than one POTW, check the box at the bottom of Section 6 and use the Part IV, the Supplemental Information Sheet to report those transfers.

On lines 6.2.1 through 6.2.3, report the amount of the chemical transferred to other off-site locations corresponding to those listed in Part II, Sections 2.1 through 2.6, including privately owned wastewater treatment facilities. In the block provided, enter the number from Part II, Section 2 corresponding to the off-site location to which the transfer is sent. For example, if the transfer is sent to the location listed in Part II, Section 2.3, enter "3" in the block provided. (The first digit of this section number has been precoded.) If you need additional space, check the box at the bottom of Section 6 and use the Supplemental Information Sheet (Part IV, Section 6) to report those transfers.

6.A Total Transfers

This column should be completed as described in the instructions for column A of Section 5 above. Enter the amount, in pounds, of the toxic chemical that is being transferred, including mixtures or trade name products containing the chemical. Do not enter the total poundage of wastes. See Section 5 for information on reporting off-site transfers of less than 1 pound. As in Section 5, if the total amount transferred is less than 1,000 pounds, you may report a range, but only for reporting years 1987, 1988, and 1989. Enter not applicable, NA, in column A.2 if you have no off-site transfers of the listed chemical.

6.B Basis of Estimate

You must identify the basis for your estimate. Enter the letter code that applies to the method by which the largest percentage of the estimate was derived. Use the same codes identified in the instructions for column B of Section 5.

6.C Type of Treatment/Disposal

Enter one of the following codes to identify the type of treatment or disposal method used by the off-site location for the chemical being reported. You should use more than one line for a single location when the toxic chemical is subject to different disposal methods; the same location code may be used more than once. You may have this information in your copy of EPA Form SO, Item S of the Annual/Biennial Hazardous Waste Treatment, Storage, and Disposal Report (RCRA). Applicable codes for Part III, Section 6(c) are as follows:

M10   Storage Only
M40   Solidification/Stabilization
M50   Incineration/Thermal Treatment
M61   Wastewater Treatment (Excluding POTW)
M69   Other Treatment

Chemical Hazard Communication Guidebook

Form R - Part III Page 28

M71 Underground Injection
M72 Landfill/Disposal Surface Impoundment
M73 Land Treatment
M79 Other Land Disposal
M90 Other Off-Site Management
M91 Transfer to Waste Broker
M99 Unknown

7. Waste Treatment Methods and Efficiency

In Section 7, you must provide the following information related to the chemical for which releases are being reported: (A) the general wastestream types containing the chemical being reported; (B) the waste treatment methods used on all wastestreams containing the chemical; (C) the range of concentrations of the chemical in the influent to the treatment method; (D) whether sequential treatment is used; (E) the efficiency or effectiveness of each treatment method in removing the chemical; and (F) whether the treatment efficiency figure was based on actual operating data. Use a separate line in Section 7 for each treatment method used on a wastestream.

In this section, report only information about treatment of wastestreams <u>at your facility</u>, not about off-site treatment. If you do not perform on-site treatment of wastes containing the chemical being reported, check the Not Applicable (NA) space at the top of Section 7.

7.A General Wastestream

For each waste treatment method, indicate the type of wastestream containing the chemical that is treated. Enter the letter code that corresponds to the general wastestream type:

A = Gaseous (gases, vapors, airborne particulates)
W = Wastewater (aqueous waste)
L = Liquid waste (non-aqueous waste)
S = Solid waste (including sludges and slurries)

If a waste is a mixture of water and organic liquid, you must report it as wastewater unless the organic content exceeds 50 percent. Slurries and sludges containing water must be reported as solid waste if they contain appreciable amounts of dissolved solids, or solids that may settle, such that the viscosity or density of the waste is considerably different from that of process wastewater.

7.B Treatment Method

Enter the appropriate code from one of the lists below for each on-site treatment method used on a wastestream containing the toxic chemical, regardless of whether the treatment method actually removes the specific chemical being reported. Treatment methods must be reported for each type of waste being treated (i.e., gaseous wastes, aqueous wastes, liquid non-aqueous wastes, and solids). The treatment codes, except for the air emission treatment codes, are not restricted to any medium.

Wastestreams containing the chemical may have a single source or may be aggregates of many sources. For example, process water from several pieces of equipment at your facility may be combined prior to treatment. Report treatment methods that apply to the aggregate wastestream, as well as treatment methods that apply to individual wastestreams. If your facility treats various wastewater streams containing the chemical in different ways, the different treatment methods must each be listed separately.

If your facility has several pieces of equipment performing a similar service, you may combine the reporting for such equipment on a single line. It is not necessary to enter four lines of data to cover four scrubber units, for example, if all four are treating wastes of similar character (e.g., sulfuric acid mist emissions), have similar influent concentrations, and have similar removal efficiencies. If, however, any of these parameters differ from one unit to the next, each scrubber must be listed separately.

<u>Air Emissions Treatment</u>

A01 Flare
A02 Condenser
A03 Scrubber
A04 Absorber
A05 Electrostatic Precipitator
A06 Mechanical Separation
A07 Other Air Emission Treatment

<u>Biological Treatment</u>

B11 Biological Treatment -- Aerobic
B21 Biological Treatment -- Anaerobic
B31 Biological Treatment -- Facultative
B99 Biological Treatment -- Other

<u>Chemical Treatment</u>

C01 Chemical Precipitation -- Lime or Sodium Hydroxide
C02 Chemical Precipitation -- Sulfide
C09 Chemical Precipitation -- Other
C11 Neutralization
C21 Chromium Reduction
C31 Complexed Metals Treatment (other than pH Adjustment)
C41 Cyanide Oxidation -- Alkaline Chlorination
C42 Cyanide Oxidation -- Electrochemical
C43 Cyanide Oxidation -- Other
C44 General Oxidation (including Disinfection) -- Chlorination

Appendix B: Regulations Reference Guide

Page 29   Form R - Part III

C45	General Oxidation (including Disinfection) -- Ozonation		R12	Solvents/Organics Recovery -- Thin-Film Evaporation
C46	General Oxidation (including Disinfection) -- Other		R13	Solvents/Organics Recovery -- Fractionation
C99	Other Chemical Treatment		R14	Solvents/Organics Recovery -- Solvent Extraction

Incineration/Thermal Treatment

- F01 Liquid Injection
- F11 Rotary Kiln with Liquid Injection Unit
- F19 Other Rotary Kiln
- F31 Two Stage
- F41 Fixed Hearth
- F42 Multiple Hearth
- F51 Fluidized Bed
- F61 Infra-Red
- F71 Fume/Vapor
- F81 Pyrolytic Destructor
- F82 Wet Air Oxidation
- F83 Thermal Drying/Dewatering
- F99 Other Incineration/Thermal Treatment

Physical Treatment

- P01 Equalization
- P09 Other Blending
- P11 Settling/Clarification
- P12 Filtration
- P13 Sludge Dewatering (non-thermal)
- P14 Air Flotation
- P15 Oil Skimming
- P16 Emulsion Breaking -- Thermal
- P17 Emulsion Breaking -- Chemical
- P18 Emulsion Breaking -- Other
- P19 Other Liquid Phase Separation
- P21 Adsorption -- Carbon
- P22 Adsorption -- Ion Exchange (other than for recovery/reuse)
- P23 Adsorption -- Resin
- P29 Adsorption -- Other
- P31 Reverse Osmosis (other than for recovery/reuse)
- P41 Stripping -- Air
- P42 Stripping -- Steam
- P49 Stripping -- Other
- P51 Acid Leaching (other than for recovery/reuse)
- P61 Solvent Extraction (other than recovery/reuse)
- P99 Other Physical Treatment

Recovery/Reuse

- R01 Reuse as Fuel -- Industrial Kiln
- R02 Reuse as Fuel -- Industrial Furnace
- R03 Reuse as Fuel -- Boiler
- R04 Reuse as Fuel -- Fuel Blending
- R09 Reuse as Fuel -- Other
- R11 Solvents/Organics Recovery -- Batch Still Distillation
- R12 Solvents/Organics Recovery -- Thin-Film Evaporation
- R13 Solvents/Organics Recovery -- Fractionation
- R14 Solvents/Organics Recovery -- Solvent Extraction
- R19 Solvents/Organics Recovery -- Other
- R21 Metals Recovery -- Electrolytic
- R22 Metals Recovery -- Ion Exchange
- R23 Metals Recovery -- Acid Leaching
- R24 Metals Recovery -- Reverse Osmosis
- R26 Metals Recovery -- Solvent Extraction
- R29 Metals Recovery -- Other
- R99 Other Reuse or Recovery

Solidification/Stabilization

- G01 Cement Processes (including Silicates)
- G09 Other Pozzolonic Processes (including Silicates)
- G11 Asphaltic Processes
- G21 Thermoplastic Techniques
- G99 Other Solidification Processes

7.C   Range of Influent Concentration

The form requires an indication of the range of concentration of the toxic chemical in the wastestream (i.e., the influent) as it typically enters the treatment equipment. Enter in the space provided one of the following code numbers corresponding to the concentration of the chemical in the influent:

1 = Greater than 1 percent
2 = 100 parts per million (0.01 percent) to 1 percent (10,000 parts per million)
3 = 1 part per million to 100 parts per million
4 = 1 part per billion to 1 part per million
5 = Less than 1 part per billion

[Note: Parts per million (ppm) is:

❏ milligrams/kilogram (mass/mass) for solids and liquids;

❏ cubic centimeters/cubic meter (volume/volume) for gases;

❏ milligrams/liter for solutions or dispersions of the chemical in water; and

❏ milligrams of chemical/kilogram of air for particulates in air. If you have particulate concentrations (at standard temperature and pressure) as grains/cubic foot of air, multiply by 1766.6 to convert to parts per million; if in milligrams/cubic meter, multiply by 0.773 to obtain parts per million. Factors are for standard conditions of 0°C (32°F) and 760 mmHg atmospheric pressure.]

Chemical Hazard Communication Guidebook

Form R - Part III   Page 30

7.D Sequential Treatment?

The blocks in this column may be used in the following case:

- ☐ Individual treatment steps are used in a series to treat the chemical, but

- ☐ You have no data on the individual efficiencies of each step, but you are able to estimate the overall efficiency of the treatment sequence.

To report sequential treatment:

- ☐ List the appropriate codes for the treatment steps in the order that they occur (in column B) and then put an "X" in the boxes in column D for all these sequential treatment steps.

- ☐ Enter the appropriate code for the influent concentration (in column C) for <u>the first treatment step</u> in the sequence. Leave this item blank for the rest of the treatment steps in the sequence.

- ☐ Provide the overall treatment efficiency (in column E) for the entire sequence by entering that value in connection with the last treatment step in the sequence only. Enter NA in column E for the efficiency of all preceding steps in the sequence.

- ☐ Mark yes or no in column F only in connection with the final step in the sequence. Do not mark in this column for preceeding steps in the sequence.

An example of how to use the sequential treatment option is provided in Appendix C.

7.E Treatment Efficiency Estimate

In the space provided, enter the number indicating the percentage of the toxic chemical removed from the wastestream through destruction, biological degradation, chemical conversion, or physical removal. The treatment efficiency (expressed as percent removal) represents the mass or weight percentage of chemical destroyed or removed, not merely changes in volume or concentration of the chemical in the wastestream. The efficiency refers only to the percent destruction, degradation, conversion, or removal <u>of the listed toxic chemical</u> from the wastestream, not the percent conversion or removal of other wastestream constituents which may occur together with the listed chemical. The efficiency also does not refer to the general efficiency of the method for any wastestream. For some treatments, the percent removal will represent removal by several mechanisms, as in as aeration basin, where a chemical may evaporate, be biodegraded, or be physically removed in the sludge.

Percent removal must be calculated as follows:

$$\frac{(I - E)}{I} \times 100$$

where I = mass of the chemical in the influent wastestream and E = mass of the chemical in the effluent wastestream.

Calculate the mass or weight of chemical in the wastestream being treated by multiplying the concentration (by weight) of the chemical in the wastestream by the flow rate. In most cases, the percent removal compares the treated effluent to the influent for the particular type of wastestream. However, for some treatment methods, such as incineration or solidification of wastewater, the percent removal of the chemical from the influent wastestream would be reported as 100 percent because the wastestream does not exist in a comparable form after treatment. Some of the treatments (e.g., fuel blending and evaporation) do not destroy, chemically convert, or physically remove the chemical from its wastestream. For these treatment methods, an efficiency of zero must be reported.

For metal compounds, the calculation of the reportable concentration and treatment efficiency is based on the weight of the parent metal, not on the weight of the metal compounds. Metals are not destroyed, only physically removed or chemically converted from one form into another. The treatment efficiency reported represents only <u>physical removal</u> of the parent metal from the wastestream, not the percent chemical conversion of the metal compound. If a listed treatment method converts but does not remove a metal (e.g., chromium reduction), the method must be reported, but the treatment efficiency must be reported as zero.

Listed toxic chemicals which are strong mineral acids or bases which are neutralized to a pH between 6-9 are considered treated at a 100 percent efficiency.

All data available at your facility must be utilized to calculate treatment efficiency and influent chemical concentration. You are <u>not</u> required to collect any new data for the purposes of this reporting requirement. If data are lacking, estimates must be made using best engineering judgment or other methods.

7.F Based on Operating Data?

This column requires you to indicate "Yes" or "No" to whether the treatment efficiency estimate is based on actual operating data. For example, you would check "Yes" if the estimate is based on monitoring of influent <u>and</u> effluent wastes under typical operating conditions. For sequential treatment, <u>do not</u> indicate "Yes" or "No" in column F for a treatment step unless you have provided a treatment estimate in column E.

If the efficiency estimate is based on published data for similar processes or on equipment supplier's literature, or if you otherwise estimated either the influent or effluent waste comparison or the flow rate, check "No."

Appendix B: Regulations Reference Guide

Form R - Part III

EXAMPLE 11: Waste Treatment Methods

One wastestream generated by your facility is aqueous waste containing lead chromate, and lead selenate as discussed in a previous example in these instructions. In this example, the waste is transferred to off-site facilities after on-site wastewater treatment. The on-site wastewater treatment plant precipitates metal sludges. The wastewater is first treated with sulfuric acid and sodium disulfate to reduce the hexavalent chromate to trivalent chromium and then treated with lime to raise the pH. This precipitates chromium hydroxide, zinc hydroxide, and lead hydroxide, but does not remove the selenium. The selenium is removed from the wastewater by an ionic exchange system. The chromium, zinc, and lead hydroxide sludge (solid) waste is transferred to an off-site land disposal facility and the selenium-containing ion exchange resin is transferred to an off-site facility for metal recovery (off-site recovery should not be reported). The treated wastewater is sent to a POTW after neutralization. You would indicate the following treatment methods for the on-site treatment of each of the lead, zinc, chromium, and selenium compounds:

C21	-	Chromium Reduction
C01	-	Chemical Precipitation -- Lime or Sodium Hydroxide
R22	-	Metals Recovery -- Ion Exchange
C11	-	Neutralization

All sequential treatment steps must be indicated for all the metal compound categories reported even if the treatment method does not affect the particular metal. For example, ionic exchange must be reported as a treatment method for lead, zinc, chromium, and selenium compounds, even though the method affects only the selenium compound.

You would indicate a discharge to a POTW in Part III, Section 6.1.1 and the location of the POTW in Part II, Section 1.1. You would also indicate the release of the metal sludge to an off-site land disposal facility in Part III, Section 6.2.1.

8. POLLUTION PREVENTION: OPTIONAL INFORMATION ON WASTE MINIMIZATION

Information provided in Part III, Section 8, of Form R is optional. In this section, you may identify waste minimization efforts relating to the reported toxic chemical. Waste minimization reduces the amount of the toxic chemical in wastes by reducing waste generation or by recycling. This can be accomplished by equipment changes, process modifications, product reformulation, chemical substitutions, or other techniques. Waste minimization refers exclusively to practices which prevent the generation of wastes. Treatment or disposal does not minimize waste and should not be reported in this section. Recycling or reuse of a toxic chemical is considered waste minimization. Waste minimization applies to air emissions and wastewater, as well as to liquid or solid materials that are released, disposed of, or treated. For example, a program to recycle material from reactor cleaning could reduce the amount of a listed chemical in wastewater prior to treatment. This reduction might not show up in annual reports of releases to receiving streams (due to effective treatment, for example) but would be captured in this section.

8.A Type of Pollution Prevention Modification

Enter the one code from the following list that best describes the type of waste minimization activity:

M1 Recycling/Reuse On-Site
(e.g., solvent recovery still; vapor recovery system; reuse of materials in a process)

M2 Recycling/Reuse Off-Site
(e.g., commercial recycler; toll recycling; at an off-site company-owned facility)

M3 Equipment/Technology Modifications
(e.g., change from solvent to mechanical stripping; modify spray systems to reduce overspray losses; install floating roofs to reduce tank emissions; install float guards to prevent tank overflow)

M4 Process Procedure Modifications
(e.g., change production schedule to minimize equipment and feedstock change-overs; improved control of operating conditions; segregation of wastes to permit recycling)

M5 Reformulation/Redesign of Product
(e.g., change in product specifications; modify design or composition; reduce or modify packaging)

M6 Substitution of Raw Materials
(e.g., change or eliminate additives; substitute water-based for solvent-based coating materials, cleaners, and pigments; increase purity of raw materials)

M7 Improved Housekeeping, Training, Inventory Control
(e.g., alter maintenance frequency; institute leak detection program; improved inventory control; institute training program on waste minimization)

M8 Other Waste Minimization Technique
(e.g., elimination of process; discontinuation of product)

8.B Quantity of the Chemical in the Wastestream Prior to Treatment/Disposal

You may report the change in the amount of the chemical generated in either of two ways. You may provide the amount of the chemical in waste produced in the reporting year and the previous year, or you may report only the percent change.

Enter the total pounds of the toxic chemical contained **in all wastes from the reporting facility (air emissions, water discharges, solid wastes** and **off-site transfers)** generated during the reporting year. This quantity may be the sum of all the release amounts reported on Form R if there is no on-site treatment of the chemical. The quantity will often be greater than the total reported release amounts because it includes waste prior to treatment.

You should consider only the quantity of the chemical in the waste. Do not report the total mass of the waste (i.e., do not include the weight of water, soil, or waste constituents which are not reportable on Form R).

Similarly, report total pounds of the toxic chemical contained in all wastes generated for the year prior to the reporting year.

Alternatively, to protect confidential information, you may wish to enter only the percentage by which the weight of the chemical in the wastes has changed. This figure may be calculated using the following formula:

$$\frac{(W_c - W_p)}{W_p} \times 100$$

where:

W_c = weight of toxic chemical in total wastes for **reporting year**

W_p = weight of toxic chemical in total wastes for the **prior year**

Note that the resulting figure will very often be negative (indicating that the total amount of waste generated has been reduced in the current year). Be sure to check-off the appropriate sign for the value where indicated on Form R.

8.C Waste Minimization Index

Enter the ratio of reporting-year production to the prior reporting-year production. This index should be calculated to most closely reflect activities involving the chemical. To determine the index, divide the production amount, which was chosen as a measure of the current reporting year's production level, by the prior year's production amount.

The index provides a means for users of the data to distinguish effects due to changes in business activity from the effects specifically due to waste minimization efforts. It is not necessary to indicate the units on which the index is based. The index should not be based on the dollar value of sales. Examples of acceptable indices include:

❑ Amount of chemical produced in 1989/amount of chemical produced in 1988. For example, a company manufactures 200,000 pounds of a chemical in 1988 and 250,000 pounds of the same chemical in 1989. The index figure to report would be 1.3 (1.25 rounded to two significant digits).

❑ Amount of paint produced in 1989/amount of paint produced in 1988.

❑ Number of appliances coated in 1989/number of appliances coated in 1988.

❑ Square feet of solar collector fabricated in 1989/square feet of solar collector fabricated in 1988.

8.D Reason for Action

Finally, enter the most appropriate code from the following list that best describes the reason for initiating the waste minimization effort:

R1   Regulatory Requirement for the Waste
R2   Reduction of Treatment/Disposal Costs
R3   Other Process Cost Reduction
R4   Self-Initiated Review
R5   Other (e.g., discontinuation of product, occupational safety).

EXAMPLE 12: WASTE MINIMIZATION (POLLUTION PREVENTION)

A facility stores toluene in a large tank, and continuously uses it as a raw material in a chemical process throughout the reporting year. Prior to the current reporting year, annual air emissions of toluene were 100,000 pounds from the tank, and another 100,000 pounds from process emissions. In addition, 150,000 pounds of sludges are created from the process and from storage tanks. The sludge contains a total of 25,000 pounds of toluene which was burned in an on-site incinerator. The Form R filed by the facility for the prior year indicated 200,000 pounds of toluene air emissions. The toluene contained in the sludge was identified as treated on-site, although the pre-treated amount of the toluene was not indicated on the Form R, since this information is not required under section 313.

At the beginning of the current reporting year, the facility installed a floating roof in its storage tank. This change reduced fugitive emissions from the tank 90 percent, from 100,000 pounds per year to 10,000 pounds. Process emissions and sludge generation remained the same.

Based on this information, Part III, Section 8 of Form R would be completed as follows:

A.   <u>Type of Modification</u>

   M3: Equipment/Technology Modification.

B. **Quantity of the Chemical in the Wastestream Prior to Treatment/Disposal**

	Tank Emissions of Toluene	Process Emissions of Toluene	Toluene in Sludges	Total Toluene Wastes
Total toluene wastes for current reporting year (pounds)	W_c = 10,000 +	100,000 +	25,000 =	135,000
Total toluene wastes for prior year (pounds)	W_p = 100,000 +	100,000 +	25,000 =	225,000

Note that only the weight of the toluene in the sludge (25,000 pounds) and not the full weight of the sludge (150,000 pounds) is included in the calculation.

The facility would record 135,000 pounds as the current reporting year waste generation (W_c), and 225,000 pounds as the prior year's waste generation (W_p).

Alternatively, the facility may opt to report only the percent change as follows:

$$\frac{(W_c - W_p)}{W_p} \times 100 = \frac{135,000 - 225,000}{225,000} \times 100$$

$$= -40\%$$

Even though the floating roof achieved a 90% reduction of toluene emissions from the tank, the overall facility-wide change in toluene waste generation is _negative_ 40% -- this is the figure that should be reported in the "or percent change" part of Section 8 of Form R.

Increases in waste generation, created by production increases that were greater than the impact of waste minimization, would be reported as a _positive_ percentage change.

C. **Index**

Usage of toluene at this facility remained the same for both years, resulting in an index of 1.0. If usage had been reduced by half, the index would have been 0.5.

D. **Reason for Action**

The facility identified code R3, Other Process Cost Reduction, as the major reason for the waste minimization action.

TABLE I

SIC CODES 20-39

20 Food and Kindred Products

- 2011 Meat packing plants
- 2013 Sausages and other prepared meat products
- 2015 Poultry slaughtering and processing
- 2021 Creamery butter
- 2022 Natural, processed, and imitation cheese
- 2023 Dry, condensed, and evaporated dairy products
- 2024 Ice cream and frozen desserts
- 2026 Fluid milk
- 2032 Canned specialties
- 2033 Canned fruits, vegetables, preserves, jams, and jellies
- 2034 Dried and dehydrated fruits, vegetables, and soup mixes
- 2035 Pickled fruits and vegetables, vegetable sauces and seasonings, and salad dressings
- 2037 Frozen fruits, fruit juices, and vegetables
- 2038 Frozen specialties, n.e.c.*
- 2041 Flour and other grain mill products
- 2043 Cereal breakfast foods
- 2044 Rice milling
- 2045 Prepared flour mixes and doughs
- 2046 Wet corn milling
- 2047 Dog and cat food
- 2048 Prepared feeds and feed ingredients for animals and fowls, except dogs and cats
- 2051 Bread and other bakery products, except cookies and crackers
- 2052 Cookies and crackers
- 2053 Frozen bakery products, except bread
- 2061 Cane sugar, except refining
- 2062 Cane sugar refining
- 2063 Beet sugar
- 2064 Candy and other confectionary products
- 2066 Chocolate and cocoa products
- 2067 Chewing gum
- 2068 Salted and roasted nuts and seeds
- 2074 Cottonseed oil mills
- 2075 Soybean oil mills
- 2076 Vegetable oil mills, except corn, cottonseed, and soybean
- 2077 Animal and marine fats and oils
- 2079 Shortening, table oils, margarine, and other edible fats and oils, n.e.c.*
- 2082 Malt beverages
- 2083 Malt
- 2084 Wines, brandy, and brandy spirits
- 2085 Distilled and blended liquors
- 2086 Bottled and canned soft drinks and carbonated waters
- 2087 Flavoring extracts and flavoring syrups, n.e.c.*
- 2091 Canned and cured fish and seafoods
- 2092 Prepared fresh or frozen fish and seafoods
- 2095 Roasted coffee
- 2096 Potato chips, corn chips, and similar snacks
- 2097 Manufactured ice
- 2098 Macaroni, spaghetti, vermicelli, and noodles
- 2099 Food preparations, n.e.c.*

21 Tobacco Products

- 2111 Cigarettes
- 2121 Cigars
- 2131 Chewing and smoking tobacco and snuff
- 2141 Tobacco stemming and redrying

22 Textile Mill Products

- 2211 Broadwoven fabric mills, cotton
- 2221 Broadwoven fabric mills, manmade fiber, and silk
- 2231 Broadwoven fabric mills, wool (including dyeing and finishing)
- 2241 Narrow fabric and other smallwares mills: cotton, wool, silk, and manmade fiber
- 2251 Women's full length and knee length hosiery, except socks
- 2252 Hosiery, n.e.c.*
- 2253 Knit outerwear mills
- 2254 Knit underwear and nightwear mills
- 2257 Weft knit fabric mills
- 2258 Lace and warp knit fabric mills
- 2259 Knitting mills, n.e.c.*
- 2261 Finishers of broadwoven fabrics of cotton
- 2262 Finishers of broadwoven fabrics of manmade fiber and silk
- 2269 Finishers of textiles, n.e.c.*
- 2273 Carpets and rugs
- 2281 Yarn spinning mills
- 2282 Yarn texturizing, throwing, twisting, and winding mills
- 2284 Thread mills
- 2295 Coated fabrics, not rubberized
- 2296 Tire cord and fabrics
- 2297 Nonwoven fabrics
- 2298 Cordage and twine
- 2299 Textile goods, n.e.c.*

23 Apparel and Other Finished Products made from Fabrics and Other Similar Materials

- 2311 Men's and boys' suits, coats, and overcoats

*"Not elsewhere classified" indicated by "n.e.c."

- 2321 Men's and boys' shirts, except work shirts
- 2322 Men's and boys' underwear and nightwear
- 2323 Men's and boys' neckwear
- 2325 Men's and boys' separate trousers and slacks
- 2326 Men's and boys' work clothing
- 2329 Men's and boys' clothing, n.e.c.*
- 2331 Women's, misses', and juniors' blouses and shirts
- 2335 Women's, misses', and juniors' dresses
- 2337 Women's, misses', and juniors' suits, skirts, and coats
- 2339 Women's, misses', and juniors', outerwear, n.e.c.*
- 2341 Women's, misses', children's, and infants' underwear and nightwear
- 2342 Brassieres, girdles, and allied garments
- 2353 Hats, caps, and millinery
- 2361 Girls', children's and infants' dresses, blouses, and shirts
- 2369 Girls', children's and infants' outerwear, n.e.c.*
- 2371 Fur goods
- 2381 Dress and work gloves, except knit and all leather
- 2384 Robes and dressing gowns
- 2385 Waterproof outerwear
- 2386 Leather and sheep lined clothing
- 2387 Apparel belts
- 2389 Apparel and accessories, n.e.c.*
- 2391 Curtains and draperies
- 2392 Housefurnishings, except curtains and draperies
- 2393 Textile bags
- 2394 Canvas and related products
- 2395 Pleating, decorative and novelty stitching, and tucking for the trade
- 2396 Automotive trimmings, apparel findings, and related products
- 2397 Schiffli machine embroideries
- 2399 Fabricated textile products, n.e.c.*

24 Lumber and Wood Products, Except Furniture

- 2411 Logging
- 2421 Sawmills and planing mills, general
- 2426 Hardwood dimension and flooring mills
- 2429 Special product sawmills, n.e.c.*
- 2431 Millwork
- 2434 Wood kitchen cabinets
- 2435 Hardwood veneer and plywood
- 2436 Softwood veneer and plywood
- 2439 Structural wood members, n.e.c.*
- 2441 Nailed and lock corner wood boxes and shook
- 2448 Wood pallets and skids
- 2449 Wood containers, n.e.c.*
- 2451 Mobile homes
- 2452 Prefabricated wood buildings and components
- 2491 Wood preserving
- 2493 Reconstituted wood products
- 2499 Wood products, n.e.c.*

25 Furniture and Fixtures

- 2511 Wood household furniture, except upholstered
- 2512 Wood household furniture, upholstered
- 2514 Metal household furniture
- 2515 Mattresses, foundations, and convertible beds
- 2517 Wood television, radio, phonograph, and sewing machine cabinets
- 2519 Household furniture, n.e.c.*
- 2521 Wood office furniture
- 2522 Office furniture, except wood
- 2531 Public building and related furniture
- 2541 Wood office and store fixtures, partitions, shelving, and lockers
- 2542 Office and store fixtures, partitions, shelving, and lockers, except wood
- 2591 Drapery hardware and window blinds and shades
- 2599 Furniture and fixtures, n.e.c.*

26 Paper and Allied Products

- 2611 Pulp mills
- 2621 Paper mills
- 2631 Paperboard mills
- 2652 Setup paperboard boxes
- 2653 Corrugated and solid fiber boxes
- 2655 Fiber cans, tubes, drums, and similar products
- 2656 Sanitary food containers, except folding
- 2657 Folding paperboard boxes, including sanitary
- 2671 Packaging paper and plastics film, coated and laminated
- 2672 Coated and laminated paper, n.e.c.*
- 2673 Plastics, foil, and coated paper bags
- 2674 Uncoated paper and multiwall bags
- 2675 Die-cut paper and paperboard and cardboard
- 2676 Sanitary paper products
- 2677 Envelopes
- 2678 Stationery tablets, and related products
- 2679 Converted paper and paperboard products, n.e.c.*

27 Printing, Publishing, and Allied Industries

- 2711 Newspapers: publishing, or publishing and printing
- 2721 Periodicals: publishing, or publishing and printing
- 2731 Books: publishing, or publishing and printing
- 2732 Book printing
- 2741 Miscellaneous publishing
- 2752 Commercial printing, lithographic
- 2754 Commercial printing, gravure
- 2759 Commercial printing, n.e.c.*
- 2761 Manifold business forms
- 2771 Greeting cards
- 2782 Blankbooks, looseleaf binders and devices

*"Not elsewhere classified" indicated by "n.e.c."

Chemical Hazard Communication Guidebook

Page 36

2789	Bookbinding and related work
2791	Typesetting
2796	Platemaking and related services

28 Chemicals and Allied Products

2812	Alkalies and chlorine
2813	Industrial gases
2816	Inorganic pigments
2819	Industrial inorganic chemicals, n.e.c.*
2821	Plastics materials, synthetic resins, and non-vulcanizable elastomers
2822	Synthetic rubber (vulcanizable elastomers)
2823	Cellulosic manmade fibers
2824	Manmade organic fibers, except cellulosic
2833	Medicinal chemicals and botanical products
2834	Pharmaceutical preparations
2835	In vitro and in vivo diagnostic substances
2836	Biological products, except diagnostic substances
2841	Soap and other detergents, except specialty cleaners
2842	Specialty cleaning, polishing, and sanitation preparations
2843	Surface active agents, finishing agents, sulfonated oils, and assistants
2844	Perfumes, cosmetics, and other toilet preparations
2851	Paints, varnishes, lacquers, enamels, and allied products
2861	Gum and wood chemicals
2865	Cyclic organic crudes and intermediates, and organic dyes and pigments
2869	Industrial organic chemicals, n.e.c.*
2873	Nitrogenous fertilizers
2874	Phosphatic fertilizers
2875	Fertilizers, mixing only
2879	Pesticides and agricultural chemicals, n.e.c.*
2891	Adhesives and sealants
2892	Explosives
2893	Printing ink
2895	Carbon black
2899	Chemicals and chemical preparations, n.e.c.*

29 Petroleum Refining and Related Industries

2911	Petroleum refining
2951	Asphalt paving mixtures and blocks
2952	Asphalt felts and coatings
2992	Lubricating oils and greases
2999	Products of petroleum and coal, n.e.c.*

30 Rubber and Miscellaneous Plastics Products

3011	Tires and inner tubes
3021	Rubber and plastics footwear
3052	Rubber and plastics hose and belting
3053	Gaskets, packing, and sealing devices
3061	Molded, extruded, and lathecut mechanical rubber products
3069	Fabricated rubber products, n.e.c.*
3081	Unsupported plastics film and sheet
3082	Unsupported plastics profile shapes
3083	Laminated plastics plate, sheet, and profile shapes
3084	Plastics pipe
3085	Plastics bottles
3086	Plastics foam products
3087	Custom compounding of purchased plastics resins
3088	Plastics plumbing fixtures
3089	Plastics products, n.e.c.*

31 Leather and Leather Products

3111	Leather tanning and finishing
3131	Boot and shoe cut stock and findings
3142	House slippers
3143	Men's footwear, except athletic
3144	Women's footwear, except athletic
3149	Footwear, except rubber, n.e.c.*
3151	Leather gloves and mittens
3161	Luggage
3171	Women's handbags and purses
3172	Personal leather goods, except women's handbags and purses
3199	Leather goods, n.e.c.*

32 Stone, Clay, Glass and Concrete Products

3211	Flat glass
3221	Glass containers
3229	Pressed and blown glass and glassware, n.e.c.*
3231	Glass products, made of purchased glass
3241	Cement, hydraulic
3251	Brick and structural clay tile
3253	Ceramic wall and floor tile
3255	Clay refractories
3259	Structural clay products, n.e.c.*
3261	Vitreous china plumbing fixtures and china and earthenware fittings and bathroom accessories
3262	Vitreous china table and kitchen articles
3263	Fine earthenware (whiteware) table and kitchen articles
3264	Porcelain electrical supplies
3269	Pottery products, n.e.c.*
3271	Concrete block and brick
3272	Concrete products, except block and brick
3273	Ready mixed concrete
3274	Lime
3275	Gypsum products
3281	Cut stone and stone products
3291	Abrasive products
3292	Asbestos products

*"Not elsewhere classified" indicated by "n.e.c."

Appendix B: Regulations Reference Guide

3295 Minerals and earths, ground or otherwise treated
3296 Mineral wool
3297 Nonclay refractories
3299 Nonmetallic mineral products, n.e.c.*

33 Primary Metal Industries

3312 Steel works, blast furnaces (including coke ovens), and rolling mills
3313 Electrometallurgical products, except steel
3315 Steel wiredrawing and steel nails and spikes
3316 Cold-rolled steel sheet, strip, and bars
3317 Steel pipe and tubes
3321 Gray and ductile iron foundries
3322 Malleable iron foundries
3324 Steel investment foundries
3325 Steel foundries, n.e.c.*
3331 Primary smelting and refining of copper
3334 Primary production of aluminum
3339 Primary smelting and refining of nonferrous metals, except copper and aluminum
3341 Secondary smelting and refining of nonferrous metals
3351 Rolling, drawing, and extruding of copper
3353 Aluminum sheet, plate, and foil
3354 Aluminum extruded products
3355 Aluminum rolling and drawing, n.e.c.*
3356 Rolling, drawing, and extruding of nonferrous metals, except copper and aluminum
3357 Drawing and insulating of nonferrous wire
3363 Aluminum die-castings
3364 Nonferrous die-castings, except aluminum
3365 Aluminum foundries
3366 Copper foundries
3369 Nonferrous foundries, except aluminum and copper
3398 Metal heat treating
3399 Primary metal products, n.e.c.*

34 Fabricated Metal Products, except Machinery and Transportation Equipment

3411 Metal cans
3412 Metal shipping barrels, drums, kegs, and pails
3421 Cutlery
3423 Hand and edge tools, except machine tools and handsaws
3425 Handsaws and saw blades
3429 Hardware, n.e.c.*
3431 Enameled iron and metal sanitary ware
3432 Plumbing fixture fittings and trim
3433 Heating equipment, except electric and warm air furnaces
3441 Fabricated structural metal
3442 Metal doors, sash, frames, molding, and trim
3443 Fabricated plate work (boiler shops)
3444 Sheet metal work
3446 Architectural and ornamental metal work
3448 Prefabricated metal buildings and components
3449 Miscellaneous structural metal work
3451 Screw machine products
3452 Bolts, nuts, screws, rivets, and washers
3462 Iron and steel forgings
3463 Nonferrous forgings
3465 Automotive stampings
3468 Crowns and closures
3469 Metal stampings, n.e.c.*
3471 Electroplating, plating, polishing, anodizing, and coloring
3479 Coating, engraving and allied services, n.e.c.*
3482 Small arms ammunition
3483 Ammunition, except for small arms
3484 Small arms
3489 Ordnance and accessories, n.e.c.*
3491 Industrial valves
3492 Fluid power valves and hose fittings
3493 Steel springs, except wire
3494 Valves and pipe fittings, n.e.c.*
3495 Wire springs
3496 Miscellaneous fabricated wire products
3497 Metal foil and leaf
3498 Fabricated pipe and pipe fittings
3499 Fabricated metal products, n.e.c.*

35 Industrial and Commercial Machinery and Computer Equipment

3511 Steam, gas and hydraulic turbines, and turbine generator set units
3519 Internal combustion engines, n.e.c.*
3523 Farm machinery and equipment
3524 Lawn and garden tractors and home lawn and garden equipment
3531 Construction machinery and equipment
3532 Mining machinery and equipment, except oil and gas field machinery and equipment
3533 Oil and gas field machinery and equipment
3534 Elevators and moving stairways
3535 Conveyors and conveying equipment
3536 Overhead traveling cranes, hoists, and monorail systems
3537 Industrial trucks, tractors, trailers, and stackers
3541 Machine tools, metal cutting types
3542 Machine tools, metal forming types
3543 Industrial patterns
3544 Special dies and tools, die sets, jigs and fixtures, and industrial molds
3545 Cutting tools, machine tool accessories, and machinists' measuring devices
3546 Power driven handtools

*"Not elsewhere classified" indicated by "n.e.c."

3547 Rolling mill machinery and equipment
3548 Electric and gas welding and soldering equipment
3549 Metalworking machinery, n.e.c.*
3552 Textile machinery
3553 Woodworking machinery
3554 Paper industries machinery
3555 Printing trades machinery and equipment
3556 Food products machinery
3559 Special industry machinery, n.e.c.*
3561 Pumps and pumping equipment
3562 Ball and roller bearings
3563 Air and gas compressors
3564 Industrial and commercial fans and blowers and air purification equipment
3565 Packaging equipment
3566 Speed changers, industrial high speed drives, and gears
3567 Industrial process furnaces and ovens
3568 Mechanical power transmission equipment, n.e.c.*
3569 General industrial machinery and equipment, n.e.c.*
3571 Electronic computers
3572 Computer storage devices
3575 Computer terminals
3577 Computer peripheral equipment, n.e.c.*
3578 Calculating and accounting machines, except electronic computers
3579 Office machines, n.e.c.*
3581 Automatic vending machines
3582 Commercial laundry, drycleaning, and pressing machines
3585 Air conditioning and warm air heating equipment and commercial and industrial refrigeration equipment
3586 Measuring and dispensing pumps
3589 Service industry machinery, n.e.c.*
3592 Carburetors, pistons, piston rings, and valves
3593 Fluid power cylinders and actuators
3594 Fluid power pumps and motors
3596 Scales and balances, except laboratory
3599 Industrial and commercial machinery and equipment, n.e.c*

36 Electronic and Other Electrical Equipment and Components, Except Computer Equipment

3612 Power, distribution, and specialty transformers
3613 Switchgear and switchboard apparatus
3621 Motors and generators
3624 Carbon and graphite products
3625 Relays and industrial controls
3629 Electrical industrial appliances, n.e.c.*
3631 Household cooking equipment
3632 Household refrigerators and home and farm freezers
3633 Household laundry equipment
3634 Electrical housewares and fans
3635 Household vacuum cleaners
3639 Household appliances, n.e.c.*
3641 Electric lampbulbs and tubes
3643 Current carrying wiring devices
3644 Noncurrent carrying wiring devices
3645 Residential electric lighting fixtures
3646 Commercial, industrial, and institutional electric lighting fixtures
3647 Vehicular lighting equipment
3648 Lighting equipment, n.e.c.*
3651 Household audio and video equipment
3652 Phonograph records and pre-recorded audio tapes and disks
3661 Telephone and telegraph apparatus
3663 Radio and television broadcasting and communications equipment
3669 Communications equipment, n.e.c.*
3671 Electron tubes
3672 Printed circuit boards
3674 Semiconductors and related devices
3675 Electronic capacitors
3676 Electronic resistors
3677 Electronic coils, transformers, and other inductors
3678 Electronic connectors
3679 Electronic components, n.e.c.*
3691 Storage batteries
3692 Primary batteries, dry and wet
3694 Electric equipment for internal combustion engines
3695 Magnetic and optical recording media
3699 Electrical machinery, equipment, and supplies, n.e.c.*

37 Transportation Equipment

3711 Motor vehicles and passenger car bodies
3713 Truck and bus bodies
3714 Motor vehicle parts and accessories
3715 Truck trailers
3716 Motor homes
3721 Aircraft
3724 Aircraft engines and engine parts
3728 Aircraft parts and auxiliary equipment, n.e.c.*
3731 Ship building and repairing
3732 Boat building and repairing
3743 Railroad equipment
3751 Motorcycles, bicycles and parts
3761 Guided missiles and space vehicles
3764 Guided missile and space vehicle propulsion units and propulsion unit parts
3769 Guided missile and space vehicle parts and auxiliary equipment, n.e.c.*
3792 Travel trailers and campers
3795 Tanks and tank components
3799 Transportation equipment, n.e.c.*

*"Not elsewhere classified" indicated by "n.e.c."

Appendix B: Regulations Reference Guide

38 Measuring, Analyzing, and Controlling Instruments; Photographic, Medical and Optical Goods; Watches and Clocks

- 3812 Search, detection, navigation, guidance, aeronautical, and nautical systems and instruments
- 3821 Laboratory apparatus and furniture
- 3822 Automatic controls for regulating residential and commercial environments and appliances
- 3823 Industrial instruments for measurement, display, and control of process variables; and related products
- 3824 Totalizing fluid meters and counting devices
- 3825 Instruments for measuring and testing of electricity and electrical signals
- 3826 Laboratory analytical instruments
- 3827 Optical instruments and lenses
- 3829 Measuring and controlling devices, n.e.c.*
- 3841 Surgical and medical instruments and apparatus
- 3842 Orthopedic, prosthetic, and surgical appliances and supplies
- 3843 Dental equipment and supplies
- 3844 X-ray apparatus and tubes and related irradiation apparatus
- 3845 Electromedical and electrotherapeutic apparatus
- 3851 Ophthalmic goods
- 3861 Photographic equipment and supplies
- 3873 Watches, clocks, clockwork operated devices, and parts

39 Miscellaneous Manufacturing Industries

- 3911 Jewelry, precious metal
- 3914 Silverware, plated ware, and stainless steel ware
- 3915 Jewelers' findings and materials, and lapidary work
- 3931 Musical instruments
- 3942 Dolls and stuffed toys
- 3944 Games, toys and children's vehicles; except dolls and bicycles
- 3949 Sporting and athletic goods, n.e.c.*
- 3951 Pens, mechanical pencils, and parts
- 3952 Lead pencils, crayons, and artists' materials
- 3953 Marking devices
- 3955 Carbon paper and inked ribbons
- 3961 Costume jewelry and costume novelties, except precious metal
- 3965 Fasteners, buttons, needles, and pins
- 3991 Brooms and brushes
- 3993 Signs and advertising specialties
- 3995 Burial caskets
- 3996 Linoleum, asphalted-felt-base, and other hard surface floor coverings, n.e.c.*
- 3999 Manufacturing industries, n.e.c.*

*"Not elsewhere classified" indicated by "n.e.c."

EPA'S TRADE SECRECY CLAIMS FOR EMERGENCY PLANNING AND COMMUNITY RIGHT-TO-KNOW INFORMATION: AND TRADE SECRET DISCLOSURES TO HEALTH PROFESSIONALS

Chemical Hazard Communication Guidebook

40 CFR Ch. I (7-1-89 Edition)

Sec.
350.13 Sufficiency of assertions.
350.15 Public petitions requesting disclosure of chemical identity claimed as trade secret.
350.16 Address to send trade secrecy claims and petitions requesting disclosure.
350.17 Appeals.
350.18 Release of chemical identity determined to be non-trade secret; notice of intent to release chemical identity.
350.19 Provision of information to States.
350.21 Adverse health effects.
350.23 Disclosure to authorized representatives.
350.25 Disclosure in special circumstances.
350.27 Substantiation form to accompany claims of trade secrecy, instructions to substantiation form.
APPENDIX A TO SUBPART A—RESTATEMENT OF TORTS SECTION 757, COMMENT b

Subpart B—Disclosure of Trade Secret Information to Health Professionals

350.40 Disclosure to health professionals.
AUTHORITY: 42 U.S.C. 11042, 11043 and 11048 Pub. L. 99-499, 100 Stat. 1747.

SOURCE: 53 FR 28801, July 29, 1988, unless otherwise noted.

Subpart A—Trade Secrecy Claims

§ 350.1 Definitions.

"Administrator" and "General Counsel" mean the EPA officers or employees occupying the positions so titled.

"Business confidentiality" or "confidential business information" includes the concept of trade secrecy and other related legal concepts which give (or may give) a business the right to preserve the confidentiality of business information and to limit its use or disclosure by others in order that the business may obtain or retain business advantages it derives from its right in the information. The definition is meant to encompass any concept which authorizes a Federal agency to withhold business information under 5 U.S.C. 552(b)(4), as well as any concept which requires EPA to withhold information from the public for the benefit of a business under 18 U.S.C. 1905.

"Claimant" means a person submitting a claim of trade secrecy to EPA in connection with a chemical otherwise required to be disclosed in a report or other filing made under Title III.

PART 350—TRADE SECRECY CLAIMS FOR EMERGENCY PLANNING AND COMMUNITY RIGHT-TO-KNOW INFORMATION: AND TRADE SECRET DISCLOSURES TO HEALTH PROFESSIONALS

Subpart A—Trade Secrecy Claims

Sec.
350.1 Definitions.
350.3 Applicability of subpart; priority where provisions conflict; interaction with 40 CFR Part 2.
350.5 Assertion of claims of trade secrecy.
350.7 Substantiating claims of trade secrecy.
350.9 Initial action by EPA.
350.11 Review of claim.

200

Appendix B: Regulations Reference Guide

Environmental Protection Agency § 350.3

"Petitioner" is any person who submits a petition under this regulation requesting disclosure of a chemical identity claimed as trade secret.

"Sanitized" means a version of a document from which information claimed as trade secret or confidential has been omitted or withheld.

"Senior management official" means an official with management responsibility for the person or persons completing the report, or the manager of environmental programs for the facility or establishments, or for the corporation owning or operating the facility or establishments responsible for certifying similar reports under other environmental regulatory requirements.

"Specific chemical identity" means the chemical name, Chemical Abstracts Service (CAS) Registry Number, or any other information that reveals the precise chemical designation of the substance. Where the trade name is reported in lieu of the specific chemical identity, the trade name will be treated as the specific chemical identity for purposes of this part.

"Submitter" means a person filing a required report or making a claim of trade secrecy to EPA under sections 303 (d)(2) and (d)(3), 311, 312, and 313 of Title III.

"Substantiation" means the written answers submitted to EPA by a submitter to the specific questions set forth in this regulation in support of a claim that chemical identity is a trade secret.

"Title III" means Title III of the Superfund Amendments and Reauthorization Act of 1986, also titled the Emergency Planning and Community Right-to-Know Act of 1986.

"Trade secrecy claim" is a submittal under sections 303 (d)(2) or (d)(3), 311, 312 or 313 of Title III in which a chemical identity is claimed as trade secret, and is accompanied by a substantiation in support of the claim of trade secrecy for chemical identity.

"Trade secret" means any confidential formula, pattern, process, device, information or compilation of information that is used in a submitter's business, and that gives the submitter an opportunity to obtain an advantage over competitors who do not know or use it. EPA intends to be guided by the Restatement of Torts, Section 757, Comment b.

"Unsanitized" means a version of a document from which information claimed as trade secret or confidential has not been withheld or omitted.

"Working day" is any day on which Federal government offices are open for normal business. Saturdays, Sundays, and official Federal holidays are not working days; all other days are.

§ 350.3 Applicability of subpart; priority where provisions conflict; interaction with 40 CFR Part 2.

(a) *Applicability of subpart.* Sections 350.1 through 350.27 establish rules governing assertion of trade secrecy claims for chemical identity information collected under the authority of sections 303 (d)(2) and (d)(3), 311, 312 and 313 of Title III of the Superfund Amendments and Reauthorization Act of 1986, and for trade secrecy or business confidentiality claims for information submitted in a substantiation under sections 303 (d)(2) and (d)(3), 311, 312, and 313 of Title III. This subpart also establishes rules governing petitions from the public requesting the disclosure of chemical identity claimed as trade secret, and determinations by EPA of whether this information is entitled to trade secret treatment. Claims for confidentiality of the location of a hazardous chemical under section 312(d)(2)(F) of Title III are not subject to the requirements of this subpart.

(b) *Priority where provisions conflict.* Where information subject to the requirements of this subpart is also collected under another statutory authority, the confidentiality provisions of that authority shall be used to claim that information as trade secret or confidential when submitting it to EPA under that statutory authority.

(c) *Interaction with 40 CFR Part 2, EPA's Freedom of Information Act procedures.* (1) No trade secrecy or business confidentiality claims other than those allowed in this subpart are permitted for information collected under sections 303 (d)(2) and (d)(3), 311, 312 and 313 of Title III.

201

§ 350.5

(2) Except as provided in § 350.25 of this subpart, request for access to chemical identities withheld as trade secret under this regulation is solely through this regulation and procedures hereunder, not through EPA's Freedom of Information Act procedures set forth at 40 CFR Part 2.

(3) Request for access to information other than chemical identity submitted to EPA under this regulation is through EPA's Freedom of Information Act regulations at 40 CFR Part 2.

§ 350.5 Assertion of claims of trade secrecy.

(a) A claim of trade secrecy may be made only for the specific chemical identity of an extremely hazardous substance under sections 303 (d)(2) and (d)(3), a hazardous chemical under sections 311 and 312, and a toxic chemical under section 313.

(b) Method of asserting claims of trade secrecy for information submitted under sections 303 (d)(2) and (d)(3).

(1) In submitting information to the local emergency planning committee under sections 303 (d)(2) or (d)(3), the submitter may claim as trade secret the specific chemical identity of any chemical subject to reporting under section 303.

(2) To make a claim, the submitter shall submit to EPA the following:

(i) A copy of the information which is being submitted under sections 303 (d)(2) or (d)(3) to the local emergency planning committee, with the chemical identity or identities claimed trade secret deleted, and the generic class or category of the chemical identity or identities inserted in its place. The method of choosing generic class or category is set forth in paragraph (f) of this section.

(ii) A sanitized and unsanitized substantiation in accordance with § 350.7 for each chemical identity claimed as trade secret.

(3) If the submitter wishes to claim information in the substantiation as trade secret or business confidential, it shall do so in accordance with § 350.7(d).

(4) Section 303 claims shall be sent to the address specified in § 350.16 of this regulation.

(c) Method of asserting claims of trade secrecy for information submitted under section 311.

(1) Submitters may claim as trade secret the specific chemical identity of any chemical subject to reporting under section 311 on the material safety data sheet or chemical list under section 311.

(2) To assert a claim for a chemical identity on a material safety data sheet under section 311, the submitter shall submit to EPA the following:

(i) One copy of the material safety data sheet which is being submitted to the State emergency response commission, the local emergency planning committee and the local fire department, which shall make it available to the public. In place of the specific chemical identity claimed as trade secret, the generic class or category of the chemical claimed as trade secret shall be inserted. The method of choosing generic class or category is set forth in paragraph (f) of this section.

(ii) A sanitized and unsanitized substantiation in accordance with § 350.7 for every chemical identity claimed as trade secret.

(3) To assert a claim for a chemical identity on a list under section 311, the submitter shall submit to EPA the following:

(i) An unsanitized copy of the chemical list under section 311. The submitter shall clearly indicate the specific chemical identity claimed as trade secret, and shall label it *"Trade Secret."* The generic class or category of the chemical claimed as trade secret shall be inserted directly below the claimed chemical identity. The method of choosing generic class or category is set forth in paragraph (f) of this section.

(ii) A sanitized copy of the chemical list under section 311. This copy shall be identical to the document in paragraph (c)(3)(i) of this section except that the submitter shall delete the chemical identity claimed as trade secret, leaving in place the generic class or category of the chemical claimed as trade secret. This copy shall be sent by the submitter to the State emergency response commission, the local emergency planning commit-

202

Environmental Protection Agency § 350.5

tee and the local fire department, which shall make it available to the public.

(iii) A sanitized and unsanitized substantiation in accordance with § 350.7 for every chemical identity claimed as trade secret.

(4) If the submitter wishes to claim information in the substantiation as trade secret or business confidential, it shall do so in accordance with § 350.7(d).

(5) Section 311 claims shall be sent to the address specified in § 350.16 of this regulation.

(d) Method of asserting claims of trade secrecy for information submitted under section 312.

(1) Submitters may claim as trade secret the specific chemical identity of any chemical subject to reporting under section 312.

(2) To assert a claim the submitter shall submit to EPA the following:

(i) An unsanitized copy of the Tier II emergency and hazardous chemical inventory form under section 312. (The Tier I emergency and hazardous chemical inventory form does not require the reporting of specific chemical identity and therefore no trade secrecy claims may be made with respect to that form.) The submitter shall clearly indicate the specific chemical identity claimed as trade secret by checking the box marked "trade secret" next to the claimed chemical identity.

(ii) A sanitized copy of the Tier II emergency and hazardous chemical inventory form. This copy shall be identical to the document in paragraph (d)(2)(i) of this section except that the submitter shall delete the chemical identity or identities claimed as trade secret and include instead the generic class or category of the chemical claimed as trade secret. The method of choosing generic class or category is set forth in paragraph (f) of this section. The sanitized copy shall be sent by the submitter to the State emergency response commission, local emergency planning committee or the local fire department, whichever entity requested the information.

(iii) A sanitized and unsanitized substantiation in accordance with § 350.7 for every chemical identity claimed as trade secret.

(3) If the submitter wishes to claim information in the substantiation as trade secret or business confidential, it shall do so in accordance with § 350.7(d).

(4) Section 312 claims shall be sent to the address specified in § 350.16 of this regulation.

(e) Method of asserting claims of trade secrecy for information submitted under section 313.

(1) Submitters may claim as trade secret the specific chemical identity of any chemical subject to reporting under section 313.

(2) To make a claim, the submitter shall submit to EPA the following:

(i) An unsanitized copy of the toxic release inventory form under section 313 with the information claimed as trade secret clearly identified. To do this, the submitter shall check the box on the form indicating that the chemical identity is being claimed as trade secret. The submitter shall enter the generic class or category that is structurally descriptive of the chemical, as specified in paragraph (f) of this section.

(ii) A sanitized copy of the toxic release inventory form. This copy shall be identical to the document in paragraph (e)(2)(i) of this section except that the submitter shall delete the chemical identity claimed as trade secret. This copy shall also be submitted to the State official or officials designated to receive this information.

(iii) A sanitized and unsanitized substantiation in accordance with § 350.7 for every chemical identity claimed as trade secret.

(3) If the submitter wishes to claim information in the substantiation as trade secret or business confidential, it shall do so in accordance with § 350.7(d).

(4) Section 313 claims shall be sent to the address specified in § 350.16 of this regulation.

(f) Method of choosing generic class or category for sections 303, 311, 312 and 313. A facility owner or operator claiming chemical identity as trade secret should choose a generic class or category for the chemical that is structurally descriptive of the chemical.

203

Chemical Hazard Communication Guidebook

§ 350.7

(g) If a specific chemical identity is submitted under Title III to EPA, or to a State emergency response commission, designated State agency, local emergency planning committee or local fire department, without asserting a trade secrecy claim, the chemical identity shall be considered to have been voluntarily disclosed, and non-trade secret.

(h) A submitter making a trade secrecy claim under this section shall submit to entities other than EPA (e.g., a designated State agency, local emergency planning committee and local fire department) only the sanitized or public copy of the submission and substantiation.

§ 350.7 Substantiating claims of trade secrecy.

(a) Claims of trade secrecy must be substantiated by providing a specific answer including, where applicable, specific facts, to each of the following questions with the submission to which the trade secrecy claim pertains. Submitters must answer these questions on the form entitled "Substantiation to Accompany Claims of Trade Secrecy" in § 350.27 of this subpart.

(1) Describe the specific measures you have taken to safeguard the confidentiality of the chemical identity claimed as trade secret, and indicate whether these measures will continue in the future.

(2) Have you disclosed the information claimed as trade secret to any other person (other than a member of a local emergency planning committee, officer or employee of the United States or a State or local government, or your employee) who is not bound by a confidentiality agreement to refrain from disclosing this trade secret information to others?

(3) List all local, State, and Federal government entities to which you have disclosed the specific chemical identity. For each, indicate whether you asserted a confidentiality claim for the chemical identity and whether the government entity denied that claim.

(4) In order to show the validity of a trade secrecy claim, you must identify your specific use of the chemical claimed as trade secret and explain why it is a secret of interest to competitors. Therefore:

(i) Describe the specific use of the chemical claimed as trade secret, identifying the product or process in which it is used. (If you use the chemical other than as a component of a product or in a manufacturing process, identify the activity where the chemical is used.)

(ii) Has your company or facility identity been linked to the specific chemical identity claimed as trade secret in a patent, or in publications or other information sources available to the public or your competitors (of which you are aware)? If so, explain why this knowledge does not eliminate the justification for trade secrecy.

(iii) If this use of the chemical claimed as trade secret is unknown outside your company, explain how your competitors could deduce this use from disclosure of the chemical identity together with other information on the Title III submittal form.

(iv) Explain why your use of the chemical claimed as trade secret would be valuable information to your competitors.

(5) Indicate the nature of the harm to your competitive position that would likely result from disclosure of the specific chemical identity, and indicate why such harm would be substantial.

(6)(i) To what extent is the chemical claimed as trade secret available to the public or your competitors in products, articles, or environmental releases?

(ii) Describe the factors which influence the cost of determining the identity of the chemical claimed as trade secret by chemical analysis of the product, article, or waste which contains the chemical (e.g., whether the chemical is in pure form or is mixed with other substances).

(b) The answers to the substantiation questions listed in paragraph (a) of this section are to be submitted on the form in § 350.27 of this subpart, and included with a submitter's trade secret claim.

(c) An owner, operator or senior official with management responsibility shall sign the certification at the end of the form contained in § 350.27. The certification in both the sanitized and

Environmental Protection Agency § 350.7

unsanitized versions of the substantiation must bear an original signature.

(d) *Claims of confidentiality in the substantiation.* (1) The submitter may claim as confidential any trade secret or confidential business information contained in the substantiation. Such claims for material in the substantiation are not limited to claims of trade secrecy for specific chemical identity, but may also include claims of confidentiality for any confidential business information. To claim this material as confidential, the submitter shall clearly designate those portions of the substantiation to be claimed as confidential by marking those portions "Confidential," or "Trade Secret." Information not so marked will be treated as public and may be disclosed without notice to the submitter.

(2) An owner, operator, or senior official with management responsibility shall sign the certification stating that those portions of the substantiation claimed as confidential would, if disclosed, reveal the chemical identity being claimed as a trade secret, or would reveal other confidential business or trade secret information. This certification is combined on the substantiation form in § 350.27 with the certification described in paragraph (c) of this section.

(3) The submitter shall submit to EPA two copies of the substantiation, one of which shall be the unsanitized version, and the other shall be the sanitized version.

(i) The unsanitized copy shall contain all of the information claimed as trade secret or business confidential, marked as indicated in paragraph (d)(1) of this section.

(ii) The second copy shall be identical to the unsanitized substantiation except that it will be a sanitized version, in which all of the information claimed as trade secret or confidential shall be deleted. If any of the information claimed as trade secret in the substantiation is the chemical identity which is the subject of the substantiation, the submitter shall include the appropriate generic class or category of the chemical claimed as trade secret. This sanitized copy shall be submitted to the State emergency response commission, a designated State agency, the local emergency planning committee and the local fire department, as appropriate, and made publicly available.

(e) *Supplemental information.* (1) EPA may request supplemental information from the submitter in support of its trade secret claim, pursuant to § 350.11(a)(1). EPA may specify the kind of information to be submitted, or the submitter may submit any additional detailed information which further supports the truth of the information previously supplied to EPA in its initial substantiation, under this section.

(2) The submitter may claim as confidential any trade secret or confidential business information contained in the supplemental information. To claim this material as confidential, the submitter shall clearly designate those portions of the supplemental information to be claimed as confidential by marking those portions "Confidential," or "Trade Secret." Information not so marked will be treated as public and may be disclosed without notice to the submitter.

(3) If portions of the supplementary information are claimed confidential, an owner, operator, or senior official with management responsibility of the submitter shall certify that those portions of the supplemental information claimed as confidential would, if disclosed, reveal the chemical identity being claimed as confidential or would reveal other confidential business or trade secret information.

(4) If supplemental information is requested by EPA and the submitter claims portions of it as trade secret or confidential, then the submitter shall submit to EPA two copies of the supplemental information, an unsanitized and a sanitized version.

(i) The unsanitized version shall contain all of the information claimed as trade secret or business confidential, marked as indicated above in paragraph (e)(2) of this section.

(ii) The second copy shall be identical to the unsanitized substantiation except that it will be a sanitized version, in which all of the information claimed as trade secret or confidential shall be deleted. If any of the information claimed as trade secret in the sup-

205

§ 350.9

plemental information is the chemical identity which is the subject of the substantiation, the submitter shall include the appropriate generic class or category of the chemical claimed as trade secret.

§ 350.9 Initial action by EPA.

(a) When a claim of trade secrecy, made in accordance with § 350.5 of this part, is received by EPA, that information is treated as confidential until a contrary determination is made.

(b) A determination as to the validity of a trade secrecy claim shall be initiated upon receipt by EPA of a petition under § 350.15 or may be initiated at any time by EPA if EPA desires to determine whether chemical identity information claimed as trade secret is entitled to trade secret treatment, even though no request for release of the information has been received.

(c) If EPA initiates a determination as to the validity of a trade secrecy claim, the procedures set forth in §§ 350.11, 350.15, and 350.17 shall be followed in making the determination.

(d) When EPA receives a petition requesting disclosure of trade secret chemical identity or if EPA decides to initiate a determination of the validity of a trade secrecy claim for chemical identity, EPA shall first make a determination that the chemical identity claimed as trade secret is not the subject of a prior trade secret determination by EPA concerning the same submitter and facility, or if it is, that the prior determination upheld the submitter's claim of trade secrecy for that chemical identity at that facility.

(1) If EPA determines that the chemical identity claimed as trade secret is not the subject of a prior trade secret determination by EPA concerning the same submitter and the same facility, or if it is, that the prior determination upheld the submitter's claim of trade secrecy, then EPA shall review the submitter's claim according to § 350.11.

(2) If such a prior determination held that the submitter's claim for that chemical identity is invalid, and such determination was not challenged by appeal to the General Counsel, or by review in the District Court, or, if challenged, was upheld, EPA shall notify the submitter by certified mail (return receipt requested) that the chemical identity claimed as trade secret is the subject of a prior, final Agency determination concerning the same facility in which it was held that such a claim was invalid. In this notification EPA shall include notice of intent to disclose chemical identity within 10 days pursuant to § 350.18(c) of this subpart. EPA shall also notify the petitioner by regular mail of the action taken pursuant to this section.

§ 350.11 Review of claim.

(a) *Determination of sufficiency.* When EPA receives a petition submitted pursuant to § 350.15, or if EPA initiates a determination of the validity of a trade secrecy claim for chemical identity, and EPA has made a determination, as required in paragraph (d)(1) of § 350.9, then EPA shall determine whether the submitter has presented sufficient support for its claim of trade secrecy in its substantiation. EPA must make such a determination within 30 days of receipt of a petition. A claim of trade secrecy for chemical identity will be considered sufficient if, assuming all of the information presented in the substantiation is true, this supporting information could support a valid claim of trade secrecy. A claim is sufficient if it meets the criteria set forth in § 350.13.

(1) *Sufficient claim.* If the claim meets the criteria of sufficiency set forth in § 350.13, EPA shall notify the submitter in writing, by certified mail (return receipt requested), that it has 30 days from the date of receipt of the notice to submit supplemental information in writing in accordance with § 350.7(e), to support the truth of the facts asserted in the substantiation. EPA will not accept any supplemental information, in response to this notice, submitted after the 30 day period has expired. The notice required by this section shall include the address to which supplemental information must be sent. The notice may specifically request supplemental information in particular areas relating to the submitter's claim. The notice must also inform the submitter of his right to

Environmental Protection Agency § 350.11

claim any trade secret or confidential business information as confidential, and shall include a reference to § 350.7(e) of this regulation as the source for the proper procedure for claiming trade secrecy for trade secret or confidential business information submitted in the supplemental information requested by EPA.

(2) *Insufficient claim.* If the claim does not meet the criteria of sufficiency set forth in § 350.13, EPA shall notify the submitter in writing of this fact by certified mail (return receipt requested). Upon receipt of this notice, the submitter may either file an appeal of the matter to the General Counsel under paragraph (a)(2)(i) of this section, or, for good cause shown, submit additional material in support of its claim of trade secrecy to EPA under paragraph (a)(2)(ii) of this section. The notice required by this section shall include the reasons for EPA's decision that the submitter's claim is insufficient, and shall inform the submitter of its rights within 30 days of receiving notice to file an appeal with EPA's General Counsel or to amend its original substantiation for good cause shown. The notice shall include the address of the General Counsel, and the address of the office to which an amendment for good cause shown should be sent. The notice shall also include a reference to § 350.11(a)(2)(i)–(iv) of this subpart as the source on the proper procedures for filing an appeal or for amending the original substantiation.

(i) *Appeal.* The submitter may file an appeal of a determination of insufficiency with the General Counsel within 30 days of receipt of the notice of insufficiency, in accordance with the procedures set forth in § 350.17.

(ii) *Good Cause.* In lieu of an appeal to the General Counsel, the submitter may send additional material in support of its trade secrecy claim, for good cause shown, within 30 days of receipt of the notice of insufficiency. To do so, the submitter shall notify EPA by letter of its contentions as to good cause, and shall include in that letter the additional supporting material.

(iii) Good cause is limited to one or more of the following reasons:

(A) The submitter was not aware of the facts underlying the additional information at the time the substantiation was submitted, and could not reasonably have known the facts at that time; or

(B) EPA regulations and other EPA guidance did not call for such information at the time the substantiation was submitted; or

(C) The submitter had made a good faith effort to submit a complete substantiation, but failed to do so due to an inadvertent omission or clerical error.

(iv) If EPA determines that the submitter has met the standard for good cause, then EPA shall decide, pursuant to paragraph (a) of this section, whether the submitter's claim meets the Agency's standards of sufficiency set forth in § 350.13.

(A) If after receipt of additional material for good cause, EPA decides the claim is sufficient, EPA will determine whether the claim presents a valid claim of trade secrecy according to the procedures set forth in paragraph (b) of this section.

(B) If after receipt of additional material for good cause, EPA decides the claim is insufficient, EPA will notify the submitter by certified mail (return receipt requested) and the submitter may seek review in U.S. District Court within 30 days of receipt of the notice. The notice required by this paragraph shall include EPA's reasons for its determination, and shall inform the submitter of its right to seek review in U.S. District Court within 30 days of receipt of the notice. The petitioner shall be notified of EPA's decision by regular mail.

(v) If EPA determines that the submitter has not met the standard for good cause, then EPA shall notify the submitter by certified mail (return receipt requested). The submitter may seek review of EPA's decision in U.S. District Court within 30 days of receipt of the notice. The notice required in this paragraph shall include EPA's reasons for its determination, and shall inform the submitter of its right to seek review in U.S. District Court within 30 days of receipt of the notice. The petitioner shall be notified of EPA's decision by regular mail.

Chemical Hazard Communication Guidebook

§ 350.13

40 CFR Ch. I (7-1-89 Edition)

(b) *Determination of trade secrecy.* Once a claim has been determined to be sufficient under paragraph (a) of this section, EPA must decide whether the claim is entitled to trade secrecy.

(1) If EPA determines that the information submitted in support of the trade secrecy claim is true and that the chemical identity is a trade secret, the petitioner shall be notified by certified mail (return receipt requested) of EPA's determination and may bring an action in U.S. District Court within 30 days of receipt of such notice. The notice required in this paragraph shall include the reasons why EPA has determined that the chemical identity is a trade secret and shall inform the petitioner of its right to seek review in U.S. District Court within 30 days of receipt of the notice. The submitter shall be notified of EPA's decision by regular mail.

(2) If EPA decides that the information submitted in support of the trade secrecy claim is not true and that the chemical identity is not a trade secret:

(i) The submitter shall be notified by certified mail (return receipt requested) of EPA's determination and may appeal to the General Counsel within 30 days of receipt of such notice, in accordance with the procedures set forth in § 350.17. The notice required by this paragraph shall include the reasons why EPA has determined that the chemical identity is not a trade secret and shall inform the submitter of its appeal rights to EPA's General Counsel. The notice shall include the address to which an appeal should be sent and the procedure for filing an appeal, as set forth in § 350.17(a) of this subpart. The petitioner shall be notified of EPA's decision by regular mail.

(ii) The General Counsel shall notify the submitter by certified mail (return receipt requested) of its decision on appeal pursuant to the requirements in § 350.17. The notice required by this paragraph shall include the reasons for EPA's determination. If the General Counsel affirms the decision that the chemical identity is not a trade secret, then the submitter shall have 30 days from the date it receives notice of the General Counsel's decision to bring an action in U.S. District Court. If the General Counsel decides that the chemical identity is a trade secret, then EPA shall follow the procedure set forth in paragraph (b)(1) of this section.

§ 350.13 Sufficiency of assertions.

(a) A substantiation submitted under § 350.7 will be determined to be insufficient to support a claim of trade secrecy unless the answers to the questions in the substantiation submitted under § 350.7 support all of the following conclusions. This substantiation must include, where applicable, specific facts.

(1) The submitter has not disclosed the information to any other person, other than a member of a local emergency planning committee, an officer or employee of the United States or a State or local government, an employee of such person, or a person who is bound by a confidentiality agreement, and such person has taken reasonable measures to protect the confidentiality of such information and intends to continue to take such measures. To support this conclusion, the facts asserted must show all of the following:

(i) The submitter has taken reasonable measures to prevent unauthorized disclosure of the specific chemical identity and will continue to take such measures.

(ii) The submitter has not disclosed the specific chemical identity to any person who is not bound by an agreement to refrain from disclosing the information.

(iii) The submitter has not previously disclosed the specific chemical identity to a local, State, or Federal government entity without asserting a confidentiality claim.

(2) The information is not required to be disclosed, or otherwise made available, to the public under any other Federal or State law.

(3) Disclosure of the information is likely to cause substantial harm to the competitive position of such person. To support this conclusion, the facts asserted must show all of the following:

(i) *Either:* (A) Competitors do not know or the submitter is not aware that competitors know that the chemi-

Environmental Protection Agency § 350.17

cal whose identity is being claimed trade secret can be used in the fashion that the submitter uses it, and competitors cannot easily duplicate the specific use of this chemical through their own research and development activities; or

(B) Competitors are not aware or the submitter does not know whether competitors are aware that the submitter is using this chemical in this fashion.

(ii) The fact that the submitter manufactures, imports or otherwise uses this chemical in a particular fashion is not contained in any publication or other information source (of which the submitter is aware) available to competitors or the public.

(iii) The non-confidential version of the submission under this title does not contain sufficient information to enable competitors to determine the specific chemical identity withheld therefrom.

(iv) The information referred to in paragraph (a)(3)(i)(A) of this section, is of value to competitors.

(v) Competitors are likely to use this information to the economic detriment of the submitter and are not precluded from doing so by a United States patent.

(vi) The resulting harm to submitter's competitive position would be substantial.

(4) The chemical identity is not readily discoverable through reverse engineering. To support this conclusion, the facts asserted must show that competitors cannot readily discover the specific chemical identity by analysis of the submitter's products or environmental releases.

(b) The sufficiency of the trade secrecy claim shall be decided entirely upon the information submitted under § 350.7, or § 350.11(a)(2)(ii).

§ 350.15 Public petitions requesting disclosure of chemical identity claimed as trade secret.

(a) The public may request the disclosure of chemical identity claimed as trade secret by submitting a written petition to the address specified in § 350.16.

(b) The petition shall include:

(1) The name, address, and telephone number of the petitioner;

(2) The name and address of the company claiming the chemical identity as trade secret; and

(3) A copy of the submission in which the submitter claimed chemical identity as trade secret, with a specific indication as to which chemical identity the petitioner seeks disclosed.

(c) EPA shall acknowledge, by letter to the petitioner, the receipt of the petition.

(d) *Incomplete petitions.* If the information contained in the petition is not sufficient to allow EPA to identify which chemical identity the petitioner is seeking to have released, EPA shall notify the petitioner that the petition cannot be further processed until additional information is furnished. EPA will make every reasonable effort to assist a petitioner in providing sufficient information for EPA to identify the chemical identity the petitioner is seeking to have released.

(e) EPA shall make a determination on a petition requesting disclosure, in accordance with § 350.11 and § 350.17, within nine months of receipt of such petition.

§ 350.16 Address to send trade secrecy claims and petitions requesting disclosure.

All claims of trade secrecy under sections 303 (d)(2), (d)(3), 311, 312, and 313 and all public petitions requesting disclosure of chemical identities claimed as trade secret should be sent to the following address: U.S. Environmental Protection Agency, Emergency Planning and Community Right-to-Know Program, P.O. Box 70266, Washington, DC 20024-0266.

§ 350.17 Appeals.

(a) *Procedure for filing appeal.* A submitter may appeal an EPA determination under §§ 350.11 (a)(2)(i) or (b)(2)(i), by filing an appeal with the General Counsel. The appeal shall be addressed to: The Office of General Counsel, U.S. Environmental Protection Agency, Contracts and Information Law Branch, Room 3600M, LE-132G, 401 M Street, SW., Washington, DC 20460.

209

§ 350.18

The appeal shall contain the following:

(1) A letter requesting review of the appealed decision; and

(2) A copy of the letter containing EPA's decision upon which appeal is requested.

(b) *Appeal of determination of insufficient claim.*

(1) Where a submitter appeals a determination by EPA under § 350.11(a)(2)(i) that the trade secrecy claim presents insufficient support for a finding of trade secrecy, the General Counsel shall make one of the following determinations:

(i) The trade secrecy claim at issue meets the standards of sufficiency set forth in § 350.13; or

(ii) The trade secrecy claim at issue does not meet the standards of sufficiency set forth in § 350.13.

(2) If the General Counsel reverses the decision made by the EPA office handling the claim, the claim shall be processed according to § 350.11(a)(1). The General Counsel shall notify the submitter of the determination on appeal in writing, by certified mail (return receipt requested). The appeal determination shall include the date the appeal was received by the General Counsel, a statement of the decision appealed from, a statement of the decision on appeal and the reasons for such decision.

(3) If the General Counsel upholds the determination of insufficiency made by the EPA office handling the claim, the submitter may seek review in U.S. District Court within 30 days after receipt of notice of the General Counsel's determination. The General Counsel shall notify the submitter of its determination on appeal in writing, by certified mail (return receipt requested). The appeal determination shall include the date the appeal was received by the General Counsel, a statement of the decision appealed from, a statement of the decision on appeal and the reasons for such decision, and a statement of the submitter's right to seek review in U.S. District Court within 30 days of receipt of such notice. The petitioner shall be notified by regular mail.

(c) *Appeal of determination of no trade secret.* (1) If a submitter appeals from a determination by EPA under § 350.11(b)(2) that the specific chemical identity at issue is not a trade secret, the General Counsel shall make one of the following determinations:

(i) The assertions supporting the claim of trade secrecy are true and the chemical identity is a trade secret; or

(ii) The assertions supporting the claim of trade secrecy are not true and the chemical identity is not a trade secret.

(2) If the General Counsel reverses the decision made by the EPA office handling the claim, the General Counsel shall notify the submitter of its determination on appeal in writing, by certified mail (return receipt requested). The appeal determination shall include the date the appeal was received by the General Counsel, a statement of the decision appealed from, a statement of the decision on appeal and the reasons for such decision. The General Counsel shall send the petitioner the notice required in § 350.11(b)(1).

(3) If the General Counsel upholds the decision of the EPA office which made the trade secret determination, the submitter may seek review in U.S. District Court within 30 days of receipt of notice of the General Counsel's decision. The General Counsel shall notify the submitter of the determination on appeal in writing, by certified mail (return receipt requested). The notice shall include the date the appeal was received by the General Counsel, a statement of the decision appealed from, the basis for the appeal determination, that it constitutes final Agency action concerning the chemical identity trade secrecy claim, and that such final Agency action may be subject to review in U.S. District Court within 30 days of receipt of such notice. The General Counsel shall notify the petitioner by regular mail.

§ 350.18 Release of chemical identity determined to be non-trade secret; notice of intent to release chemical identity.

(a) Where a submitter fails to seek review within U.S. District Court within 20 days of receiving notice of a

210

Environmental Protection Agency

§ 350.21

determination of the General Counsel under § 350.17(b)(3) of this subpart that the trade secrecy claim is insufficient, or under § 350.17(c)(3) of this subpart that chemical identity claimed as trade secret is not entitled to trade secret protection, EPA may furnish notice of intent to disclose the chemical identity claimed as trade secret within 10 days by furnishing the submitter with the notice set forth in paragraph (d) of this section by certified mail (return receipt requested).

(b) Where a submitter fails to seek review within U.S. District Court within 20 days of receiving notice of an EPA determination under § 350.11(a)(2)(iv)(B), or § 350.11(a)(2)(v) of this regulation, or fails to pursue appeal to the General Counsel within 20 days after being notified of its right to do so under § 350.11(a)(2)(i) or § 350.11(b)(2)(i), EPA may furnish notice of intent to disclose the chemical identity claimed as trade secret within 10 days by furnishing the submitter with the notice set forth in paragraph (d) of this section by certified mail (return receipt requested).

(c) Where EPA, upon initial review under § 350.9(d), determines that the chemical identity claimed as trade secret in a submittal submitted pursuant to this part is the subject of a prior final Agency determination concerning a claim of trade secrecy for the same chemical identity for the same facility, in which such claim was held invalid, EPA shall furnish notice of intent to disclose chemical identity within 10 days by furnishing the submitter with the notice set forth in paragraph (d) of this section by certified mail (return receipt requested).

(d) EPA shall furnish notice of its intent to release chemical identity claimed as trade secret by sending the following notification to submitters, under the circumstances set forth in paragraphs (a), (b), and (c) of this section. The notice shall state that EPA will make the chemical identity available to the petitioner and the public on the tenth working day after the date of the submitter's receipt of written notice (or on such later date as the Office of General Counsel may establish), unless the Office of General Counsel has first been notified of the submitter's commencement of an action in Federal court to obtain judicial review of the determination at issue, and to obtain preliminary injunctive relief against disclosure, or, where applicable, as described in paragraph (b) of this section, of commencement of an appeal to the General Counsel. The notice shall further state that if Federal court action is timely commenced, EPA may nonetheless make the information available to the petitioner and the public (in the absence of an order by the court to the contrary), once the court has denied a motion for a preliminary injunction in the action or has otherwise upheld the EPA determination, or, that if Federal court action or appeal to the General Counsel is timely commenced, EPA may nonetheless make the information available to the petitioner and the public whenever it appears to the General Counsel, after reasonable notice to the submitter, that the submitter is not taking appropriate measures to obtain a speedy resolution of the action.

§ 350.19 Provision of information to States.

(a) Any State may request access to trade secrecy claims, substantiations, supplemental substantiations, and additional information submitted to EPA. EPA shall release this information, even if claimed confidential, to any State requesting access if:

(1) The request is in writing;

(2) The request is from the Governor of the State; and

(3) The State agrees to safeguard the information with procedures equivalent to those which EPA uses to safeguard the information.

(b) The Governor of a State which receives access to trade secret information under this section may disclose such information only to State employees.

§ 350.21 Adverse health effects.

The Governor or State emergency response commission shall identify the adverse health effects associated with each of the chemicals claimed as trade secret and shall make this information

211

available to the public. The material safety data sheets submitted to the State emergency response commissions may be used for this purpose.

§ 350.23 Disclosure to authorized representatives.

(a) Under section 322(f) of the Act, EPA possesses the authority to disclose to any authorized representative of the United States any information to which this section applies, notwithstanding the fact that the information might otherwise be entitled to trade secret or confidential treatment under this part. Such authority may be exercised only in accordance with paragraph (b) of this section.

(b)(1) A person under contract or subcontract to EPA or a grantee who performs work for EPA in connection with Title III or regulations which implement Title III may be considered an authorized representative of the United States for purposes of this § 350.23. Subject to the limitations in this § 350.23(b), information to which this section applies may be disclosed to such a person if the EPA program office managing the contract, subcontract, or grant first determines in writing that such disclosure is necessary in order that the contractor, subcontractor or grantee may carry out the work required by the contract, subcontract or grant.

(2) No information shall be disclosed under this § 350.23(b) unless this contract, subcontract, or grant in question provides:

(i) That the contractor, subcontractor or the grantee and the contractor's, subcontractor's, or grantee's employees shall use the information only for the purpose of carrying out the work required by the contract, subcontract, or grant, and shall refrain from disclosing the information to anyone other than EPA without the prior written approval of each affected submitter or of an EPA legal office, and shall return to EPA all copies of the information (and any abstracts or extracts therefrom) upon request by the EPA program office, whenever the information is no longer required by the contractor, subcontractor or grantee for the performance of the work required under the contract, subcontract or grant, or upon completion of the contract, subcontract or grant;

(ii) That the contractor, subcontractor or grantee shall obtain a written agreement to honor such terms of the contract or subcontract from each of the contractor's, subcontractor's or grantee's employees who will have access to the information, before such employee is allowed such access; and

(iii) That the contractor, subcontractor or grantee acknowledges and agrees that the contract, subcontract or grant provisions concerning the use and disclosure of confidential business information are included for the benefit of, and shall be enforceable by, both EPA and any covered facility having an interest in information concerning it supplied to the contractor, subcontractor or grantee by EPA under the contract or subcontract or grant.

(3) No information shall be disclosed under this § 350.23(b) until each affected submitter has been furnished notice of the contemplated disclosure by the EPA program office and has been afforded a period found reasonable by that office (not less than 5 working days) to submit its comments. Such notice shall include a description of the information to be disclosed, the identity of the contractor, subcontractor or grantee, the contract, subcontract or grant number, if any, and the purposes to be served by the disclosure. This notice may be published in the FEDERAL REGISTER or may be sent to individual submitters.

(4) The EPA program office shall prepare a record of disclosures under this § 350.23(b). The EPA program office shall maintain the record of disclosure and the determination of necessity prepared under paragraph (b)(1) of this section for a period of not less than 36 months after the date of the disclosure.

§ 350.25 Disclosure in special circumstances.

Other disclosure of specific chemical identity may be made in accordance with 40 CFR 2.209.

Appendix B: Regulations Reference Guide

Environmental Protection Agency § 350.27

§ 350.27 Substantiation form to accompany claims of trade secrecy, instructions to substantiation form.

(a) The form in paragraph (b) of this section must be completed and submitted as required in § 350.7(a).

(b) Substantiation form to accompany claims of trade secrecy.

213

Chemical Hazard Communication Guidebook

§ 350.27 40 CFR Ch. I (7-1-89 Edition)

United States Environmental Protection Agency
Washington, DC 20460

EPA — Substantiation To Accompany Claims of Trade Secrecy Under the Emergency Planning and Community Right-To-Know Act of 1986

Form Approved
OMB No. 2050-0078
Approval expires 10-31-90

Paperwork Reduction Act Notice

Public reporting burden for this collection of information is estimated to vary from 27.7 hours to 33.2 hours per response, with an average of 28.8 hours per response, including time for reviewing instructions, searching existing data sources, gathering and maintaining the data needed, and completing and reviewing the collection of information. Send comments regarding the burden estimate or any other aspect of this collection of information, including suggestions for reducing this burden, to Chief, Information Policy Branch, PM-223, U.S. Environmental Protection Agency, 401 M Street, SW, Washington, DC 20460; and to the Office of Information and Regulatory Affairs, Office of Management and Budget, Washington, DC 20503.

Part 1. Substantiation Category

1.1 Title III Reporting Section (check only one)

☐ 303   ☐ 311   ☐ 312   ☐ 313

1.2 Reporting Year 19____

1.3 Indicate Whether This Form Is (check only one)

1.3a. ☐ **Sanitized** (answer 1.3.1a below)

1.3.1a. Generic Class or Category

1.3b. ☐ **Unsanitized** (answer 1.3.1b. and 1.3.2b. below)

1.3.1b. CAS Number

☐☐☐☐☐☐-☐☐-☐

1.3.2b. Specific Chemical Identity

Part 2. Facility Identification Information

2.1 Name

2.2 Street Address

2.3 City, State, and ZIP Code

2.4 Dun and Bradstreet Number

☐☐-☐☐☐-☐☐☐☐

EPA Form 9510-1 (7-88) Page 1 of 5

214

Appendix B: Regulations Reference Guide

Environmental Protection Agency § 350.27

Part 3. Responses to Substantiation Questions

3.1 Describe the specific measures you have taken to safeguard the confidentiality of the chemical identity claimed as trade secret, and indicate whether these measures will continue in the future.

3.2 Have you disclosed the information claimed as trade secret to any other person (other than a member of a local emergency planning committee, officer or employee of the United States or a State or local government, or your employee) who is not bound by a confidentiality agreement to refrain from disclosing this trade secret information to others?

☐ Yes ☐ No

3.3 List all local, State, and Federal government entities to which you have disclosed the specific chemical identity. For each, indicate whether you asserted a confidentiality claim for the chemical identity and whether the government entity denied that claim.

Government Entity	Confidentiality Claim Asserted		Confidentiality Claim Denied	
	Yes	No	Yes	No

EPA Form 9510-1 (7-88)   Page 2 of 5

215

Chemical Hazard Communication Guidebook

§ 350.27   **40 CFR Ch. I (7-1-89 Edition)**

3.4   In order to show the validity of a trade secrecy claim, you must identify your specific use of the chemical claimed as trade secret and explain why it is a secret of interest to competitors. Therefore:

(i)   Describe the specific use of the chemical claimed as trade secret, identifying the product or process in which it is used. (If you use the chemical other than as a component of a product or in a manufacturing process, identify the activity where the chemical is used.)

(ii)   Has your company or facility identity been linked to the specific chemical identity claimed as trade secret in a patent, or in publications or other information sources available to the public or your competitors (of which you are aware)?

☐ Yes   ☐ No

If so, explain why this knowledge does not eliminate the justification for trade secrecy.

(iii)   If this use of the chemical claimed as trade secret is unknown outside your company, explain how your competitors could deduce this use from disclosure of the chemical identity together with other information on the Title III submittal form.

EPA Form 9510-1 (7-88)   Page 3 of 5

216

Appendix B: Regulations Reference Guide

Environmental Protection Agency §350.27

3.4 (iv) Explain why your use of the chemical claimed as trade secret would be valuable information to your competitors.

3.5 Indicate the nature of the harm to your competitive position that would likely result from disclosure of the specific chemical identity, and indicate why such harm would be substantial.

3.6 (i) To what extent is the chemical claimed as trade secret available to the public or your competitors in products, articles, or environmental releases?

EPA Form 9510-1 (7-88)

Chemical Hazard Communication Guidebook

§ 350.27   **40 CFR Ch. I (7-1-89 Edition)**

3.6 (ii) Describe the factors which influence the cost of determining the identity of the chemical claimed as trade secret by chemical analysis of the product, article, or waste which contains the chemical (e.g., whether the chemical is in pure form or is mixed with other substances).

Part 4. Certification (Read and sign after completing all sections)

I certify under penalty of law that I have personally examined the information submitted in this and all attached documents. Based on my inquiry of those individuals responsible for obtaining the information, I certify that the submitted information is true, accurate, and complete, and that those portions of the substantiation claimed as confidential would, if disclosed, reveal the chemical identity being claimed as a trade secret, or would reveal other confidential business or trade secret information. I acknowledge that I may be asked by the Environmental Protection Agency to provide further detailed factual substantiation relating to this claim of trade secrecy, and certify to the best of my knowledge and belief that such information is available. I understand that if it is determined by the Administrator of EPA that this trade secret claim is frivolous, EPA may assess a penalty of up to $25,000 per claim.

I acknowledge that any knowingly false or misleading statement may be punishable by fine or imprisonment or both under applicable law.

4.1 Name and official title of owner or operator or senior management official

4.2 Signature (All signatures must be original)

4.3 Date Signed

EPA Form 9510-1 (7-88)

218

Appendix B: Regulations Reference Guide

Environmental Protection Agency § 350.27

INSTRUCTIONS FOR COMPLETING THE EPA TRADE SECRET SUBSTANTIATION FORM

General Information

EPA requires that the information requested in a trade secret substantiation be completed using this substantiation form in order to ensure that all facility and chemical identifier information, substantiation questions, and certification statements are completed. Submitter-devised forms will not be accepted. Incomplete substantiations will in all likelihood be found insufficient to support the claim, and the claim will be denied. *Moreover, the statute provides that a submitter who fails to provide information required will be subject to a $10,000 fine.* For the submitter's own protection, therefore, the EPA form must be used and completed in its entirety.

The statute for section 322 establishes a two-phase process in which the submitter must do the following:

1. At the time a report is submitted, the submitter must present a complete set of assertions that (if true) would be sufficient to justify the claim of trade secrecy; and

2. If the claim is reviewed by EPA, the submitter will be asked to provide additional factual information sufficient to establish the truthfulness of the assertions made at the time the claim was made.

In making its assertions of trade secrecy, a submitter should provide, where applicable, descriptive factual statements. Conclusory statements of compliance (such as positive or negative restatements of the questions) may not provide EPA with enough information to make a determination and may be found insufficient to support a claim.

What May Be Withheld

Only the specific chemical identity required to be disclosed in sections 303, 311, 312, and 313 submissions may be claimed trade secret on the Title III submittal itself. (Other trade secret or confidential business information included in answer to a question on the substantiation may be claimed trade secret or confidential, as described below.)

Location information claimed as confidential under section 312(d)(2)(F) should *not* be sent to EPA; this should only be sent to the SERC, LEPC, and the fire department, as requested.

Sanitized and Unsanitized Copies

You must submit this form to EPA in sanitized and unsanitized versions, along with the sanitized and unsanitized copies of the submittal that gives rise to this trade secrecy claim (except for the section 303 submittal, and for MSDSs under section 311). The *unsanitized* version of this form contains specific chemical identity and CAS number and may contain other trade secret or confidential business information, which should be clearly labeled as such. Failure to claim other information trade secret or confidential will make that information publicly available. In the *sanitized* version of this form, the specific chemical identity and CAS number must be replaced with the chemical's generic class or category and any other trade secret or confidential business information should be deleted. *You should also send sanitized copies of the submittal and this form to relevant State and local authorities.*

Each question on this form must be answered. *Submitters are encouraged to answer in the space provided.* If you need more space to answer a particular question, please use additional sheets. If you use additional sheets, be sure to include the number (and if applicable, the subpart) of the question being answered and write your facility's Dun and Bradstreet Number on the lower right-hand corner of each sheet.

When the Forms Must be Submitted

The sanitized and unsanitized report forms and trade secret substantiations must be submitted to EPA by the normal reporting deadline for that section (e.g., section 313 submissions for any calendar year must be submitted on or before July 1 of the following year).

219

Chemical Hazard Communication Guidebook

§ 350.27

Where to Send the Trade Secrecy Claim

All trade secrecy claims should be sent to the following address: U.S. Environmental Protection Agency, Emergency Planning and Community Right-to-Know Program, P.O. Box 70266, Washington, DC 20024-0266.

In addition, you must send sanitized copies of the report form and substantiation to relevant State and local authorities. States will provide addresses where the copies of the reports are to be sent.

Packaging of Claim(s)

A completed section 322 claim package must include four items, packaged in the following order:

1. An unsanitized trade secret substantiation form.
2. A sanitized trade secret substantiation form.
3. An unsanitized 312 or 313 report (it is not necessary to create an unsanitized section 303 submittal or MSDS for submission under section 311).
4. A sanitized (public) section 303, 311, 312, or 313 or report.

It is important to securely fasten together (binder clip or rubber band) each of the reporting forms and substantiations for the particular chemical being claimed trade secret. This process will make it clear that a claim is physically complete when submitted. When submitters submit claims for more than one chemical, EPA requests that the four parts associated with each chemical be assembled as a set and each set for different chemicals be kept separate within the package sent to EPA. Following these guidelines permits the Agency to make the appropriate determinations of trade secrecy, and to make public only those portions of each submittal required to be disclosed.

How to Obtain Forms and Other Information

Additional copies of the Trade Secret Substantiation Form may be obtained by writing to: Emergency Planning and Community Right-to-Know Program, U.S. Environmental Protection Agency, WH-562A, 401 M Street, SW., Washington, DC 20460.

40 CFR Ch. I (7-1-89 Edition)

Instructions for Completing Specific Sections of the Form

Part 1. Substantiation Category

1.1 Title III Reporting Section. Check the box corresponding to the section for which this particular claim of trade secrecy is being made. Checking off more than one box for a claim is *not* permitted.

1.2 Reporting Year. Enter the year to which the reported information applies, not the year in which you are submitting the report.

1.3a Sanitized. If this copy of the submission is the "public" or sanitized version, check this box and complete 1.3.1a. which asks for generic class or category. Do *not* complete the information required in the unsanitized box (1.3b.).

1.3.1a Generic Class or Category. You must complete this if you are claiming the specific chemical identity as a trade secret and have marked the box in 1.3a. The generic chemical name must be structurally descriptive of the chemical.

1.3b Unsanitized. Check the box if this version of the form contains the specific chemical identity or any other trade secret or confidential business information.

1.3.1b CAS Number. You must enter the Chemical Abstract Service (CAS) registry number that appears in the appropriate section of the rule for the chemical being reported. Use leading place holding zeros. If you are reporting a chemical category (e.g., copper compounds), enter N/A in the CAS number space.

1.3.2b Specific Chemical Identity. Enter the name of the chemical or chemical category as it is listed in the appropriate section of the reporting rule.

Part 2. Facility Identification Information

2.1-2.3 Facility Name and Location. You must enter the name of your facility (plant site name or appropriate facility designation), street address, city, State and ZIP Code in the space provided. You may not use a post office box number for this location.

220

Environmental Protection Agency

2.4 Dun and Bradstreet Number. You must enter the number assigned by Dun and Bradstreet for your facility or each establishment within your facility. If the establishment does not have a D & B number, enter N/A in the boxes reserved for those numbers. Use leading place holding zeros.

Part 3. Responses to Substantiation Questions

The six questions posed in this form are based on the four statutory criteria found in section 322(b) of Title III. The information you submit in response to these questions is the basis for EPA's initial determination as to whether the substantiation is sufficient to support a claim of trade secrecy. EPA has indicated in § 350.13 of the final rule the specific criteria that it regards as the legal basis for evaluating whether the answers you have provided are sufficient to warrant protection of the chemical identity. You are urged to review those criteria before preparing answers to the questions on the form.

Part 4. Certification

An *original* signature is required for each trade secret substantiation submitted to EPA, both sanitized and unsanitized. It indicates the submitter is certifying that the particular substantiation provided to EPA is complete, true, and accurate, and that it is intended to support the specific trade secret claim being made. Noncompliance with this certification requirement may jeopardize the trade secret claim.

4.1 Name and Official Title. Print or type the name and title of the person who signs the statement at 4.2.

4.2 Signature. This certification must be signed by the owner or operator, or a senior official with management responsibility for the person (or persons) completing the form. An *original* signature is required for each trade secret substantiation submitted to EPA, both sanitized and unsanitized. Since the certification applies to all information supplied on the forms, it should be signed only after the substantiation has been completed.

4.3 Date. Enter the date when the certification was signed.

Part 350, Subpt. A, App. A

APPENDIX A TO SUBPART A—RESTATEMENT OF TORTS SECTION 757, COMMENT b

b. Definition of trade secret. A trade secret may consist of any formula, pattern, device or compilation of information which is used in one's business, and which gives him an opportunity to obtain an advantage over competitors who do not know or use it. It may be a formula for a chemical compound, a process of manufacturing, treating or preserving materials, a pattern for a machine or other device, or a list of customers. It differs from other secret information in a business (see section 759) in that it is not simply information as to single or ephemeral events in the conduct of the business, as, for example, the amount or other terms of a secret bid for a contract or the salary of certain employees, or the security investments made or contemplated, or the date fixed for the announcement of a new policy or for bringing out a new model or the like. A trade secret is a process or device for continuous use in the operation of the business. Generally it relates to the production of goods, as, for example, a machine or formula for the production of an article. It may, however, relate to the sale of goods or to other operations in the business, such as a code for determining discounts, rebates or other concessions in a price list or catalogue, or a list of specialized customers, or a method of bookkeeping or other office management.

Secrecy. The subject matter of a trade secret must be secret. Matters of public knowledge or of general knowledge in an industry cannot be appropriated by one as his secret. Matters which are completely disclosed by the goods which one markets cannot be his secret. Substantially, a trade secret is known only in the particular business in which it is used. It is not requisite that only the proprietor of the business know it. He may, without losing his protection, communicate it to employees involved in its use. He may likewise communicate it to others pledged to secrecy. Others may also know of it independently, as, for example, when they have discovered the process or formula by independent invention and are keeping it secret. Nevertheless, a substantial element of secrecy must exist, so that, except by the use of improper means, there would be difficulty in acquiring the information. An exact definition of a trade secret is not possible. Some factors to be considered in determining whether given information is one's trade secret are: (1) The extent to which the information is known outside of his business; (2) the extent to which it is known by employees and others involved in his business; (3) the extent of

221

§ 350.40

measures taken by him to guard the secrecy of the information; (4) the value of the information to him and to his competitors; (5) the amount of effort or money expended by him in developing the information; (6) the ease or difficulty with which the information could be properly acquired or duplicated by others.

Novelty and prior art. A trade secret may be a device or process which is patentable; but it need not be that. It may be a device or process which is clearly anticipated in the prior art or one which is merely a mechanical improvement that a good mechanic can make. Novelty and invention are not requisite for a trade secret as they are for patentability. These requirements are essential to patentability because a patent protects against unlicensed use of the patented device or process even by one who discovers it properly through independent research. The patent monopoly is a reward to the inventor. But such is not the case with a trade secret. Its protection is not based on a policy of rewarding or otherwise encouraging the development of secret processes or devices. The protection is merely against breach of faith and reprehensible means of learning another's secret. For this limited protection it is not appropriate to require also the kind of novelty and invention which is a requisite of patentability. The nature of the secret is, however, an important factor in determining the kind of relief that is appropriate against one who is subject to liability under the rule stated in this section. Thus, if the secret consists of a device or process which is a novel invention, one who acquires the secret wrongfully is ordinarily enjoined from further use of it and is required to account for the profits derived from his past use. If, on the other hand, the secret consists of mechanical improvements that a good mechanic can make without resort to the secret, the wrongdoer's liability may be limited to damages, and an injunction against future use of the improvements made with the aid of the secret may be inappropriate.

Subpart B—Disclosure of Trade Secret Information to Health Professionals

§ 350.40 Disclosure to health professionals.

(a) *Definitions.* "Medical emergency" means any unforeseen condition which a health professional would judge to require urgent and unscheduled medical attention. Such a condition is one which results in sudden and/or serious symptom(s) constituting a threat to a person's physical or psychological well-being and which requires immediate medical attention to prevent possible deterioration, disability, or death.

(b) The specific chemical identity, including the chemical name of a hazardous chemical, extremely hazardous substance, or a toxic chemical, is made available to health professionals, in accordance with the applicable provisions of this section.

(c) *Diagnosis or Treatment by Health Professionals in Non-Emergency Situations.* (1) An owner or operator of a facility which is subject to the requirements of sections 311, 312, and 313, shall, upon request, provide the specific chemical identity, if known, of a hazardous chemical, extremely hazardous substance, or a toxic chemical to a health professional if:

(i) The request is in writing;

(ii) The request describes why the health professional has a reasonable basis to suspect that:

(A) The specific chemical identity is needed for purposes of diagnosis or treatment of an individual,

(B) The individual or individuals being diagnosed or treated have been exposed to the chemical concerned, and

(C) Knowledge of the specific chemical identity of such chemical will assist in diagnosis or treatment.

(iii) The request contains a confidentiality agreement which includes:

(A) A description of the procedures to be used to maintain the confidentiality of the disclosed information; and

(B) A statement by the health professional that he will not use the information for any purpose other than the health needs asserted in the statement of need authorized in paragraph (c)(1)(ii) of this section and will not release the information under any circumstances, except as authorized by the terms of the confidentiality agreement or by the owner or operator of the facility providing such information.

(iv) The request includes a certification signed by the health professional stating that the information contained in the statement of need is true.

(2) Following receipt of a written request, the facility owner or operator to whom such request is made shall pro-

Environmental Protection Agency § 350.40

vide the requested information to the health professional promptly.

(d) *Preventive Measures and Treatment by Local Health Professionals.* (1) An owner or operator of a facility subject to the requirements of sections 311, 312, or 313 shall provide the specific chemical identity, if known, of a hazardous chemical, an extremely hazardous substance, or a toxic chemical to any health professional (such as a physician, toxicologist, epidemiologist, or nurse) if:

(i) The requester is a local government employee or a person under contract with the local government;

(ii) The request is in writing;

(iii) The request describes with reasonable detail one or more of the following health needs for the information:

(A) To assess exposure of persons living in a local community to the hazards of the chemical concerned.

(B) To conduct or assess sampling to determine exposure levels of various population groups.

(C) To conduct periodic medical surveillance of exposed population groups.

(D) To provide medical treatment to exposed individuals or population groups.

(E) To conduct studies to determine the health effects of exposure.

(F) To conduct studies to aid in the identification of chemicals that may reasonably be anticipated to cause an observed health effect.

(iv) The request contains a confidentiality agreement which includes:

(A) A description of the procedures to be used to maintain the confidentiality of the disclosed information; and

(B) A statement by the health professional that he will not use the information for any purpose other than the health needs asserted in the statement of need authorized in paragraph (d)(1)(iii) of this section and will not release the information under any circumstances except as may otherwise be authorized by the terms of such agreement or by the owner or operator of the facility person providing such information.

(v) The request includes a certification signed by the health professional stating that the information contained in the statement of need is true.

(2) Following receipt of a written request, the facility owner or operator to whom such request is made shall promptly provide the requested information to the local health professional.

(e) *Medical Emergency.* (1) An owner or operator of a facility which is subject to the requirements of sections 311, 312, or 313 must provide a copy of a material safety data sheet, an inventory form, or a toxic chemical release form, including the specific chemical identity, if known, of a hazardous chemical, extremely hazardous substance, or a toxic chemical, to any treating physician or nurse who requests such information if the treating physician or nurse determines that:

(i) A medical emergency exists as to the individual or individuals being diagnosed or treated;

(ii) The specific chemical identity of the chemical concerned is necessary for or will assist in emergency or first-aid diagnosis or treatment; and,

(iii) The individual or individuals being diagnosed or treated have been exposed to the chemical concerned.

(2) Owners or operators of facilities must provide the specific chemical identity to the requesting treating physician or nurse immediately following the request, without requiring a written statement of need or a confidentiality agreement in advance.

(3) The owner or operator may require a written statement of need and a written confidentiality agreement as soon as circumstances permit. The written statement of need shall describe in reasonable detail the factors set forth in paragraph (e)(1) of this section. The written confidentiality agreement shall be in accordance with paragraphs (c)(1)(iii) and (f) of this section.

(f) *Confidentiality Agreement.* The confidentiality agreement authorized in paragraphs (c)(1)(iii), (d)(1)(iv) and (e)(3) of this section:

(i) May restrict the use of the information to the health purposes indicated in the written statement of need;

(ii) May provide for appropriate legal remedies in the event of a breach of the agreement; and

223

(iii) May not include requirements for the posting of a penalty bond.

(g) Nothing in this regulation is meant to preclude the parties from pursuing any non-contractual remedies to the extent permitted by law, or from pursuing the enforcement remedy provided in section 325(e) of Title III.

(h) The health professional receiving the trade secret information may disclose it to EPA only under the following circumstances: The health professional must believe that such disclosure is necessary in order to learn from the Agency additional information about the chemical necessary to assist him in carrying out the responsibilities set forth in paragraphs (c), (d), and (e) of this section. Such information comprises facts regarding adverse health and environmental effects.

DOT REQUIREMENTS APPLICABLE TO SHIPPERS OF HAZARDOUS MATERIALS

A. DETERMINING THE PROPER SHIPPING NAME AND CLASSIFICATION

1. Shippers must meet the general obligations for classification and description.

The shipper must class and describe the hazardous material according to the hazardous materials tables and hazardous materials communications regulations (Part 172) and the general requirements for shipments and packagings for shippers (Part 173).

The shipper must determine that the packaging or container has been manufactured, assembled, and marked in accordance with specifications for transportation. A shipper must comply with any applicable specification or exemption.

Refer To: 49 CFR § 173.22

2. Shippers must classify materials having more than one hazard in sequence according to the specified order of hazard.

A hazardous material, having more than one hazard must be classed according to the following order of hazards (See Appendix C for definitions):

(1) Radioactive material (except a limited quantity)
(2) Poison A
(3) Flammable gas
(4) Non-flammable gas
(5) Flammable liquid
(6) Oxidizer
(7) Flammable Solid
(8) Corrosive material (liquid)
(9) Poison B
(10) Corrosive material (solid)
(11) Irritating materials
(12) Combustible liquid (in containers having capacities exceeding 110 gallons)
(13) ORM-B
(14) ORM-A
(15) Combustible liquid (in containers having capacities of 110 gallons or less)

Chemical Hazard Communication Guidebook

(16) ORM-E
Refer To: 49 CFR § 173.2

CLASSIFICATION/NAME - CHECKLIST FOR COMMON VIOLATIONS

Improper description and/or proper shipping
name for material being shipped
Refer To: 49 CFR § 172.101,102

Improper classification of hazardous material
Refer To: 49 CFR 172.101,102

Failure to properly classify material having
more than one hazard
Refer To: 49 CFR § 173.2

Omission of technical name of material
following n.o.s. description of material offered
for export by vessel
Refer To: 49 CFR § 172.203(i)

The letters "RQ" not displayed in association
with the proper shipping name when required
(See Discussion in Chapter 4).
Refer To: 49 CFR § 172.203(c)(2)

B. COMPLYING WITH GENERAL RESTRICTIONS ON TRANSPORTING HAZARDOUS MATERIALS

1. Shippers must ensure that all shipments are in compliance with the general regulations in order to be offered for transportation.

Each person who offers a hazardous material for transportation, each carrier by air, highway, rail, or water who transports a hazardous material, or a person who otherwise performs a packaging, labeling, or marking function must perform the function in accordance with the applicable requirements for shipping papers, package marking, labeling, and transport vehicle

Appendix B: Regulations Reference Guide

placarding of those hazardous materials as described in Part 172.

A shipment must be prepared properly to be offered for transportation; each person who offers hazardous materials for transportation must:

a. Instruct each of his officers, agents, and employees having any responsibility for preparing hazardous materials for shipment as to applicable regulations (173.1);

b. Must classify and describe it in accordance with Parts 172 and 173 of Subchapter C; and

c. Determine that the packaging or container has been manufactured, assembled, and marked in accordance with the regulation (173.22).

Refer To: 49 CFR § 173.1,3,22

---

2. Shippers must comply with any special restrictions on the modes of transportation permitted, such as restriction on shipments by air.

Shippers must meet special restrictions pertaining to specified modes of transportation, for example:

When the regulations indicate a hazardous material is forbidden aboard cargo aircraft, the material is also forbidden aboard passenger-carrying aircraft.

Refer To: 49 CFR § 173.6

---

3. Certain materials cannot be shipped and some methods of shipment are prohibited.

Unless otherwise provided in Subchapter C, shippers may not offer certain materials, or packages for transportation. These include:

a. A hazardous material in the same packaging, freight container, or overpack with another hazardous material that, if mixed, would be liable to cause a dangerous evolution of heat or gas, have corrosive tendencies;

Chemical Hazard Communication Guidebook

b. A package containing a material which is liable to decompose or polymerize at a specified temperature with evolution of a dangerous quantity of heat or gas, unless it is stabilized or inhibited in such a manner as to preclude this danger;

c. Packages which evolve a dangerous quantity of flammable gas or vapor released from a material which would not otherwise be subject to Subchapter C, i.e., the release of flammable vapor or gas in such quantities that a flammable mixture with air would be created within a transport vehicle;

d. Packages containing materials (other than explosives) which will detonate in a fire; and

e. Packages containing a device with fuel and equipped with an ignition element (such as a cigarette lighter) unless the device design and packaging have been approved as specified.

Refer To: 49 CFR § 173.21

SHIPMENT - CHECKLIST FOR COMMON VIOLATIONS

Failure to instruct each officer, agents, and employees as to the applicable Hazardous Materials Regulations
Refer To: 49 CFR § 173.1

C. DETERMINING THE PROPER PACKAGING

1. Shippers must meet the general requirements that apply to packaging and containers.

Shippers must ensure that the commodity is shipped in the proper container or package. However, when a package or container is required to be marked with a DOT specification, compliance with that requirement is the responsibility of the packaging or container manufacturer. Marking the packaging or container with the DOT specification is understood to certify compliance by the manufacturer.

Appendix B: Regulations Reference Guide

	The manufacturer of a packaging or container who transfers it to some other person must inform them of any specification requirements that have not been met at the time of transfer, unless they are specifically excepted from that responsibility by the regulation (178.0-2).
	If containers are reused, the original manufacturer will not be responsible for markings certifying that the packaging is in compliance (173.28).
Refer To: 49 CFR § 178.0-2	

---

2. Shippers must ensure that packages meet the standard requirements.

Each package used for shipping hazardous materials under Subchapter C must be designed and constructed and its contents so limited that under conditions normally incident to transportation:

a. There will be no significant release of the hazardous materials to the environment (See Chapters 3-4 regarding spill reporting);

b. The effectiveness of the packaging will not be substantially reduced; and

c. There will be no mixture of gases or vapors in the package which could, through any credible spontaneous increase of heat or pressure, or through an explosion, significantly reduce the effectiveness of the package.

Packaging used for the shipment of hazardous materials must meet all of the specified design and construction criteria from Parts 172 and 173.

Refer To: 49 CFR § 173.24

Chemical Hazard Communication Guidebook

3. Shippers must fulfill any special packaging requirements for specific materials, such as for certain poisonous materials.

Shippers must adhere to special packaging requirements for specific materials, e.g., poisons. Persons who offer for transportation a liquid material, other than a liquefied compressed gas, addressed in the Hazardous Materials Table by specified prescribed packaging requirements and having a saturated vapor concentration exceeding thresholds set forth in 173.3a(b)(2), must meet special criteria for packaging in addition to general packaging requirements. The packaging criteria for these poisonous materials that exceed the vapor concentration threshold relate to:

a. Packaging which is chemically compatible with the materials being transported;

b. Containment capacity; and

c. Whether authorized approval for the level of safety is obtained.

Refer To: 49 CFR § 173.3a

---

4. Shippers must fulfill any additional packaging requirements that apply to modes of transportation, such as shipment by air.

Shippers are required to make special provisions to ensure secure packaging based on the mode of transportation. For air shipment, for example:

a. Packaging must be constructed or equipped to secure contents despite changes in altitude and temperature;

b. Closures must be secured against loosening due to vibration;

c. Bags must be water resistant; and

d. Valves must be protected from damage.

Refer To: 49 CFR § 173.6

Appendix B: Regulations Reference Guide

5. Shippers must correct, in the manner specified, leaking containers of hazardous materials being transported (See Chapters 3-4 for Discussion of Spill Reporting).

Packages of hazardous materials that are damaged or found leaking and hazardous materials that have been spilled or leaked may be placed in a metal, removable head, salvage drum that is compatible with the lading and shipped for repackaging or disposal in compliance with conditions described in section 173.3(c).

Refer To: 49 CFR § 173.3(c)

---

6. Shippers must fulfill special requirements when reusing packagings or containers.

Containers used more than once must be in such condition, including closure devices and cushioning materials, that they comply in all respects with the prescribed requirements for those containers. Repairs must be made in an efficient manner in accordance with requirements for materials and construction as prescribed in Parts 178 and 179 for new containers. Parts that are weak, broken, or otherwise deteriorated must be replaced. If testing is prescribed for packaging, retesting may be required if packaging or containers are to be reused.

Refer To: 49 CFR § 173.28

---

7. Shippers may find exceptions to certain packaging requirements, based on quantity of the material shipped.

Small quantities of Flammable liquids, Flammable solids, Oxidizers, Organic Peroxides, Corrosive materials, Poison B, and ORM A, B, C, and Radioactive materials that also meet definitions of at least one of these hazard classes are not subject to any other requirement of Subchapter C if the material is contained in an inner receptacle which meets criteria specified for:

a. Quantity in volume or weight (maximum of 30 ml for liquids or 30 grams for solids, except poisons, 1 gram for Poison B or Inhalation hazards, specified for radioactive material);

Chemical Hazard Communication Guidebook

                                              b. Construction;

                                              c. Closure devices;

                                              d. Cushioning;

                                              e. Testing by drop and by load; and

                                              f. Placement within the full package.

Refer To: 49 CFR § 173.4

---

8. Shippers must permit inspection of various transportation procedures and products that affect safety.

        Methods of packing, and storage of hazardous materials that affect safety in transportation must be open to inspection by a duly authorized representative of the initial carrier or a representative of DOT.

Refer To: 49 CFR § 173.3(a)

---

9. Shippers must ensure that the commodity being shipped is in containers authorized for that use by DOT specification requirements.

PACKAGING (GENERAL)
CHECKLIST FOR COMMON VIOLATIONS

Offering for shipment improperly packaged material

Use of DOT specification containers which are not authorized for the commodity being shipped

---

Use of containers that are leaking
(See Discussion in Chapters 3-4 of Spill Reporting)
Refer To: 49 CFR § 173.3(C)

Appendix B: Regulations Reference Guide

Manufacturing and marking containers as meeting a DOT specification when they do not meet the specification

---

Packages exceeding maximum quantity limitations for materials

---

Omission of identification numbers on packagings
Refer To: 49 CFR § 172.301

D. SELECTING THE PROPER LABEL(S)

1. Shippers must fulfill the general requirements for labeling.

Each person who offers a package, overpack, or freight container containing a hazardous material for transportation must label it, when required, according to Part 172, Subpart E and with labels prescribed for the material in the Hazardous Materials Table and Optional Hazardous Materials Table. Additional requirements may apply for specific materials or quantities, containers, conditions, or modes of transportation; there may also be exceptions to the coverage. (For Hazardous Materials Table, see below)

Refer To: 49 CFR § 172.400(a)

---

2. Shippers must ensure that labels conform to the specifications of Subpart E.

Packages must be labeled in accordance with Subpart E. Some examples of criteria for labeling are that:

a. Each label affixed, printed on, or attached to a package must be durable and weather resistant;

b. Any color on a label must be able to withstand, without substantial change, exposure as specified in section 172.407; and

Chemical Hazard Communication Guidebook

c. Labels must contain the requisite wording and numbers, if required, and be of the prescribed dimensions, colors, and shapes

Refer To: 49 CFR § 172.407

---

3. Shippers must place labels on packages or containers in the manner required.

Each label required by Subpart E must be printed on or affixed to the surface of the package near the marked proper shipping name required by Subpart D. When two or more different labels are required, they must be displayed or affixed next to each other.

Each label must be affixed to a background of contrasting color, or must have a dotted or solid line outer border.

Labels must not be obscured by markings or attachments.

In certain cases there are exceptions to the standard placement of labels. For example, labels may be printed on or placed on a securely affixed tag or affixed by other suitable means to:

a. A compressed gas cylinder;

b. A package which has such an irregular surface that a label cannot be satisfactorily attached; or

c. A package that contains no radioactive material and which has dimensions less than those of the required label.

Additional labels may be required on certain packages. For example, on each package of hazardous materials for which labeling is required, and which has a volume of 64 cubic feet or more, labels must be displayed on at least two sides or two ends (excluding the bottom) of containers.

Refer To: 49 CFR § 172.406

Appendix B: Regulations Reference Guide

4. Shippers must fulfill additional labelling requirements that apply for types or conditions of packaging, such as when materials are mixed or consolidated.

Shippers must fulfill additional requirements that may pertain to specific types of packaging or containers, or specific ways in which hazardous materials are packaged.

For example, the packaging, outside container, or overpack must be labeled as required for each hazard class of hazardous material contained within when:

a. Hazardous materials having different hazard classes are packed within the same packaging, or within the same outside container or overpack; and

b. Two or more packages containing compatible hazardous material are packaged together within the same outside container or overpack (172.404).

To take another example of an additional labeling requirement related to the container used, each metal barrel or drum containing a flammable liquid at a specified vapor pressure must have affixed to it a "BUNG" label in addition to the "FLAMMABLE LIQUID" labels required (172.402(e)).

Refer To: e.g. 49 CFR § 172.402

---

5. Shippers must fulfill additional labelling requirements that apply for specific materials, such as laboratory samples.

Shippers may have to provide additional information on labels or additional labels based on the hazardous contents of a package. For example, in the case of samples which must be transported for laboratory analysis, where there is reasonable doubt as to the class and labeling requirements for the material being shipped, the shipper can make a tentative class assignment based on defining criteria and hazard precedence prescribed in Subchapter C and the shipper's knowledge of the material (172.402(h)).

Refer To: 49 CFR § 172.402

Chemical Hazard Communication Guidebook

6. Shippers must fulfill any additional labeling requirements that apply for specific modes of transportation, such as by cargo aircraft.

Shippers may be required to provide additional labels based on the mode used to transport a hazardous material. For example, each person who offers for transportation by air a package containing a hazardous material authorized only on cargo aircraft shall affix a "CARGO AIRCRAFT ONLY" label to the package.

Refer To: 49 CFR § 172.402

---

7. Shippers must fulfill additional labelling requirements that apply to specific conditions, such as when a material is dangerous when wet.

Additional requirements may pertain to conditions under which the material must be transported. For instance, a "DANGEROUS WHEN WET" label is required when a package contains a material fitting that description (172.402(c)).

Refer To: 49 CFR § 172.402

---

8. Shippers may find exceptions to the standard requirements for labeling, such as for certain packages handled under supervision of DOD personnel.

Exceptions to labeling requirements may exist for certain materials under certain conditions or shipped by a specified mode of transportation. Among the many exceptions are the following examples. A label is not required on:

a. A package containing a hazardous material other than ammunition that is loaded and unloaded under the supervision of DOD personnel and escorted by DOD personnel in a separate vehicle;

b. A cargo tank or tank car other than a multi-unit tank car tank; and

c. A package containing a material classed as ORM-A, B, C, D, or E if that package does not contain any other material classed as a hazardous material that requires labeling.

However, caution should be used in determining whether an exemption applies. For example, CARGO AIRCRAFT

Appendix B: Regulations Reference Guide

ONLY labels may still be required on these packages which would be otherwise exempted from labeling.
Refer To: 49 CFR § 172.400(b)

LABELING - CHECKLIST FOR COMMON VIOLATIONS

Label on the container not consistent with the hazard class on the shipping papers when appropriate
Refer To: 49 CFR § 172.400

No labels on outer container to represent mixed packaging of hazardous materials (materials with more than one hazard - dual labeling)
Refer To: 49 CFR § 172.404

Use of obsolete labels
Refer To: 49 CFR § 172.400

Labeling containers not authorized to be labeled
Refer To: 49 CFR § 172.401

Color and/or size of label does not meet the standards of Subpart E
Refer To: 49 CFR § 172.407

No label on "LIMITED QUANTITIES" offered for air transportation
Refer To: 49 CFR §

No label on shipments destined for air transport
Refer To: 49 CFR § 172.402

Chemical Hazard Communication Guidebook

Less than two Radioactive Materials labels
(White I, Yellow II or Yellow III) on containers
(two opposite sides)
Refer To: 49 CFR § 172.403,402

E. MARKING THE PACKAGING IN THE MANNER REQUIRED

1. Shippers must fulfill the general requirements for marking materials for transportation.

Each person who offers a hazardous material for transportation, or each carrier that transports a hazardous material must mark each package, freight container, and transport vehicle containing the hazardous material in the manner required.

Specific additional marking requirements apply to different types of hazardous materials, such as radioactive materials, liquid hazardous materials, and packagings containing materials classed as ORM, and to different holding containers, such as portable tanks, cargo tanks, and tank cars and multi-unit tank car tanks.

Refer To: 49 CFR § 172.300

---

2. Shippers must construct and display markings properly.

Each person who offers a hazardous material for transportation must mark each package, freight container, and transport vehicle containing the hazardous material in the manner specified in Part 172, Subpart D. Additional requirements may apply to specific substances, such as liquid hazardous materials, or specific modes of transport, such as air shipment.

Packages containing hazardous materials must have markings, for example:

a. With the proper shipping name and identification number;

b. Constructed and displayed properly; and

c. With the consignee's or consignor's name and address.

Appendix B: Regulations Reference Guide

Markings must be, for example:

a. Durable;

b. In English;

c. Printed on or affixed to the surface of a package or on a label, tag, or sign;

d. Displayed on a background of sharply contrasting color;

e. Unobscured by labels or attachments; and

f. Must be located away from any other marking (such as advertising) that could substantially reduce its effectiveness.

Refer To: 49 CFR § 172.300,304

---

3. Shippers must fulfill any additional marking requirements for transport under certain conditions or modes, such as export shipments by water

Shippers must mark packages according to specifications for different modes of transportation or different conditions.

For example:

a. Each package of hazardous material offered for export by water and described by a "n.o.s." (Not Otherwise Specified) entry in the Hazardous Materials Table or Optional Hazardous Materials Table(when authorized) must have the technical name or names of the material added in parentheses immediately following the proper shipping name. For a mixture of two or more hazardous materials, the technical name of at least two components most predominantly contributing to the hazard(s) of the mixture must be added in parentheses immediately following the proper shipping name (172.302).

b. ORM-D that is prepared for air shipment and packaged in accordance with 173.6 must be marked "ORM-D-AIR" on the packaging (172.316(a)(5)).

Refer To: 49 CFR § 172.302

Chemical Hazard Communication Guidebook

4. Shippers must mark identification numbers on placards and orange panels when required.

Identification numbers must be displayed on orange panels or placards when required by the sections on portable tanks (172.326), cargo tanks (172.328), tank cars and multi-unit tank car tanks (172.330) and other bulk packagings (172.331), in the manner described in this section. Special provisions and exceptions are listed in section 172.336. For example, for hazardous materials in hazard classes for which hazard warning placards are not specified, identification numbers, when required, must be displayed on an orange panel or a on a plain white square-on-point display configuration having the same outside dimensions as a placard. The latter is not, however, considered to be a placard.

Refer To: 49 CFR § 172.332

---

5. Shippers must replace missing or damaged identification numbers.

If more than one of the identification number markings on placards, orange panels, or white square-on-point display configurations that are required to be displayed are lost, damaged or destroyed during transportation, the carrier must replace all the missing or damaged identification numbers as soon as practicable. However, in such a case, the numbers may be entered by hand on the appropriate placard, orange panel, or white display configuration providing the correct identification numbers are entered legibly using an indelible marking material. When entered by hand, the numbers must be located in the white display area specified in 172.332. This does not preclude required compliance with the placarding requirements of Subpart F.

Refer To: 49 CFR § 172.338

Appendix B: Regulations Reference Guide

6. Shippers must fulfill additional requirements that apply to specific containers, such as portable or cargo tanks, tank cars, and bulk packagings.

Shippers transporting hazardous substances must fulfill additional requirements for marking specific types of containers, such as portable tanks, cargo tanks, tank car tanks, and bulk packagings.

A person using portable tanks, for example:

a. Must mark the proper shipping name of the material on two opposing sides in the manner specified; and

b. Must mark each side and each end with the identification number if the tank has a capacity of 1,000 gallons or more.

A portable tank marked to identify a specific hazardous substance may not be used to transport any other material unless the marking is removed, or changed to identify the hazardous material actually in the portable tank, as appropriate (172.326). (See also 178.245-272 for additional requirements.)

A person using cargo tanks, for example:

a. Must use special markings for gases; and

b. Must indicate if the tanks are constructed of quenched and tempered steel (172.328). (See also 178.315-343 for additional requirements.)

For portable tanks, cargo tanks, tank cars and multi-unit tank car tanks, and other bulk packagings, there is a general requirement that the container that is marked as specified must remain marked unless it is reloaded with a material not subject to Subchapter C or sufficiently cleaned of residue and purged of vapor to remove any potential

Chemical Hazard Communication Guidebook

Refer To: 49 CFR § 172.326-331

hazards. These exceptions may not apply to tank cars and multi-unit tank car tanks containing a combustible liquid.

---

7. Shippers must fulfill any additional requirements for marking containers encasing certain materials, such as liquid hazardous matter, ORM, and wastes.

Shippers must mark containers of hazardous materials in the manner specified in Subpart D. Packagings may require additional or specialized markings based on the hazardous contents. For example:

a. Packages containing liquid hazardous materials must, among other specifications, be marked "THIS END UP" or "THIS SIDE UP" to indicate the upward position of the inside packaging, an arrow symbol indicating "This Way Up" as specified in ANSI's "Pictorial Marking for Handling of Goods" should be used in addition to the marking required by this section and 173.25, and arrows for purposes other than indicating proper package orientation may not be displayed on the packaging (172.312).

b. Markings on packaging containing material classed as ORM ("Other Regulated Material") must include the appropriate ORM designation (such as ORM-A for an ORM-A, or ORM-B-KEEP DRY for an ORM-B solid and corrosive only to aluminum when wet), and that designation is the certification by the person offering the package for transportation that the material is properly described, classed, packaged, marked, and labeled (when appropriate) and in proper condition for transportation according to the applicable regulations. This certification does not preclude the certification required on shipping papers in section 172.204 (172.316).

c. If the word "Waste" is not included in the hazardous material description in the Table, the proper shipping name for a hazardous waste must include the word "Waste" preceding the shipping name of the material. The proper shipping name

Appendix B: Regulations Reference Guide

is not required to include the word if the package bears the special markings designated by EPA for hazardous wastes in 40 CFR § 262.32.

Refer To: e.g., 49 CFR § 172.301,312

8. Shippers may find exceptions to the standard requirements for marking packagings, such as cases where identification numbers are not required.

There are some exceptions to general marking requirements. For example, identification numbers are not required:

a. On the ends of a portable tank, cargo tank or tank car having more than one compartment if hazardous materials having different identification numbers are being transported therein (in such a circumstance the identification numbers on the sides of the tank must be displayed in the same sequence as the compartments containing the materials they identify);

b. On a cargo tank containing only gasoline or only fuel oil if the cargo tank is marked "Gasoline" or "Fuel Oil" respectively on each side and rear as specified or is placarded as specified; or

c. For each of different liquid petroleum distillate fuels transported in cargo tanks, compartmented cargo tanks, or tank cars, if the identification number is displayed for the fuel having the lowest flash point.

Refer To: 49 CFR § 172.336

MARKING OF CONTAINERS - CHECKLIST FOR COMMON VIOLATIONS

No commodity description (proper shipping name) on the container
Refer To: 49 CFR § 172.301

Gross weight not marked on Radioactive Materials packages weighing over 110 pounds
Refer To: 49 CFR § 172.310

Chemical Hazard Communication Guidebook

No name and address of consignee or consignor
on the container
Refer To: 49 CFR § 172.306

Container markings not in a contrasting color
Refer To: 49 CFR § 172.304(a)(2)

Container of liquid hazardous material not
marked on outside "THIS END UP" or "THIS SIDE UP"
Refer To: 49 CFR § 172.312

Reconditioned drum improperly marked

Portable tanks not marked with proper name of
the hazardous material
Refer To: 49 CFR § 172.326

Omission of identification numbers (when
required) on placard or orange panel
Refer To: 49 CFR § 172.332

Omission of marking of INHALATION HAZARD when
required
Refer To: 49 CFR § 172.301

No DOT Exemption number on containers shipped
under DOT Exemption

USA not included as part of the DOT
Specification markings for Radioactive Materials
packages destined for export
Refer To: 49 CFR § 172.310(a)(3)

Appendix B: Regulations Reference Guide

F. DETERMINING THE PROPER PLACARD(S)

1. Shippers must meet the general placarding requirements of Part 172, Subpart F.

Each person who offers for transportation or transports any hazardous material subject to Subchapter C must comply with the placarding requirements of this subpart.

For example, each transport vehicle and freight container containing any quantity of a hazardous material is generally required to be placarded on each end and each side with the type of placards specified in the tables and placarding requirements of Subpart F. There may be exceptions, or additional placarding requirements, associated with specific materials, containers or vehicles, conditions, or modes of transport (172.504,516).

Refer To: 49 CFR § 172.500,504

---

2. Shippers must fulfill any additional placarding requirements that apply for specific materials, such as certain poisons.

Shippers must provide additional placards or additional information on placards if required by Subpart F for specific materials. For example, each transport vehicle, portable tank, or freight container that contains a material subject to the "Poison-Inhalation Hazard" shipping paper description of 172.203(k)(4) must be placarded "POISON" on each side and each end in addition to the placards necessary under the general requirements. Duplication of "POISON" placards or display of UN class numbers at the bottom of additional placards is unnecessary under this additional requirement for poisons.

Refer To: 49 CFR § 172.505

Chemical Hazard Communication Guidebook

3. Shippers must fulfill any additional placarding requirements that apply for containers, such as cargo tanks or rail cars, under certain conditions.

Additional placarding requirements apply for certain types of containers under certain conditions. For example:

a. Each cargo tank and portable tank that is required to be placarded when it contains a hazardous material must remain placarded when it is emptied unless it is reloaded with a material not subject to Subchapter C or is sufficiently cleaned and purged of vapors to remove any potential hazard.

b. Rail cars, truck bodies or trailers containing lading which has been fumigated or treated with poisonous liquid, solid, or gas must be placarded on or near each door with a placard signifying the danger and name of the flammable or poisonous material used for treatment as specified in 173.9.

c. Delivery of rail cars, freight containers, or trailers containing lading, fumigated or treated with flammable liquid or gas for transportation by rail carrier is prohibited until 48 hours have elapsed since the treatment, or until sufficient ventilation has re moved the danger due to the presence of flammable vapors of fire or explosion.

Refer To: e.g., 49 CFR § 173.9, 172.51

---

4. Shippers must provide identification numbers on placards.

Specific rules pertain to marking identification numbers on placards, described in section 172.334. For example:

a. The identification number of a material may be displayed only on the placards required by the tables in section 172.504, except as required for a combustible liquid.

b. If a placard is required by 172.504, an identification number may not be displayed on an orange panel unless it is displayed in proximity to the placard

Refer To: 49 CFR § 172.503

Appendix B: Regulations Reference Guide

5. Shippers may find exceptions to the general placarding requirements, for certain hazardous materials.

The placarding requirements of Subpart F do not apply to etiologic agents or to hazardous materials classed as ORM-A, B, C, D, or E, or to hazardous materials authorized under Subchapter C to be offered as Limited Quantities when identified as such on shipping papers in accordance with 172.203(b).

Refer To: 49 CFR § 172.500

---

6. Shippers must fulfill any additional placarding requirements that apply to specific modes of transport, such as by highway or rail.

Special placarding provisions apply to certain modes of transportation.

For example:

a. Each person offering a motor carrier a hazardous material for transportation by highway must provide to the motor carrier the required placards for the material being offered prior to or at the same time the material is offered for transportation, unless the carrier's motor vehicle is already placarded for the materials as required

b. Each person offering a hazardous material for transportation by rail must affix the specified placards to the rail car unless the placards already displayed on motor vehicles, transport containers, or port able tanks on a rail car are already in compliance with the requirements

c. Each placard on a motor vehicle or rail car must be readily visible from the direction it faces except from the direction of another vehicle or car to which it is coupled. This requirement may be met by the placards displayed on the freight containers or portable tanks loaded on a motor vehicle or rail car (172.516).

Refer To: 49 CFR § 172.506-516

Chemical Hazard Communication Guidebook

PLACARDING - CHECKLIST FOR COMMON VIOLATIONS

Failure to placard vehicle requiring placarding
Refer To: 49 CFR § 172.500

_____

Failure to use more than one kind of placard to indicate more than one hazard class of material loaded within vehicle
Refer To: 49 CFR § 172.504

_____

Placards not applied to both sides of cargo tank
Refer To: 49 CFR § 172.504,514

_____

Freight container containing hazardous material over 640 cubic feet not placarded
Refer To: 49 CFR § 172.512

_____

Placarding material not authorized to be placarded
Refer To: 49 CFR § 172.502

_____

Omission of identification number (when required) on placard or orange panel
Refer To: 49 CFR § 172.503 (172.334)

_____

Appendix B: Regulations Reference Guide

G. PREPARING SHIPPING PAPERS

1. Shippers must describe the hazardous material properly on the shipping papers.

The shipping description of a hazardous material on the shipping paper must include, in sequence:

a. The proper shipping name prescribed for the material in section 172.101 or 172.101 (when authorized);

b. The hazard class, except as noted in section 172.202(2);

c. The identification number (preceded by "UN" or "NA" as appropriate; and

d. The total quantity (by weight, volume, or as otherwise appropriate).

General description requirements apply as specified in 172.201. Additional requirements also apply to certain quantities or types of substances, containers, conditions, and modes of transport as specified in 172.203 and elsewhere.

Refer To: 49 CFR § 172.202

---

2. Shippers must meet the general requirements for preparation of shipping papers of Part 172, Subpart C.

Each person who offers a hazardous material for transportation shall describe it on shipping papers in the manner specified. Additional requirements may apply to specific materials or quantities, containers, conditions, or modes of transportation; there may also be exceptions to the coverage.

Refer To: 49 CFR § 172.200

Chemical Hazard Communication Guidebook

3. Shippers must fulfill additional requirements that apply to preparation of shipping papers for specific modes of transportation, such as by water.

Shippers must provide additional information on shipping papers for moving hazardous materials by certain modes of transportation, such as transportation by rail, by highway, and by water. For example, each shipment by water must have as additional shipping paper entries:

a. Identification of the type of packages such as barrels, drums, cylinders, and boxes;

b. The number of each type of package including those in a freight container or on a pallet;

c. The gross weight of each type of package or the individual gross weight of each package; and

d. The name of the shipper (172.201(b)).

The shipping paper for a hazardous material offered for transportation by vessel to any country outside the U.S.A. must include:

a. The technical name of the material immediately following the proper shipping name when the material is described by an n.o.s. entry in the Hazardous Materials Table or the Optional Hazardous Materials Table; and

b. The names of at least two components most predominately contributing to the hazard or hazards of the mixture if the material is a mixture of two or more hazardous materials.

Refer To: 49 CFR § 172.203

Appendix B: Regulations Reference Guide

4. Shippers must fulfill any additional shipping paper requirements that apply to specific materials, such as poisonous materials, or wastes.

Shippers must provide additional information on shipping papers for certain hazardous materials, such as poisonous materials, liquified petroleum gas, hazardous wastes, and other materials.

For example:

a. For poisonous materials, if the name of the compound or principal constituent that causes a material to meet the definition of a poison is not included in the proper shipping name, the compound or constituent name must be entered on the shipping paper in association with the shipping description; additional requirements and exclusions may also apply, as specified in 172.203(k);

b. For liquified petroleum gas, the shipper must indicate suitability for transportation by highway, as specified in 172.203(h)(2); and

c. For hazardous wastes, a hazardous waste manifest, EPA Form 8700-22 and 8700-22A (when necessary), must be prepared by the person responsible in accordance with 40 CFR § 262.20, in order to offer, transport, transfer, or deliver, and signed, carried, and given as required of that person by section 172.205.

Refer To: 49 CFR § 172.203, 205

---

5. Shippers must fulfill additional requirements that apply to specific containers, such as cargo tanks under some circumstances, or empty packaging.

Shippers must provide additional information on shipping papers for certain types of containers or packagings or conditions that affect them. For example, the description on the shipping paper for a packaging containing:

a. The residue of a hazardous material may include the words "RESIDUE: Last contained ***" in association with the

Chemical Hazard Communication Guidebook

	basic description of the hazardous material last contained in the packaging (172.203(e)); and
	b. Materials such as anhydrous ammonia or liquefied petroleum gas transported by highway in cargo tanks made of quenched and tempered steel as authorized by 173.315(a)(1) must indicate suitability of the tank for shipping as specified in 172.203(h)
Refer To: 49 CFR § 172.203	
6. Shippers must fulfill additional requirements that apply to shipping papers for specific conditions, such as material that is dangerous when wet.	Shippers must provide additional information on shipping papers for certain conditions. For example, the words "Dangerous When Wet" must be entered on the shipping paper in association with the basic description when a package must be labeled as such due to the properties of its contents (172.203(j)).
Refer To: e.g., 49 CFR § 172.203	
7. Shippers may find exceptions to the standard requirements on shipping papers.	Exceptions to shipping paper requirements may exist for certain materials, under certain conditions, or by certain modes of transportation. For example, Subpart C does not apply to any material, other than a hazardous waste or a hazardous substance, that is:

a. An ORM-A, B, or C, unless it is offered or intended for transportation by air when it is subject to the regulations pertaining to transportation by air as specified in 172.101; or

b. An ORM-A, B, or C, unless it is offered or intended for transportation by water when it is subject to the regulations pertaining to transportation by water as specified in 172.101; or

c. An ORM-D unless it is offered or intended for transportation by air. |

Appendix B: Regulations Reference Guide

Each shipping paper issued in connection with a shipment made under an exemption must bear the notation "DOT-E" followed by the exemption number (172.203(a)).

Refer To: 49 CFR § 172.200

---

7. Shippers must certify a shipment before offering it for transportation.

Each person who offers a hazardous material for transportation must certify that it is properly classified, described, packaged, marked and labeled, and is in proper condition for transportation according to the requirements of Hazardous Materials Regulations (Subchapter C) by making in writing the declaration or an approved alternate specified in section 172.204.

Refer To: 49 CFR § 172.204

SHIPPING PAPERS - CHECKLIST FOR COMMON VIOLATIONS

No proper shipping name and/or classification of hazardous material entered on shipping papers.
Refer To: 49 CFR § 172.202

Proper shipping name and/or classification abbreviated
Refer To: 49 CFR § 172.201(a)(3)

Improper format for hazardous materials description on shipping papers, e.g. HM entries not first, highlighted, or no HM column
Refer To: 49 CFR § 172.201

No certification for shipment
Refer To: 49 CFR § 172.204

No identification number (UN or NA) on shipping paper
Refer To: 49 CFR § 172.202(a)(3)

Chemical Hazard Communication Guidebook

No DOT Exemption number on shipments moving
under DOT Exemption
Refer To: 49 CFR § 172.203(a)

No indication for "LIMITED QUANTITY" on shipments
excepted from specification packaging and labeling
Refer To: 49 CFR § 172.203(b)

Color of label indicated in lieu of the proper
hazard class
Refer To: 49 CFR § 172.202(a)(2)

DOT HAZARDOUS MATERIALS TRANSPORTATION TABLE

Chemical Hazard Communication Guidebook

§172.101   Title 49—Transportation

§172.101 Hazardous Materials Table

(1)	(2)	(3)	(3A)	(4)	(5) Packaging		(6) Maximum net quantity in one package		(7) Water shipments		
+/A/W	Hazardous materials descriptions and proper shipping names	Hazard class	Identification number	Label(s) required (if not excepted)	(a) Exceptions	(b) Specific requirements	(a) Passenger carrying aircraft or railcar	(b) Cargo only aircraft	(a) Cargo vessel	(b) Passenger vessel	(c) Other requirements
	Accumulator, pressurized (pneumatic or hydraulic), containing nonflammable gas	Nonflammable gas	NA1956	Nonflammable gas	173.306		No limit	No limit	1,2	1,2	
	Acetal	Flammable liquid	UN1088	Flammable liquid	173.118	173.119	1 quart	10 gallons	1,3	4	
	Acetaldehyde (ethyl aldehyde)	Flammable liquid	UN1089	Flammable liquid	None	173.119	Forbidden	10 gallons	1,3	5	
A	Acetaldehyde ammonia	ORM-A	UN1841	None	173.505	173.510	No limit	No limit	1,2	1,2	
	Acetic acid (aqueous solution)	Corrosive material	UN2790	Corrosive	173.244	173.245	1 quart	10 gallons	1,2	1,2	Stow separate from nitric acid or oxidizing materials
	Acetic acid, glacial	Corrosive material	UN2789	Corrosive	173.244	173.245	1 quart	10 gallons	1,2	1,2	Stow separate from nitric acid or oxidizing materials
	Acetic anhydride	Corrosive material	UN1715	Corrosive	173.244	173.245	1 quart	1 gallon	1,2	1,2	
	Acetone	Flammable liquid	UN1090	Flammable liquid	173.118	173.119	1 quart	10 gallons	1,3	4	Shade from radiant heat. Stow away from corrosive materials
	Acetone cyanohydrin	Poison B	UN1541	Poison	None	173.346 173.3a	Forbidden	55 gallons	1	5	Shade from radiant heat. Stow away from corrosive materials
	Acetone oil	Flammable liquid	UN1091	Flammable liquid	173.118	173.119	1 quart	10 gallons	1,2	1	
	Acetonitrile	Flammable liquid	NA1648	Flammable liquid	173.118	173.119	1 quart	10 gallons	1	4	Shade from radiant heat
	Acetyl acetone peroxide, in solution with not more than 9% by weight active oxygen. See Organic peroxide, liquid or solution, n.o.s.		UN2080								
	Acetyl acetone peroxide with more than 9% by weight active oxygen	Forbidden									
	Acetyl benzoyl peroxide, not more than 40% in solution. See Acetyl benzoyl peroxide solution, not over 40% peroxide		UN2081								
	Acetyl benzoyl peroxide, solid, or more than 40% in solution	Forbidden									
	Acetyl benzoyl peroxide solution, not over 40% peroxide	Organic peroxide	UN2081	Organic peroxide	None	173.222	Forbidden	1 quart	1,2	1	Keep dry. Glass carboys not permitted on passenger vessels
	Acetyl bromide	Corrosive material	UN1716	Corrosive	173.247	173.247	1 quart	1 gallon	1	1	Stow away from alcohols. Keep cool and dry. Separate longitudinally by an intervening complete compartment or hold from explosives
	Acetyl chloride	Flammable liquid	UN1717	Flammable liquid	173.244	173.247	1 quart	1 gallon	1	1	

112

Appendix B: Regulations Reference Guide

Chapter I—Research and Special Programs Administration   **§172.101**

Material	Class	ID No.	Label	Packaging §173	Max qty passenger	Max qty cargo	Vessel stowage	Other
Acetyl cyclohexanesulfonyl peroxide, more than 82%, wetted with less than 12% water	Forbidden							
Acetyl cyclohexanesulphonyl peroxide, not more than 82%, wetted with not less than 12% water. See Organic peroxide, solid, n.o.s.		UN2082						
Acetyl cyclohexanesulphonyl peroxide, not more than 32% in solution. See Organic peroxide, liquid or solution, n.o.s.		UN2083						
Acetylene	Flammable gas	UN1001	Flammable gas	None	173.303	Forbidden	300 pounds	1	Shade from radiant heat
Acetylene (liquid)	Forbidden							
Acetylene silver nitrate	Forbidden							
Acetylene tetrabromide	ORM-A	UN2504	None	173.505	173.510	10 gallons	55 gallons	1	
Acetyl iodide	Corrosive material	UN1898	Corrosive	173.244	173.247	1 quart	1 gallon		Keep dry. Glass carboys not permitted on passenger vessels
Acetyl peroxide, not more than 25% in solution. See Acetyl peroxide solution, not over 25% peroxide		UN2084							
Acetyl peroxide, solid, or more than 25% in solution	Forbidden							
Acetyl peroxide solution, not over 25% peroxide	Organic peroxide	UN2084	Organic peroxide	173.153	173.222	Forbidden	1 quart	1,2	1	
Acid butyl phosphate	Corrosive material	UN1718	Corrosive	173.244	173.245	1 quart	5 gallons	1,2	1,2	Glass carboys in hampers not permitted under deck
Acid carboy, empty. See Carboy, empty.									
Acid, liquid, n.o.s.	Corrosive material	NA1760	Corrosive	173.244	173.245	1 quart	5 pints	1	4	Keep cool
Acrolein, inhibited	Flammable liquid	UN1092	Flammable liquid and Poison	None	173.122	Forbidden	1 quart	1,2	5	Keep cool. Stow away from living quarters
Acrylic acid	Corrosive material	UN2218	Corrosive	173.244	173.245	1 quart	5 pints	1	1	Keep cool
Acrylonitrile	Flammable liquid	UN1093	Flammable liquid and Poison	None	173.119	Forbidden	1 quart	1,2	5	Keep cool and dry
Actuating cartridge, explosive (fire extinguisher, or valve)	Class C explosive			173.114		50 pounds	150 pounds	1,2	1,2	
Adhesive	Combustible liquid	UN1133	None	173.118a	None	No limit	No limit	1,2	1,2	
Adhesive	Flammable liquid	UN1133	Flammable liquid	173.118	173.132	1 quart	10 gallons	1,2	1	
Aerosol product. See Compressed gas, n.o.s.										
Air, compressed	Nonflammable gas	UN1002	Nonflammable gas	173.306	173.302	150 pounds	300 pounds	1,2	1,2	
Air conditioning machine. See Refrigerating machine										
Air, refrigerated liquid (cryogenic liquid)	Nonflammable gas	UN1003	Nonflammable gas		173.316 173.318	Forbidden	300 pounds	1,2	1,2	Stow separate from flammables. Do not overstow with other cargo
Airplane flare. See Fireworks, special										
Alcoholic beverage	Flammable liquid	UN1170	Flammable liquid	173.118	173.125	See 173.118(*)	10 gallons (*)	1,2	1	
Alcoholic beverage	Combustible liquid	UN1170	None	173.118a	None	No limit	No limit	1,2	1	
Alcohol, n.o.s.	Flammable liquid	UN1987	Flammable liquid	173.118	173.125	1 quart	10 gallons	1,2	1	

113

Chemical Hazard Communication Guidebook

§172.101   Title 49—Transportation

§172.101 Hazardous Materials Table

(1) +/A/W	(2) Hazardous materials descriptions and proper shipping names	(3) Hazard class	(3A) Identification number	(4) Label(s) required (if not excepted)	(5)(a) Packaging Exceptions	(5)(b) Packaging Specific requirements	(6)(a) Passenger carrying aircraft or railcar	(6)(b) Cargo only aircraft	(7)(a) Cargo vessel	(7)(b) Passenger vessel	(7)(c) Other requirements
	Alcohol, n.o.s.	Combustible liquid	UN1987	None	173.118a	None	No limit	No limit	1,2	1,2	
	Aldrin	Poison B	NA2761	Poison	173.364	173.376	50 pounds	200 pounds	1,2	1,2	
	Aldrin, cast solid	ORM-A	NA2761	None	173.505	173.510	No limit	No limit	1,2	1,2	
	Aldrin mixture, dry (*with more than 65% aldrin*)	Poison B	NA2761	Poison	173.364	173.376	50 pounds	200 pounds	1,2	1,2	
A	Aldrin mixture, dry, with 65% or less aldrin	ORM-A	NA2761	None	173.505	173.510	No limit	55 gallons	1,2	1,2	If flash point less than 141 deg. F, segregation same as for flammable liquids
A	Aldrin mixture, liquid (*with more than 60% aldrin*)	Poison B	NA2762	Poison	173.345	173.361	1 quart		1,2	1,2	
A	Aldrin mixture, liquid, with 60% or less aldrin	ORM-A	NA2762	None	173.505	173.510	No limit	No limit	1,2	1,2	
	Alkaline (*corrosive*) liquid, n.o.s.	Corrosive material	NA1719	Corrosive	173.244	173.249	1 quart	5 gallons	1,2	1,2	
	Alkanesulfonic acid	Corrosive material	UN2584	Corrosive	173.244	173.245	5 pints	1 gallon	1,2	1	
	Alkyl aluminum halides. See Pyrophoric liquid, n.o.s.										
A	Allethrin	ORM-A	NA2902	None	173.505	173.510	No limit	No limit	1,2	1	
	Allyl alcohol	Flammable liquid	UN1098	Flammable liquid and Poison	None	173.119 173.3a	1 quart	10 gallons	1,2	1	
	Allyl bromide	Flammable liquid	UN1099	Flammable liquid	173.118	173.119	Forbidden	10 gallons	1,2	1	
	Allyl chloride	Flammable liquid	UN1100	Flammable liquid	None	173.119	Forbidden	10 gallons	1,3	5	
	Allyl chlorocarbonate	Flammable liquid	UN1722	Flammable liquid	None	173.288	Forbidden	5 pints	1	5	Keep dry. Separate longitudinally by an intervening complete hold or compartment from explosives. Segregation same as for corrosive materials
	Allyl chloroformate. See Allyl chlorocarbonate										
	Allyl trichlorosilane	Corrosive material	UN1724	Corrosive	None	173.280	Forbidden	10 gallons	1	1	Keep dry
	Aluminum alkyl. See Pyrophoric liquid, n.o.s.										
	Aluminum bromide, anhydrous	Corrosive material	UN1725	Corrosive	173.244	173.245b	25 pounds	100 pounds	1,2	1,2	Keep dry
	Aluminum chloride, anhydrous	Corrosive material	UN1726	Corrosive	173.244	173.245b	25 pounds	100 pounds	1,2	1,2	Keep dry
	Aluminum dross, wet or hot. See 173.173	Forbidden									

114

Chapter I—Research and Special Programs Administration §172.101

	Name	Hazard class	ID No.	Label(s) required	(ref)	(ref)	Passenger aircraft	Cargo aircraft	(col)	(col)	Other requirements
	Aluminum hydride	Flammable solid	UN2463	Flammable solid and Dangerous When wet	None	173.206	Forbidden	25 pounds	1,2	5	Segregation same as for flammable solid labeled Dangerous When Wet
	Aluminum, metallic, powder	Flammable solid	UN1396	Flammable solid	173.232	173.232	25 pounds	100 pounds	1,2	1,2	Keep dry. Segregation same as for flammable solids labeled dangerous when wet
	Aluminum nitrate	Oxidizer	UN1438	Oxidizer	173.153	173.182	25 pounds	100 pounds	1,2	1,2	
	Aluminum phosphate solution	Corrosive material	NA1760	Corrosive	173.244	173.245	1 quart	10 gallons	1,2	1,2	
	Aluminum phosphide	Flammable solid	UN1397	Flammable solid and Dangerous when wet	None	173.154	Forbidden	25 pounds	1,2	1,2	Stow away from acids and oxidizing materials
A	Aluminum sulfate solution	ORM-B	NA1760	None	173.505	173.510	25 pounds	100 pounds	1,2	1,2	
	Amatol. See High explosive										
	2-(2-Aminoethoxy) ethanol	Corrosive material	NA1760	Corrosive	173.244	173.245	1 quart	10 gallons	1,2	1,2	
	N-Aminoethylpiperazine	Corrosive material	UN2815	Corrosive	173.244	173.245	1 quart	10 gallons	1,2	1,2	
	Aminopropyldiethanolamine	Corrosive material	NA1760	Corrosive	173.244	173.245	1 quart	10 gallons	1,2	1,2	
	N-Aminopropylmorpholine	Corrosive material	NA1760	Corrosive	173.244	173.245	1 quart	10 gallons	1,2	1,2	
	bis (Aminopropyl) piperazine	Corrosive material	NA1760	Corrosive	173.244	173.245	1 quart	10 gallons	1,2	1,2	
	Ammonia, anhydrous	Nonflammable gas	UN1005	Nonflammable gas	173.306	173.304 173.314 173.315	Forbidden	300 pounds	1,2	4	Stow in well ventilated space
	Ammonia solution (containing more than 44% ammonia)	Nonflammable gas	UN2073	Nonflammable gas	173.306	173.304 173.314 173.315	Forbidden	300 pounds	1,2	4	Stow in well ventilated space
	Ammonia solution (containing 44% or less ammonia in water). See Ammonium hydroxide										
	Ammonium arsenate, solid	Poison B	UN1546	Poison	173.364	173.365	50 pounds	200 pounds	1,2	1,2	Stow away from alkaline corrosives
A	Ammonium azide	Forbidden									
	Ammonium bifluoride, solid or solution. See Ammonium hydrogen fluoride, solid or solution										
	Ammonium bisulfite, solid	ORM-B	NA2693	None	173.505	173.510	25 pounds	100 pounds	1,2	1,2	
	Ammonium bisulfite solution	Corrosive material	NA2693	Corrosive	173.244	173.245	1 quart	5 gallons	1,2	1,2	
	Ammonium bromate	Forbidden									
A	Ammonium carbamate	ORM-A	NA9083	None	173.505	173.510	50 pounds	No limit	1,2	1,2	Keep away from heat
A	Ammonium carbonate	ORM-A	NA9084	None	173.505	173.510	50 pounds	No limit	1,2	1,2	Keep away from heat, acids, alum and salts of iron or zinc
	Ammonium chlorate	Forbidden									
	Ammonium dichromate (ammonium bichromate)	Oxidizer	UN1439	Oxidizer	173.153	173.154 173.235	25 pounds	100 pounds	1,2	1,2	
A	Ammonium fluoborate	ORM-B	NA9088	None	None	173.510	25 pounds	100 pounds	1,2	1,2	
A	Ammonium fluoride	ORM-B	UN2505	None	173.505	173.800	25 pounds	100 pounds	1,2	1,2	
	Ammonium fulminate	Forbidden									
	Ammonium hydrogen fluoride, solid	Corrosive material	UN1727	Corrosive	173.244	173.245b	25 pounds	100 pounds	1,2	1,2	Keep dry
	Ammonium hydrogen fluoride solution	Corrosive material	UN2817	Corrosive	173.244	173.245	1 quart	5 gallons	1,2	1,2	Keep dry
A	Ammonium hydrogen sulfate	ORM-B	UN2506	None	173.505	173.800	25 pounds	100 pounds	1,2	1,2	

115

Chemical Hazard Communication Guidebook

§172.101   Title 49—Transportation

§172.101 Hazardous Materials Table

(1) +/A/W	(2) Hazardous materials descriptions and proper shipping names	(3) Hazard class	(3A) Identification number	(4) Label(s) required (if not excepted)	(5) Packaging		(6) Maximum net quantity in one package				(7) Water shipments	
					(a) Exceptions	(b) Specific requirements	(a) Passenger carrying aircraft or railcar	(b) Cargo only aircraft	(a) Cargo vessel	(b) Passenger vessel		(c) Other requirements
A	Ammonium hydrosulfide solution	ORM-A	NA2683	None	173.505	173.605	10 gallons	55 gallons	1			
	Ammonium hydroxide (containing notless than 12% but not more than 44% ammonia)	Corrosive material	NA2672	Corrosive	173.244	173.245	2 gallons	2 gallons		4		
AW	Ammonium hydroxide (containing less than 12% ammonia)	ORM-A	NA2672	None	173.505	173.510	10 gallons	55 gallons	1	1		
	Ammonium nitrate-carbonate mixture	Oxidizer	UN2068	Oxidizer	173.153	173.182	25 pounds	100 pounds	1,2	1,2		
	Ammonium nitrate fertilizer, containing no more than 0.2% carbon	Oxidizer	UN2067	Oxidizer	173.153	173.182	25 pounds	100 pounds	1,2	1,2		
	Ammonium nitrate - fuel oil mixture. See High explosive											
	Ammonium nitrate-fuel oil mixture (containing only prilled ammonium nitrate and fuel oil)	Blasting agent		Blasting agent	None	173.114a	Forbidden	100pounds	1,2	1,2		
	Ammonium nitrate mixed fertilizer	Oxidizer	UN2069	Oxidizer	173.153	173.182	25 pounds	100 pounds	1,2	1,2		
	Ammonium nitrate (no organic coating)	Oxidizer	UN1942	Oxidizer	173.153	173.182	25 pounds	100 pounds	1,2	1,2		
	Ammonium nitrate (organic coating)	Oxidizer	NA1942	Oxidizer	173.153	173.182	25 pounds	100 pounds	1,2	1,2		
	Ammonium nitrate-phosphate	Oxidizer	UN2070	Oxidizer	173.153	173.182	25 pounds	100 pounds	1,2	1,2		
	Ammonium nitrate, solution (containing not less than 15% water). See 173.154(a)(17) and 173.154(a)(18)	Oxidizer	UN2426	Oxidizer								
	Ammonium nitrite	Forbidden										
	Ammonium oxalate	ORM-A	NA2449	None	173.505	173.510	50 pounds	200 pounds	1,2	1,2	Stow away from powdered metals	
	Ammonium perchlorate	Oxidizer	UN1442	Oxidizer	173.153	173.239a	25 pounds	100 pounds	1,2	4		
	Ammonium perchlorate. See High explosive											
	Ammonium permanganate	Oxidizer	NA9190	Oxidizer	None	173.154	Forbidden	Forbidden	1,2	1,2	Separate from ammonium compounds and hydrogen peroxide. This material may be forbidden in water transportation by certain countries	
	Ammonium persulfate	Oxidizer	UN1444	Oxidizer	173.153	178.154	50 pounds	200 pounds	1,2	1,2		
	Ammonium picrate, dry. See High explosive											
	Ammonium picrate, wet (with 10% or more water)	Flammable solid	UN1310	Flammable solid	173.192		1 pound	1 pound	1	4	Stow away from heavy metals and their compounds	
	Ammonium picrate, wet, with 10% or more water, over 16 ounces in one outside packaging. See High explosive											
	Ammonium polysulfide solution	ORM-A	UN2818	None	173.505	173.605	10 gallons	55 gallons	1,2	1,2		
A	Ammonium silicofluoride	ORM-B	UN2854	None	173.505	173.510	25 pounds	100 pounds	1,2	1,2		

116

Appendix B: Regulations Reference Guide

Chapter I—Research and Special Programs Administration   §172.101

w	Ammonium sulfate nitrate	ORM-C	NA1477	None	173.505	173.910	1 quart	10 gallons	1,2	1,2	Must not be accepted for transportation while hot. Separate by an intervening hold or compartment from Class A explosives. Separate from other explosives, corrosive materials, flammable solids, liquids, or gases, oxidizing materials, organic peroxides, or organic materials
	Ammonium sulfide solution	Flammable liquid	UN2683	Flammable liquid	173.118	173.119					
	Ammonium, chemical (containing a Poison A liquid or gas). See Chemical ammunition, nonexplosive (containing a Poison A material)										
	Ammunition, chemical (containing a Poison B material). See Chemical ammunition, nonexplosive (containing a Poison B material)										
	Ammunition, chemical (containing an irritating liquid or solid). See Chemical ammunition, nonexplosive (containing an irritating material)										
	Ammunition, chemical, explosive, with Poison A material	Class A explosive		Explosive A and Poison gas	None	173.59	Forbidden	Forbidden	6	5	No other cargo may be stowed in the same hold with these items
	Ammunition, chemical, explosive, with Poison B material	Class A explosive		Explosive A and Poison gas	None	173.59	Forbidden	Forbidden	6	5	No other cargo may be stowed in the same hold with these items
	Ammunition, chemical, explosive, with irritant	Class A explosive		Explosive A and Irritant	None	173.59	Forbidden	Forbidden	6	5	No other cargo may be stowed in the same hold with these items
	Ammunition for cannon with empty projectile	Class B explosive		Explosive B	None	173.89	Forbidden	Forbidden	1,2	5	
	Ammunition for cannon with explosive projectile	Class A explosive		Explosive A	None	173.54	Forbidden	Forbidden	6	5	
	Ammunition for cannon with gas projectile	Class A explosive		Explosive A	None	173.54	Forbidden	Forbidden	6	5	
	Ammunition for cannon with illuminating projectile	Class A explosive		Explosive A	None	173.54	Forbidden	Forbidden	6	5	
	Ammunition for cannon with incendiary projectile	Class B explosive		Explosive B	None	173.89	Forbidden	Forbidden	1,2	5	
	Ammunition for cannon with inert loaded projectile	Class B explosive		Explosive B	None	173.89	Forbidden	Forbidden	1,2	5	
	Ammunition for cannon without projectile	Class A explosive		Explosive A	None	173.54	Forbidden	Forbidden	6	5	
	Ammunition for cannon with smoke projectile	Class B explosive		Explosive B	None	173.89	Forbidden	Forbidden	1,2	5	
	Ammunition for cannon with solid projectile	Class A explosive		Explosive A	None	173.54	Forbidden	Forbidden	6	5	
	Ammunition for cannon with tear gas projectile	Class B explosive		Explosive B	None	173.89	Forbidden	Forbidden	1,2	5	
	Ammunition for small-arms with explosive projectile	Class A explosive		Explosive A	None	173.58	Forbidden	Forbidden	6	5	
	Ammunition for small-arms with incendiary projectile	Class A explosive		Explosive A	None	173.58	Forbidden	Forbidden	6	5	
	Ammunition, non-explosive. See 173.55										
	Ammunition, rocket. See Rocket ammunition with										

117

Chemical Hazard Communication Guidebook

§172.101   Title 49—Transportation

§172.101 Hazardous Materials Table

(1)	(2) Hazardous materials descriptions and proper shipping names	(3) Hazard class	(3A) Identification number	(4) Label(s) required (if not excepted)	(5) Packaging		(6) Maximum net quantity in one package		(7) Water shipments		
+/A/W					(a) Exceptions	(b) Specific requirements	(a) Passenger carrying aircraft or railcar	(b) Cargo only aircraft	(a) Cargo vessel	(b) Passenger vessel	Other requirements
	Ammunition, small-arms. See Small-arms ammunition										
	Amyl acetate	Flammable liquid	UN1104	Flammable liquid	173.118	173.119	1 quart	10 gallons	1,2	1,2	
	Amyl acid phosphate	Corrosive material	UN2819	Corrosive	173.244	173.245	1 quart	10 gallons	1,2	1,2	
	Amylamine	Flammable liquid	UN1106	Flammable liquid	173.118	173.119	1 quart	10 gallons	1,2	1	
	Amyl chloride	Flammable liquid	UN1107	Flammable liquid	173.118	173.119	1 quart	10 gallons	1,2	1	
	Amylene, normal	Flammable liquid	UN1108	Flammable liquid	173.118	173.119	1 quart	10 gallons	1,3	1,3	
	Amyl formate	Flammable liquid	UN1109	Flammable liquid	173.118	173.119	1 quart	10 gallons	1,2	1	
	Amyl mercaptan	Flammable liquid	UN1111	Flammable liquid	None	173.141	Forbidden	10 gallons	1,2	1	
	Amyl nitrite	Flammable liquid	UN1113	Flammable liquid	173.118	173.119	1 quart	10 gallons	1,3	4	
	tert-Amyl peroxy-2-ethylhexanoate, technically pure. See Organic peroxide, liquid or solution, n.o.s.		UN2898								
	tert-Amyl peroxyneodecanoate, not more than 75% with phlegmatizer. See Organic peroxide, liquid or solution, n.o.s.		UN2891								
	Amyl trichorosilane	Corrosive material	UN1728	Corrosive	None	173.280	Forbidden	10 gallons	1	5	Keep dry
	Anhydrous ammonia. See Ammonia, anhydrous										
	Anhydrous hydrazine. See Hydrazine, anhydrous										
	Anhydrous hydrofluoric acid. See Hydrogen fluoride										
	Aniline oil drum, empty See 173.347(d)								1,2	1	Do not accept unless returnable package notice is on drum and the instructions thereon have been carried out
+	Aniline oil, liquid	Poison B	UN1547	Poison	None	173.347	Forbidden	55 gallons	1,2	1,2	Stow away from oxidizing materials and acids
	Anisoyl chloride	Corrosive material	UN1729	Corrosive	173.244	173.279	1 quart	1 quart	1	1	Keep dry
	Antimonous chloride. See Antimony trichloride										
A	Antimony lactate, solid	ORM-A	UN1550	None	173.505	173.510	No limit	No limit			

118

Appendix B: Regulations Reference Guide

Chapter I—Research and Special Programs Administration §172.101

Material	Class	ID No.	Label(s)	Packaging (exceptions)	Packaging (specific)	Max qty (passenger)	Max qty (cargo)	(col a)	(col b)	Other requirements
Antimony pentachloride	Corrosive material	UN1730	Corrosive	None	173.247	1 quart	1 quart	1	1	Keep dry. Glass carboys not permitted on passenger vessels
Antimony pentachloride solution	Corrosive material	UN1731	Corrosive	173.244	173.245	1 quart	5 pints	1	1	Keep dry. Glass carboys not permitted on passenger vessels
Antimony pentafluoride	Corrosive material	UN1732	Corrosive	None	173.246	Forbidden	25 pounds	1	5	Keep dry
Antimony potassium tartrate, solid	ORM-A	UN1551	None	173.505	173.510	No limit	No limit	1,2	1,2	
Antimony sulfide and a chlorate, mixtures of	Forbidden									
Antimony sulfide, solid	ORM-A	NA1325	None	173.505	173.510	No limit	No limit			
Antimony tribromide, solid	Corrosive material	NA1549	Corrosive	173.244	173.245b	25 pounds	100 pounds	1,2	1,2	Keep dry
Antimony tribromide solution	Corrosive material	NA1549	Corrosive	173.244	173.245	1 quart	5 pints	1	1	Keep dry
Antimony trichloride, solid	Corrosive material	UN1733	Corrosive	173.244	173.245b	25 pounds	100 pounds	1,2	1,2	Keep dry
Antimony trichloride solution	Corrosive material	UN1733	Corrosive	173.244	173.245	1 quart	5 pints	1	1	Keep dry
Antimony trifluoride, solid	Corrosive material	NA1549	Corrosive	173.244	173.245b	25 pounds	100 pounds	1,2	1,2	Keep dry
Antimony trifluoride solution	Corrosive material	NA1549	Corrosive	173.244	173.245	1 quart	5 pints	1	1	Keep dry
Aqua ammonia solution *(containing 44% or less ammonia). See Ammonium hydroxide*										
Argon or Argon, compressed	Nonflammable gas	UN1006	Nonflammable gas	173.306	173.302	150 pounds	300 pounds	1,2	1,3	
Argon, refrigerated liquid *(cryogenic liquid)*	Nonflammable gas	UN1951	Nonflammable gas	173.320	173.314 173.316 173.318	100 pounds	1,100 pounds	1,3	1,3	
Arsenic acid, solid	Poison B	UN1554	Poison	173.364	173.366	50 pounds	200 pounds	1,2	1,2	
Arsenic acid solution	Poison B	NA1553	Poison	173.345	173.348	1 quart	55 gallons	1,2	1,2	
Arsenical compound, liquid, n.o.s., or Arsenical mixture, liquid, n.o.s.	Poison B	UN1556	Poison	173.345	173.346	1 quart	55 gallons	1,2	1,2	
Arsenical compound, solid, n.o.s., or Arsenical mixture, solid, n.o.s.	Poison B	UN1557	Poison	173.364	173.367	50 pounds	200 pounds	1,2	1,2	Keep dry
Arsenical dip, liquid *(sheep dip)*	Poison B	NA1557	Poison	173.345	173.346	1 quart	55 gallons	1,2	1,2	
Arsenical dust	Poison B	UN1562	Poison	173.364	173.368	50 pounds	200 pounds	1,2	1,2	
Arsenical pesticide, liquid, n.o.s.*(compounds and preparations)*	Flammable liquid	UN2760	Flammable liquid	173.118	173.119	1 quart	10 gallons	1	1	
Arsenical pesticide, liquid, n.o.s.*(compounds and preparations)*	Poison B	UN2759	Poison	173.345	173.346	1 quart	55 gallons	1,2	1,2	
Arsenical pesticide, solid n.o.s *(compounds and preparations)*	Poison B	UN2759	Poison	173.364	173.367	50 pounds	200 pounds	1,2	1,2	
Arsenic bromide, solid	Poison B	UN1555	Poison	173.364	173.365	50 pounds	200 pounds	1,2	1,2	
Arsenic chloride, liquid. See Arsenic trichloride.										
Arsenic disulfide. See Arsenic sulfide, solid										
Arsenic iodide, solid	Poison B	NA1557	Poison	173.364	173.365	50 pounds	200 pounds	1,2	1,2	
Arsenic pentoxide, solid	Poison B	UN1559	Poison	173.364	173.365	50 pounds	200 pounds	1,2	1,2	
Arsenic, solid	Poison B	UN1558	Poison	173.364	173.365	50 pounds	200 pounds	1,2	1,2	
Arsenic sulfide and a chlorate, mixtures of	Forbidden									
Arsenic sulfide, solid	Poison B	NA1557	Poison	173.364	173.365	50 pounds	200 pounds	1,2	1,2	
Arsenic trichloride, liquid	Poison B	UN1560	Poison	173.345	173.346	1 quart	55 gallons	1,2	1,2	
Arsenic trioxide, solid	Poison B	UN1561	Poison	173.364	173.366	50 pounds	200 pounds	1,2	1,2	
Arsenic trisulfide	Poison B	NA1557	Poison	173.364	173.365	50 pounds	200 pounds	1,2	2	
Arsenic, white, solid. See Arsenic trioxide, solid										

119

Chemical Hazard Communication Guidebook

§172.101   **Title 49—Transportation**

§172.101 Hazardous Materials Table

(1)	(2)	(3)	(3A)	(4)	(5) Packaging		(6) Maximum net quantity in one package		(7) Water shipments		
	Hazardous materials descriptions and proper shipping names	Hazard class	Identification number	Label(s) required (if not excepted)	(a) Exceptions	(b) Specific requirements	(a) Passenger carrying aircraft or railcar	(b) Cargo only aircraft	(a) Cargo vessel	(b) Passenger vessel	(c) Other requirements
+/ A/ W	Arsenious acid, solid. See Arsenic trioxide, solid										
	Arsenious and mercuric iodide solution	Poison B	NA2810	Poison	173.345	173.346	1 quart	55 gallons	1,2	1,2	Segregation same as for flammable gases
	Arsine	Poison A	UN2188	Poison gas and Flammable gas	None	173.328	Forbidden	Forbidden	1	5	
	Asbestos	ORM-C		None	173.1090	173.1090	No limit	No limit	1,2	1,2	Stow and handle to avoid airborne particles.
	Ascaridole (organic peroxide)	Forbidden								5	
W	Asphalt, at or above its flashpoint	ORM-C	NA1999	None	None	None	Forbidden	Forbidden	1		When applicable, no fire or residue thereof may be present in the furnace heating the substance while the vehicle is on board a cargo vehicle
	Asphalt, cut back	Flammable liquid	NA1999	Flammable liquid	173.118	173.131	1 quart	10 gallons	1,2	1	
	Asphalt, cut back	Combustible liquid	NA1999	None	173.118a	None	No limit	No limit	1,2	1,2	
	Automobile, motorcycle, tractor, or other self-propelled vehicle. See Motor vehicle										
	Automobile, motorcycle, tractor, or other self-propelled vehicle, engine, or other mechanical apparatus, with charged electric storage battery, wet. See Motor vehicle										
	Azaurolic acid (salt of). (dry)	Forbidden									
	3-Azido-1,2-Propylene glycol dinitrate	Forbidden									
	5-Azido-1-hydroxy tetrazole	Forbidden									
	Azidodithiocarbonic acid	Forbidden									
	Azidoethyl nitrate	Forbidden									
	Azido guanidine picrate (dry)	Forbidden									
	Azido hydroxy tetrazole (mercury and silver salts)	Forbidden									
	Azinphos methyl	Poison B	NA2783	Poison	173.364	173.365	50 pounds	200 pounds	1,2	4	
	Azinphos methyl mixture, liquid	Poison B	NA2783	Poison	173.345	173.346	½ pint	1 quart	1,2	5	
	1-Aziridinyl phosphine oxide (tris). See Tris-(1-aziridinyl) phosphine oxide										
	Azoteterazole (dry)	Forbidden									
	Bags, burlap, used, must be classed for the hazardous material previously contained in bag. See 173.28, 173.29										
	Barium azide, wet, 50% or more water	Flammable solid	UN1571	Flammable solid	None	173.239	Forbiddenn	1 pound	1,2	1,2	Stow away from heavy metals

120

Chapter I—Research and Special Programs Administration §172.101

Name	Class	ID	Ref	Ref2	Limit1	Limit2	Col	Notes
Barium chlorate	Oxidizer	UN1445	Oxidizer	173.153	173.153	25 pounds	100 pounds	1,2	Separate from ammonium compounds. Stow away from powdered metals
Barium chlorate, wet	Oxidizer	NA1445	Oxidizer	173.153	173.153	25 pounds	200 pounds	1,2	Separate from ammonium compounds. Stow away from powdered metals
Barium cyanide, solid	Poison B	UN1565	Poison	173.370	173.370	25 pounds	200 pounds	1,2	Stow away from acids
Barium nitrate	Oxidizer	NA1446	Oxidizer	173.153	173.182	25 pounds	100 pounds	1,2	
Barium oxide	ORM-B	UN1884	None	173.505	173.800	25 pounds	100 pounds	1,2	
Barium perchlorate	Oxidizer	UN1447	Oxidizer	173.153	173.154	25 pounds	100 pounds	1,2	Stow away from powdered metals
Barium permanganate	Oxidizer	UN1448	Oxidizer	173.153	172.154	25 pounds	100 pounds	1,2	Separate from ammonium compounds and hydrogen peroxide
Barium peroxide	Oxidizer	UN1449	Oxidizer	173.153	173.156	25 pounds	100 pounds	1,2	Keep dry
Barium styphnate, monohydrate. See Initiating explosive									
Barrel, empty. See Drum, empty									
Battery, dry. Not subject to Parts 170-189 of this subchapter									
Battery, electric storage, dry (containing potassium hydroxide, dry solid, flake, bead, or granular)	Corrosive material	NA1813	Corrosive	173.244	173.245b	25 pounds	100 pounds	1,2	Keep dry
Battery, electric storage, wet, filled with acid	Corrosive material	UN2794	Corrosive	173.260	173.260	Forbidden	No limit	1,2	
Battery, electric storage, wet, filled with acid, with automobile (or specifically named self-propelled vehicle or mechanical apparatus)	Corrosive material	NA2794	Corrosive	173.250	173.260	No limit	No limit	1,2	Keep dry
Battery, electric storage, wet, filled with alkali, with automobile (or specifically named self-propelled vehicle or mechanical apparatus)	Corrosive material	NA2795	Corrosive	173.250	173.260	No limit	No limit	1,2	Keep dry
Battery, electric storage, wet, with wheelchair	Corrosive material		Corrosive	173.250	173.250 175.10	No limit	No limit	1,2	Keep dry
Battery, electric storage, wet, nonspillable. See §173.260(d)									
Battery, electric storage, wet, filled with alkali	Corrosive material	UN2795	Corrosive	173.260	173.260	Forbidden	No limit	1,2	
Battery fluid, acid	Corrosive material	UN2796	Corrosive	173.244	173.257	1 quart	5 gallons	1,2	
Battery fluid, acid, with electronic equipment or actuating device	Corrosive material	NA2796	Corrosive	None	173.259	Forbidden	5 pints	1,2	
Battery fluid, acid, with battery, electric storage, wet, empty, or dry	Corrosive material	NA2796	Corrosive	None	173.258	Forbidden	5 pints	1,2	
Battery fluid, alkali	Corrosive material	UN2797	Corrosive	173.244	173.257	1 quart	5 gallons	1,2	
Battery fluid, alkali, with electronic equipment or actuating device	Corrosive material	NA2797	Corrosive	None	173.259	Forbidden	5 pints	1,2	
Battery fluid, alkali, with battery, electric storage wet, empty or dry	Corrosive material	UN2797	Corrosive	None	173.258	Forbidden	5 pints	1,2	
Battery, lithium. See 173.206(f)									
Battery parts (plates, grids, etc. unwashed, exhausted)	ORM-C		None	173.505	173.915	No limit	No limit	1,2	
Benzaldehyde	Combustible liquid	NA1989	None	173.118a	None	1 quart	No limit	4	
Benzene (benzol)	Flammable liquid	UN1114	Flammable liquid	173.118	178.119	1 quart	10 gallons	1,2	
Benzene diazonium chloride (dry)	Forbidden							1	
Benzene diazonium nitrate (dry)	Forbidden								

121

§172.101 Hazardous Materials Table

(1) +/A/W	(2) Hazardous materials descriptions and proper shipping names	(3) Hazard class	(3A) Identification number	(4) Label(s) required (if not excepted)	(5) Packaging (a) Exceptions	(5) Packaging (b) Specific requirements	(6) Maximum net quantity in one package (a) Passenger carrying aircraft or railcar	(6) Maximum net quantity in one package (b) Cargo only aircraft	(7) Water shipments (a) Cargo vessel	(7) Water shipments (b) Passenger vessel	(7) Water shipments (c) Other requirements
	Benzene phosphorus dichloride	Corrosive material	UN2798	Corrosive	None	173.250a	Forbidden	5 pints	1	5	
	Benzene phosphorus thiodichloride	Corrosive material	UN2799	Corrosive	None	173.250a	Forbidden	5 pints	1	5	
	Benzenethiol. See Phenyl mercaptan										
	Benzene triozonide	Forbidden									
	Benzidine	Poison B	UN1885	Poison	173.364	173.365	50 pounds	200 pounds	1,2	1	
	Benzine	Flammable liquid	UN1115	Flammable liquid	173.118	173.119	1 quart	10 gallons	1,2	1	
	Benzoic derivative pesticide, liquid, n.o.s. (compounds and preparations)	Flammable liquid	UN2770	Flammable liquid	173.118	173.119	1 quart	10 gallons	1,2	1	
	Benzoic derivative pesticide, liquid, n.o.s. (compounds and preparations)	Poison B	UN2769	Poison	173.345	173.346	1 quart	55 gallons	1,2	1,2	
	Benzoic derivative pesticide, solid, n.o.s. (compounds and preparations), solid	Poison B	UN2769	Poison	173.364	173.365	50 pounds	200 pounds	1,2	1,2	
	Benzonitrile	Combustible liquid	UN2224	None	173.118a		No limit	No limit	1,2	1,2	
	Benzoxidiazoles (dry)	Forbidden									
	Benzoyl azide	Forbidden									
	Benzoyl chloride	Corrosive material	UN1736	Corrosive	173.244	173.247	1 quart	1 quart	1	1	Keep dry. Glass carboys not permitted on passenger vessels
	Benzoyl peroxide	Organic peroxide	NA2085	Organic peroxide	None	173.157 173.158	Forbidden	25 pounds	1,2	1	
	Benzoyl peroxide, more than 77% but less than 95% with water. See Benzoyl peroxide.		UN2088								
	Benzoyl peroxide, not less than 30% but not more than 52% with inert solid. See Organic peroxide, solid, n.o.s.		UN2089								
	Benzoyl peroxide, not more than 72% as a paste. See Organic peroxide, n.o.s.		UN2087								
	Benzoyl peroxide, not more than 77% with water. See Benzoyl peroxide.		UN2090								
	Benzoyl peroxide, technically pure or Benzoyl peroxide, more than 52% with inert solid. See Benzoyl peroxide.		UN2085								
	Benzyl bromide (bromotoluene, alpha)	Corrosive material	UN1737	Corrosive	None	173.281	Forbidden	5 pints	1	5	Keep dry
	Benzyl chloride	Corrosive material	UN1738	Corrosive	173.244	173.295	Forbidden	1 quart	1	4	Keep dry

Appendix B: Regulations Reference Guide

Chapter I—Research and Special Programs Administration §172.101

Material	Class	ID No.	Label	Packaging (Exceptions)	Packaging (Specific)	Max. Pkg. (Passenger)	Max. Pkg. (Cargo)	Vessel	Vessel Stow	Other
Benzyl chloroformate (benzyl chlorocarbonate)	Corrosive material	UN1739	Corrosive	None	173.298	Forbidden	5 pints	1	5	Keep dry
Beryllium chloride	Poison B	NA1566	Poison	173.364	173.365	50 pounds	200 pounds	1,2	1,2	
Beryllium compound, n.o.s.	Poison B	NA1566	Poison	173.364	173.365	50 pounds	200 pounds	1,2	1,2	
Beryllium fluoride	Poison B	NA1566	Poison	173.364	173.365	50 pounds	200 pounds	1,2	1,2	
Beryllium nitrate	Oxidizer	UN2464	Oxidizer	173.153	173.182	25 pounds	100 pounds	1,2	1,2	
Biphenyl triazonium	Forbidden									
Bipyridilium pesticide, liquid, n.o.s. (compounds and preparations)	Flammable liquid	UN2782	Flammable liquid	173.118	173.119	1 quart	10 gallons	1,2	1	
Bipyridilium pesticide, liquid, n.o.s. (compounds and preparations)	Poison B	UN2781	Poison	173.345	173.346	1 quart	55 gallons	1,2	1,2	
Bipyridilium pesticide, solid, n.o.s. (compounds and preparations)	Poison B	UN2781	Poison	173.364	173.365	50 pounds	200 pounds	1,2	1,2	
Black powder	Class A explosive		Explosive A	None	173.60	Forbidden	Forbidden	6	5	
Black powder igniter with empty cartridge bag	Class C explosive		Explosive C	None	173.106	50 pounds	150 pounds	1,3	1,3	
Blasting agent, n.o.s.	Blasting agent		Blasting agent	None	173.114a	Forbidden	100 pounds	1,2	1,2	
Blasting caps. See Detonators, Class A or Class C explosives										
Blasting caps, electric. See Detonators, Class A or Class C explosives										
Blasting caps, percussion activated. See Detonators, Class A or Class C explosives										
Blasting caps with detonating cord. See Detonators, Class A or Class C explosives and Detonating primers, Class A or Class C explosives.										
Blasting caps with metal clad mild detonating fuze. See Detonators, Class A or Class C explosives										
Blasting caps with safety fuse. See Detonators, Class A or Class C explosives										
Blasting caps with shock tubes. See Detonators, Class A or Class C explosives										
Blasting gelatin. See High explosive										
Blasting powder. See Black powder										
Bleaching powder, containing 39% or less available chlorine	ORM-C	UN2208	None	173.505	173.220			1,2	1,2	Keep dry. Stow separate from flammable liquids and acids. (Stow away from oils, grease, and similar organic materials.)
Bomb, explosive. See Explosive bomb										
Bomb, explosive with gas, smoke, or incendiary material. See Explosive bomb										
Bomb, fireworks. See Fireworks, special										
Bomb, gas, smoke, or incendiary, nonexplosive. See Chemical ammunition, nonexplosive										
Bomb, incendiary, or smoke without bursting charge. See Fireworks, special										
Bomb, practice, with electric primer or electric squib (non-explosive). See 173.55										
Bomb, sand-loaded or empty (non-explosive). See 173.55										
Booster, explosive	Class A explosive		Explosive A	None	173.69	Forbidden	Forbidden	6	5	
Bordeaux arsenite, liquid	Poison B	NA2759	Poison	173.345	173.346	1 quart	55 gallons	1,2	1,2	
Bordeaux arsenite, solid	Poison B	NA2759	Poison	173.364	173.365	50 pounds	200 pounds	1,2	1,2	

123

Chemical Hazard Communication Guidebook

§172.101   Title 49—Transportation

§172.101 Hazardous Materials Table

(1)	(2)	(3)	(3A)	(4)	(5) Packaging		(6) Maximum net quantity in one package		(7) Water shipments		
+/A/W	Hazardous materials descriptions and proper shipping names	Hazard class	Identification number	Label(s) required (if not excepted)	(a) Exceptions	(b) Specific requirements	(a) Passenger carrying aircraft or railcar	(b) Cargo only aircraft	(a) Cargo vessel	(b) Passenger vessel	(c) Other requirements
	Boron tribromide	Corrosive material	UN2692	Corrosive	None	173.251	Forbidden	1 quart	1	5	Stow in well ventilated space. Shade from radiant heat. Segregation same as for nonflammable gases.
	Boron trichloride	Corrosive material	UN1741	Corrosive	None	173.251	Forbidden	1 quart	1,2	5	Stow away from living quarters and foodstuffs
	Boron trifluoride	Nonflammable gas	UN1008	Nonflammable gas and Poison	None	173.302	Forbidden	Forbidden	1	5	
	Boron trifluoride-acetic acid complex	Corrosive material	UN1742	Corrosive	173.244	173.247	1 quart	1 gallon	1,2	1,2	Keep cool
	Bromine	Corrosive material	UN1744	Corrosive	None	173.252	Forbidden	1 quart	1	5	
	Bromine azide	Forbidden									
	Bromine pentafluoride	Oxidizer	UN1745	Oxidizer	None	173.246	Forbidden	100 pounds	1	5	Shade from radiant heat. Segregation same as for corrosives
	Bromine trifluoride	Oxidizer	UN1746	Oxidizer and Poison	None	173.246	Forbidden	100 pounds	1	5	Shade from radiant heat. Segregation same as for corrosives
	4-*Bromo-1,2-dinitrobenzene (unstable at 59 deg C.)*	Forbidden									
	Bromoacetic acid, solid	Corrosive material	UN1938	Corrosive	173.244	173.245b	25 pounds	100 pounds	1,2	1,2	Keep dry
	Bromoacetic acid solution	Corrosive material	UN1938	Corrosive	173.244	173.245	1 quart	1 quart	1,2	1,2	Glass carboys in hampers not permitted under deck
	Bromoacetone, liquid	Poison A	UN1569	Poison gas	None	173.329	Forbidden	Forbidden	1	5	Segregation same as for flammable liquids
	Bromobenzene	Combustible liquid	UN2514	None	173.118a	None	No limit	No limit	1,2	1,2	
	Bromochloromethane	ORM-A	UN1887	None	173.505	173.605	10 gallons	55 gallons			
	Bromosilane	Forbidden									
A	*Bromotoluene, alpha. See Benzyl bromide*										
	Bromotrifluoromethane (R-13B1 or H-1301)	Nonflammable gas	UN1009	Nonflammable gas	173.306	173.304 173.314 173.315	150 pounds	300 pounds	1,2	1,2	
	Brucine, solid (*dimethoxy strychnine*)	Poison B	UN1570	Poison	173.364	173.365	50 pounds	200 pounds	1,2	1,2	Separate from flammable gases or liquids, oxidizing materials, or organic peroxides
	Burnt cotton, not repicked	Flammable solid	NA1325	Flammable solid	None	173.159	Forbidden	Forbidden	1	5	
	Burster, explosive	Class A explosive		Explosive A	None	173.69	Forbidden	Forbidden	6	5	
+	Butadiene, inhibited	Flammable gas	UN1010	Flammable gas	173.306	173.304 173.314 173.315	Forbidden	300 pounds	1,2	1	Stow away from living quarters

124

Appendix B: Regulations Reference Guide

Chapter I—Research and Special Programs Administration §172.101

Description	Hazard class	ID number	Packaging (§173)	Packaging exceptions	Max net quantity per package (passenger)	Max net quantity per package (cargo)	Label(s)
Butane or Liquefied petroleum gas. See Liquefied petroleum gas							
1,2,4-Butanetriol trinitrate	Forbidden						
tert-Butoxycarbonyl azide	Forbidden						
n-Butyl-4,4-di-(tert-butylperoxy)valerate, technically pure. See Organic peroxide, liquid or solution, n.o.s.		UN2140					
n-Butyl-4,4-di-(tert-butylperoxy)valerate, not more than 52% with inert solid. See Organic peroxide, solid, n.o.s.		UN2141					
Butyl acetate	Flammable liquid	UN1123	173.118	173.119	1 quart	10 gallons	1,2	1
n-Butyl acid phosphate. See Acid butyl phosphate							
Butyl alcohol	Flammable liquid	NA1120	173.118	173.125	1 quart	10 gallons	1,2	1
Butylamine	Flammable liquid	UN1125	173.118	173.119	1 quart	10 gallons	1,2	1
Butyl bromide, normal	Flammable liquid	UN1126	173.118	173.119	1 quart	10 gallons	1,2	1
Butyl chloride	Flammable liquid	UN1127	173.118	173.119	1 quart	10 gallons	1,2	1
tert-Butyl cumyl peroxide, technically pure or tert-Butyl cumene peroxide, technically pure. See Organic peroxide, liquid or solution, n.o.s.		UN2091					
Butyl ether	Flammable liquid	UN1149	173.118	173.119	1 quart	10 gallons	1,2	1,2
Butyl formate	Flammable liquid	UN1128	173.118	173.119	1 quart	10 gallons	1,2	1
tert-Butyl hydroperoxide, more than 72% but not more than 90% with water. See Organic peroxide, liquid or solution, n.o.s.		UN2094					
tert-Butyl hydroperoxide, not more than 72% with water. See Organic peroxide, liquid or solution, n.o.s.		UN2093					
tert-Butyl hydroperoxide, not more than 80% in di-tert-butyl peroxide and solvent. See Organic peroxide, liquid or solution, n.o.s.		UN2092					
tert-Butyl hydroperoxide, not more than 80% in di-tert-butyl peroxide or solvent. See Organic peroxide, liquid or solution, n.o.s.		UN2092					
tert-Butyl hydroperoxide, more than 90% with water	Forbidden						
n-Butyl isocyanate	Flammable liquid	UN2485	None	173.119 173.3a	1 quart	10 gallons	1,2	1
tert-Butyl isopropyl benzene hydroperoxide	Organic peroxide	NA2091	173.153	173.224	1 quart	1 quart	1,2	4
Butyl mercaptan	Flammable liquid and Poison	UN2347	None	173.141	Forbidden	10 gallons	1,3	5
tert-Butyl peroxy-2-ethylhexanoate, technically pure. See Organic peroxide, liquid or solution, n.o.s.	Organic peroxide Flammable liquid	UN2143					

125

Chemical Hazard Communication Guidebook

§172.101   **Title 49—Transportation**

§172.101 Hazardous Materials Table

(1) +/A/W	(2) Hazardous materials descriptions and proper shipping names	(3) Hazard class	(3A) Identification number	(4) Label(s) required (if not excepted)	(5)(a) Packaging Exceptions	(5)(b) Specific requirements	(6)(a) Passenger carrying aircraft or railcar	(6)(b) Cargo only aircraft	(7)(a) Cargo vessel	(7)(b) Passenger vessel	(7)(c) Other requirements
	tert-Butyl peroxy-2-ethylhexanoate, *not more than 30% with 2,2-Di-(tert-butylperoxy)butane, not more than 35%, with not less than 35% phlegmatizer. See Organic peroxide, liquid or solution, n.o.s.*		UN2886								
	tert-Butyl peroxy-2-ethylhexanoate, *not more than 12% with 2,2-Di-(tert-butylperoxy)butane, not more than 14% with not less than 14% phlegmatizer and 60% inert inorganic solid. See Organic peroxide, solid, n.o.s.*		UN2887								
	tert-Butyl peroxy-2-ethylhexanoate, *not more than 50% with phlegmatizer. See Organic peroxide, liquid or solution, n.o.s.*		UN2888								
	tert-Butyl peroxy-3,5,5-trimethylhexanoate or tert-Butyl peroxyisomonanoate, *technically pure. See Organic peroxide, liquid or solution, n.o.s.*		UN2104								
	3-tert-Butyl peroxy-3-phenylphthalide, *technically pure. See Organic peroxide, solid, n.o.s.*		UN2596								
	tert-Butyl peroxyacetate, *not more than 76% in solution. See Organic peroxide, liquid or solution, n.o.s.*		UN2095								
	tert-Butyl peroxyacetate, *not more than 52% in solution. See Organic peroxide, liquid or solution, n.o.s.*		UN2096								
	tert-Butyl peroxyacetate, *more than 76% in solution*	Forbidden									
	tert-Butyl peroxybenzoate, *not more than 75% in solution. See Organic peroxide, liquid or solution, n.o.s.*		UN2098								
	tert-Butyl peroxybenzoate, *not more than 50% with inert inorganic solid. See Organic peroxide, solid, n.o.s.*		UN2890								
	tert-Butyl peroxybenzoate, *technically pure or tert-Butyl peroxybenzoate, more than 75% in solution. See Organic peroxide, liquid or solution, n.o.s.*		UN2097								

126

Appendix B: Regulations Reference Guide

Chapter I—Research and Special Programs Administration   **§172.101**

tert-Butyl peroxycrotonate, not more than 76% in solution. See Organic peroxide, liquid or solution, n.o.s.		UN2183					
n-Butyl peroxydicarbonate, not more than 52% in solution. See Organic peroxide, liquid or solution, n.o.s.		UN2169					
n-Butyl peroxydicarbonate, not more than 27% in solution. See Organic peroxide, liquid or solution, n.o.s.		UN2170					
n-Butyl peroxydicarbonate, more than 52% in solution	Forbidden						
tert-Butyl peroxydiethylacetate, 33% with tert-Butyl peroxybenzoate, 33%, and solvent. See Organic peroxide, liquid or solution, n.o.s.		UN2551					
tert-Butyl peroxydiethylacetate, technically pure. See Organic peroxide, liquid or solution, n.o.s.		UN2144					
tert-Butyl peroxyisobutyrate, more than 52% but not more than 77% in solution. See Organic peroxide, liquid or solution, n.o.s.		UN2142					
tert-Butyl peroxyisobutyrate, not more than 52% in solution. See Organic peroxide, liquid or solution, n.o.s.		UN2562					
tert-Butyl peroxyisobutyrate, more than 77% in solution	Forbidden						
tert-Butyl peroxyisopropyl carbonate, technically pure. See Organic peroxide, liquid or solution, n.o.s.		UN2103					
tert-Butyl peroxymaleate, not more than 55% in solution. See Organic peroxide, liquid or solution, n.o.s.		UN2100					
tert-Butyl peroxymaleate, not more than 55% as a paste. See Organic peroxide, solid, n.o.s.		UN2101					
tert-Butyl peroxymaleate, technically pure. See Organic peroxide, solid, n.o.s.		UN2099					
tert-Butyl peroxyneodecanoate, not more than 77% in solution. See Organic peroxide, liquid or solution, n.o.s.		UN2177					
tert-Butyl peroxyneodecanoate, technically pure. See Organic peroxide, liquid or solution, n.o.s.		UN2594					
tert-Butyl peroxyphthalate, technically pure. See Organic peroxide, solid, n.o.s.		UN2105					
tert-Butyl peroxypivalate, not more than 77% in solution. See Organic peroxide, liquid or solution, n.o.s.		UN2110					
Butyl phosphoric acid. See Acid butyl phosphate							
Butyl trichlorosilane	Corrosive material	UN1747	Corrosive	None	173.280	Forbidden	10 gallons	1	1
Butyraldehyde	Flammable liquid	UN1129		173.118	173.119	1 quart	10 gallons	1,2	1
Butyric acid	Corrosive material	UN2820	Corrosive	173.244	173.245	1 quart	10 gallons	1,2	1,2	Keep dry

127

Chemical Hazard Communication Guidebook

§172.101   Title 49—Transportation

§172.101 Hazardous Materials Table

(1)	(2)	(3)	(3A)	(4)	(5) Packaging		(6) Maximum net quantity in one package		(7) Water shipments		
+/A/W	Hazardous materials descriptions and proper shipping names	Hazard class	Identification number	Label(s) required (if not excepted)	(a) Exceptions	(b) Specific requirements	(a) Passenger carrying aircraft or railcar	(b) Cargo only aircraft	(a) Cargo vessel	(b) Passenger vessel	(c) Other requirements
	Cabazide	Forbidden									
	Calcium arsenate, solid	Poison B	UN1573	Poison	173.364	173.367	50 pounds	200 pounds	1,2	1,2	
	Calcium arsenite, solid	Poison B	NA1574	Poison	173.364	173.368 173.365	50 pounds	200 pounds	1,2	1,2	
	Calcium bisulfite solution. *See* Calcium hydrogen sulfite solution										
	Calcium carbide	Flammable solid	UN1402	Flammable solid and Dangerous when wet	None	173.178	Forbidden	25 pounds	1,2	1,2	Keep dry. Stow away from copper, its alloys, and salts
	Calcium chlorate	Oxidizer	UN1452	Oxidizer	173.153	173.163	25 pounds	100 pounds	1,2	1,2	Separate from ammonium compounds. Stow away from powdered metals and cyanide
	Calcium chlorite	Oxidizer	UN1453	Oxidizer	None	173.160	Forbidden	100 pounds	1,2	1,2	Separate from ammonium compounds, powdered materials, and cyanides
AW	Calcium cyanamide, not hydrated *(containing more than 0.1% calcium carbide)*	ORM-C	UN1403	None	None	173.945	25 pounds	200 pounds	1,2	1,2	Segregation same as for flammable solids labeled Dangerous When Wet
	Calcium cyanide, solid *or* Calcium cyanide mixture, solid	Poison B	UN1575	Poison	173.370	173.370	25 pounds	200 pounds	1,2	1,2	Stow away from corrosive liquids. Keep dry
	Calcium hydrogen sulfite solution	Corrosive material	NA2693	Corrosive	173.244	173.245	1 quart	5 gallons	1,2	1,2	
	Calcium hypochlorite, hydrated *(minimum 5.5% but not more than 10% water, and containing more than 39% available chlorine)*	Oxidizer	UN2880	Oxidizer	173.153	173.217	50 pounds	100 pounds	1,2	1,2	
	Calcium hypochlorite mixture, dry. *(Containing more than 39% available chlorine)*	Oxidizer	UN1748	Oxidizer	173.153	173.217	50 pounds	100 pounds	1,2	1,2	Keep cool and dry
	Calcium, metal	Flammable solid	UN1401	Flammable solid and Dangerous when wet	173.153	173.154	25 pounds	100 pounds	1,2	4	Keep cool and dry. Segregation same as for flammable solids labeled Dangerous When Wet
	Calcium, metal, crystalline	Flammable solid	NA1401	Flammable solid and Dangerous when wet	None	173.231	Forbidden	25 pounds	1,2	5	Keep cool and dry. Segregation same as for flammable solids labeled Dangerous When Wet
AW	Calcium nitrate *(See 173.182(a) Note)*	Oxidizer ORM-B	UN1454 UN1910	Oxidizer None	173.153 173.505	173.182 173.850	25 pounds 25 pounds	100 pounds 100 pounds	1,2 1,2	1,2 1,2	Keep dry. Stow away from explosives, acids, combustible materials, and ammonium salts
	Calcium permanganate	Oxidizer	UN1456	Oxidizer	173.153	173.154	25 pounds	100 pounds	1,2	1,2	Separate from ammonium compounds and hydrogen peroxide

128

Chapter I—Research and Special Programs Administration §172.101

	Name	Class	ID	Label(s)	Packaging (Exceptions)	Packaging (Specific)	Max qty (passenger)	Max qty (cargo)	Vessel stowage	Other stowage req.	Other requirements
	Calcium peroxide	Oxidizer	UN1457	Oxidizer	173.153	173.156	25 pounds	100 pounds	1,2	1,2	Keep dry
	Calcium phosphide	Flammable solid	UN1360	Flammable solid and Dangerous when wet	None	173.161	Forbidden	25 pounds	1	5	Keep cool and dry. Segregation same as for flammable solids labeled Dangerous When Wet
	Calcium resinate	Flammable solid	UN1313	Flammable solid	None	173.166	Forbidden	125 pounds	1	5	
	Calcium resinate, fused	Flammable solid	UN1314	Flammable solid	None	173.166	Forbidden	125 pounds	1	5	Segregation same as for flammable solids labeled Dangerous When Wet
	Calcium silicon (*powder*)	Flammable solid	UN1406	Flammable solid and Dangerous when wet	173.153	173.178	Forbidden	25 pounds	1,2	4	
AW	Camphene	ORM-A	NA9011	None	173.505	173.610	No limit	No limit	1,3	1,3	
	Camphor oil	Combustible liquid	UN1130	None	173.118a	None	No limit	No limit	1,2	1,2	Stow away from foodstuffs and living quarters
	Cannon primer	Class C explosive		None	None	173.107	50 pounds	150 pounds	1,3	5	
	Caprylyl peroxide solution	Organic peroxide	NA2129	Organic peroxide	173.153	173.221	1 quart	1 quart	1,2	4	Keep cool. Stow separate from combustible materials, explosives, or acids
	Caps, blasting. *See* Detonators, Class A or Class C explosives										
	Caps, toy. *See* Toy caps										
	Carbamate pesticide, liquid, n.o.s. (*compounds and preparations*),	Flammable liquid	UN2758	Flammable liquid	173.118	173.119	1 quart	10 gallons	1,2	1	
	Carbamate pesticide, liquid, n.o.s. (*compounds and preparations*),	Poison B	UN2757	Poison	173.345	173.346	1 quart	55 gallons	1,2	1,2	
	Carbamate pesticide, solid, n.o.s. (*compounds and preparations*),	Poison B	UN2757	Poison	173.364	173.365	50 pounds	200 pounds	1,2	1,2	
A	Carbaryl	ORM-A		None	173.505	173.510	No limit	No limit	1,2	1,2	
	Carbofuran	Poison B	NA2757	Poison	173.364	173.364	50 pounds	200 pounds	1,2	1,2	
+	Carbofuran mixture, liquid	Poison B	NA2757	Poison	173.345	173.346	1 quart	55 gallons	1,2	1,2	
	Carbolic acid. *See* Phenol										
	Carbolic acid, liquid (*liquid tar acid containing over 50% phenol*). *See* Phenol, liquid										
	Carbon bisulfide, *or* Carbon disulfide	Flammable liquid	UN1131	Flammable liquid	None	173.121	Forbidden	Forbidden	1	5	Keep cool. Not permitted on any vessel transporting explosives, except that quantities not exceeding 200 pounds may be transported on such vessels under conditions approved by the Captain of the Port
	Carbon dioxide	Nonflammable gas	UN1013	Nonflammable gas	173.306	178.302 178.304	150 pounds	300 pounds	1,2	1,2	
	Carbon dioxide-nitrous oxide mixture	Nonflammable gas	UN1015	Nonflammable gas	173.306	178.304	150 pounds	300 pounds	1,2	1,2	
	Carbon dioxide-oxygen mixture	Nonflammable gas	UN1014	Nonflammable gas	173.306	178.304	150 pounds	300 pounds	1,2	1,2	
	Carbon dioxide, refrigerated liquid	Nonflammable gas	UN2187	Nonflammable gas	173.306	178.314 178.315	150 pounds	300 pounds	1,2	1,2	
AW	Carbon dioxide, solid, *or* Dry ice, *or* Carbonice	ORM-A	UN1845	None	None	173.615	440 pounds	440 pounds	1	1	Stow away from open ventilators. Stow away from cyanides or cyanide mixtures, liquid or dry
	Carbon monoxide	Flammable gas	UN1016	Flammable gas	173.306	173.302	Forbidden	150 pounds	1	4	Stow away from living quarters
+	Carbon monoxide, cryogenic liquid	Flammable gas	NA9202	Flammable gas	None	173.318	Forbidden	Forbidden	1	5	Stow away from living quarters
AW	Carbon tetrachloride	ORM-A	UN1846	None	173.505	173.620	1 quart	55 gallons	1,2	1,2	

129

Chemical Hazard Communication Guidebook

§172.101   Title 49—Transportation

§172.101 Hazardous Materials Table

(1)	(2)	(3)	(3A)	(4)	(5) Packaging		(6) Maximum net quantity in one package		(7) Water shipments		
+/A/W	Hazardous materials descriptions and proper shipping names	Hazard class	Identification number	Label(s) required (if not excepted)	(a) Exceptions	(b) Specific requirements	(a) Passenger carrying aircraft or railcar	(b) Cargo only aircraft	(a) Cargo vessel	(b) Passenger vessel	(c) Other requirements
	Carbonyl chloride. See Phosgene.										
	Carboys, empty, material previously contained in carboy. See 173.29										
	Cartridge bags, empty, with black powder igniter	Class C explosive	NA1133	Explosive C	None	173.106	50 pounds	150 pounds	1,3	1,3	
	Cartridge, practice ammunition	Class C explosive	NA1133	Explosive C	None	173.101a	50 pounds	150 pounds	1,2	1,2	
	Case oil. See Gasoline or Naptha										
	Casinghead gasoline. See Gasoline										
W	Castor Beans	ORM-C		None	173.505	173.952			1,2	1,2	Stow away from living quarters and foodstuffs. Bulk shipments permitted in tight vans or containers only on cargo vessels (Castor beans only)
	Castor pomace. See Castor beans										
W	*Caustic, potash, dry, solid, flake, bead, or granular. See Potassium hydroxide, dry, etc.*										
	Caustic potash, liquid or solution. See Potassium hydroxide solution										
	Caustic soda, dry, solid, flake, bead, or granular. See Sodium hydroxide, dry, etc.										
	Caustic soda, liquid or solution. See Sodium hydroxide solution										
W	*Cellosolve. See Ethylene glycol monoethyl ether*										
W	*Cellosolve acetate. See Ethylene glycol monoethyl ether acetate*										
	Cement	Combustible Liquid		None	173.118a	None	No limit	No limit	1,2	1,2	
	Cement	Flammable liquid	NA1133	Flammable liquid	173.118	173.132	1 quart	10 gallons	1,2	1	
	Cement, adhesive, n.o.s. See Cement										
	Cement, container, linoleum, tile, or wallboard, liquid	Flammable liquid	NA1133	Flammable liquid	173.118	173.132	1 quart	15 gallons	1,2	1	
	Cement, leather	Flammable liquid	NA1133	Flammable liquid	173.118	173.119	1 quart	10 gallons	1,2	1	
	Cement, pyroxylin	Flammable liquid	NA1133	Flammable liquid	173.118	173.132	1 quart	15 gallons	1,2	1	
	Cement, roofing, liquid	Flammable liquid	NA1133	Flammable liquid	173.118	173.119	1 quart	10 gallons	1,2	1	

130

Chapter I—Research and Special Programs Administration §172.101

Name	Class	ID No.	Label(s)	Packaging (Exceptions)	Packaging (Specific)	Max Qty Passenger	Max Qty Cargo	Vessel Stowage	Other
Cement, rubber	Flammable liquid	NA1133	Flammable liquid	173.118	173.152	1 quart	15 gallons	1,2	1
Cesium metal	Flammable solid	UN1407	Flammable solid and Dangerous when wet	None	173.206	Forbidden	25 pounds	1,2	5	Segregation same as for flammable solids labeled Dangerous When Wet
Charcoal briquettes or briquets	Flammable solid	NA1361	Flammable solid	173.162	173.162	50 pounds	50 pounds	1,2	1,2
Charcoal screenings, made from 'piñon' wood	Flammable solid	NA1361	Flammable solid	173.162	173.162	25 pounds	200 pounds	1,2	1,
Charcoal, shell	Flammable solid	NA1361	Flammable solid	173.162	173.162	25 pounds	200 pounds	1,2	1,2
Charcoal, wood, ground, crushed, granulated, or pulverized	Flammable solid	NA1361	Flammable solid	173.162	173.162	25 pounds	200 pounds	1,2	1,2
Charcoal, wood, lump	Flammable solid	NA1361	Flammable solid	173.162	173.162	50 pounds	50 pounds	1,2	1,2
Charcoal wood screenings, other than 'piñon' wood screenings	Flammable solid	NA1361	Flammable solid	None	173.162	Forbidden	Forbidden	1	1
Charged well casing jet perforating gun (total explosive contents in guns 20 pounds or more per motor vehicle)	Class A explosive		Explosive A	None	178.53 178.80	Forbidden	Forbidden			Forbidden
Charged well casing jet perforating gun (total explosive contents in guns not exceeding 20 pounds per motor vehicle or special offshore down hole tool pallet	Class C explosive		Explosive C	None	173.53 173.110	Forbidden	Forbidden	1,2	5
Chemical ammunition, explosive. See Ammunition, chemical, explosive, with ...									
Chemical ammunition, nonexplosive (containing a Poison B material)	Poison B	UN2016	Poison	173.345	173.350	Forbidden	55 gallons			See correct shipping name of applicable Poison B material for storage, special handling, and special segregation requirements
Chemical ammunition, nonexplosive (containing an irritating material)	Irritating material	UN2017	Irritant	None	173.368	Forbidden	20 pounds			See correct shipping name of applicable Irritant material for storage, special handling, and special segregation requirements
Chemical ammunition, nonexplosive (containing a Poison A material)	Poison A	UN2016	Poison gas	None	173.350	Forbidden	Forbidden			See correct shipping name of applicable Poison A material for storage, special handling, and special segregation requirements
Chemical kit	Corrosive material	NA1760	Corrosive	173.296	173.229	1 quart	1 quart	1,3	1,3
Chlorate and borate mixture (containing more than 28% chlorate)	Oxidizer	UN1458	Oxidizer	173.153	173.229	25 pounds	100 pounds	1,2	4	Stow away from ammonium compounds and away from powdered metals
Chlorate and magnesium chloride mixture (containing more than 28% chlorate)	Oxidizer	UN1459	Oxidizer	173.153	173.229	25 pounds	100 pounds	1,2	4	Stow away from ammonium compounds and away from powdered metals
Chlorate explosive, dry. See High explosive									
Chlorate, n.o.s.	Oxidizer	UN1461	Oxidizer	173.153	173.163	25 pounds	100 pounds	1,2	4	Stow away from ammonium compounds and away from powdered metals
Chlorate, n.o.s., wet	Oxidizer	NA1461	Oxidizer	173.153	173.163	25 pounds	200 pounds	1,2	4	Stow away from ammonium compounds and away from powdered metals
Chlorate of potash. See Potassium chlorate									
Chlorate of soda. See Sodium chlorate									
Chlorate powder. See High explosive									
Chlordane, liquid	Flammable liquid Combustible liquid	NA2762 NA2762	Flammable liquid None	173.118 173.118a	173.119 None	1 quart No limit	10 gallons No limit	1,2	1
Chloric acid	Oxidizer	NA2626	Oxidizer and Poison	None	178.237	Forbidden	Forbidden	1,2	1,2	Forbidden

131

Chemical Hazard Communication Guidebook

§172.101   Title 49—Transportation

§172.101 Hazardous Materials Table

(1)	(2)	(3)	(3A)	(4)	(5) Packaging		(6) Maximum net quantity in one package			(7) Water shipments		
+/A/W	Hazardous materials descriptions and proper shipping names	Hazard class	Identification number	Label(s) required (if not excepted)	(a) Exceptions	(b) Specific requirements	(a) Passenger carrying aircraft or railcar	(b) Cargo only aircraft	(a) Cargo vessel	(b) Passenger vessel	Other requirements	
	Chloride of phosphorus. See Phosphorus trichloride											
	Chloride of sulfur. See Sulfur chloride											
	Chlorinated lime (chloride of lime.) See Bleaching powder											
W	Chlorine	Nonflammable gas	UN1017	Nonflammable gas and Poison	None	173.304 173.314 173.315	Forbidden	Forbidden	1,2	5	Stow in a well-ventilated space. Stow away from organic materials	
	Chlorine azide	Forbidden										
+	Chlorine dioxide hydrate, frozen	Oxidizer	NA9191	Oxidizer and Poison	None	173.237	Forbidden	Forbidden			Forbidden	
	Chlorine dioxide (not hydrate)	Forbidden										
	Chlorine trifluoride	Oxidizer	UN1749	Oxidizer and Poison	None	173.246	Forbidden	100 pounds	1,3	5	Stow in well ventilated area away from organic material	
	Chloroacetic acid, liquid or solution	Corrosive material	UN1750	Corrosive	173.244	173.294	1 quart	1 quart	1,2	1,2	Glass carboys in hampers not permitted under deck	
	Chloroacetic acid, solid	Corrosive material	UN1751	Corrosive	173.244	173.245b	25 pounds	100 pounds	1,2	1,2	Keep dry	
	Chloroacetophenone, gas, liquid, or solid (CN)	Irritating material	UN1697	Irritant	None	173.382	Forbidden	75 pounds	1	5		
	Chloroacetyl chloride	Corrosive material	UN1752	Corrosive	None	173.253	Forbidden	1 quart	1	5	Keep dry	
	Chlorobenzene	Flammable liquid	UN1134	Flammable liquid	173.118	173.119	1 quart	10 gallons	1,2	1,2		
	Chlorobenzol. See Chlorobenzene											
	p-Chlorobenzoyl peroxide	Organic peroxide	UN2113	Organic peroxide	None	173.157 173.158	Forbidden	25 pounds	1	1		
	p-Chlorobenzoyl peroxide, *not more than 75% with water.* See p-Chlorobenzoyl peroxide.		UN2113									
	p-Chlorobenzoyl peroxide, *not more than 52% as a paste.* See Organic peroxide, solid, n.o.s.		UN2114									
	p-Chlorobenzoyl peroxide, *not more than 52% in solution.* See Organic peroxide, liquid or solution, n.o.s.		UN2115									
	Chlorodinitrobenzene. See Dinitrochlorobenzene											
	1-Chloro-1,1-difluoroethane. See Chlorodifluoroethane (R-142b)											

132

Chapter I—Research and Special Programs Administration §172.101

	Material	Class	ID No.	Label	(a)	(b)	(c) Passenger	(d) Cargo	Col 7	Stowage
+	Chlorodifluoroethane (R-142b) or (1-Chloro-1,1-difluoroethane).	Flammable gas	UN2517	Flammable gas	173.306	173.304	Forbidden	300 pounds	1,2	1	Stow away from living quarters and foodstuffs
	Chlorodifluoromethane (R-22)	Nonflammable gas	UN1018	Nonflammable gas	173.306	173.304 173.314 173.315	150 pounds	300 pounds	1,2	1	
	Chlorodifluoro-methane and chloropentafluoroethane mixture (constant boiling mixture) (R-502). See Refrigerant gas, n.o.s.										
AW	Chloroform	ORM-A	UN1888	None	173.505	173.620	10 gallons	55 gallons	1,2	1	
	4-Chloro-o-toluidine hydrochloride	Poison B	UN1579	Poison	None	173.362	Forbidden	Forbidden	1,2	5	Keep cool
+	Chloropentafluoroethane (R-115)	Nonflammable gas	UN1020	Nonflammable gas	173.306	173.304 173.314 173.315	150 pounds	300 pounds	1,2	5	Keep cool. Segregation same as for flammable gases
	3-Chloroperoxybenzoic acid, not more than 86% with 3-chlorobenzoic acid See Organic peroxide, solid, n.o.s.		UN2755								
	Chlorophenyltrichlorosilane	Corrosive material	UN1753	Corrosive	None	173.290	Forbidden	10 gallons	1	1	Keep cool
++	Chloropicrin, absorbed	Poison B	NA1583	Poison	None	173.357	Forbidden	Forbidden	1	5	Keep cool
++	Chloropicrin and methyl chloride mixture	Poison A	UN1582	Poison gas and Flammable gas	None	173.329	Forbidden	Forbidden	1	5	Keep cool
+	Choloropicrin and nonflammable, nonliquefied compressed gas mixture	Poison A	NA1955	Poison gas and nonflammable gas	None	173.329	Forbidden	Forbidden	1	5	Keep cool
	Chloropicrin mixture, flammable (pressure not exceeding 14.7 psia; flash point below 100°F)	Poison B	NA2929	Poison and flammable liquid	None	173.357	Forbidden	Forbidden	1	5	Keep cool
++	Chloropicrin, liquid	Poison B	UN1580	Poison	None	173.357	Forbidden	Forbidden	1	5	Keep cool
++	Chloropicrin mixture (containing no compressed gas or Poison A liquid)	Poison B	UN1583	Poison	None	173.357	Forbidden	Forbidden	1	5	Keep cool
<	Chloroplatinic acid, solid	ORM-B	UN2507	None	173.505	173.800	25 pounds	100 pounds	1,2	1	
	Chloroprene, inhibited	Flammable liquid	UN1991	Flammable liquid	173.118	173.119	1 quart	10 gallons	1,2	1	
	Chloroprene, uninhibited	*Forbidden*									
	2-Chloropropene	*Flammable liquid*	UN2456	Flammable liquid	None	173.254	1 quart	10 gallons	1	5	
	Chlorosulfonic acid	Corrosive material	UN1754	Corrosive	173.244	173.254	1 quart	1 quart	1	1	Keep dry. Glass carboys not permitted on passenger vessels
	Chlorosulfonic acid-sulfur trioxide mixture	Corrosive material	UN1754	Corrosive	173.144	173.254	1 quart	1 quart	1	1	Keep dry. Glass carboys not permitted on passenger vessels
	Chlorotetrafluoroethane (R-124)	Nonflammable gas	UN1021	Nonflammable gas	173.306	173.304	150 pounds	300 pounds	1,2	1,2	
	Chlorotrifluoromethane (R-13)	Nonflammable gas	UN1022	Nonflammable gas	173.306	173.304 173.314 173.315	150 pounds	300 pounds	1,2	1,2	
<	Chlorpyrifos	ORM-A	NA2783	None	173.505	173.510	100 pounds	No limit	1,2	1,2	Stow away from foodstuffs
	Chromic acid mixture, dry	Oxidizer	NA1463	Oxidizer	173.153	173.164	25 pounds	100 pounds	1,2	1,2	Stow away from foodstuffs. Stow separate from flammable liquids and solids
	Chromic acid, solid	Oxidizer	NA1463	Oxidizer	173.153	173.164	25 pounds	100 pounds	1,2	1,2	
	Chromic acid solution	Corrosive material	UN1755	Corrosive	173.244	173.287	1 quart	1 gallon	1	1	
	Chromic anhydride. See Chromic acid, solid										

133

Chemical Hazard Communication Guidebook

§172.101   **Title 49—Transportation**

§172.101 Hazardous Materials Table

(1) +/A/W	(2) Hazardous materials descriptions and proper shipping names	(3) Hazard class	(3A) Identification number	(4) Label(s) required (if not excepted)	(5) Packaging (a) Exceptions	(5) Packaging (b) Specific requirements	(6) Maximum net quantity in one package (a) Passenger carrying aircraft or railcar	(6) Maximum net quantity in one package (b) Cargo only aircraft	(7) Water shipments (a) Cargo vessel	(7) Water shipments (b) Passenger vessel	(7) Water shipments (c) Other requirements
	Chromic fluoride, solid	Corrosive material	UN1756	Corrosive	173.244	173.245b	25 pounds	100 pounds	1,2	1,2	
	Chromic fluoride solution	Corrosive material	UN1757	Corrosive	173.244	173.245	1 quart	1 gallon	1,2	1,2	
	Chromic trioxide. *See* Chromic acid, solid										
	Chromium oxychloride or Chromyl chloride	Corrosive material	UN1758	Corrosive	None	173.247	Forbidden	1 gallon	1	1	Keep dry. Glass carboys not permitted on passenger vessels
	Cigarette lighter (*or other similar ignition device*)	Flammable gas	UN1057	Flammable gas	173.21 175.10	173.308	21 ounces	25 pounds	1	1	
	Cigarette lighter (*or other similar ignition device*)	Flammable liquid	UN1226	Flammable liquid	173.21 175.10	173.118	Forbidden	Forbidden	1	1	
	Cigarette load	Class C explosive		Explosive C	None	173.111	50 pounds	150 pounds	1,2	1,2	
	Cloud gas cylinder. *See* Chemical ammunition, nonexplosive										
	Coal briquettes, hot	Forbidden									
	Coal facings. *See* Coal ground bituminous, etc.										
	Cool gas. *See* Hydrocarbon gas, nonliquefied										
	Coal, ground bituminous, sea coal or coal facings	Flammable solid	NA1361	Flammable solid	173.165	173.165	Forbidden	Forbidden	1	1	Separate from flammable gases or liquids, oxidizing materials, or organic peroxides
	Coal oil (export shipment only). *See* Kerosene										
	Coal tar distillate	Combustible liquid	UN1137	None	173.118a	None	No limit	No limit	1,2	1,2	
	Coal tar distillate	Flammable liquid	UN1136	Flammable liquid	173.118	173.119	1 quart	10 gallons	1,2	1	
	Coal tar dye, liquid (*not otherwise specifically named in 172.101*)	Corrosive material	NA2801	Corrosive	173.244	173.249a	1 quart	10 gallons	1,2	1,2	
	Coating solution	Flammable liquid	UN1139	Flammable liquid	173.118	173.132	1 quart	15 gallons	1,2	1	
	Cobalt resinate, precipitated	Flammable solid	UN1318	Flammable solid	None	173.166	Forbidden	125 pounds	1,2	1,2	
	Cocculus, solid (*fishberry*)	Poison B	UN1584	Poison	173.364 173.505	173.365 173.955	50 pounds	200 pounds	1,2 1,2	1,2 4	Keep dry
W	Coconut meal pellets containing at least 6% and not more than 13% moisture and not more than 10% residual fat content	ORM-C		None							
	Coke, hot	Forbidden									
	Collodion	Flammable liquid	NA2059	Flammable liquid	173.118	173.119	1 quart	10 gallons	1,2	1	

Chapter I—Research and Special Programs Administration　　　　**§172.101**

Item	Class	ID No.	Label	Ref 1	Ref 2	Qty A	Qty B	Col 1	Col 2
Combination fuze	Class C explosive		Explosive C	None	173.106	50 pounds	150 pounds	1,3	1,3
Combination primer	Class C explosive		None	None	173.107	50 pounds	150 pounds	1,3	5
Combustible liquid, n.o.s.	Combustible liquid	NA1993	None	173.118a	None	No limit	No limit	1,3	1,2
Commercial shaped charge. See High explosive									
Common fireworks. See Fireworks, common									
Compound, cleaning, liquid	Flammable liquid	NA1993	Flammable liquid	173.118	173.119	1 quart	10 gallons	1,2	1
Compound, cleaning, liquid	Corrosive material	NA1760	Corrosive	173.244	173.245	1 quart	1 quart	1,2	1,2
Compound, cleaning, liquid	Combustible liquid	NA1993	None	173.118a	None	No limit	No limit	1,2	1,2
Compound, cleaning, liquid	Corrosive material	NA1760	Corrosive	173.244	173.240a	1 quart	1 quart	1,2	1,2
Compound, cleaning, liquid (*containing phosphoric acid, acetic acid, sodium hydroxide or potassium hydroxide*)	Corrosive material	NA1789	Corrosive	173.244	173.263	1 quart	1 gallon	1	1
Compound, cleaning, liquid (*containing hydrochloric (muriatic) acid*)	Corrosive material	NA1789	Corrosive	173.244	173.256	1 quart	1 gallon	1	4
Compound, cleaning, liquid (*containing hydrofluoric acid*)	Corrosive material	NA1142	Corrosive	173.118	173.119	1 quart	55 gallons	1,2	1
Compound, polishing, liquid	Flammable liquid	NA1142	Flammable liquid	173.244	173.245	1 quart	1 gallon	1,2	1,2
Compound, rust preventing *or* Compound, rust removing	Corrosive material	NA1760	Corrosive	173.118a	None	No limit	No limit	1,2	1,2
Compound, tree or weed killing, liquid	Combustible liquid	NA1993	None	173.244	173.245	1 quart	1 quart	1,2	1,2
Compound, tree or weed killing, liquid	Corrosive material	NA1760	Corrosive	173.244	173.245	1 quart	1 quart	1	1
Compound, tree or weed killing, liquid	Flammable liquid	NA1993	Flammable liquid	173.118	173.119	1 quart	10 gallons	1,2	1,2
Compound, tree or weed killing, liquid	Poison B	NA2810	Poison	173.345	173.346	1 quart	55 gallons	1,2	1,2
Compound, tree or weed killing, solid	Oxidizer	NA1479	Oxidizer	173.153	173.154 173.229	25 pounds	100 pounds		
Compound, vulcanizing, liquid	Corrosive material	NA1760	Corrosive	173.244	173.245	1 quart	1 quart	1,2	1,2
Compound, vulcanizing, liquid	Flammable liquid	NA1142	Flammable liquid	173.118	172.119	1 quart	10 gallons	1,2	1
Compressed gas, n.o.s.	Flammable gas	UN1964	Flammable gas	173.306	173.302 173.304 173.305	Forbidden	300 pounds	1	4
Compressed gas, n.o.s.	Nonflammable gas	UN1956	Nonflammable gas	173.306 173.307	173.302 173.304 173.305	150 pounds	300 pounds	1,2	1,2
Consumer commodity	ORM-D		None	173.500a(b)	173.510 173.1200	65 pounds gross	65 pounds gross		Not subject to requirements of Part 175
Container, reused or empty, must be classed for the hazardous material previously contained. See 173.28, 173.29									
Copper acetoarsenite, solid	Poison B	UN1585	Poison	173.364	173.367	50 pounds	200 pounds	1,2	1,2
Copper acetylide	Forbidden								
Copper amine azide	Forbidden								
Copper arsenite, solid	Poison B	UN1586	Poison	173.364	173.365	50 pounds	200 pounds	1,2	1,2
Copper based pesticide, liquid, n.o.s. (*compounds and preparations*),	Flammable liquid	UN2776	Flammable liquid	173.118	173.119	1 quart	10 gallons	1,2	1,2

135

Chemical Hazard Communication Guidebook

§172.101   Title 49—Transportation

§172.101 Hazardous Materials Table

(1) +/A/W	(2) Hazardous materials descriptions and proper shipping names	(3) Hazard class	(3A) Identification number	(4) Label(s) required (if not excepted)	(5) Packaging (a) Exceptions	(5) Packaging (b) Specific requirements	(6) Maximum net quantity in one package (a) Passenger carrying aircraft or railcar	(6) (b) Cargo only aircraft	(7) Water shipments (a) Cargo vessel	(7) (b) Passenger vessel	(7) (c) Other requirements
	Copper based pesticide, liquid, n.o.s. (compounds and preparations).	Poison B	UN2775	Poison	173.345	173.346	1 quart	55 gallons	1,2	1,2	
	Copper based pesticide, solid, n.o.s. (compounds and preparations).	Poison B	UN2775	Poison	173.364	173.365	50 pounds	200 pounds	1,2	1,2	
A	Copper chloride	ORM-B	UN2802	None	173.505	173.800	25 pounds	100 pounds	1,2	1,2	Stow away from acids
	Copper cyanide	Poison B	UN1587	Poison	173.370	173.370	25 pounds	200 pounds	1,2	1,2	
	Copper tetramine nitrate	Forbidden									
W	Copra	ORM-C	UN1363	None	173.505	173.960			1,2	1,2	Segregation same as for flammable solids. Separate from flammable gases or liquids, oxidizing materials, or organic peroxides
	Copra pellets. See Coconut meal pellets										
	Cord, detonating *flexible*	Class A explosive		Explosive A	173.81	173.81	Forbidden	Forbidden	6	5	
	Cord, detonating *flexible*	Class C explosive		Explosive C	None	173.104	Forbidden	150 pounds	1,3	1,3	
	Corrosive liquid, n.o.s.	Corrosive material	UN1760	Corrosive	173.244	173.245 173.245a	1 quart	1 quart	1	4	For material that meets only the corrosion to skin criteria of 49 CFR 173.240(a)(1), "under deck" stowage is also authorized if the description includes the additional entry specified by 172.203(i)(2)
	Corrosive liquid, poisonous, n.o.s.	Corrosive material	UN2922	Corrosive and Poison	173.244	173.245	1 quart	1 quart	1	4	
	Corrosive solid, n.o.s.	Corrosive material	UN1759	Corrosive	173.244	173.245b	25 pounds	100 pounds	1	4	For material that meets only the corrosion to skin criteria of 49 CFR 173.240(a)(1), "under deck" stowage is also authorized if the description includes the additional entry specified by 172.203(i)(2)
	Cosmetics, liquid, n.o.s.	Corrosive material	NA1760	Corrosive	173.244	173.245	1 quart	1 quart	1,2	1,2	
	Cosmetics, n.o.s.	Combustible liquid	NA1993	None	173.118a	None	No limit	No limit	1,2	1,2	
	Cosmetics, n.o.s.	Flammable liquid	NA1993	Flammable liquid	173.118	173.119	1 quart	10 gallons	1,2	1	
	Cosmetics, n.o.s.	Flammable solid	NA1325	Flammable solid	173.153	173.154	25 pounds	100 pounds	1,2	1,2	
	Cosmetics, n.o.s.	Oxidizer	NA1479	Oxidizer	173.153	173.154	25 pounds	100 pounds	1,2	1,2	
	Cosmetics, solid, n.o.s.	Corrosive material	NA1759	Corrosive	173.244	173.245b	25 pounds	100 pounds	1,2	1,2	
W	Cotton	ORM-C		None	173.505	173.965			1,2	1,2	Keep dry
	Coumaphos	Poison B	NA2783	Poison	173.364	173.365	50 pounds	200 pounds	1,2	1,2	Segregation same as for flammable solids. See 176.900 to 176.904

136

Chapter I—Research and Special Programs Administration §172.101

Material	Hazard Class	ID No.	Label	Ref 1	Ref 2	Pass Limit	Cargo Limit	Col A	Col B	Notes
Coumaphos mixture, liquid	Poison B	NA2783	Poison	173.345	173.346	1 pint	1 quart	1,2	1,2	
Cresol	Corrosive material	UN2076	Corrosive	173.244	173.245	1 quart	10 gallons	1,2	1,2	
Crotonaldehyde	Flammable liquid	UN1143	Flammable liquid and Poison	None	173.119 173.3a	1 quart	1 gallon	1,2	1	
Crotonic acid	Corrosive material	UN2823	Corrosive	173.244	173.245	1 quart	10 gallons	1,2	1,2	
Crotonylene	Flammable liquid	UN1144	Flammable liquid	173.118	173.119	1 quart	10 gallons	1,3	4	
Crude oil, petroleum	Combustible liquid	UN1267	None	173.118a	None	No limit	No limit	1,2	1,2	
Crude oil, petroleum	Flammable liquid	UN1267	Flammable liquid	173.118	173.119	1 quart	10 gallons	1,2	1	
Cumene hydroperoxide	Organic peroxide	UN2116	Organic peroxide	173.153	173.224	1 quart	1 quart	1,2	4	
Cumene hydroperoxide, technically pure. See Cumene hydroperoxide.		UN2116								
Cupric cyanide. See Copper cyanide.										
Cupric nitrate	Oxidizer	NA1479	Oxidizer	173.153	173.182	25 pounds	100 pounds	1,2	1,2	Keep dry. Stow away from acids
Cupriethylene-diamine solution	Corrosive material	UN1761	Corrosive	173.244	173.249	1 quart	1 gallon	1,2	1,2	Stow away from acids
Cyanide or cyanide mixture, dry	Poison B	UN1588	Poison	173.364	173.370	25 pounds	200 pounds	1,2	1,2	Shade from radiant heat
Cyanide solution, n.o.s.	Poison B	UN1935	Poison	173.345	173.352	1 quart	55 gallons	1,2	1,2	Shade from corrosive materials
Cyanogen bromide	Poison B	UN1889	Poison	None	173.379	Forbidden	25 pounds	1	5	Shade from radiant heat
Cyanogen chloride containing less than 0.9% water	Poison A	UN1589	Poison gas and Flammable gas	None	173.328	Forbidden	Forbidden	1	5	Segregation same as for flammable gas
Cyanogen gas	Poison A	UN1026	Poison gas and Flammable gas	None	173.328	Forbidden	Forbidden	1	5	
Cyanuric triazide	Forbidden									
Cyclohexane	Flammable liquid	UN1145	Flammable liquid	173.118	173.119	1 quart	10 gallons	1,3	4	
Cyclohexanone peroxide, 50 to 85%	Organic peroxide	UN2119	Organic peroxide	173.157	173.156	Forbidden	25 pounds	1	1	
Cyclohexanone peroxide, as a paste with not more than 9% by weight active oxygen. See Cyclohexanone peroxide, 50 to 85% peroxide.		UN2896								
Cyclohexanone peroxide, in solution with not more than 9% by weight active oxygen. See Cyclohexanone peroxide, 50 to 85% peroxide		UN2118								
Cyclohexanone peroxide, not over 50% peroxide	Organic peroxide	UN2896	Organic peroxide	173.153	173.154	2 pounds	25 pounds	1,2	1,2	
Cyclohexanone peroxide and di-(1-hydroxy cyclohexyl) peroxide mixture. See appropriate cyclohexanone peroxide entry										
Cyclohexenyl trichlorosilane	Corrosive material	UN1762	Corrosive	None	173.280	Forbidden	10 gallons	1	1	Keep dry
Cyclohexylamine	Flammable liquid	UN2357	Flammable liquid	173.118	173.119	1 quart	10 gallons	1,2	1	
Cyclohexyl trichlorosilane	Corrosive material	UN1763	Corrosive	None	173.280	Forbidden	10 gallons	1	1	Keep dry

Chemical Hazard Communication Guidebook

§172.101   Title 49—Transportation

§172.101 Hazardous Materials Table

(1)	(2)	(3)	(3A)	(4)	(5) Packaging		(6) Maximum net quantity in one package		(7) Water shipments		
+/A/W	Hazardous materials descriptions and proper shipping names	Hazard class	Identification number	Label(s) required (if not excepted)	(a) Exceptions	(b) Specific requirements	(a) Passenger carrying aircraft or railcar	(b) Cargo only aircraft	(a) Cargo vessel	(b) Passenger vessel	(c) Other requirements
	Cyclopentane	Flammable liquid	UN1146	Flammable liquid	173.118	173.119	1 quart	10 gallons	1,3	4	
	Cyclopentane, methyl	Flammable liquid	UN2298	Flammable liquid	173.118	173.119	1 quart	10 gallons	1,3	4	
+	Cyclopropane	Flammable gas	UN1027	Flammable gas	173.306	173.304	Forbidden	300 pounds	1,2	1	
	Cyclotetramethylene tetranitramine (dry) (HMX)	Forbidden									
	Cyclotetramethylene tetranitramine, wet with not less than 10% water. See High explosive.										
	Cyclotrimethylene trinitramine, desensitized. See High explosive										
	Cyclotrimethylene trinitramine, wet with not less than 10% water. See High explosive										
	Cylinder, empty, including ton tanks, must be classed for the hazardous material previously contained in cylinder. See 173.29										
	2,4-D. See 2,4-Dichlorophenoxyacetic acid										
A	DDT or Dichlorodiphenyltrichloroethane	ORM-A	NA2761	None	173.505	173.510	No limit	No limit	1,2	1,2	
A	Decaborane	Flammable solid	UN1868	Flammable solid and Poison	None	173.236	Forbidden	25 pounds	1,2	1,2	
	Decahydronaphthalene	Combustible liquid	UN1147	None	173.118a	None	No limit	No limit	1,2	1,2	
	Decalin. See Decahydronaphthalene										
	Decanoyl peroxide, technically pure. See Organic peroxide, solid, n.o.s.										
	Delay connectors. See Detonators, Class A or Class C explosives and Detonating primers, Class A or Class C explosives										
	Delay electric igniter	Class C explosive	UN2120	Explosive C	None	173.106	50 pounds	150 pounds	1,3	1,3	
	Denatured alcohol	Flammable liquid		Flammable liquid	173.118	173.125	1 quart	10 gallons	1,2	1	
	Depth bomb. See Explosive bomb										
	Detonating fuze, Class A, with or without radioactive components	Class A explosive	NA1986	Explosive A	None	173.69	Forbidden	Forbidden	6	5	
	Detonating fuze, Class C explosive	Class C explosive		Explosive C	None	173.113	50 pounds	150 pounds	1,3	1,3	

Chapter I—Research and Special Programs Administration §172.101

Detonating primers, Class A explosives. See 173.53	Class A explosive		Explosive A	None	173.63	Forbidden	Forbidden	5
Detonating primers, Class C explosives. See 173.100	Class C explosive		Explosive C	None	173.66	173.103(d)	150 pounds	6	1,2
Detonators, Class A explosives. See 173.53	Class A explosive		Explosive A	None	173.66	Forbidden	Forbidden		1,2
Detonators, Class C explosives. See 173.100	Class C explosive		Explosive C	None	173.66	173.103(d)	150 pounds	6	1,2
Detonators, commercial. See Detonators, Class A or Class C explosives								
Di-(1-hydroxycyclohexyl) peroxide, technically pure. See Organic peroxide, solid, n.o.s.	Forbidden	UN2148						
Di-(1-hydroxytetrazole) (dry)	Forbidden							
Di-(1-naphthoyl)peroxide		UN2122						
Di-(2-ethylhexyl) peroxydicarbonate, technically pure. See Organic peroxide, liquid or solution, n.o.s.		UN2123						
Di-(2-ethylhexyl) peroxydicarbonate, not more than 77% in solution. See Organic peroxide, liquid or solution, n.o.s.								
Di-(2-ethylhexyl) phosphoric acid	Corrosive material	NA1902	Corrosive	173.244	173.245	1 quart	10 gallons	1,2
Di-(2-methylbenzoyl)peroxide, not more than 85% with water. See Organic peroxide, solid, n.o.s.		UN2593						
1,3-Di-(2-tert-butylperoxyisopropyl) benzene, technically pure or more than 40% with inert solid. See Organic peroxide, solid, n.o.s.		UN2112						
1,3-Di-(2-tert-butylperoxyisopropyl) benzene and 1,4-Di-(2-tert-butylperoxyisopropyl) benzene mixture, technically pure or more than 40% with inert solid. See Organic peroxide, solid, n.o.s.		UN2112						
1,4-Di-(2-tert-butylperoxyisopropyl) benzene, technically pure or more than 40% with inert solid. See Organic peroxide, solid, n.o.s.		UN2112						

139

Chemical Hazard Communication Guidebook

§172.101   Title 49—Transportation

§172.101 Hazardous Materials Table

(1) +/A/W	(2) Hazardous materials descriptions and proper shipping names	(3) Hazard class	(3A) Identification number	(4) Label(s) required (if not excepted)	(5) Packaging (a) Exceptions	(5) Packaging (b) Specific requirements	(6) Maximum net quantity in one package (a) Passenger carrying aircraft or railcar	(6) Maximum net quantity in one package (b) Cargo only aircraft	(7) Water shipments (a) Cargo vessel	(7) Water shipments (b) Passenger vessel	(7) Water shipments (c) Other requirements
	Di-(3,5,5-trimethyl-1,2-dioxolanyl-3)peroxide, *not more than 50% as a paste, with phlegmatizer.* See Organic peroxide, solid, n.o.s.		UN2597								
	2,2-Di-(4,4-di-tert-butylperoxycyclohexyl)propane, *not more than 42% with inert solid.* See Organic peroxide, solid, n.o.s.		UN2168								
	2,2-Di-(4,4-di-tert-butylperoxycyclohexyl)propane, *more than 42% with inert solid*	Forbidden	UN2154								
	Di-(4-tert-butylcyclohexyl)peroxydicarbonate, *technically pure.* See Organic peroxide, solid, n.o.s.										
	Di-(4-tert-butylcyclohexyl)peroxydicarbonate, *not more than 42%, stable dispersion, in water.* See Organic peroxide, liquid or solution, n.o.s.		UN2894								
	Diacetone alcohol	Combustible liquid	UN1148	None	173.118a	None	No limit	No limit	1,2	1,2	
	Diacetone alcohol	Flammable liquid	UN1148	Flammable liquid	173.118	173.119	1 quart	10 gallons	1,2	1	
	Diacetone alcohol peroxide, *not more than 57% in solution with not more than 9% hydrogen peroxide, not less than 26% diacetone alcohol and not less than 9% water; total active oxygen content not more than 9%.* See Organic peroxide, liquid or solution, n.o.s.		UN2163								
	Diacetone alcohol peroxides, *more than 57% in solution with more than 9% hydrogen peroxide, less than 26% diacetone alcohol and less than 9% water; total active oxygen content more than 9% by weight*	Forbidden									
	Diacetyl	Flammable liquid	UN2346	Flammable liquid	173.118	173.119	1 quart	10 gallons	1,2	1	
	p-Diazidobenzene	Forbidden									
	1,2-Diazidoethane	Forbidden									
A	Diazinon	ORM-A	NA2783	None	173.505	173.510	No limit	No limit	1,2	1,2	
	1,1'-Diazoaminonaphthalene	Forbidden									

140

Appendix B: Regulations Reference Guide

Chapter I—Research and Special Programs Administration　　　**§172.101**

	Name	Classification	ID No.	Label	Ref 1	Ref 2	Limit 1	Limit 2	Col A	Col B	Notes
	Diazoaminotetrazole (dry)	Forbidden									
	Diazodinitrophenol. See Initiating explosive										
	Diazodinitrophenol (dry)	Forbidden									
	Diazodiphenylmethane	Forbidden									
	Diazonium nitrates (dry)	Forbidden									
	Diazonium perchlorates (dry)	Forbidden									
	1,3-Diazopropane	Forbidden									
	Dibenzyl peroxydicarbonate, not more than 87% with water. See Organic peroxide, solid, n.o.s.		UN2149								
	Dibenzyl peroxydicarbonate, more than 87% with water	Forbidden									
	Di-(beta-nitroxyethyl)ammonium nitrate	Forbidden									
+	Diborane or diborane mixtures	Flammable gas	UN1911	Flammable gas and Poison	None	173.302	Forbidden	Forbidden	1	5	Separate from Chlorine and materials bearing the oxidizer label.
	Dibromoacetylene	Forbidden									
A	Dibromodifluoromethane	ORM-A	UN1941	None	173.505	173.605	10 gallons	55 gallons			
AW	1,2-Dibromoethane. See Ethylene dibromide										
	Dicetyl peroxydicarbonate, not more than 42%, stable dispersion, in water. See Organic peroxide, liquid or solution, n.o.s.		UN2895								
	Dicetyl peroxydicarbonate, technically pure. See Organic peroxide, solid, n.o.s.		UN2164								
	N,N'-Dichlorazodicarbonamidine (salts of) (dry)	Forbidden									
	1,1'-Dichloro-2,2-bis(parachlorophenyl)ethane. See TDE										
	Dichloroacetic acid	Corrosive material	UN1764	Corrosive	173.244	173.245	1 quart	1 quart	1,2	1,2	
	Dichloroacetyl chloride	Corrosive material	UN1765	Corrosive	173.244	173.247	1 quart	1 gallon	1	4	
	Dichloroacetylene	Forbidden									
A	Dichlorobenzene, ortho, liquid	ORM-A	UN1591	None	173.505	173.510	No limit	No limit	1,2	1,2	
A	Dichlorobenzene, para, solid	ORM-A	UN1592	None	173.505	173.510	No limit	No limit	1,2	1,2	
	2,4-Dichlorobenzoyl peroxide, not more than 75% with water. See Organic peroxide, solid, n.o.s.		UN2137								
	2,4-Dichlorobenzoyl peroxide, not more than 52% as a paste. See Organic peroxide, solid, n.o.s.		UN2138								
	2,4-Dichlorobenzoyl peroxide, not more than 52% in solution. See Organic peroxide, liquid or solution, n.o.s.		UN2139								
	2,4-Dichlorobenzoyl peroxide, more than 75% with water	Forbidden									
	Dichlorobutene	Flammable liquid	NA2924	Flammable liquid	173.118	173.119	1 quart	10 gallons	1,2	1	Glass carboys in hampers not permitted under deck
	Dichlorobutene	Corrosive material	NA2924	Corrosive	173.244	173.245 173.245a	1 quart	10 gallons	1	4	Keep dry
	Dichlorodifluoroethylene	ORM-A	NA9018	None	173.505	173.605 173.304	10 gallons	55 gallons			
A	Dichlorodifluoromethane (R-12)	Nonflammable gas	UN1028	Nonflammable gas	173.308	173.314 173.315	150 pounds	300 pounds	1,2	1,2	

141

Chemical Hazard Communication Guidebook

§172.101   Title 49—Transportation

§172.101 Hazardous Materials Table

(1) +/A/W	(2) Hazardous materials descriptions and proper shipping names	(3) Hazard class	(3A) Identification number	(4) Label(s) required (if not excepted)	(5) Packaging (a) Exceptions	(5) Packaging (b) Specific requirements	(6) Maximum net quantity in one package (a) Passenger carrying aircraft or railcar	(6) (b) Cargo only aircraft	(7) Water shipments (a) Cargo vessel	(7) (b) Passenger vessel	(7) (c) Other requirements
	Dichlorodifluoromethane and difluoroethane mixture (*constant boiling mixture*) (*R-500*). *See Refrigerant gas, n.o.s. or Dispersant gas, n.o.s.*										
	Dichlorodifluoromethane (*R-12*) and dichlorotetrafluoroethane (*R-114*) mixture. *See Refrigerant gas, n.o.s. or Dispersant gas, n.o.s.*										
	Dichlorodifluoromethane (*R-12*) and chlordifluoromethane (*R-22*) mixture. *See Refrigerant gas, n.o.s. or Dispersant gas, n.o.s.*										
	Dichlorodifluoromethane (*R-12*) and trichlorofluoromethane (*R-11*) mixture. *See Refrigerant gas, n.o.s. or Dispersant gas, n.o.s.*										
	Dichlorodifluoromethane (*R-12*) and trichlorofluoromethane (*R-11*) and chlorodifluoromethane (*R-22*) mixture. *See Refrigerant gas, n.o.s. or Dispersant gas, n.o.s.*										
	Dichlorodifluoromethane (*R-12*) and trichlorotrifluoroethane (*R-113*) mixture. *See Refrigerant gas, n.o.s. or Dispersant gas, n.o.s.*										
	Dichlorodiphenyltrichloroethane. *See DDT*										
A	Dichloroethylene	Flammable liquid	UN1150	Flammable liquid	173.118	173.119	1 quart	10 gallons	1,2	1	
	Dichloroisopropyl ether	Corrosive material	UN2490	Corrosive	173.244	173.245	1 quart	10 gallons	1,2	1,2	
A	Dichloromethane *or* Methylene chloride	ORM-A	UN1593	None	173.505	173.605	10 gallons	55 gallons	1,2	1,2	
	Dichloropentane	Flammable liquid	UN1152	Flammable liquid	173.118	173.119	1 quart	10 gallons	1,2	1,2	
A	2,4-Dichlorophenoxyacetic acid	ORM-A	NA2765	None	173.505	173.510	50 pounds	No limit	1,2	1,2	
	Dichlorophenyltrichlorosilane	Corrosive material	UN1766	Corrosive	None	173.280	Forbidden	10 gallons	1	1	Keep dry
	Dichloropropane. *See Propylene dichloride*										
	Dichloropropene	Flammable liquid	UN2047	Flammable liquid	173.118	173.119	1 quart	10 gallons	1,2	1	
	Dichloropropene and propylene dichloride mixture	Flammable liquid	NA2047	Flammable liquid	173.118	173.119	1 quart	10 gallons	1,2	1	

142

Appendix B: Regulations Reference Guide

Chapter I—Research and Special Programs Administration §172.101

									Keep dry. Segregation same as for corrosives
									Stow away from living quarters.
2,2-Dichloropropionic acid	Corrosive material	NA1760	Corrosive	173.244	173.245	1 quart	10 gallons	1,2	1,2
Dichlorvos	Poison B	NA2783	Poison	173.345	173.346	Forbidden	1 quart	1,2	1,2
Dichlorvos mixture, dry	Poison B	NA2783	Poison	173.364	173.365	50 pounds	200 pounds	1,2	1,2
Dicumyl peroxide 50% solution	Organic peroxide	NA3121	Organic peroxide	173.153	173.224	1 quart	1 quart	1,2	4
Dicumyl peroxide, technically pure or Dicumyl peroxide, with inert solid. See Dicumyl peroxide, dry.									
Dicumyl peroxide, dry	Organic peroxide	UN2121	Organic peroxide	173.153	173.154	2 pounds	25 pounds	1,2	1,2
Dicyclohexyl peroxydicarbonate, technically pure. See Organic peroxide, solid, n.o.s.		UN2121							
Dicyclohexyl peroxydicarbonate, not more than 91% with water. See Organic peroxide, solid, n.o.s.		UN2153							
Dieldrin	ORM-A	NA2761	None	173.505	173.510	No limit	No limit	1,2	1,2
Diesel fuel. See Fuel oil	Forbidden								
Diethanol nitrosamine dinitrate (dry)	Forbidden								
Diethylamine	Flammable liquid	UN1154	Flammable liquid	173.118	173.119	Forbidden	5 pints	1,2	4
Diethyl cellosolve. See Ethylene glycol diethyl ether	Flammable liquid								
Diethyl dichlorosilane	Flammable liquid	UN1767	Flammable liquid	None	173.125	Forbidden	10 gallons	1	1
Diethylene glycol dinitrate. See 173.51	Forbidden								
Diethylgold bromide	Forbidden								
Diethyl ketone	Flammable liquid	UN1156	Flammable liquid	173.118	173.119	1 quart	10 gallons	1,2	1
Diethyl peroxydicarbonate, not more than 27% in solution. See Organic peroxide, liquid or solution, n.o.s.		UN2175							
Diethyl peroxydicarbonate, more than 27% in solution	Forbidden			173.306	173.304 173.314 173.315	Forbidden	200 pounds	1,2	1,2
Diethylzinc. See Pyrophoric liquid, n.o.s.									
Difluoroethane (R-152a)	Flammable gas	UN1030	Flammable gas	173.306	173.304	Forbidden	300 pounds	1,2	5
1,1-Difluoroethylene (R-1132A)	Flammable gas	UN1768	Corrosive	None	173.275	Forbidden	1 gallon	1,2	1,2
Difluorophosphoric acid, anhydrous	Corrosive material	UN2178							
2,2-Dihydroperoxy propane, not more than 25% with inert organic solid. See Organic peroxide, solid, n.o.s.		UN2376							
Dihydropyran	Flammable liquid	UN1157	Flammable liquid	173.118	173.119	1 quart	10 gallons	1,2	4
1,8-Dihydroxy-2,4,5,7-tetranitroanthraquinone (chrysamminic acid)	Forbidden								
Diiodoacetylene	Forbidden								
Diisobutyl ketone	Combustible liquid	UN2163	None	173.118a	None	No limit	No limit	1,2	1,2
Diisobutyryl peroxide, not more than 32% in solution. See Organic peroxide, liquid or solution, n.o.s.									

143

Chemical Hazard Communication Guidebook

§172.101   Title 49—Transportation

§172.101 Hazardous Materials Table

(1) +/A/W	(2) Hazardous materials descriptions and proper shipping names	(3) Hazard class	(3A) Identification number	(4) Label(s) required (if not excepted)	(5) Packaging (a) Exceptions	(5) Packaging (b) Specific requirements	(6) Maximum net quantity in one package (a) Passenger carrying aircraft or railcar	(6) (b) Cargo only aircraft	(7) Water shipments (a) Cargo vessel	(7) (b) Passenger vessel	(7) (c) Other requirements
	Diisooctyl acid phosphate	Corrosive material	UN1902	Corrosive	173.244	173.296	1 quart	1 quart	1,2	1,2	Glass carboys in hampers not permitted under deck
	Diisopropylamine	Flammable liquid	UN1158	Flammable liquid	173.118	173.119	1 quart	10 gallons	1,2	1	
	Diisopropylbenzene hydroperoxide solution, not over 72% peroxide	Organic peroxide	UN2171	Organic peroxide	173.153	173.224	1 quart	1 quart	1,2	4	
	Diisopropylbenzene hydroperoxide, not more than 72% in solution. See Diisopropylbenzene hydroperoxide solution, not more than 72% peroxide.		UN2171								
	Diisopropylbenzene hydroperoxide, more than 72% in solution	Forbidden									
	Diisopropyl ether	Flammable liquid	UN1159	Flammable liquid	173.118	173.119	1 quart	10 gallons	1,3	4	
	Diisotridecyl peroxydicarbonate, technically pure. See Organic peroxide, liquid or solution, n.o.s.		UN2889								
	2,5-Dimethyl-2,5-di-(2-ethylhexanoylperoxy)hexane, technically pure. See Organic peroxide, liquid or solution, n.o.s.		UN2157								
	2,5-Dimethyl-2,5-di-(benzoylperoxy)hexane, technically pure. See Organic peroxide, solid, n.o.s.		UN2172								
	2,5-Dimethyl-2,5-di-(benzoylperoxy)hexane, not more than 82% with inert solid. See Organic peroxide, solid, n.o.s.		UN2173								
	2,5-Dimethyl-2,5-dihydroperoxy hexane, not more than 82% with water. See Dimethylhexane dihydroperoxide, with 18% or more water.		UN2174								
	2,5-Dimethyl-2,5-dihydroperoxy hexane, more than 82% with water	Forbidden									
	2,5-Dimethyl-2,5-di-(tert-butylperoxy)hexane, technically pure. See Organic peroxide, liquid or solution, n.o.s.		UN2155								

144

Chapter I—Research and Special Programs Administration §172.101

Article		ID No.	Class	(6)	(7)	Passenger	Cargo	(10)	(11)	Notes
2,5-Dimethyl-2,5-di-(tert-butylperoxy)hexane, not more than 52% with inert solid. See Organic peroxide, solid, n.o.s.										
2,5-Dimethyl-2,5-di-(tert-butylperoxy)hexyne-3, technically pure. See Organic peroxide, liquid or solution, n.o.s.										
2,5-Dimethyl-2,5-di-(tert-butylperoxy)hexyne-3, not more than 52% with inert solid. See Organic peroxide, solid, n.o.s.										
Dimethylamine, anhydrous	Flammable gas	UN1032	Flammable gas	173.306	173.304 173.314 173.315	Forbidden	300 pounds	1,2	4	
Dimethylamine, aqueous solution	Flammable liquid	UN1160	Flammable liquid	173.118	173.119	1 quart	10 gallons	1,2	1	
2,3-Dimethylbutane	Flammable liquid	UN2457	Flammable liquid	173.118	173.119	1 quart	10 gallons	1,3	4	
Dimethyl carbonate	Flammable liquid	UN1161	Flammable liquid	173.118	173.119	1 quart	10 gallons	1,2	1	
Dimethyl chlorothiophosphate, see Dimethyl phosphorochloridothioate.										
1,4-Dimethylcyclohexane	Flammable liquid	UN2263	Flammable liquid	173.118	173.119	1 quart	10 gallons	1,2	1	
Dimethyldichlorosilane	Flammable liquid	UN1162	Flammable liquid	None	173.135	Forbidden	5 pints	1,2	1	
Dimethyl ether	Flammable gas	UN1033	Flammable gas	173.306	173.304 173.314 173.315	Forbidden	300 pounds	1,2	1	
Dimethylhexane dihydroperoxide (dry)	Forbidden									
Dimethylhexane dihydroperoxide, (with 18% or more water)	Organic peroxide	UN2174	Organic peroxide	None	173.157	Forbidden	25 pounds	1	1	Keep dry. Separate from corrosive and oxidizing materials, and organic peroxides.
Dimethylhydrazine, unsymmetrical (UDMH)	Flammable liquid	UN1163	Flammable liquid and Poison	None	173.145	Forbidden	5 pints	1,2	1	Keep cool
Dimethyl phosphorochloridothioate. See Dimethyl clorothiophosphate.										
Dimethyl phosphorochloridothioate	Corrosive material	NA2267	Corrosive and poison	173.244	173.245	1 quart	1 quart	1,2	4	
Dimethyl sulfate	Corrosive material	UN1595	Corrosive	None	173.255	Forbidden	1 quart	1	5	
Dimethyl sulfide	Flammable liquid	UN1164	Flammable liquid	None	173.119	Forbidden	10 gallons	1,2	5	
Dimyristyl peroxydicarbonate, technically pure. See Organic peroxide, solid, n.o.s.		UN2506								
Dimyristyl peroxydicarbonate, not more than 22%, stable dispersion, in water. See Organic peroxide, liquid or solution, n.o.s.		UN2892								
1,4-Dinitro-1,1,4,4-tetramethylolbutanetetranitrate (dry)	Forbidden									
2,4-Dinitro-1,3,5-trimethylbenzene	Forbidden									
1,3-Dinitro-4,5-dinitrosobenzene	Forbidden									
1,3-Dinitro-5,5-dimethyl hydantoin	Forbidden									
Dinitro-7,8-dimethylgcoluril (dry)	Forbidden									

145

Chemical Hazard Communication Guidebook

§172.101   Title 49—Transportation

§172.101 Hazardous Materials Table

(1) +/A/W	(2) Hazardous materials descriptions and proper shipping names	(3) Hazard class	(3A) Identification number	(4) Label(s) required (if not excepted)	(5) Packaging (a) Exceptions	(5) Packaging (b) Specific requirements	(6) Maximum net quantity in one package (a) Passenger carrying aircraft or railcar	(6) Maximum net quantity in one package (b) Cargo only aircraft	(7) Water shipments (a) Cargo vessel	(7) Water shipments (b) Passenger vessel	(7) Water shipments (c) Other requirements
	Dinitrobenzene, solid, or Dinitrobenzol, solid	Poison B	UN1597	Poison	173.364	173.371	50 pounds	200 pounds	1,2	1,2	
	Dinitrobenzene solution	Poison B	UN1597	Poison	173.345	173.346	1 quart	55 gallons	1,2	1,2	
	Dinitrochlorobenzene	Poison B	UN1577	Poison	173.364	173.365	50 pounds	200 pounds	1,2	1,2	
A	Dinitrocyclohexylphenol	ORM-A	NA9026	None	173.505	173.510	No limit	No limit	1,2	1,2	
	1,2-Dinitroethane	Forbidden									
	1,1-Dinitroethane (dry)	Forbidden									
	Dinitroglycoluril	Forbidden									
	Dinitromethane	Forbidden									
	Dinitrophenol solution	Poison B	UN1599	Poison	173.345	173.362a	1 quart	65 pounds	1,2	1,2	Stow away from heavy metals and their compounds. If flash point is 141 deg F or less segregation same as for flammable liquids
	Dinitropropylene glycol	Forbidden									
	2,4-Dinitroresorcinol (heavy metal salts of) (dry)	Forbidden									
	4,6-Dinitroresorcinol (heavy metal salts of) (dry)	Forbidden									
	3,5-Dinitrosalicylic acid (lead salt) (dry)	Forbidden									
	Dinitrosobenzylamidine and salts of (dry)	Forbidden									
	2,2-Dinitrostilbene	Forbidden									
	a,a'-Di-(nitroxy)methylether	Forbidden									
	1,9-Dinitroxy pentamethylene-2,4,6,8-tetramine (dry)	Forbidden									
	Di-n-propyl peroxydicarbonate, *technically pure. See* Organic peroxide, liquid, n.o.s.		UN2176								
	Dioxane	Flammable liquid	UN1165	Flammable liquid	173.118	173.119	1 quart	10 gallons	1,2	1	
	Dioxolane	Flammable liquid	UN1166	Flammable liquid	173.118	173.119	1 quart	10 gallons	1,2	1	
	Diphenylaminechloroarsine (DM)	Irritating material	UN1698	Irritant	None	173.382	Forbidden	75 pounds	1	5	
	Diphenyl dichlorosilane	Corrosive material	UN1769	Corrosive	None	173.280	Forbidden	10 gallons	1	1	
	Diphenyl methyl bromide, solid	Corrosive material	UN1770	Corrosive	173.244	173.245b	25 pounds	100 pounds	1	4	
	Diphenyl methyl bromide solution	Corrosive material	UN1770	Corrosive	173.244	173.247	1 quart	1 gallon	1,2	1,2	
	Diphosgene. *See* Phosgene										
	Di-sec-butyl peroxydicarbonate, *technically pure. See* Organic peroxide, liquid or solution, n.o.s.		UN2150								

146

Appendix B: Regulations Reference Guide

Chapter I—Research and Special Programs Administration §172.101

Di-sec-butyl peroxydicarbonate, not more than 52% in solution. See Organic peroxide, liquid or solution, n.o.s.								
Disinfectant, liquid	Corrosive material	UN2151	Corrosive	173.244	173.245	1 quart	10 gallons	1	4
Disinfectant, liquid	Poison B	UN1903	Poison	173.345	173.346	1 quart	55 gallons	1,2	1
Disinfectant, liquid, n.o.s.	Combustible liquid	NA1993	None	173.118a	None	No limit	No limit	1,2	1,2
Disinfectant, solid	Poison B	UN1601	Poison	173.364	173.365	50 pounds	200 pounds	1,2	1
Dispersant gas, n.o.s. See Refrigerant gas, n.o.s.									
Distearyl peroxydicarbonate, not more than 85% with stearyl alcohol. See Organic peroxide, solid, n.o.s.									
Disulfoton	Poison B	NA2783	Poison	None	173.358	Forbidden	1 quart	1,2	5
Disulfoton mixture, dry	Poison B	NA2783	Poison	173.377	173.377	Forbidden	200 pounds	1,2	4
Disulfoton mixture, liquid	Poison B	NA2783 UN2102	Poison	173.359	173.359	1 pint	1 quart	1,2	5
Di-tert-butyl peroxide, technically pure. See Organic peroxide, liquid or solution, n.o.s.									
1,1-Di-(tert-butylperoxy)-3,3,5-trimethyl cyclohexane, technically pure. See Organic peroxide, liquid or solution n.o.s.		UN2145							
1,1-Di-(tert-butylperoxy)-3,3,5-trimethyl cyclohexane, not more than 57% in solution. See Organic peroxide, liquid or solution, n.o.s.		UN2146							
1,1-Di-(tert-butylperoxy)-3,3,5-trimethyl cyclohexane, not more than 58% with inert solid. See Organic peroxide, solid, n.o.s.		UN2147							
2,2-Di-(tert-butylperoxy)butane, not more than 55% in solution. See Organic peroxide, liquid or solution, n.o.s.		UN2111							
2,2-Di-(tert-butylperoxy)butane, more than 55% in solution	Forbidden								
1,1-Di-(tert-butylperoxy)cyclohexane, technically pure. See Organic peroxide, liquid or solution, n.o.s.		UN2179							
1,1-Di-(tert-butylperoxy)cyclohexane, not more than 77% in solution. See Organic peroxide, liquid or solution, n.o.s.		UN2180							
1,2-Di-(tert-butylperoxy)cyclohexane, not more than 77% in solution. See Organic peroxide, liquid or solution, n.o.s.		UN2181							
1,1-Di-(tert-butylperoxy)cyclohexane, not more than 40% with inert inorganic solid, with not less than 13% phlegmatizer. See Organic peroxide, solid, n.o.s.		UN2265							
1,1-Di-(tert-butylperoxy)cyclohexane, not more than 50% with phlegmatizer. See Organic peroxide, solid, n.o.s.		UN2297							
Di-(t,n-butylperoxy) phthalate, more than 55% in solution	Forbidden								
Di-(tert-butylperoxy)phthalate, technically pure. See Organic peroxide, solid, n.o.s.		UN2106							

147

Chemical Hazard Communication Guidebook

§172.101 Title 49—Transportation

§172.101 Hazardous Materials Table

(1) +/A/W	(2) Hazardous materials descriptions and proper shipping names	(3) Hazard class	(3A) Identification number	(4) Label(s) required (if not excepted)	(5) Packaging (a) Exceptions	(5) Packaging (b) Specific requirements	(6) Maximum net quantity in one package (a) Passenger carrying aircraft or railcar	(6) (b) Cargo only aircraft	(7) Water shipments (a) Cargo vessel	(7) (b) Passenger vessel	(7) (c) Other requirements
	Di-(tert-butylperoxy)phthalate, *not more than 55% in solution.* See Organic peroxide, liquid or solution, n.o.s.		UN2107								
	Di-(tert-butylperoxy)phthalate, *not more than 55% as a paste.* See Organic peroxide, solid, n.o.s.		UN2108								
	2,2-Di-(tert-butylperoxy)propane, *not more than 50% with phlegmatizer.* See Organic peroxide, liquid or solution, n.o.s.		UN2883								
	2,2-Di-(tert-butylperoxy)propane, *not more than 40% with inert inorganic solid with not less than 13% phlegmatizer.* See Organic peroxide, solid, n.o.s.		UN2884								
	Dithiocarbamate pesticide, liquid, n.o.s. *(compounds and preparations),*	Flammable liquid	UN2772	Flammable liquid	173.118	173.119	1 quart	10 gallons	1,2	1	
	Dithiocarbamate pesticide , liquid, n.o.s.*(compounds and preparations),*	Poison B	UN2771	Poison	173.345	173.346	1 quart	55 gallons	1,2	1,2	
	Dithiocarbamate pesticide, solid, n.o.s. *(compounds and preparations),*	Poison B	UN2771	Poison	173.364	173.365	50 pounds	200 pounds	1,2	1,2	
	Divinyl ether	Flammable liquid	UN1167	Flammable liquid	None	173.119	Forbidden	10 gallons	1,3	5	
	Dodecylbenzenesulfonic acid	Corrosive material	NA2584	Corrosive	173.244	173.245	1 quart	10 gallons	1,2	1,2	
	Dodecyl trichlorosilane	Corrosive material	UN1771	Corrosive	None	173.280	Forbidden	10 gallons	1	1	Keep dry
	Driers, paint or varnish, liquid, n.o.s	Combustible liquid	UN1168	None	173.118a	None	No limit	No limit	1,2	1,2	
	Driers, paint or varnish, liquid, n.o.s	Flammable liquid	UN1168	Flammable liquid	173.118 173.128	173.128	1 quart	55 gallons	1,2	1	
	Drill cartridge. See 173.55										
	Drugs, n.o.s.	Combustible liquid	NA1993	None	173.118a	None	No limit	No limit	1,2	1,2	
	Drugs, n.o.s.	Flammable solid	NA1325	Flammable solid	173.153	173.154	25 pounds	100 pounds	1,2	1,2	
	Drugs, n.o.s,	Oxidizer	NA1479	Oxidizer	173.153	173.154	25 pounds	100 pounds	1,2	1,2	
	Drugs, n.o.s,	Flammable liquid	NA1993	Flammable liquid	173.118	173.119	1 quart	10 gallons	1,2	1	
	Drugs, liquid, n.o.s.	Corrosive material	NA1760	Corrosive	173.244	173.245	1 quart	1 quart	1,2	1,2	
	Drugs, liquid,n.o.s.	Poison B	NA2810	Poison	173.345	173.346	1 quart	55 gallons	1,3	1	

148

Appendix B: Regulations Reference Guide

Chapter I—Research and Special Programs Administration §172.101

Drugs, solid, n.o.s.	Corrosive material Poison B	NA1759	Corrosive Poison	173.244 173.364	173.345 173.365	25 pounds 50 pounds	100 pounds 200 pounds	1,2 1,3	1,2 1,3
Drums, empty, must be classed for the hazardous material previously contained in drum. See 173.29									
Dry ice. See Carbon dioxide, solid									
Dye intermediate, liquid	Corrosive material	UN2801	Corrosive	173.244	173.249a	1 quart	10 gallons	1,2	1,2	Stow away from foodstuffs and living quarters
Dynamite. See High explosive									
Electric blasting caps. See Detonators, Class A or Class C explosives									
Electric squib	Class C explosive		Explosive C	None	173.106	50 pounds	150 pounds	1,3	1,3	Stow away from foodstuffs and living quarters
Electrolyte (acid) battery fluid (not over 47% acid). See Battery fluid, acid.									
Empty cartridge bag with black powder igniter	Class C explosive		Explosive C	None	173.106	50 pounds	150 pounds	1,3	1,3
Endosulfan	Poison B	NA2761	Poison	173.364	173.365	1 pound	10 pounds	1,2	1,2	If stowed under deck, must be stowed in a recoverable location.
Endosulfan mixture, liquid	Poison B	NA2761	Poison	173.345 173.364	173.346 173.365	1 quart 1 pound	55 gallons 10 pounds	1,2 1,2	1,2 1,2	If stowed under deck, must be stowed in a recoverable location.
Endrin	Poison B	NA2761	Poison	173.345 173.364	173.346 173.365	1 quart 1 pound	55 gallons 10 pounds	1,2 1,2	1,2 1,2	
Endrin mixture, liquid	Poison B	NA2761	Poison	173.345	173.346	1 quart	55 gallons	1,2	1,2
Engine, internal combustion	Flammable gas	UN1950	Flammable gas	173.150	173.304	Forbidden	60 pounds	1,2	5
Engine starting fluid	Flammable liquid	UN2023	Flammable liquid	None	173.119	1 quart	10 gallons	1,2	1,2
Epichlorohydrin	Corrosive material	NA1790	Corrosive	None	173.299	Forbidden	10 pounds	1	5
Escape or Evacuation slide, inflatable. See Life rafts, inflatable									
Etching acid, liquid, n.o.s.	Flammable gas	UN1035	Flammable gas	173.306	173.304	Forbidden	300 pounds	1,2	4	Stow away from living quarters
Ethane or Ethane, compressed	Flammable gas	UN1961	Flammable gas	None	173.315	Forbidden	Forbidden	1	5	Stow away from living quarters
Ethane, refrigerated liquid	Flammable gas	UN1961	Flammable gas	None	173.315	Forbidden	Forbidden	1	5	Stow away from living quarters
Ethane-Propane mixture, refrigerated liquid	Forbidden								
Ethanol. See Ethyl alcohol									
Ethanol amine dinitrate	Forbidden								
Ethion	Poison B	NA2783	Poison	173.345	173.346	Forbidden	1 quart	1,2	1,2
Ethion mixture, dry	Poison B	NA2783	Poison	173.364	173.365	50 pounds	200 pounds	1,2	1,2
Ethyl-3,3-di-(tert-butylperoxy)butyrate, technically pure. See Organic peroxide, liquid or solution, n.o.s.		UN2184							
Ethyl-3,3-di-(tert-butylperoxy)butyrate, not more than 77% in solution. See Organic peroxide, liquid or solution, n.o.s.		UN2185							
Ethyl-3,3-di-(tert-butylperoxy)butyrate, not more than 50% with inert inorganic solid. See Organic peroxide, solid, n.o.s.		UN2596							
Ethyl acetate	Flammable liquid	UN1173	Flammable liquid	173.118	173.119	1 quart	10 gallons	1,2	1
Ethyl acrylate, inhibited	Flammable liquid	UN1917	Flammable liquid	173.118	173.119	1 quart	10 gallons	1,2	1

149

§172.101 Hazardous Materials Table

(1) +/A/W	(2) Hazardous materials descriptions and proper shipping names	(3) Hazard class	(3A) Identification number	(4) Label(s) required (if not excepted)	(5) Packaging (a) Exceptions	(5) Packaging (b) Specific requirements	(6) Maximum net quantity in one package (a) Passenger carrying aircraft or railcar	(6) Maximum net quantity in one package (b) Cargo only aircraft	(7) Water shipments (a) Cargo vessel	(7) Water shipments (b) Passenger vessel	(7) Water shipments (c) Other requirements
	Ethyl alcohol	Flammable liquid	UN1170	Flammable liquid	173.118	173.125	1 quart	10 gallons	1,2	1	
	Ethyl aldehyde. See Acetaldehyde										
	Ethyl benzene	Flammable liquid	UN1175	Flammable liquid	173.118	173.119	1 quart	10 gallons	1,2	1	
	Ethyl borate	Flammable liquid	UN1176	Flammable liquid	173.118	173.119	1 quart	10 gallons	1,2	1	Keep dry
	Ethyl butyl acetate	Combustible liquid	UN1177	None	173.118a	None	No limit	No limit	1,2	1,2	
	Ethyl butyl ether	Flammable liquid	UN1179	Flammable liquid	173.118	173.119	1 quart	10 gallons	1,2	1	
	Ethyl butyraldehyde	Flammable liquid	UN1178	Flammable liquid	173.118	173.119	1 quart	10 gallons	1,2	1	
	Ethyl butyrate	Flammable liquid	UN1180	Flammable liquid	173.118	173.119	1 quart	10 gallons	1,2	1,2	
	Ethyl chloride	Flammable liquid	UN1037	Flammable liquid	None	173.123	Forbidden	See 173.123	1,2	1	Segregation same as for flammable gases
	Ethyl chloroacetate	Combustible liquid	UN1181	None	173.118a	None	No limit	No limit	1,2	1,2	
	Ethyl chloroformate (chlorocarbonate)	Flammable liquid	UN1182	Flammable liquid and Poison	None	173.288	Forbidden	5 pints	1,2	1	
	Ethyl chlorothioformate	Corrosive material	UN2826	Corrosive	173.244	173.245 173.245a	1 quart	1 quart	1,2	1	
	Ethyl crotonate	Flammable liquid	UN1862	Flammable liquid	173.118	173.119	1 quart	10 gallons	1,2	1	
	Ethyl dichlorosilane	Flammable liquid	UN1183	Flammable liquid	None	173.135	Forbidden	5 pints	1,2	1	
	Ethyl phosphonothioic dichloride, anhydrous	Corrosive material	NA1760	Corrosive	173.244	173.245 173.245a	1 quart	1 quart	1	4	
	Ethylene or Ethylene, compressed	Flammable gas	UN1962	Flammable gas	173.306	173.304	Forbidden	300 pounds	1,2	4	Stow away from living quarters
	Ethylene, refrigerated liquid (cryogenic liquid)	Flammable gas	UN1038	Flammable gas	None	173.318 173.319	Forbidden	Forbidden	1	5	Stow away from living quarters
	Ethylene chlorohydrin	Poison B	UN1135	Poison	173.345	173.346 173.3a	1 quart	55 gallons	1,2	1	Segregation same as for flammable liquids
	Ethylenediamine	Corrosive material	UN1604	Corrosive	173.244	173.245	1 quart	1 quart	1,2	1,2	
+	Ethylene diamine diperchlorate	Forbidden									
	Ethylene dibromide	Poison B	UN1605	Poison	173.345	173.346	1 quart	55 gallons	1,2	1,2	Stow away from living quarters

Chapter I—Research and Special Programs Administration §172.101

Ethylene dichloride	Flammable liquid	UN1184	Flammable liquid	173.118	173.119	1 quart	10 gallons	1,2	1
Ethylene glycol diethyl ether (diethyl Cellosolve)	Flammable liquid	UN1153	Flammable liquid	173.118	173.119	1 quart	10 gallons	1,2	1,2
Ethylene glycol dinitrate	Forbidden								
Ethylene glycol monoethyl ether (Cellosolve)	Combustible liquid	UN1171	None	173.118a	None	No limit	No limit	1,2	1,2
Ethylene glycol monoethyl ether acetate (Cellosolve acetate)	Combustible liquid	UN1172	None	173.118a	None	No limit	No limit	1,2	1,2
Ethylene glycol monomethyl ether (methyl Cellosolve)	Combustible liquid	UN1188	None	173.118a	None	No limit	No limit	1,2	1,2
Ethylene glycol monomethyl ether acetate (methyl Cellosolve acetate)	Combustible liquid	UN1189	None	173.118a	None	No limit	No limit	1,2	1,2
Ethylene imine, inhibited	Flammable liquid	UN1185	Flammable liquid and Poison	None	173.139	Forbidden	5 pints	1,2	1
Ethylene oxide	Flammable liquid	UN1040	Flammable liquid	None	173.124	Forbidden	See 173.124	1,2	1	Segregation same as for flammable gases
Ethyl ether	Flammable liquid	UN1155	Flammable liquid	None	173.119	Forbidden	10 gallons	1,3	5
Ethyl formate	Flammable liquid	UN1190	Flammable liquid	173.118	173.119	1 quart	10 gallons	1,3	4
Ethylhexaldehyde	Combustible liquid	UN1191	None	173.118a	None	No limit	No limit	1,2	1,2
Ethyl hydroperoxide (explodes above 100 deg C)	Forbidden								
Ethyl lactate	Combustible liquid	UN1192	None	173.118a	None	No limit	No limit	1,2	1,2
Ethyl mercaptan	Flammable liquid	UN2363	Flammable liquid	None	173.141	Forbidden	10 gallons	1,2	1	Segregation same as for flammable gases
Ethyl methyl ether	Flammable liquid	UN1039	Flammable liquid	None	173.119	Forbidden	10 gallons	1,3	1
Ethyl methyl ketone	Flammable liquid	UN1193	Flammable liquid	173.118	173.119	1 quart	10 gallons	1,2	1
Ethyl nitrate (nitric ether)	Flammable liquid	NA1993	Flammable liquid	173.118a	173.119	Forbidden	Forbidden	1,2	1
Ethyl nitrite (nitrous ether)	Flammable liquid	UN1194	Flammable liquid	None	173.119	Forbidden	Forbidden	1,3	5
Ethyl perchlorate	Forbidden								
Ethyl phenyl dichlorosilane	Corrosive material	UN2435	Corrosive	None	173.290	Forbidden	10 gallons	1	5
Ethyl phosphonous dichloride. See Pyroforic liquid, n.o.s.									
Ethyl phosphonothioic dichloride, anhydrous	Corrosive material	NA 1760	Corrosive	173.244	173.245 178.245a	1 quart	1 quart	1	4
Ethyl phosphorodichloridate	Corrosive material	NA1760	Corrosive	173.244	173.245 178.245a	1 quart	1 quart	1	4
Ethyl propionate	Flammable liquid	UN1195	Flammable liquid	173.118	173.119	1 quart	10 gallons	1,2	1
Ethyl silicate (tetraethyl orthosilicate)	Combustible liquid	UN1292	None	173.118a	None	No limit	No limit	1,2	1,2
Ethyl trichlorosilane	Flammable liquid	UN1196	Flammable liquid	None	173.135	Forbidden	5 pints	1,2	1
Etiologic agent, n.o.s.	Etiologic agent	NA2814	Etiologic agent	173.386	173.387	See 173.386	4 liters		
Explosive auto alarm	Class C explosive		Explosive C	None	173.111	50 pounds	150 pounds	1,2	1,2	Not permitted except under specific conditions approved by the Department

Chemical Hazard Communication Guidebook

§172.101   **Title 49—Transportation**

§172.101 Hazardous Materials Table

(1)	(2)	(3)	(3A)	(4)	(5) Packaging		(6) Maximum net quantity in one package		(7) Water shipments		
+/A/W	Hazardous materials descriptions and proper shipping names	Hazard class	Identification number	Label(s) required (if not excepted)	(a) Exceptions	(b) Specific requirements	(a) Passenger carrying aircraft or railcar	(b) Cargo only aircraft	(a) Cargo vessel	(b) Passenger vessel	(c) Other requirements
	Explosive bomb	Class A explosive		Explosive A	None	173.56	Forbidden	Forbidden	1,2	5	Magazine stowage authorized. No other cargo may be stowed in the same hold with these items
	Explosive cable cutter	Class C explosive		Explosive C	None	173.102	50 pounds	150 pounds	1,3	1,3	
	Explosive, forbidden. See Sec. 173.51	Forbidden									
	Explosive mine	Class A explosive		Explosive A	None	173.56	Forbidden	Forbidden	1,2	5	Magazine stowage authorized. No other cargo may be stowed in the same hold with these items
	Explosive, new approval, and evaluation. See 173.86										
	Explosive pest control devices	Class C explosive		Explosive C	None	173.100	50 pounds	150 pounds	1,3	1,3	
	Explosive power device, Class B	Class B explosive		Explosive B	None	173.94	Forbidden	150 pounds	1,2	5	
	Explosive power device, Class C	Class C explosive		Explosive C	None	173.102	50 pounds	150 pounds	1,3	1,3	
	Explosive projectile	Class A explosive		Explosive A	None	173.56	Forbidden	Forbidden	1,2	5	Magazine stowage authorized. No other cargo may be stowed in the same hold with this material
	Explosive release device	Class C explosive		Explosive C	None	173.102	50 pounds	150 pounds	1,3	1,3	
	Explosive rivet	Class C explosive		Explosive C	None	173.100	50 pounds	150 pounds	1,2	1,2	
	Explosive, sample for laboratory examination				173.86		Forbidden	See 173.86			
	Explosive torpedo	Class A explosive		Explosive A	None	173.56	Forbidden	Forbidden	1,2	5	Magazine stowage authorized. No other cargo may be stowed in the same hold with this material
	Extract, liquid, flavoring	Flammable liquid	UN1197	Flammable liquid	173.118	173.119	1 quart	10 gallons	1,2	1	
	Ferric arsenate, solid	Poison B	UN1606	Poison	173.364	173.365	50 pounds	200 pounds	1,2	1,2	
	Ferric arsenite, solid	Poison B	UN1607	Poison	173.364	173.365	50 pounds	200 pounds	1,2	1,2	
A	Ferric chloride, solid, *anhydrous*	ORM-B	UN1773	None	173.505	173.510	25 pounds	100 pounds	1,2	1,2	
	Ferric chloride solution	Corrosive material	UN2582	Corrosive	173.244	173.245	1 quart	10 quarts	1,2	1,2	
	Ferric nitrate	Oxidizer	UN1466	Oxidizer	173.153	173.182	25 pounds	100 pounds	1,2	1,2	
W	Ferrochrome, exothermic	ORM-C		None	173.505	173.985			1	1	
W	Ferromanganese, exothermic. *See Ferrochrome, exothermic*										
W	Ferrophosphorus	ORM-A		None	173.505	173.635			1,2	1,2	Keep dry. Stow away from living quarters

152

Chapter I—Research and Special Programs Administration §172.101

	Article	Class	ID No.	Label(s) required	Packaging exceptions	Packaging specific requirements	Max pkg passenger	Max pkg cargo	Passenger vessel	Cargo vessel	Other requirements
AW	Ferrosilicon, containing 30% or more but not more than 70% silicon	ORM-A	UN1408	None	173.505	173.645	Forbidden	25 pounds	1,2	1,2	Keep dry. Stow away from living quarters. Segregation same as for flammable solids labeled Dangerous When Wet.
A	Ferrous arsenate, solid	Poison B	UN1608	Poison	173.364	173.365	50 pounds	200 pounds	1,2	1,2	
	Ferrous chloride, solid	ORM-B	NA1759	None	173.505	173.510	No limit	No limit	1,2	1,2	
	Ferrous chloride, solution	Corrosive material	NA1760	Corrosive	173.244	173.245	1 quart	5 gallons	1,2	1,2	
	Fertilizer ammoniating solution containing free ammonia (more than 25.3 p.s.i.g.)	Nonflammable gas	UN1043	Nonflammable gas	173.306	173.304 173.314	Forbidden	300 pounds	1,2	4	
W	Fibers (jute, hemp, flax, sisal, coir, kapok, and similar vegetable fibers)	ORM-C	NA1372	None	173.505	173.965				1,2	Stow away from animal or vegetable oils. Segregation same as for flammable solids.
	Film (nitrocellulose)	Flammable solid	NA1324	Flammable solid	None	173.177	50 pounds	200 pounds	1,3	1,3	Stow away from other flammable cargo or substances
	Firecracker. See Fireworks, common or special										
	Firecracker salute. See Fireworks, common or special										
	Fire extinguisher	Nonflammable gas	UN1044	Nonflammable gas	173.308		150 pounds	300 pounds	1,2	1,2	
	Fire extinguisher charge containing sulfuric acid	Corrosive material	UN1774	Corrosive	173.261		1 quart	1 gallon	1,2	1,2	
	Fireworks, common	Class C explosive		Explosive C	None	173.100 173.108	50 pounds	200 pounds	1,3	1,3	Passenger vessels in metal lockers only
	Fireworks, exhibition display piece. See Fireworks, special										
	Fireworks, special	Class B explosive	NA2216	Explosive B	None	173.88 173.91			3	3	Passenger vessels in metal lockers only. Toy torpedoes must not be packed with other special fireworks
W	Fish meal or fish scrap containing 6% to 12% water	ORM-C		None	173.505	173.995	Forbidden	200 pounds	1,2	1,2	Segregation same as for flammable solids. Separate from flammable gases or liquids, oxidizing materials, or organic peroxides. *Use double strip stowage* for cargo 6-12 percent moisture containing not more than 12 percent fat. *Use single strip stowage* for cargo 6-12 percent moisture containing 12-15 percent fat.
	Fish meal or fish scrap containing less than 6% or more than 12% water	Flammable solid	NA1374	Flammable solid	None	173.171	Forbidden	Forbidden	1,2	1,2	
	Fissile radioactive material. See Radioactive material, fissile										
	Flame retardant compound liquid	Corrosive material	NA1760	Corrosive	173.244	173.291	1 quart	10 gallons	1,2	1,2	Separate from flammable gases or liquids, oxidizing materials, or organic peroxides
+	Flammable gas n.o.s. *See* Compressed gas, n.o.s.										
	Flammable liquid, corrosive, n.o.s.	Flammable liquid	UN2924	Flammable liquid and Corrosive	None	173.119	1 quart	1 quart	1,2	1	
	Flammable liquid, n.o.s.	Flammable liquid	UN1993	Flammable liquid	173.118	173.119	1 quart	10 gallons	1,2	1	
	Flammable liquid, poisonous, n.o.s.	Flammable liquid	UN1992	Flammable liquid and Poison	None	173.119	1 quart	10 gallons	1,2	1	

153

Chemical Hazard Communication Guidebook

§172.101 Title 49—Transportation

§172.101 Hazardous Materials Table

(1) +/A/W	(2) Hazardous materials descriptions and proper shipping names	(3) Hazard class	(3A) Identification number	(4) Label(s) required (if not excepted)	(5) Packaging (a) Exceptions	(5) Packaging (b) Specific requirements	(6) Maximum net quantity in one package (a) Passenger carrying aircraft or railcar	(6) Maximum net quantity in one package (b) Cargo only aircraft	(7) Water shipments (a) Cargo vessel	(7) Water shipments (b) Passenger vessel	(7) Water shipments (c) Other requirements
	Flammable solid, corrosive, n.o.s.	Flammable solid	UN2925	Flammable solid and Corrosive	173.153	173.154	25 pounds	25 pounds	1	4	
	Flammable solid, n.o.s.	Flammable solid	UN1325	Flammable solid	173.153	173.154	25 pounds	25 pounds	1,2	1,2	
	Flammable solid, poisonous, n.o.s.	Flammable solid	UN2926	Flammable solid and Poison	173.153	173.154	25 pounds	25 pounds	1,2	1	
	Flare. See Fireworks, common										
	Flare, airplane. See Fireworks, special										
	Flash cartridge. See Fireworks, special or Low explosives										
	Flash cracker. See Fireworks, common or special										
	Flash powder. See Fireworks, special or Low explosives										
	Flexible linear shaped charge, metal clad	Class C explosive		Explosive C	None	173.104	50 pounds	300 pounds	1,3	1,3	
	Flowers of sulfur. See Sulfur										
	Flue dust, poisonous	Poison B	NA2811	Poison	173.364	173.368	50 pounds	200 pounds	1,2	1,2	
	Fluoboric acid	Corrosive material	UN1775	Corrosive	173.244	173.283	1 quart	1 gallon	1,2	1,2	
	Fluoric acid. See Hydrofluoric acid										
	Fluorine	Nonflammable gas	UN1045	Poison and Oxidizer	None	173.302	Forbidden	Forbidden	1	5	Stow in well ventilated space away from organic materials
	Fluorophosphoric acid, anhydrous. See Monofluorophosphoric acid, anhydrous										
	Fluorosilicic acid. See Hydrofluorosilicic acid										
	Fluorosulfonic acid or Fluosulfonic acid	Corrosive material	UN1777	Corrosive	None	173.274	Forbidden	1 gallon	1	5	Keep dry
	Fluosilicic acid	Corrosive material	UN1778	Corrosive	None	173.265	1 quart	1 gallon	1,2	1,2	
	Forbidden explosives. See 173.51	Forbidden									
	Forbidden materials. See 173.21	Forbidden									
	Formaldehyde solution (Flash point more than 141 deg F.; in containers of 110 gallons or less)	ORM-A	UN2209	None	173.505	173.510	10 gallons	55 gallons	1,2	4	
AW	Formaldehyde solution (flash point not more than 141 deg F.; in containers over 110 gallons)	Combustible liquid	UN1198	None	173.118a	None	10 gallons	55 gallons	1,2	1,2	

154

Chapter I—Research and Special Programs Administration §172.101

AW	Formaldehyde solution (flash point not more than 141 deg F.; in containers of 110 gallons or less)	ORM-A	UN1198	None	173.505	173.510	10 gallons	55 gallons	1,2	4
	Formaldehyde solution (flash point more than 141 deg F.; in containers over 110 gallons)	Combustible liquid	UN2209	None	173.118a	None	10 gallons	55 gallons	1,2	1,2
	Formalin. See Formaldehyde solution									
	Formic acid	Corrosive material	UN1779	Corrosive	173.244	173.289	1 quart	5 gallons	1,2	1,2
	Formic acid solution	Corrosive material	UN1779	Corrosive	173.244	173.289	1 quart	5 gallons	1,2	1,2
	Fuel, aviation, turbine engine	Flammable liquid	UN1863	Flammable liquid	173.118	173.119	1 quart	10 gallons	1,2	1
	Fuel, aviation, turbine engine	Combustible liquid	UN1863	None	173.118a	None	No limit	No limit	1,2	1,2
	Fuel oil	Combustible liquid	NA1993	None	173.118a	None	No limit	No limit	1,2	1,2
	Fuel oil, diesel. See Fuel oil									
	Fuel oil, No. 1, 2, 4, 5 or 6	Combustible liquid	NA1993	None	173.118a	None	No limit	No limit	1,2	1,2
	Fulminate of mercury (dry)	Forbidden								
	Fulminate of mercury, wet. See Initiating explosive									
	Fulminating gold	Forbidden								
	Fulminating mercury	Forbidden								
	Fulminating platinum	Forbidden								
	Fulminating silver	Forbidden								
	Fulminic acid	Forbidden								
	Fumaryl chloride	Corrosive material	UN1780	Corrosive	173.244	173.245	1 quart	1 quart	1	1
	Fumigant. See 173.152(a) Note 1									
	Furan	Flammable liquid	UN2389	Flammable liquid	173.118	173.119	1 quart	10 gallons	1,2	1
	Furfural	Combustible liquid	UN1199	None	173.118a	None	No limit	No limit	1,2	1
	Fusee (railway or highway)	Flammable solid	NA1325	Flammable solid	None	173.154a	50 pounds	200 pounds	1,3	1,3
	Fuse igniter	Class C explosive		Explosive C	None	173.106	50 pounds	150 pounds	1,3	1,3
	Fuse, instantaneous	Class C explosive		Explosive C	173.100		50 pounds	150 pounds	1,2	1,2
	Fuse lighter	Class C explosive		Explosive C	None	173.106	50 pounds	150 pounds	1,3	1,3
	Fuel oil	Combustible liquid	UN1201	None	173.118a	None	No limit	No limit	1,2	1,2
	Fuse, mild detonating, metal clad	Class C explosive		Explosive C	None	173.104	50 pounds	300 pounds	1,2	1,2
	Fuse, safety	Class C explosive		Explosive C	173.100	173.100	50 pounds	300 pounds	1,2	1,2
	Fuze, combination	Class C explosive		Explosive C	None	173.105	50 pounds	150 pounds	1,3	1,3
	Fuze, detonating	Class A explosive		Explosive A	None	173.69	Forbidden	Forbidden	6	5
	Fuze, detonating, Class C explosive	Class C explosive		Explosive C	None	173.113	50 pounds	150 pounds	1,3	1,3
	Fuze, detonating, radioactive	Class A explosive		Explosive A	None	173.69	Forbidden	Forbidden	6	5

Notes (sideways):
- Glass carboys in hampers not permitted under deck
- Glass carboys not permitted

155

Chemical Hazard Communication Guidebook

§172.101   Title 49—Transportation

§172.101 Hazardous Materials Table

(1) +/A/W	(2) Hazardous materials descriptions and proper shipping names	(3) Hazard class	(3A) Identification number	(4) Label(s) required (if not excepted)	(5) Packaging (a) Exceptions	(5) Packaging (b) Specific requirements	(6) Maximum net quantity in one package (a) Passenger carrying aircraft or railcar	(6) Maximum net quantity in one package (b) Cargo only aircraft	(7) Water shipments (a) Cargo vessel	(7) Water shipments (b) Passenger vessel	(7) Water shipments (c) Other requirements
	Fuze, percussion	Class C explosive		Explosive C	None	173.105	50 pounds	150 pounds	1,3	1,3	
	Fuze, time	Class C explosive		Explosive C	None	173.105	50 pounds	150 pounds	1,3	1,3	
	Fuze, tracer	Class C explosive		Explosive C	None	173.105	50 pounds	150 pounds	1,3	1,3	
	Galactan trinitrate	Forbidden									
	Gallium metal, liquid	ORM-B	UN2803	None	None	173.861	Forbidden	Forbidden	1	5	None
	Gallium metal, solid	ORM-B	UN2803	None	None	173.862	40 pounds	40 pounds	1,3	1	Shade from radiant heat
	Gas cylinder, empty. See Cylinder, empty										
	Gas drips, hydrocarbon	Combustible liquid	UN1864	None	173.118a	None	No limit	No limit	1,2	1,2	
	Gas drips, hydrocarbon	Flammable liquid	UN1864	Flammable liquid	173.118	173.119	1 quart	10 gallons	1,2	1	
	Gas identification set	Poison A	NA9035	Poison gas	None	173.331	Forbidden	Forbidden	1	5	
	Gas identification set	Irritating material	NA9035	Irritant	None	173.331	Forbidden	Forbidden	1	5	
	Gas mine. See Explosive mine										
	Gasohol (gasoline mixed with ethyl alcohol containing 20% maximum alcohol). See Gasoline										
	Gasoline (including casing-head and natural)	Flammable liquid	UN1203	Flammable liquid	173.118	173.119	1 quart	10 gallons	1,2	4	
	Gelatin Dynamite. See High explosive										
	Germane	Poison A	UN2192	Poison gas and Flammable gas	None	173.328	Forbidden	Forbidden	1	5	Segregation same as for flammable gases
	Glycerol-1,3-dinitrate	Forbidden									
	Glycerol monogluconate trinitrate	Forbidden									
	Glycerol monolactate trinitrate	Forbidden									
	Grenade without bursting charge. (With incendiary material)	Class B explosive		Explosive B	None	173.91	Forbidden	Forbidden	3	3	Passenger vessels in metal lockers only
	Grenade without bursting charge. (With smoke charge) (Smoke grenade)	Class C explosive		Explosive C	None	173.108	50 pounds	150 pounds	1,3	1,3	
	Grenade without bursting charge. (With Poison A gas charge)	Poison A	NA2016	Poison gas	None	173.330	Forbidden	Forbidden			See correct shipping name of applicable Poison A material for stowage, special handling, and special segregation requirements

156

Chapter I—Research and Special Programs Administration §172.101

Grenade without bursting charge. (With Poison B charge)	Poison B	NA2016	Poison	None	173.350	Forbidden	Forbidden		See correct shipping name of applicable Poison B material for stowage, special handling, and special segregation requirements
Grenade, empty, primed	Class C explosive		None	None	173.107	50 pounds	150 pounds	1,3 1,3	
Grenade, hand or rifle, explosive (with or without gas, smoke, or incendiary material)	Class A explosive		Explosive A	None	173.56	Forbidden	Forbidden	1,2 5	No other cargo may be stowed in the same hold with these items
Grenade, tear gas	Irritating material	NA2017	Irritant	None	173.385	Forbidden	75 pounds	1,2 1	
Guanidine nitrate	Oxidizer	UN1467	Oxidizer	173.153	173.182	25 pounds	100 pounds	1,2 1,2	Separate from nitro-compounds, chlorates, and acids
Guanyl nitrosamino guanylidene hydrazine. See Initiating explosive									
Guanyl nitrosamino guanylidene hydrazine (dry)	Forbidden								
Guanyl nitrosamino guanyl tetrazene. See Initiating explosive									
Guided missile, without warhead. See Rocket motor, Class A explosive or Rocket motor, Class B explosive									
Guided missile with warhead. See Rocket ammunition with explosive, illuminating, gas, incendiary, or smoke projectile									
Guncotton. See High explosive									
Guthion mixture, liquid. See Azinphos methyl mixture, liquid									
Hafnium metal, dry. (See 173.214 Note 3)	Flammable solid	UN2545	Flammable solid	None	173.214	Forbidden	75 pounds	1 5	
Hafnium metal, wet	Flammable solid	UN1326	Flammable solid	None	173.214	Forbidden	150 pounds	1,2 5	
Hand signal device	Class C explosive		Explosive C	None	173.108	50 pounds	200 pounds	1,2 1,2	
Hazardous substance, liquid or solid, n.o.s. or ORM-E, liquid or solid, n.o.s.	ORM-E	NA9188	None	None	173.1300	No limit	No limit	1,2 1,2	
Hazardous waste, liquid or solid, n.o.s.	ORM-E	NA9189	None	None	173.1300	Forbidden	550 pounds	1,2 1,2	
Hazardous waste, meeting the definition of a hazard class other than ORM-E See 172.101(c)(10)									
Heater for refrigerator car, liquid fuel type (containing fuel)	Flammable liquid	NA1993	Flammable liquid	173.146	173.302	Forbidden	Forbidden	1,2 1	
Helium or Helium, compressed	Nonflammable gas	UN1046	Nonflammable gas	173.306	173.302 173.314	150 pounds	300 pounds	1,2 1,2	
Helium-oxygen mixture	Nonflammable gas	NA1980	Nonflammable gas	173.306	173.302	150 pounds	300 pounds	1,2 1,2	
Helium, refrigerated liquid (cryogenic liquid)	Nonflammable gas	UN1963	Nonflammable gas	173.320	173.316 173.318	100 pounds	1,100 pounds	1,3 1,3	
Heptane	Flammable liquid	UN1206	Flammable liquid	173.118	173.119	1 quart	10 gallons	1,2 1	
Hexachlorocyclopentadiene	Corrosive material	UN2646	Corrosive	173.244	173.245	1 quart	10 gallons	1,2 1,2	
W Hexachloroethane	ORM-A	NA9037	None	178.505	173.650	Forbidden	Forbidden	1,2 1,2	
Hexadecyltrichlorosilane	Corrosive material	UN1781	Corrosive	None	173.280	Forbidden	10 gallons	1 1	Keep dry
Hexadiene	Flammable liquid	UN2458	Flammable liquid	None	173.119	Forbidden	10 gallons	1,2 5	

157

Chemical Hazard Communication Guidebook

§172.101   Title 49—Transportation

§172.101 Hazardous Materials Table

(1) +/A/W	(2) Hazardous materials descriptions and proper shipping names	(3) Hazard class	(3A) Identification number	(4) Label(s) required (if not excepted)	(5) Packaging (a) Exceptions	(5) Packaging (b) Specific requirements	(6) Maximum net quantity in one package (a) Passenger carrying aircraft or railcar	(6) Maximum net quantity in one package (b) Cargo only aircraft	(7) Water shipments (a) Cargo vessel	(7) Water shipments (b) Passenger vessel	(7) Water shipments (c) Other requirements
	Hexaethyl tetraphosphate and compressed gas mixture	Poison A	UN1612	Poison gas	None	173.334	Forbidden	Forbidden	1	5	Shade from radiant heat
	Hexaethyl tetraphosphate, liquid	Poison B	UN1611	Poison	None	173.358	Forbidden	1 quart	1	4	
	Hexaethyl tetraphosphate mixture, dry (containing more than 2% hexaethyl tetraphosphate)	Poison B	NA2783	Poison	None	173.377	Forbidden	200 pounds	1,2	5	
	Hexaethyl tetraphosphate mixture, dry (containing not more than 2% hexaethyl tetraphosphate)	Poison B	NA2783	Poison	173.377	173.377	50 pounds	200 pounds	1,2	4	
	Hexaethyl tetraphosphate mixture, liquid (containing more than 25% hexaethyl tetraphosphate)	Poison B	UN2783	Poison	None	173.359	Forbidden	1 quart	1,2	5	
	Hexaethyl tetraphosphate mixture, liquid (containing not more than 25% hexaethyl tetraphosphate)	Poison B	UN2783	Poison	173.359	173.359	1 quart	1 quart	1,2	4	
	Hexafluorophosphoric acid	Corrosive material	UN1782	Corrosive	None	173.275	Forbidden	1 gallon	1,2	1,2	
	Hexafluoropropylene	Nonflammable gas	UN1858	Nonflammable gas	173.306	173.304 173.314 173.315	150 pounds	300 pounds	1	4	
	Hexafluoropropylene oxide	Nonflammable gas	NA1956	Nonflammable gas	173.306	173.304 173.314	150 pounds	300 pounds	1,2	1,2	
	Hexaldehyde	Flammable liquid	UN1207	Flammable liquid	173.118	173.119	1 quart	10 gallons	1,2	1,2	
	3,3,6,6,9,9-Hexamethyl-1,2,4,5-tetraoxocyclononane, technically pure. See Organic peroxide, solid, n.o.s.		UN2165								
	3,3,6,6,9,9-Hexamethyl-1,2,4,5-tetraoxocyclononane, not more than 52% with inert solid. See Organic peroxide, solid, n.o.s.		UN2166								
	3,3,6,6,9,9-Hexamethyl-1,2,4,5-tetraoxocyclononane, not more than 52% in solution. See Organic peroxide, liquid or solution, n.o.s.		UN2167								
	Hexamethylenediamine, solid	Corrosive material	UN2280	Corrosive	173.244	173.245b	25 pounds	100 pounds	1,2	1,2	
	Hexamethylenediamine, solution	Corrosive material	UN1783	Corrosive	173.244	173.292	1 quart	10 gallons	1,2	1,2	

158

Appendix B: Regulations Reference Guide

Chapter I—Research and Special Programs Administration §172.101

Material	Hazard Class	ID No.	Packaging (Exceptions)	Packaging (Specific)	Max Qty Passenger	Max Qty Cargo	Label	Water Shipment	Other Requirements
Hexamethyleneimine	Flammable liquid	UN2493	Flammable liquid and Corrosive	None	173.119	1 quart	1 gallon	1,2	1
Hexamethylene triperoxide diamine (dry)	Forbidden								
Hexamethylol benzene hexanitrate (dry)	Forbidden								
Hexane	Flammable liquid	UN1208	Flammable liquid	173.118	173.119	1 quart	10 gallons	1,3	4
2,2',4,4',6,6'-Hexanitro-3,3'-dihydroxyazobenzene (dry)	Forbidden								
Hexanitroazoxy benzene	Forbidden								
2,2',3',4,4',6-Hexanitrodiphenylamine	Forbidden								
2,3',4,4',6,6'-Hexanitrodiphenylether	Forbidden								
N,N'-(hexanitrodiphenyl)ethylene dinitramine (dry)	Forbidden								
Hexanitrodiphenyl urea	Forbidden								
Hexanitroethane	Forbidden								
Hexanitrooxanilide	Forbidden								
Hexanoic acid	Corrosive material	NA1760	Corrosive	173.244	173.245	1 quart	10 gallons	1,2	1,2
Hexyltrichlorosilane	Corrosive material	UN1784	Corrosive	None	173.280	Forbidden	10 gallons	1	1
High explosive	Class A explosive			173.65	173.61 to 173.87	Forbidden	Forbidden	6	5	Keep dry
High explosive, liquid	Class A explosive			None	173.62	Forbidden	Forbidden	6	5
Hydraulic accumulator. See Accumulator, pressurized									
Hydrazine, anhydrous	Flammable liquid	UN2029	Flammable liquid and Poison	None	173.276	Forbidden	5 pints	1	5	Segregation same as for corrosives
Hydrazine, aqueous solution	Corrosive material	UN2030	Corrosive	None	173.276	Forbidden	5 pints	1	5
Hydrazine azide	Forbidden								
Hydrazine chlorate	Forbidden								
Hydrazine dicarbonic acid diazide	Forbidden								
Hydrazine perchlorate	Forbidden								
Hydrazine selenate	Forbidden								
Hydriodic acid	Corrosive material	UN1787	Corrosive	173.244	173.245	1 quart	1 gallon	1	1	Glass carboys not permitted on passenger vessel
Hydrobromic acid, more than 49% strength	Corrosive material	UN1788	Corrosive	None	173.262	Forbidden	Forbidden	1	1	Glass carboys not permitted on passenger vessel
Hydrobromic acid not more than 49% strength	Corrosive material	UN1788	Corrosive	173.244	173.262	1 quart	1 gallon	1	1
Hydrocarbon gas, liquefied	Flammable gas	UN1965	Flammable gas	173.306	173.304 173.314	Forbidden	300 pounds	1,2	1
Hydrocarbon gas, nonliquefied	Flammable gas	UN1964	Flammable gas	173.306	173.302	Forbidden	300 pounds	1,2	1
Hydrochloric acid	Corrosive material	UN1789	Corrosive	173.244	173.263	1 quart	1 gallon	1	1	Glass carboys not permitted on passenger vessel
Hydrochloric acid, anhydrous. See Hydrogen chloride									
Hydrochloric acid mixture	Corrosive material	NA1789	Corrosive	173.244	173.263	1 quart	1 gallon	1	1	Glass carboys not permitted on passenger vessel
Hydrochloric acid solution, inhibited	Corrosive material	UN1789	Corrosive	173.244	173.263	1 quart	1 gallon	1	1	Glass carboys not permitted on passenger vessel

159

§172.101 Hazardous Materials Table

(1) +/A/W	(2) Hazardous materials descriptions and proper shipping names	(3) Hazard class	(3A) Identification number	(4) Label(s) required (if not excepted)	(5) Packaging (a) Exceptions	(5) Packaging (b) Specific requirements	(6) Maximum net quantity in one package (a) Passenger carrying aircraft or railcar	(6) (b) Cargo only aircraft	(7) Water shipments (a) Cargo vessel	(7) (b) Passenger vessel	(7) (c) Other requirements
	Hydrocyanic acid (prussic) solution (5% or more hydrocyanic acid)	Poison A	UN1613	Poison gas and Flammable gas	None	173.332	Forbidden	Forbidden	1	5	Shade from radiant heat. Aqueous solutions containing more than 20 percent hydrogen cyanide are not permitted in transportation by water. Segregation same as for flammable gases.
	Hydrocyanic acid, liquefied	Poison A	NA1051	Poison gas and Flammable gas	None	173.332	Forbidden	Forbidden	1	5	Segregation same as for flammable gases.
	Hydrocyanic acid (prussic), unstabilized	Forbidden									
+	Hydrocyanic acid solution, less than 5% hydrocyanic acid	Poison B	UN1613	Poison	None	173.351	Forbidden	25 pounds	1	5	Shade from radiant heat
	Hydrofluoric acid, anhydrous. See Hydrogen fluoride										
	Hydrofluoric acid solution	Corrosive material	UN1790	Corrosive	173.244	173.264	1 quart	1 gallon	1	4	
	Hydrofluoric and sulfuric acid mixture	Corrosive material	UN1786	Corrosive	None	173.290	Forbidden	1 gallon	1	5	
	Hydrofluoroboric acid. See Fluoboric acid										
	Hydrofluorosilicic acid or Hydrofluosilicic acid	Corrosive material	NA1778	Corrosive	None	173.265	1 quart	1 gallon	1,2	1,2	
+	Hydrogen or Hydrogen, compressed	Flammable gas	UN1049	Flammable gas	173.306	173.302 173.314 173.316 173.318 173.319	Forbidden	300 pounds	1,2	4	Stow away from living quarters
+	Hydrogen, refrigerated liquid (cryogenic liquid)	Flammable gas	UN1966	Flammable gas	None	173.304	Forbidden	Forbidden	5	5	
	Hydrogen bromide	Nonflammable gas	UN1048	Nonflammable gas	173.306	173.304	Forbidden	300 pounds	1	4	Stow away from living quarters
	Hydrogen chloride or Hydrogen chloride, anhydrous	Nonflammable gas	UN1050	Nonflammable gas	173.306	173.304	Forbidden	300 pounds	1	4	Stow away from living quarters
	Hydrogen chloride, refrigerated liquid	Nonflammable gas	UN2186	Nonflammable gas	None	173.314 173.315	Forbidden	300 pounds	1,2	4	Stow in well ventilated space
	Hydrogen fluoride	Corrosive material	UN1052	Corrosive	None	173.264	Forbidden	110 pounds	1	5	Segregation same as for nonflammable gases
	Hydrogen iodide solution. See Hydriodic acid										
	Hydrogen peroxide solution (40% to 52% peroxide)	Oxidizer	UN2014	Oxidizer	173.244	173.266	Forbidden	Forbidden	1	4	Shade from radiant heat. Separate from permanganates. Keep away from powdered metals
	Hydrogen peroxide solution (8% to 40% peroxide)	Oxidizer	UN2014	Oxidizer	173.244	173.266	1 quart	1 gallon	1,2	1	Shade from radiant heat. Separate from permanganates. Keep away from powdered metals

Chapter I—Research and Special Programs Administration §172.101

Article	Class	UN#	Label	(col)	(col)	Limit 1	Limit 2	(col)	(col)	Notes
Hydrogen peroxide solution (over 52% peroxide)	Oxidizer	UN2015	Oxidizer and Corrosive	None	173.266	Forbidden	Forbidden	1	5	Shade from radiant heat. Separate from permanganates. Keep away from powdered metals. Concentrations greater than 60% hydrogen peroxide not permitted on any vessel except under conditions approved by the Department.
+ Hydrogen selenide	Poison A	UN 2202	Poison gas & Flammable gas	None	173.328	Forbidden	Forbidden	1	5	Stow away from living quarters.
Hydrogen sulfate. See Sulfuric acid										
+ Hydrogen sulfide	Flammable gas	UN1053	Flammable gas and Poison	None	173.304 173.314	Forbidden	300 pounds	1	5	
Hydrosilicofluoric acid. See Hydrofluorosilicic acid										
Hydroxyl amine iodide	Forbidden									
Hypochlorite solution containing more than 7% available chlorine by weight	Corrosive material	UN1791	Corrosive	173.244	173.277	1 quart	4 gallons	1,2	1	Glass carboys in hampers not permitted under deck
△ Hypochlorite solution containing not more than 7% available chlorine by weight	ORM-B	NA1791	None	173.505	173.510	No limit	No limit			
Hyponitrous acid	Forbidden									
Igniter	Class C explosive		Explosive C	None	173.106	50 pounds	150 pounds	1,3	1,3	
Igniter cord	Class C explosive		Explosive C	None	173.100	50 pounds	150 pounds	1,3	1,3	
Igniter fuse, metal clad	Class C explosive		Explosive C	None	173.106	50 pounds	150 pounds	1,3	1,3	
Igniter, jet thrust (jato)	Class A explosive		Explosive A	None	173.79	Forbidden	Forbidden	6	5	
Igniter, jet thrust (jato)	Class B explosive		Explosive B	None	173.92	Forbidden	550 pounds	1,3	5	
Igniter, rocket motor	Class A explosive		Explosive A	None	173.79	Forbidden	Forbidden	6	5	
Igniter, rocket motor	Class B explosive		Explosive B	None	173.92	Forbidden	550 pounds	1,3	5	
Illuminating projectile. See Fireworks, special										
Iminobispropylamine	Corrosive material	UN2269	Corrosive	173.244	173.245	1 quart	10 gallons	1,2	1,2	
Infectious substance, human, n.o.s. See Etiologic Agent n.o.s.										
Initiating explosive (diazodinitrophenol)	Class A explosive		Explosive A	None	173.70	Forbidden	Forbidden	6	5	
Initiating explosive (fulminate of mercury)	Class A explosive		Explosive A	None	173.71	Forbidden	Forbidden	6	5	
Initiating explosive (guanyl nitrosamino guanylidene hydrazine)	Class A explosive		Explosive A	None	173.72	Forbidden	Forbidden	6	5	
Initiating explosive (lead azide, dextrinated type only)	Class A explosive		Explosive A	None	173.73	Forbidden	Forbidden	6	5	
Initiating explosive (lead mononitroresorcinate)	Class A explosive		Explosive A	None	173.70	Forbidden	Forbidden	6	5	
Initiating explosive barium styphnate, monohydrate, lead styphnate (lead trinitroresorcinate)	Class A explosive		Explosive A	None	173.74	Forbidden	Forbidden	6	5	
Initiating explosive (nitro mannite)	Class A explosive		Explosive A	None	173.75	Forbidden	Forbidden	6	5	

161

Chemical Hazard Communication Guidebook

§172.101   Title 49—Transportation

§172.101 Hazardous Materials Table

(1)	(2)	(3)	(3A)	(4)	(5) Packaging		(6) Maximum net quantity in one package		(7) Water shipments		
+/A/W	Hazardous materials descriptions and proper shipping names	Hazard class	Identification number	Label(s) required (if not excepted)	(a) Exceptions	(b) Specific requirements	(a) Passenger carrying aircraft or railcar	(b) Cargo only aircraft	(a) Cargo vessel	(b) Passenger vessel	(c) Other requirements
	Initiating explosive (nitrosoguanidine)	Class A explosive		Explosive A	None	173.76	Forbidden	Forbidden	6	5	
	Initiating explosive (pentaerythrite teranitrate)	Class A explosive		Explosive A	None	173.77	Forbidden	Forbidden	6	5	
	Initiating explosive (tetrazene (guanyl nitrosamine guanyl tetrazene))	Class A explosive		Explosive A	None	173.78	Forbidden	Forbidden	6	5	
	Initiating explosives (dry)	Forbidden									
	Ink	Combustible liquid	UN1210	None	173.118a	None	No limit	No limit	1,2	1,2	
	Ink	Flammable liquid	UN1210	Flammable liquid	173.118	173.144	1 quart	10 gallons	1,2	1	
	Inositol hexanitrate (dry)	Forbidden									
	Insecticide, dry, n.o.s.	Poison B	NA2588	Poison	173.364	173.365	50 pounds	200 pounds	1,2	1,2	Shade from radiant heat
	Insecticide, liquefied gas (containing no Poison A or B material)	Nonflammable gas	NA1968	Nonflammable gas	173.306	173.304	150 pounds	300 pounds	1,3	1,3	
	Insecticide, liquefied gas, containing Poison A material or Poison B material	Poison A	NA1967	Poison gas	None	173.329 173.334	Forbidden	Forbidden	1	5	
	Insecticide, liquid, n.o.s.	Combustible liquid	NA1993	None	173.118a	None	No limit	No limit	1,2	1,2	
	Insecticide, liquid, n.o.s.	Flammable liquid	NA1993	Flammable liquid	173.118	173.119	1 quart	10 gallons	1,2	1	
	Insecticide, liquid, n.o.s.	Poison B	NA2902	Poison	173.345	173.346	1 quart	55 gallons	1,2	1,2	
	Inulin trinitrate (dry)	Forbidden									
	Iodine azide (dry)	Forbidden									
	Iodine monochloride	Corrosive material	UN1792	Corrosive	None	173.293	Forbidden	1 quart	1	5	Keep dry
	Iodine pentafluoride	Oxidizer	UN2495	Oxidizer and Poison	None	173.246	Forbidden	100 pounds	1	1	Keep dry
	Iodoxy compounds (dry)	Forbidden									
	Indium nitratopentamine indium nitrate	Forbidden									
	Iron chloride, solid. See Ferric chloride, solid										
	Iron mass or sponge, not properly oxidized	Flammable solid	NA1383	Flammable solid	None	173.174	Forbidden	Forbidden	1,2	5	Separate from flammable gases or liquids, oxidizing materials, or organic peroxides
	Iron mass or sponge, spent	Flammable solid	UN1376	Flammable solid	None	173.174	Forbidden	Forbidden	1,2	5	Separate from flammable gases or liquids, oxidizing materials, or organic peroxides
	Iron oxide, spent. See Iron mass or sponge, spent										
	Iron sesquichloride, solid. See Ferric chloride										

162

Appendix B: Regulations Reference Guide

Chapter I—Research and Special Programs Administration §172.101

								Stow away from living quarters
Irritating agent, n.o.s.	Irritating material	NA1693	Irritant	None	173.382	Forbidden	75 pounds	1	1
Isobutane or Liquefied petroleum gas. See Liquefied petroleum gas									
Isobutyl acetate	Flammable liquid	UN1213	Flammable liquid	173.118	173.119	1 quart	10 gallons	1,2	1
Isobutylamine	Flammable liquid	UN1214	Flammable liquid	173.118	173.119	1 quart	10 gallons	1,2	1
Isobutylene or Liquefied petroleum gas. See Liquefied petroleum gas									
Isobutyric acid	Corrosive material	UN2529	Corrosive	173.244	173.245	1 quart	10 gallons	1,2	1,2
Isobutyric anhydride	Corrosive material	UN2530	Corrosive	173.244	173.245	1 quart	10 gallons	1,2	1,2
Isomanyl peroxide, *technically pure or* Isomanyl peroxide, *in solution. See* Organic peroxide, liquid or solution, n.o.s.		UN2128							
Isooctane	Flammable liquid	UN1262	Flammable liquid	173.118	173.119	1 quart	10 gallons	1,2	1
Isooctene	Flammable liquid	UN1216	Flammable liquid	173.118	173.119	1 quart	10 gallons	1,3	4
Isopentane	Flammable liquid	UN1265	Flammable liquid	173.118	173.119	Forbidden	10 gallons	1,3	4
Isopentanoic acid	Corrosive material	NA1760	Corrosive	173.244	173.245	1 quart	10 gallons	1,2	1,2
Isoprene	Flammable liquid	UN1218	Flammable liquid	173.118	173.119	Forbidden	10 gallons	1,3	4
Isopropanol	Flammable liquid	UN1219	Flammable liquid	173.118	173.125	1 quart	10 gallons	1,2	1
Isopropyl acetate	Flammable liquid	UN1220	Flammable liquid	173.118	173.119	1 quart	10 gallons	1,2	1
Isopropyl acid phosphate, solid	Corrosive material	UN1793	Corrosive	173.244	173.245b	25 pounds	100 pounds	1,2	1,2
Isopropyl alcohol. See Isopropanol									
Isopropylamine	Flammable liquid	UN1221	Flammable liquid	None	173.119	Forbidden	10 gallons	1,3	5
Isopropyl mercaptan	Flammable liquid	NA 2402	Flammable liquid	None	173.141	Forbidden	10 gallons	1,3	5
Isopropyl nitrate	Flammable liquid	UN1222	Flammable liquid	173.118	173.119	1 quart	10 gallons	1,2	1
Isopropyl percarbonate, stabilized	Organic peroxide	NA2134	Organic peroxide	None	173.292	Forbidden	Forbidden	5	5
Isopropyl percarbonate, unstabilized	Organic peroxide	NA2133	Organic peroxide	None	173.218	Forbidden	Forbidden	5	5
Isopropyl peroxydicarbonate, *technically pure. See* Isopropyl percarbonate, unstabilized		UN1793							
Isopropyl peroxydicarbonate, *not more than 52% in solution. See* Organic peroxide, liquid or solution, n.o.s.		UN2134							
Isopropyl phosphoric acid, solid. *See* Isopropyl acid phosphate, solid									
Isothiocyanic acid (polymerization hazard)	Forbidden								
Jet thrust igniter. *See* Igniter, jet thrust									
Jet thrust unit (*jato*)	Class A explosive		Explosive A	None	173.79	Forbidden	Forbidden	6	5

163

Chemical Hazard Communication Guidebook

§172.101   Title 49—Transportation

§172.101 Hazardous Materials Table

(1)	(2) Hazardous materials descriptions and proper shipping names	(3) Hazard class	(3A) Identification number	(4) Label(s) required (if not excepted)	(5) Packaging (a) Exceptions	(5) Packaging (b) Specific requirements	(6) Maximum net quantity in one package (a) Passenger carrying aircraft or railcar	(6) (b) Cargo only aircraft	(7) Water shipments (a) Cargo vessel	(7) (b) Passenger vessel	(7) (c) Other requirements
	Jet thrust unit (jato)	Class B explosive	UN1223	Explosive B	None	173.92	Forbidden	550 pounds	1,3	5	
	Kerosene	Combustible liquid		None	173.118a	None	No limit	No limit	1,2	1,2	
	Lacquer base or Lacquer chips, plastic (wet with alcohol or solvent)	Flammable liquid	UN1263	Flammable liquid	173.118	173.127	1 quart	25 pounds	1,2	1	
	Lacquer base, or Lacquer chips, dry	Flammable solid	NA2557	Flammable solid	173.153	173.175	25 pounds	100 pounds	1	1	
	Lauroyl peroxide	Organic peroxide	UN2124	Organic peroxide	173.153	173.157 173.158	2 pounds	25 pounds	1,2	1	
	Lauroyl peroxide, not more than 42%, stable dispersion, in water. See Organic peroxide, liquid or solution, n.o.s.										
	Lauroyl peroxide, technically pure. See Lauroyl peroxide.		UN2124								
	Lead arsenate, solid	Poison B	UN1617	Poison	173.364	173.367	50 pounds	200 pounds	1,2	1,2	Stow away from acids
	Lead arsenite, solid	Poison B	UN1618	Poison	173.364	173.365	50 pounds	200 pounds	1,2	1,2	Segregation same as for corrosive materials
	Lead azide. See Initiating explosive										
	Lead azide (dry)	Forbidden									
A	Lead chloride	Poison B	NA2291	None	173.505	173.800	25 pounds	100 pounds	1,2	1,2	
	Lead cyanide	Poison B	UN1620	Poison	173.370	173.1010	25 pounds	No limit	1,2	1,2	
W	Lead dross (containing 3% or more free acid)	ORM-B	NA1794	None	173.505	173.510	25 pounds	100 pounds	1,2	1,2	
A	Lead fluoborate	ORM-C	NA2291	None	173.505	173.510	25 pounds	100 pounds	1,2	1,2	
A	Lead fluoride	ORM-B	NA2811	None	173.505		25 pounds	100 pounds	1,2	1,2	
	Lead mononitroresorcinate. See Initiating explosive										
	Lead mononitroresorcinate (dry)	Forbidden									
	Lead nitrate	Oxidizer	UN1469	Oxidizer	173.153	173.182	25 pounds	100 pounds	1,2	1,2	Stow away from foodstuffs
	Lead peroxide	Oxidizer	UN1872	Oxidizer	173.153	173.154	25 pounds	100 pounds	1,2	1,2	Stow away from foodstuffs
	Lead picrate (dry)	Forbidden									
	Lead scrap. See Lead dross										
	Lead styphnate (dry)										
	Lead styphnate (lead trinitroresorcinate). See Initiating explosive										
W	Lead sulfate, solid (containing more than 3% free acid)	Corrosive material	UN1794	Corrosive	173.244	173.245b	25 pounds	100 pounds	1,2	1,2	

164

Chapter I—Research and Special Programs Administration §172.101

						1 per inaccessible cargo compartment	No limit	1,2	1,2	
Life rafts, inflatable	ORM-C		None	None	173.906	1 per inaccessible cargo compartment	No limit	1,2	1,2	
Life-saving appliances, self-inflating	ORM-C	UN 2990	None	None	173.906	1 per inaccessible cargo compartment	No limit	1,2	1,2	
Lime-nitrogen. See Calcium cyanamide, not hydrated										
Lime, unslaked. See Calcium oxide										
Lindane	ORM-A	NA2761	None	173.505	173.510	No limit	No limit	1,2	1,2	
Liquefied hydrocarbon gas. See Hydrocarbon gas, liquefied										
Liquefied nonflammable gas (*charged with nitrogen, carbon dioxide, or air*)	Nonflammable gas	NA1058	Nonflammable gas	173.306	173.304	300 pounds	300 pounds	1,2	1,2	
Liquefied petroleum gas	Flammable gas	UN1075	Flammable gas	173.306	173.304 173.314 173.315	Forbidden	300 pounds	1,2	1	
Liquid other than one classed as flammable, corrosive, poison or irritant, charged with nitrogen, carbon dioxide, or air. See Compressed gas n.o.s.										
Lithium acetylide-ethylene diamine complex	Flammable solid	NA2813	Flammable solid and Dangerous when wet	None	173.206	Forbidden	25 pounds	1,2	5	Segregation same as for flammable solid labeled Dangerous When Wet
Lithium aluminum hydride	Flammable solid	UN1410	Flammable solid and Dangerous when wet	None	173.206	Forbidden	25 pounds	1,2	5	Segregation same as for flammable solid labeled Dangerous When Wet
Lithium aluminum hydride, ethereal	Flammable liquid	UN1411	Flammable liquid	None	173.137	Forbidden	1 quart	1	5	Segregation same as for flammable solid labeled Dangerous When Wet
Lithium amide, powdered	Flammable solid	UN1412	Flammable solid	173.153	173.168	25 pounds	100 pounds	1,2	4	Segregation same as for flammable solid labeled Dangerous When Wet
Lithium batteries, for disposal *Lithium battery. See 173.206(f).*	ORM-C	UN1413	None	None	173.1015	Forbidden	Forbidden			
Lithium borohydride	Flammable solid	UN1413	Flammable solid and Dangerous when wet	None	173.206	Forbidden	25 pounds	1,2	5	Segregation same as for flammable solid labeled Dangerous When Wet
Lithium ferrosilicon	Flammable solid	UN2830	Flammable solid and Dangerous when wet	None	173.206	Forbidden	25 pounds	1,2	5	Segregation same as for flammable solid labeled Dangerous When Wet
Lithium hydride	Flammable solid	UN1414	Flammable solid and Dangerous when wet	None	173.206	Forbidden	25 pounds	1,2	5	Segregation same as for flammable solid labeled Dangerous When Wet
Lithium hydride in fused solid form	Flammable solid	UN2805	Flammable solid and Dangerous when wet	None	173.206	Forbidden	100 pounds	1,2	5	Segregation same as for flammable solid labeled Dangerous When Wet

165

Chemical Hazard Communication Guidebook

§172.101   Title 49—Transportation

§172.101 Hazardous Materials Table

(1)	(2)	(3)	(3A)	(4)	(5) Packaging		(6) Maximum net quantity in one package		(7) Water shipments		
+/A/W	Hazardous materials descriptions and proper shipping names	Hazard class	Identification number	Label(s) required (if not excepted)	(a) Exceptions	(b) Specific requirements	(a) Passenger carrying aircraft or railcar	(b) Cargo only aircraft	(a) Cargo vessel	(b) Passenger vessel	(c) Other requirements
	Lithium hypochlorite compound, dry (containing more than 39% available chlorine)	Oxidizer	UN1471	Oxidizer	173.153	173.217	50 pounds	100 pounds	1,2	1,2	
	Lithium metal	Flammable solid	UN1415	Flammable solid and Dangerous when wet	None	173.206	Forbidden	25 pounds	1,2	5	Segregation same as for flammable solids labeled Dangerous When Wet
	Lithium metal, in cartridges	Flammable solid	UN1415	Flammable solid and Dangerous when wet	173.206	173.206	1 pound	25 pounds	1,2	4	Segregation same as for flammable solids labeled Dangerous When Wet
	Lithium nitride	Flammable solid	UN2806	Flammable solid and Dangerous when wet	None	173.206	Forbidden	25 pounds	1,2	5	Segregation same as for flammable solids labeled Dangerous When Wet
	Lithium peroxide	Oxidizer	UN1472	Oxidizer	173.153	173.154	25 pounds	100 pounds	1,2	1,2	Keep dry
	Lithium silicon	Flammable solid	UN1417	Flammable solid and Dangerous when wet	None	173.206	Forbidden	25 pounds	1,2	1,2	Segregation same as for flammable solids labeled Dangerous When Wet
	London purple, solid	Poison B	UN1621	Poison	173.364	173.365	50 pounds	200 pounds	1,2	1,2	
	Low blasting explosive. See Low explosive										
	Low explosive	Class A explosive		Explosive A	None	173.60	Forbidden	Forbidden	6	5	
	Lye. See Sodium hydroxide, solid										
	Magnesium aluminum phosphide	Flammable solid	UN1419	Flammable solid and Dangerous when wet	None	173.206	Forbidden	25 pounds	1,2	1,2	Segregation same as for flammable solids labeled Dangerous When Wet
	Magnesium arsenate, solid	Poison B	UN1622	Poison	173.364	173.367	50 pounds	200 pounds	1,2	1,2	
	Magnesium dross, wet or hot. See 173.173	Forbidden									
	Magnesium granules coated, particle size not less than 149 microns	Flammable solid	UN2950	Flammable solid and Dangerous when wet	173.153	173.178	25	100	1,2	1,2	Segregation same as for flammable solids labeled Dangerous When Wet
+	Magnesium metal (powder, pellets, turnings, or ribbon) or Magnesium aluminum powder	Flammable solid	UN1869	Flammable solid and Dangerous when wet	173.153	173.220	25	100	1,2	1,2	Segregation same as for flammable solids labeled Dangerous When Wet
	Magnesium nitrate	Oxidizer	UN1474	Oxidizer	173.153	173.182	25 pounds	100 pounds	1,2	1,2	
	Magnesium perchlorate	Oxidizer	UN1475	Oxidizer	173.153	173.154	25 pounds	100 pounds	1,3	1,3	Stow away from powdered metals

166

Appendix B: Regulations Reference Guide

Chapter I—Research and Special Programs Administration §172.101

	Magnesium peroxide, solid	Oxidizer	UN1476		173.153	173.154	25 pounds	100 pounds	1,2	1,2 Keep dry
	Magnesium scrap (borings, clippings, shavings, sheet, turnings, or scalpings)	Flammable solid	NA1869		173.153	173.220	Forbidden	Forbidden	1,2	1,3 Segregation same as for flammable solids labeled Dangerous When Wet
A	*Magnetized material. See 173.21(f)*									
A	Malathion	ORM-A	NA2783	None	1,3,510	1,3,510	No limit	No limit	1,2	1,2
A	Maleic acid	ORM-A	NA2215	None	173.505	173.510	50 pounds	200 pounds	1,2	1,2 Keep tightly closed.
AW	Maleic anhydride	ORM-A	UN2215	None	173.505	173.510	50 pounds	200 pounds	1,2	1,2 Stow away from foodstuffs
	Mannitan tetranitrate	Forbidden								
	Matches, block. See Matches, strike anywhere									
	Matches, safety, book, card, or strike-on box	Flammable solid	UN1944	None	173.176	None	50 pounds	50 pounds	1,2	1,2
	Matches, strike anywhere	Flammable solid	UN1331	Flammable solid	None	173.176a	Forbidden	Forbidden	1,2	1
	Matting acid. See Sulfuric acid									
	Medicines, n.o.s.	Combustible liquid	UN1851	None	173.118a	None	No limit	No limit	1,2	1,2
	Medicines, n.o.s.	Flammable liquid	UN1851	Flammable liquid	173.118	173.119	1 quart	10 gallons	1,2	1
	Medicines, n.o.s.	Flammable solid	UN1851	Flammable solid	173.153	173.154	25 pounds	100 pounds	1,2	1,2
	Medicines, n.o.s.	Oxidizer	UN1851	Oxidizer	173.153	173.154	25 pounds	100 pounds	1,2	1,2
	Medicines, liquid, n.o.s.	Corrosive material	UN1851	Corrosive	173.244	173.245	1 quart	1 quart	1,2	1,2
	Medicines, liquid, n.o.s.	Poison B	UN1851	Corrosive	173.345	173.346	1 quart	55 gallons	1,3	1
	Medicines, solid, n.o.s.	Corrosive material	UN1851	Corrosive	173.244	173.245b	25 pounds	100 pounds	1,2	1,2 Keep dry
	Medicines, solid, n.o.s., p-Menthane hydroperoxide, technically pure. See Paramenthane hydroperoxide	Poison B	UN1851	Poison	173.364	173.365	50 pounds	200 pounds	1,3	1,3
	Mercaptan mixture, aliphatic	Flammable liquid	UN1851 UN2125	Flammable liquid	None	173.141	Forbidden	10 gallons	1,3	5 Stow in well ventilated space away from living quarters
	Mercaptan mixture, aliphatic *(in containers over 110 gallons)*	Combustible liquid	NA1228	None	173.118a	None	Forbidden	10 gallons	1,2	1,2
AW	Mercaptan mixture, aliphatic *(in containers of 110 gallons or less). See 173.141(b)*	ORM-A	NA1228		173.505	173.510	Forbidden	10 gallons	1,3	5
+	Mercuric acetate	Poison B	UN1629	Poison	173.364	173.365	50 pounds	200 pounds	1,2	1,2
+	Mercuric ammonium chloride, solid	Poison B	UN1630	Poison	173.364	173.365	50 pounds	200 pounds	1,2	1,2
+	Mercuric benzoate, solid	Poison B	UN1631	Poison	173.364	173.365	50 pounds	200 pounds	1,2	1,2
+	Mercuric bromide, solid	Poison B	UN1634	Poison	173.364	173.365	Forbidden	25 pounds	1,2	1,2
+	Mercuric chloride, solid	Poison B	UN1624	Poison	173.364	173.372	Forbidden	25 pounds	1,2	1,2
+	Mercuric cyanide, solid	Poison B	UN1636	Poison	173.370	173.370	50 pounds	200 pounds	1,2	1,2 *Stow away from acids*
+	Mercuric iodide, solid	Poison B	UN1638	Poison	173.364	173.365	50 pounds	200 pounds	1,2	1,2
+	Mercuric iodide, solution	Poison B	UN1638	Poison	173.345	173.346	1 quart	55 gallons	1,2	1,2 If stowed under deck, must be stowed in a recoverable location.
	Mercuric nitrate	Oxidizer	UN1625	Oxidizer	173.153	173.182	25 pounds	100 pounds	1,2	1,2
+	Mercuric oleate, solid	Poison B	UN1640	Poison	173.364	173.365	50 pounds	200 pounds	1,2	1,2 *Stow away from acids*
+	Mercuric oxide, solid	Poison B	UN1641	Poison	173.364	173.365	50 pounds	200 pounds	1,2	1,2 *Stow away from acids*
++	Mercuric oxycyanide, solid (desensitized)	Poison B	UN1642	Poison	173.364	173.365	25 pounds	200 pounds	1,2	1,2
++	Mercuric potassium cyanide, solid	Poison B	UN1626	Poison	173.364	173.365 178.370	25 pounds	200 pounds	1,2	1,2
++	Mercuric potassium iodide, solid	Poison B	UN1643	Poison	173.364	173.365	50 pounds	200 pounds	1,2	1,2
++	Mercuric salicylate, solid	Poison B	UN1644	Poison	173.364	173.365	50 pounds	200 pounds	1,2	1,2
++	Mercuric subsulfate, solid	Poison B	NA9025	Poison	173.364	173.365	50 pounds	200 pounds	1,2	1,2
+	Mercuric sulfate, solid	Poison B	UN1645	Poison	173.364	173.365	50 pounds	200 pounds	1,2	1,2

167

Chemical Hazard Communication Guidebook

§172.101 **Title 49—Transportation**

§172.101 Hazardous Materials Table

(1) +/A/W	(2) Hazardous materials descriptions and proper shipping names	(3) Hazard class	(3A) Identification number	(4) Label(s) required (if not excepted)	(5) Packaging (a) Exceptions	(5) Packaging (b) Specific requirements	(6) Maximum net quantity in one package (a) Passenger carrying aircraft or railcar	(6) Maximum net quantity in one package (b) Cargo only aircraft	(7) Water shipments (a) Cargo vessel	(7) Water shipments (b) Passenger vessel	(7) Water shipments (c) Other requirements
+	Mercuric sulfocyanate, solid *or* Mercuric thiocyanate, solid	Poison B	UN1646	Poison	173.364	173.365	50 pounds	200 pounds	1,2	1,2	
+	Mercurol *or* Mercury nucleate, solid	Poison B	UN1639	Poison	173.364	173.365	50 pounds	200 pounds	1,2	1,2	
+	Mercurous acetate, solid	Poison B	UN1629	Poison	173.364	173.365	50 pounds	200 pounds	1,2	1,2	
	Mercurous azide	Forbidden									
+	Mercurous bromide, solid	Poison B	UN1634	Poison	173.364	173.365	50 pounds	200 pounds	1,2	1,2	
+	Mercurous gluconate, solid	Poison B	UN1637	Poison	173.364	173.365	50 pounds	200 pounds	1,2	1,2	
+	Mercurous iodide, solid	Poison B	UN1638	Poison	173.153	173.154	50 pounds	100 pounds	1,2	1,2	
	Mercurous nitrate, solid	Oxidizer	UN1627	Oxidizer	173.364	173.365	50 pounds	200 pounds	1,2	1,2	
+	Mercurous oxide, black, solid	Poison B	UN1641	Poison	173.364	173.365	50 pounds	200 pounds	1,2	1,2	
+	Mercurous sulfate, solid	Poison B	UN1628	Poison	173.364	173.365	50 pounds	200 pounds	1,2	1,2	
	Mercury acetylide	Forbidden									
	Mercury based pesticide, liquid, n.o.s. (compounds and preparations)	Flammable liquid	UN2778	Flammable liquid	173.118	173.119	1 quart	10 gallons	1,2	1	
	Mercury based pesticide, liquid, n.o.s. (compounds and preparations)	Poison B	UN2777	Poison	173.345	173.346	1 quart	55 gallons	1,2	1,2	
	Mercury based pesticide, solid, n.o.s. (compounds and preparations)	Poison B	UN2777	Poison	173.364	173.365	50 pounds	200 pounds	1,2	1,2	
	Mercury compound, solid, n.o.s.	Poison B	UN2025	Poison	173.364	173.365	50 pounds	200 pounds	1,2	1,2	
	Mercury fulminate. See Initiating explosive	Forbidden									
	Mercury iodide aquabasic ammonobasic (Iodide of Millon's base)										
A	Mercury, metallic	ORM-B	NA2809	None	None	173.860	173.860	See 173.860	1,2	1,2	
	Mercury nitride	Forbidden									
	Mercury oxycyanide	Forbidden									
	Mesityl oxide	Flammable liquid	UN1229	Flammable liquid	None	173.119	1 quart	10 gallons	1,2	1	
	Metal alkyl, solution, n.o.s.	Flammable liquid	NA9195	Flammable liquid	173.118	173.119	1 quart	1 gallon	1,2	1	
W	Metal borings, shavings, turnings, or cuttings (*ferrous metals only, except stainless steel*)	ORM-C	UN2793	None	173.505	173.1025	50 pounds	200 pounds	1,2	1,2	Keep dry, not permitted if temperature of material is at or above 130 deg F
	Metal salts of methyl nitramine (dry)	Forbidden									
+	Methane *or* Methane, compressed	Flammable gas	UN1971	Flammable gas	173.306	173.302	Forbidden	300 pounds	1,2	4	Stow away from living quarters
	Methane, refrigerated liquid (cryogenic liquid)	Flammable gas	UN1972	Flammable gas	None	173.318	Forbidden	Forbidden	1	5	Stow away from living quarters
	Methanol. *See* Methyl alcohol										
	Methazoic acid	Forbidden									

168

Chapter I—Research and Special Programs Administration §172.101

	Description		UN/NA	Class						
	Methyl acetate	Flammable liquid	UN1231	Flammable liquid	173.118	173.119	1 quart	10 gallons	1,2	1	
	Methyl acetone	Flammable liquid	UN1232	Flammable liquid	173.118	173.119	1 quart	10 gallons	1,2	1	
+	Methylacetylene-propadiene, stabilized	Flammable gas	UN1060	Flammable gas	173.306	173.304 173.314 173.315	Forbidden	300 pounds	1,2	1	
	Methyl acrylate, inhibited	Flammable liquid	UN1919	Flammable liquid	173.118	173.119	1 quart	10 gallons	1,2	1	
	Methylal	Flammable liquid	UN1234	Flammable liquid	None	173.119	Forbidden	10 gallons	1,3	5	
	Methyl alcohol	Flammable liquid	UN1230	Flammable liquid	173.306	173.119	1 quart	10 gallons	1,2	1	
+	Methylamine, anhydrous	Flammable gas	UN1061	Flammable gas	173.306	173.304 173.314 173.315	Forbidden	300 pounds	1	4	Stow away from mercury and its compounds
	Methylamine, aqueous solution	Flammable liquid	UN1235	Flammable liquid	173.118	173.119	1 quart	10 gallons	1,3	4	
	Methylamine dinitramine and dry salts thereof	Forbidden									
	Methylamine nitroform	Forbidden									
	Methylamine perchlorate (dry)	Forbidden									
	Methylamyl acetate	Flammable liquid	UN1233	Flammable liquid	173.118	173.119	1 quart	10 gallons	1,2	1,2	
	Methyl amyl ketone	Combustible liquid	UN1110	None	173.118a	None	No limit	No limit	1	5	Shade from radiant heat
+	Methyl bromide and more than 2% chloropicrin mixture, liquid	Poison B	NA1581	Poison	None	173.353	Forbidden	Forbidden	1	5	Stow away from living quarters
+	Methyl bromide and nonflammable, nonliquefied compressed gas mixture, liquid (including up to 2% chloropicrin)	Poison B	NA1955	Poison	None	173.353a	Forbidden	300 pounds	1	1	
+	Methyl bromide - ethylene dibromide mixture, liquid	Poison B	UN1647	Poison	None	173.353	Forbidden	55 gallons	1	5	Stow away from living quarters. Segregation same as for nonflammable gas.
+	Methyl bromide, liquid (including up to 2% chloropicrin)	Poison B	UN1062	Poison	None	173.353	Forbidden	55 gallons	1	5	
	Methyl butene	Flammable liquid	UN2460	Flammable liquid	None	173.119	Forbidden	10 gallons	1,2	5	
	Methyl butyrate	Flammable liquid	UN1237	Flammable liquid	173.118	173.119	1 quart	10 gallons	1,2	1	
	Methyl cellosolve. See Ethylene glycol monomethyl ether										
	Methyl cellosolve acetate. See Ethylene glycol monomethyl ether acetate										
+	Methyl chloride	Flammable gas	UN1063	Flammable gas	173.306	173.304 173.314 173.315	Forbidden	300 pounds	1,2	4	
+	Methyl chloride-methylene chloride mixture	Flammable gas	UN1912	Flammable gas	173.306	173.304 173.314	Forbidden	300 pounds	1,2	4	
	Methyl chlorocarbonate. See Methyl chloroformate										
▲	Methyl chloroform. See 1,1,1-Trichloroethane										
	Methyl chloroformate	Flammable liquid	UN1238	Flammable liquid and Poison	None	173.288	Forbidden	5 pints	1,2	1	

169

Chemical Hazard Communication Guidebook

§172.101   **Title 49—Transportation**

§172.101 Hazardous Materials Table

(1) +/A/W	(2) Hazardous materials descriptions and proper shipping names	(3) Hazard class	(3A) Identification number	(4) Label(s) required (if not excepted)	(5)(a) Packaging Exceptions	(5)(b) Specific requirements	(6)(a) Max net qty — Passenger carrying aircraft or railcar	(6)(b) Max net qty — Cargo only aircraft	(7)(a) Water — Cargo vessel	(7)(b) Water — Passenger vessel	(7)(c) Other requirements
	Methylchloromethyl ether, anhydrous	Flammable liquid	UN1239	Flammable liquid and Poison	None	173.143	Forbidden	Forbidden	1	5	Shade from radiant heat
	Methyl cyanide	Flammable liquid	UN1648	Flammable liquid	173.118	173.119	1 quart	10 gallons	1	4	Shade from radiant heat
	Methylcyclohexane	Flammable liquid	UN2296	Flammable liquid	173.118	173.119	1 quart	10 gallons	1,2	1	
	Methylcyclopentane	Flammable liquid	UN2298	Flammable liquid	173.118	173.119	1 quart	10 gallons	1,3	4	
	Methyl dichloroacetate	Corrosive material	UN2299	Corrosive	173.244	173.245	1 quart	1 quart	1,2	1,2	
	Methyldichloroarsine	Poison A	NA1556	Poison gas	None	173.328	Forbidden	Forbidden	1	5	Shade from radiant heat
	Methyl dichlorosilane	Flammable liquid	UN1242	Flammable liquid	None	173.136	Forbidden	5 pints	1,2	1	
	Methylene chloride. *See* Dichloromethane										
	Methylene glycol dinitrate	Forbidden									
	Methyl ethyl ether. *See* Ethyl methyl ether										
	Methyl ethyl ketone	Flammable liquid	UN1193	Flammable liquid	173.118	173.119	1 quart	10 gallons	1,2	1	
	Methyl ethyl ketone peroxide, in solution with not more than 9% by weight active oxygen. *See* Organic peroxide, liquid, or solution, n.o.s.		UN2550								
	Methyl ethyl ketone peroxide, in solution with more than 9% by weight active oxygen	Forbidden									
	Methyl ethyl pyridine	Corrosive material	UN2300	Corrosive	173.244	173.245	1 quart	10 gallons	1,2	1,2	
	Methyl formate	Flammable liquid	UN1243	Flammable liquid	allons		1 quart	10 gallons	1,3	4	
	Methylfuran	Flammable liquid	UN2301	Flammable liquid	173.118	173.119	1 quart	10 gallons	1,3	4	
	a-Methylglucoside tetranitrate	Forbidden									
	a-Methylglycerol trinitrate	Forbidden									
	Methylhydrazine	Flammable liquid	UN1244	Flammable liquid and Poison	None	173.145	Forbidden	5 pints	1,2	1	Stow separate from oxidizing materials and corrosives
	Methyl isobutyl ketone peroxide, in solution with not more than 9% by weight active oxygen. *See* Organic peroxide, liquid or solution, n.o.s.		UN2126								

170

Chapter I—Research and Special Programs Administration §172.101

Material	Class	ID No.	Label	Packaging (Exceptions)	Packaging (Specific)	Max Pass.	Max Cargo	Stowage	Other
Methyl isobutyl ketone peroxide, in solution with more than 9% by weight active oxygen	Forbidden								
Methyl isocyanate	Flammable liquid	UN2480	Flammable liquid and Poison	None	173.119 173.*a	Forbidden	10 gallons	1	5	Keep cool. Stow away from living quarters and sources of heat
Methyl isopropenyl ketone, inhibited	Flammable liquid	UN1246	Flammable liquid	173.118	173.119	1 quart	10 gallons	1,2	1	
Methyl magnesium bromide in ethyl ether not over 40% concentration	Flammable liquid	UN1928	Flammable liquid	None	173.149	Forbidden	Forbidden	1	1	Segregation same as for flammable solids. Separate from flammable gases or liquids, oxidizing materials or organic peroxides
Methyl mercaptan	Flammable gas	UN1064	Flammable gas	173.306	173.304 173.814 173.315	Forbidden	300 pounds	1,2	1	
Methyl methacrylate monomer, inhibited	Flammable liquid	UN1247	Flammable liquid	173.118	173.119	1 quart	10 gallons	1,2	1	
Methyl methacrylate monomer, uninhibited (high-purity, if acceptable under 173.21 of this subchapter)	Flammable liquid	NA1247	Flammable liquid	173.118	173.119	Forbidden	Forbidden	1,2	1	
Methyl nitrate	Forbidden								
N-Methyl-N'-nitro-N-nitrosoguanidine (not exceeding 25 grams in one outside packaging)	Flammable solid	NA1325	Flammable solid	None	173.179	Forbidden	Forbidden	4	5	
Methyl parathion, liquid	Poison B	NA2783	Poison	None	173.358	Forbidden	1 quart	1,3	1,3	
Methyl parathion mixture, dry	Poison B	NA2783	Poison	173.377	173.377	50 pounds	200 pounds	1,2	1,2	
Methyl parathion mixture, liquid, (containing 25% or less methyl parathion)	Poison B	NA2783	Poison	None	173.359	1/2 pint	1 quart	1,2	1,2	
Methyl parathion mixture, liquid, (containing over 25% methyl parathion)	Poison B	NA2783	Poison	None	173.359	Forbidden	1 quart	1,2	1,2	
Methylpentadiene	Flammable liquid	UN2461	Flammable liquid	173.118	173.119	1 quart	10 gallons	1,2	1	
Methyl pentane	Flammable liquid	UN2462	Flammable liquid	173.118	173.119	1 quart	10 gallons	1,2	1	
Methyl phosphonothioic dichloride, anhydrous	Corrosive material	NA1760	Corrosive	173.244	173.245 173.245a	1 quart	1 quart	1	4	
Methyl phosphonic dichloride	Corrosive material	NA9206	Corrosive and Poison	None	173.271	Forbidden	Forbidden	1	1	Keep dry. Glass carboys not permitted on passenger vessels
Methyl phosphonous dichloride. See Pyrofonic liquid, n.o.s										
Methyl picric acid (heavy metal salts of)	Forbidden									
Methyl propionate	Flammable liquid	UN1248	Flammable liquid	173.118	173.119	1 quart	10 gallons	1,2	1	
Methyl propyl ketone	Flammable liquid	UN1249	Flammable liquid	173.118	173.119	1 quart	10 gallons	1,2	1	
Methyl sulfate. See Dimethyl sulfate										
Methyl sulfide. See Dimethyl sulfide										
Methyltrichlorosilane	Flammable liquid	UN1250	Flammable liquid	None	173.135	Forbidden	10 gallons	1,2	1	
Methyl trimethylol methane trinitrate	Forbidden									
Methyl vinyl ketone, inhibited	Flammable liquid	UN1251	Poison	173.147	173.147	4 ounces	10 gallons	1,2	1	
Mevinphos	Poison B	NA2783	Poison	None	173.358	Forbidden	1 quart	1,2	5	
Mevinphos mixture, dry	Poison B	NA2783	Poison	173.377	173.377	Forbidden	200 pounds	1,2	4	
Mevinphos mixture, liquid	Poison B	NA2783	Poison	173.359	173.359	1 pint	1 quart	1,2	5	
Mexacarbate	Poison B	NA2757	Poison	173.364	173.365	50 pounds	200 pounds	1,2	1,2	
Mild detonating fuse, metal clad. See Fuse, mild detonating, metal clad										

171

Chemical Hazard Communication Guidebook

§172.101   Title 49—Transportation

§172.101 Hazardous Materials Table

(1)	(2) Hazardous materials descriptions and proper shipping names	(3) Hazard class	(3A) Identification number	(4) Label(s) required (if not excepted)	(5) Packaging (a) Exceptions	(5) Packaging (b) Specific requirements	(6) Max net quantity (a) Passenger carrying aircraft or railcar	(6) Max net quantity (b) Cargo only aircraft	(7) Water shipments (a) Cargo vessel	(7) Water shipments (b) Passenger vessel	(7) Water shipments (c) Other requirements
	Mine, empty. See 173.55										
	Mine, explosive, with gaseous material. See Explosive mine										
+/A/W	Mine rescue equipment containing carbon dioxide	Nonflammable gas	NA1956	Nonflammable gas	173.306		150 pounds	300 pounds	1,2	1,2	
	Mining reagent, liquid (containing 20% or more cresylic acid)	Corrosive material	NA2022	Corrosive	173.244	173.249a	1 quart	10 gallons	1,2	1,2	
	Mipafox	ORM-A	UN2783	None	173.505	173.510	No limit	No limit			
A	Mixed acid. *See Nitrating acid*										
A	Molybdenum pentachloride	ORM-B	UN2508	None	173.505	173.800	25 pounds	100 pounds	1	1	Stow away from living quarters
	Monochloroacetone, stabilized or inhibited	Irritating material	UN1695	Irritant	None	173.384	Forbidden	5 gallons			
	Monochloroacetone (unstabilized)	Forbidden									
	Monochloroethylene. See Vinyl chloride										
	Monoethanolamine	Corrosive material	UN2491	Corrosive	173.244	173.245	1 quart	10 gallons	1,2	1,2	Segregation same as for flammable gas
	Monoethylamine	Flammable liquid	UN1036	Flammable liquid	None	173.148	Forbidden	5 pints	1,2	5	Keep dry
	Monofluorophosphoric acid, anhydrous	Corrosive material	UN1776	Corrosive	None	173.275	Forbidden	1 gallon	1,2	1,2	
	Morpholine	Flammable liquid	UN2054	Flammable liquid	173.118	173.119	1 quart	10 gallons	1,2	1	
	Morpholine, aqueous, mixture	Flammable liquid	NA2054	Flammable liquid	173.118	173.119	1 quart	10 gallons	1,2	1	
	Morpholine, aqueous, mixture	Corrosive material	NA1760	Corrosive	173.244	173.245	1 quart	10 gallons	1	4	
	Moth balls. See Naphthalene										
	Motion picture film. See Film										
+	Motor fuel antiknock compound or Antiknock compound	Poison B	UN1649	Poison	None	173.354	Forbidden	55 gallons	1	5	If flashpoint less than 141 deg F. segregation same as for flammable liquids
	Motor, internal combustion				173.120						
	Motor vehicle, etc., including automobile, motorcycle, truck, tractor, and other self-propelled vehicle or equipment powered by internal combustion engine, when offered new or used for transportation and which contain fuel in the engine or fuel tank or the electric storage battery is connected to either terminal of the electrical system	ORM-C		None	173.120 173.250 173.257 173.306 175.305 176.905				1,2	1,2	
	Muriatic acid. See Hydrochloric acid										

172

Chapter I—Research and Special Programs Administration §172.101

	Naphtha	Combustible liquid	UN2553	None	173.118a	None	No limit	No limit	1,2	1,2	
	Naphtha	Flammable liquid	UN2553	Flammable liquid	173.118	173.119	1 quart	10 gallons	1,2	1	Segregation same as for flammable solids
AW	Naphthalene or Naphthalin	ORM-A	UN1334	None	173.505	173.655	25 pounds	300 pounds	1,2	1	Stow away from living quarters
	Naphthalene diazonide	Forbidden									
	Naphtha petroleum. See Petroleum naphtha										
	Naphthyl amineperchlorate	Forbidden									
	Natural gas, refrigerated liquid (with methane content) (cryogenic liquid)	Flammable gas	UN1972	Flammable gas	None	173.318	Forbidden	Forbidden	1	5	
	Natural gasoline. See Gasoline										
	Neohexane	Flammable liquid	UN1208	Flammable liquid	173.118	173.119	1 quart	10 gallons	1,3	4	
	Neon or Neon, compressed	Nonflammable gas	UN1065	Nonflammable gas	173.306	173.302	150 pounds	300 pounds	1,2	1,2	
	Neon, refrigerated liquid (cryogenic liquid)	Nonflammable gas	UN1913	Nonflammable gas	173.320	173.316	100 pounds	1,100 pounds	1,3	1,3	
	New explosive or explosive device. See 173.51 and 173.86										
	Nickel carbonyl	Flammable liquid	UN1259	Flammable liquid and Poison	None	173.126	Forbidden	Forbidden	1	5	Shade from radiant heat. Segregation same as for flammable liquids. Not permitted on a vessel transporting explosives, except that quantities not exceeding 200 pounds may be transported on such vessels under conditions approved by the Captain of the Port
	Nickel cyanide, solid	Poison B	UN1653	Poison	173.370	173.346	25 pounds	200 pounds	1,2	1,2	Stow away from acids
+	Nickel nitrate	Oxidizer	UN2725	Oxidizer	173.153	173.346	25 pounds	100 pounds	1,2	1,2	
	Nickel picrate	Forbidden									
	Nicotine hydrochloride	Poison B	UN1656	Poison	173.345	173.346	1 quart	55 gallons	1,2	1,2	
	Nicotine, liquid	Poison B	UN1654	Poison	None	173.346	Forbidden	55 gallons	1,2	1,2	
	Nicotine salicylate	Poison B	UN1657	Poison	173.364	173.365	50 pounds	200 pounds	1,2	1,2	
	Nicotine sulfate, liquid	Poison B	UN1658	Poison	173.345	173.346	1 quart	55 gallons	1,2	1,2	
	Nicotine sulfate, solid	Poison B	UN1658	Poison	173.364	173.365	50 pounds	200 pounds	1,2	1,2	
	Nicotine tartrate	Poison B	UN1659	Poison	173.364	173.365	50 pounds	200 pounds	1,2	1,2	
	Nitrated paper (unstable)	Forbidden									
++	Nitrate, n.o.s.	Oxidizer	NA1477	Oxidizer	173.153	173.182	25 pounds	100 pounds	1,2	1,2	
	Nitrate of ammonia explosives. See High explosive										
	Hydrogen selenide	Flammable gas	UN2202	Flammable gas and Poison	None	173.328	Forbidden	Forbidden	1	5	
	Nitrates of diazonium compounds	Forbidden									
	Nitrating acid, mixture (with not more than 50% nitric acid)	Corrosive material	UN1796	Corrosive	None	173.267	Forbidden	1 quart	1	5	Stow away from fluorides
	Nitrating acid, mixture (with more than 50% nitric acid)	Oxidizer	UN1796	Oxidizer and Corrosive	None	173.267	Forbidden	1 quart	1	5	Segregation same as for corrosive material. Stow away from fluorides
	Nitrating acid, spent	Corrosive material	NA1826	Corrosive	None	173.248	Forbidden	1 quart	1	5	
	Nitric acid (over 40%)	Oxidizer	UN2031	Oxidizer and Corrosive	None	173.266	Forbidden	5 pints	1	5	Segregation same as for corrosive materials, separate from diethylenetriamine
	Nitric acid, 40% or less	Corrosive material	NA1760	Corrosive	None	173.268	Forbidden	5 pints	1	5	Stow away from hydrazine, separate from diethylenetriamine
	Nitric acid, fuming	Oxidizer	UN2032	Oxidizer and Poison	None	173.268	Forbidden	Forbidden	1	5	Segregation same as for corrosive materials. Stow away from hydrazine, separate from diethylenetriamine

173

Chemical Hazard Communication Guidebook

§172.101   Title 49—Transportation

§172.101 Hazardous Materials Table

(1) +/A/W	(2) Hazardous materials descriptions and proper shipping names	(3) Hazard class	(3A) Identification number	(4) Label(s) required (if not excepted)	(5) Packaging (a) Exceptions	(5) Packaging (b) Specific requirements	(6) Maximum net quantity in one package (a) Passenger carrying aircraft or railcar	(6) Maximum net quantity in one package (b) Cargo only aircraft	(7) Water shipments (a) Cargo vessel	(7) Water shipments (b) Passenger vessel	(7) Water shipments (c) Other requirements
	Nitric ether. See Ethyl nitrate										
	Nitric oxide	Poison A	UN1660	Poison gas	None	173.337	Forbidden	Forbidden	1	5	
	2-Nitro-2-methylpropanol nitrate	Forbidden									
	6-Nitro-4-diazotoluene-3-sulfonic acid (dry)	Forbidden									
	p-Nitroaniline. See Nitroaniline										
	N-Nitroaniline										
+	Nitroaniline	Poison B	UN1661	Poison	173.364	173.373	50 pounds	200 pounds	1,2	1,2	
	m-Nitrobenzene diazonium perchlorate	Forbidden									
	Nitrobenzene, liquid *or* Nitrobenzol, liquid (oil of mirbane)	Poison B	UN1662	Poison	173.345	173.346	1 quart	55 gallons	1,2	1,2	
	Nitro carbonitrate. See Blasting agent, n.o.s.										
	Nitrocellulose, colloided, granular *or* flake, wet with not less than 20% alcohol *or* solvent, *or* block, wet with not less than 25% alcohol	Flammable liquid	NA2059	Flammable liquid	173.118	173.127	1 quart	25 pounds	1,3	1	
	Nitrocellulose, colloided, granular *or* flake, wet with not less than 20% water	Flammable solid	NA2555	Flammable solid	173.153	173.184	25 pounds	100 pounds	1,3	1	
	Nitrocellulose, dry. See High explosive										
	Nitrocellulose, wet with not less than 30% alcohol *or* solvent	Flammable liquid	NA2556	Flammable liquid	173.118	173.127	1 quart	25 pounds	1,3	1	
	Nitrocellulose, wet with not less than 20% water	Flammable solid	NA2555	Flammable solid	173.153	173.184	25 pounds	100 pounds	1,3	1	
	Nitrochlorobenzene, *meta or para*, solid	Poison B	UN1578	Poison	173.364	173.374	50 pounds	200 pounds	1,2	1,2	
	Nitrochlorobenzene, *ortho*, liquid	Poison B	UN1578	Poison	173.345	173.346	1 quart	55 gallons	1,2	1,2	
	Nitroethane	Flammable liquid	UN2842	Flammable liquid	173.118	173.119	15 gallon	55 gallons	1,2	1	
	Nitroethylene polymer	Forbidden									
	Nitroethyl nitrate	Forbidden									
	Nitrogen *or* Nitrogen, compressed	Nonflammable gas	UN1066	Nonflammable gas	173.306	173.302 173.314	150 pounds	300 pounds	1,2	1,2	Segregation same as for nonflammable gases. Stow away from organic materials
	Nitrogen, refrigerated liquid *(cryogenic liquid)*	Nonflammable gas	UN1977	Nonflammable gas	173.320	173.316 173.318	100 pounds	1,100 pounds	1,3	1,3	Segregation same as for nonflammable gases. Stow away from organic materials
	Nitrogen dioxide, liquid	Poison A	UN1067	Poison gas and Oxidizer	None	173.336	Forbidden	Forbidden	1	5	Segregation same as for nonflammable gases. Stow away from organic materials
	Nitrogen peroxide, liquid	Poison A	NA1067	Poison gas and oxidizer	None	173.336	Forbidden	Forbidden	1	5	Segregation same as for nonflammable gases. Stow away from organic materials
	Nitrogen tetroxide, liquid	Poison A	NA1067	Poison gas and oxidizer	None	173.336	Forbidden	Forbidden	1	5	Segregation same as for nonflammable gases. Stow away from organic materials
	Nitrogen trichloride	Forbidden									

174

Chapter I—Research and Special Programs Administration §172.101

Name	Class	UN No.	Label	Packaging (exceptions)	Packaging (specific)	Max qty passenger	Max qty cargo	Vessel stowage	Other requirements
Nitrogen trifluoride	Nonflammable gas	UN2451	Nonflammable gas	None	173.302	Forbidden	300 pounds	1	5	Stow away from living quarters and organic materials
Nitrogen triiodide	Forbidden								
Nitrogen triiodide monoamine	Forbidden								
Nitroglycerin, liquid, desensitized. See High explosive, liquid									
Nitroglycerin, liquid, not desensitized. See 173.51	Forbidden								
Nitroglycerin, spirits of. See Spirits of nitroglycerin									
Nitroguanidine, dry. See High explosive	Forbidden								
Nitroguanidine nitrate	Flammable solid	UN1336	Flammable solid	173.153	173.184	25 pounds	100 pounds	1,2	4
Nitroguanidine, wet with not less than 20% water	Forbidden								
1-Nitro hydantoin	Corrosive material	UN1798	Corrosive	None	173.278	Forbidden	5 pints	1	5
Nitrohydrochloric acid	Corrosive material	UN1798	Corrosive	None	173.278	Forbidden	5 pints	1	5
Nitrohydrochloric acid, diluted	Forbidden								
Nitro isobutane triol trinitrate	Forbidden								
Nitromannite. See High explosive									
Nitromannite (dry)	Forbidden								
Nitromethane	Flammable liquid	UN1261	Flammable liquid	173.118	173.149a	1 quart	10 gallons	1,2	1,2
Nitromuriatic acid. See Nitrohydrochloric acid									
N-Nitro-N-methylglycolamide nitrate	Forbidden								
Nitrophenol pesticide, substituted, liquid or solid, n.o.s. (compounds and preparations). See Substituted nitrophenol pesticide, liquid or solid, n.o.s. (compounds and preparations)									
m-Nitrophenyldinitro methane	Forbidden								
Nitropropane	Flammable liquid	UN2608	Flammable liquid	173.118	173.119	15 gallon	55 gallons	1,2	1
Nitrosoguanidine. See Initiating explosive									
Nitrostarch, dry. See High explosive									
Nitrostarch, wet with not less than 30% alcohol or solvent	Flammable liquid	UN1337	Flammable liquid	173.118	173.127	1 quart	25 pounds	1,2	1
Nitrostarch, wet with not less than 20% water	Flammable solid	UN1337	Flammable solid	173.153	173.184	25 pounds	100 pounds	1	4
Nitrosugars (dry)	Forbidden								
Nitrosyl chloride	Nonflammable gas	UN1069	Nonflammable gas	173.306	173.304 173.314	Forbidden	300 pounds	1	4
Nitrourea. See High explosive									
Nitrous oxide or Nitrous oxide, compressed	Nonflammable gas	UN1070	Nonflammable gas	173.306	173.304	150 pounds	300 pounds	1,2	1,2	Under deck stowage must be in well-ventilated space Stow away from flammables. Do not overstow with other cargo
Nitrous oxide, refrigerated liquid	Nonflammable gas	UN2201	Nonflammable gas	173.306	173.315	Forbidden	Forbidden	1	1
Nitroxylol	Poison B	NA1665	Poison	173.345	173.346	1 quart	55 gallons	1,2	1
Nonflammable gas, n.o.s. See Compressed gas, n.o.s.									
Nonliquefied hydrocarbon gas. See Hydrocarbon gas, nonliquefied									
Nonyltrichlorosilane	Corrosive material	UN1799	Corrosive	None	173.280	Forbidden	10 gallons	1	1	Keep dry
Nordhausen acid. See Sulfuric acid									

175

Chemical Hazard Communication Guidebook

§172.101 Title 49—Transportation

§172.101 Hazardous Materials Table

(1)	(2)	(3)	(3A)	(4)	(5) Packaging		(6) Maximum net quantity in one package		(7) Water shipments		
+/A/W	Hazardous materials descriptions and proper shipping names	Hazard class	Identification number	Label(s) required (if not excepted)	(a) Exceptions	(b) Specific requirements	(a) Passenger carrying aircraft or railcar	(b) Cargo only aircraft	(a) Cargo vessel	(b) Passenger vessel	(c) Other requirements
	Octadecyltrichlorosilane	Corrosive material	UN1800	Corrosive	None	173.280	Forbidden	10 gallons	1	1	Keep dry
	1,7-Octadiene-3,5-diyne-1,8-dimethoxy-9-octadecynoic acid	Forbidden									
	Octane	Flammable liquid	UN1262	Flammable liquid	173.118	173.119	1 quart	10 gallons	1,2	1	
	n-Octanoyl peroxide, technically pure. See Organic peroxide, liquid or solution, n.o.s.		UN2129								
	Octyltrichlorosilane	Corrosive material	UN1801	Corrosive	None	173.280	Forbidden	10 gallons	1	1	Keep dry
	Oil, described as oil, Oil, n.o.s. See Petroleum oil, or Petroleum oil, n.o.s.	Combustible liquid	NA1270	None	173.118a	None	No limit	No limit	1,2	1,2	
	Oil, described as oil, Oil, n.o.s. See Petroleum oil, or Petroleum oil, n.o.s.	Flammable liquid	NA1270	Flammable liquid	173.118	173.119	1 quart	10 gallons	1,2	1	
	Oil of mirbane. See Nitrobenzene, liquid										
	Oil of vitriol. See Sulfuric acid										
	Oil well cartridge	Class C explosive	NA1831	Class C explosive	None	173.112	50 pounds	150 pounds	1,3	1,3	Under deck stowage must be in metal drums only. Keep dry.
	Oleum (fuming sulfuric acid)	Corrosive material		Corrosive	None	173.272	Forbidden	5 pints	1,2	1	Stow separate from combustible materials, explosives, or acids
	Organic peroxide, liquid or solution, n.o.s.	Flammable liquid	NA1993	Flammable liquid and Organic peroxide	None	173.119	Forbidden	1 quart	1,2	5	
	Organic peroxide, liquid or solution, n.o.s.	Organic peroxide	NA9183	Organic peroxide	173.153	173.221	Forbidden	1 quart	1,2	1,2	Stow separate from combustible materials, explosives, or acids.
	Organic peroxide, mixture. See Organic peroxide, solid, n.o.s. or Organic peroxide, liquid or solution, n.o.s., as appropriate.		UN2756								
	Organic peroxide, sample, solid, n.o.s. or Organic peroxide, liquid or solution, n.o.s., as appropriate.		UN2255								
	Organic peroxide, solid, n.o.s.	Organic peroxide	NA9187	Organic peroxide	173.153	173.154	Forbidden	25 pounds	1,2	1,2	Stow separate from combustible materials, explosives, or acids.
	Organic peroxide, trial quantity, n.o.s. See Organic peroxide, solid, n.o.s. or Organic peroxide, liquid or solution, n.o.s., as appropriate.		UN2899								

176

Chapter I—Research and Special Programs Administration §172.101

Description	Class	ID No.	Label	Packaging (Exceptions)	Packaging (Specific)	Max qty (passenger)	Max qty (cargo)	Vessel stowage	Vessel req.	Other
Organic phosphate mixture, Organic phosphate compound mixture, or Organic phosphorus compound mixture; liquid	Poison B	NA2783	Poison	173.359	173.359	1/2 pint	1 quart	1,2	5	
Organic phosphate mixture, Organic phosphate compound mixture, or Organic phosphorus compound mixture; solid or dry	Poison B	NA2783	Poison	173.377	173.377	50 pounds	200 pounds	1,2	4	
Organic phosphate, Organic phosphorus compound, or Organic phosphorus compound; mixed with compressed gas	Poison A	NA1955	Poison gas	None	173.334	Forbidden	Forbidden	1	5	Shade from radiant heat
Organic phosphate, Organic phosphate compound, or Organic phosphorus compound; liquid	Poison B	NA2783	Poison	None	173.358	Forbidden	1 quart	1,2	5	
Organic phosphate, Organic phosphate compound, or Organic phosphorus compound; solid or dry	Poison B	NA2783	Poison	None	173.377	Forbidden	200 pounds	1,2	4	
Organochlorine pesticide, liquid, n.o.s. (compounds and preparations),	Flammable liquid Poison B	UN2762 UN2761	Flammable liquid Poison	173.118 173.345	173.119 173.346	1 quart 1 quart	10 gallons 55 gallons	1,2 1,2	1 1,2	
Organochlorine pesticide, solid, n.o.s (compounds and preparations),	Poison B	UN2761	Poison	173.364	173.365	50 pounds	200 pounds	1,2	1,2	
Organophosphorus pesticide, liquid, n.o.s. (compounds and preparations),	Flammable liquid Poison B	UN2784 UN2783	Flammable liquid Poison	None 173.359	173.119 173.359	Forbidden Forbidden	1 quart 1 quart	1,2 1,2	5 5	
Organophosphorus pesticide, solid, n.o.s. (compounds and preparations),	Poison B	UN2783	Poison	173.377	173.377	Forbidden	200 pounds	1,2	4	
Organotin pesticide, liquid, n.o.s. (compounds and preparations),	Flammable liquid Poison B	UN2787 UN2786	Flammable liquid Poison	173.118 173.345	173.119 173.346	1 quart 1 quart	10 gallons 55 gallons	1,2 1,2	1 1,2	
Organotin pesticide, solid, n.o.s. (compounds and preparations),	Poison B	UN2786	Poison	173.364	173.365	50 pounds	200 pounds	1,2	1,2	
ORM-A, n.o.s.	ORM-A	NA1693	None	173.505	173.510	No limit	No limit			
ORM-B, n.o.s.	ORM-B	NA1760	None	173.505	173.510	No limit	No limit			
ORM-C. See 173.500 and 176.900										
ORM-E, liquid or solid, n.o.s. See Hazardous substance, liquid or solid, n.o.s.										
Orthonitroaniline. See Nitroaniline.										
Oxidizer, corrosive, liquid, n.o.s.	Oxidizer	NA9193	Oxidizer and Corrosive	None	173.245	Forbidden	1 quart	1	5	
Oxidizer, corrosive, solid, n.o.s.	Oxidizer	NA9194	Oxidizer and Corrosive	173.153	173.154	25 pounds	25 pounds	1	4	
Oxidizer material packed with other articles. See 173.152										
Oxidizer, n.o.s. or Oxidizing material, n.o.s.	Oxidizer Oxidizer	UN1479 NA9199	Oxidizer Oxidizer and Poison	173.153 None	173.154 173.154	25 pounds Forbidden	25 pounds 1 quart	1,2 1	1,2 5	
Oxidizer, poisonous, liquid, n.o.s.	Oxidizer	NA9200	Oxidizer and Poison	173.153	173.154	25 pounds	25 pounds	1,2	4	
Oxygen or Oxygen, compressed	Nonflammable gas	UN1072	Oxidizer	173.306	173.302 173.314 173.316	150 pounds	300 pounds	1,2	1,2	Under deck stowage must be in a well ventilated space. Stow separate from flammables. Do not overstow with other cargo
Oxygen, refrigerated liquid (cryogenic liquid)	Nonflammable gas	UN1073	Oxidizer	173.320	173.318	Forbidden	Forbidden	1	1	

177

Chemical Hazard Communication Guidebook

§172.101   Title 49—Transportation

§172.101 Hazardous Materials Table

(1)	(2)	(3)	(3A)	(4)	(5) Packaging		(6) Maximum net quantity in one package		(7) Water shipments		
+/A/W	Hazardous materials descriptions and proper shipping names	Hazard class	Identification number	Label(s) required (if not excepted)	(a) Exceptions	(b) Specific requirements	(a) Passenger carrying aircraft or railcar	(b) Cargo only aircraft	(a) Cargo vessel	(b) Passenger vessel	(c) Other requirements
	Paint	Combustible liquid	UN1263	None	173.118a	None	No limit	No limit	1,2	1,2	
	Paint	Flammable liquid	UN1263	Flammable liquid	173.118	173.128	1 quart	55 gallons	1,2	1	
	Paint or paint related material	Corrosive material	NA1760	Corrosive	173.244	173.245	1 quart	1 gallon	1,2	1,2	
	Paint related material	Combustible liquid	NA1263	None	173.118a	None	No limit	No limit	1,2	1,2	
	Paint related material	Flammable liquid	NA1263	Flammable liquid	173.118 173.128	173.128	1 quart	55 gallons	1,2	1	
	Paper caps. See Toy caps										
	Paraformaldehyde	ORM-A	UN2213	None	173.505	173.510	50 pounds	200 pounds	1,2	1,2	
	Paraldehyde	Flammable liquid	UN1264	Flammable liquid	173.118	173.119	1 quart	10 gallons	1,2	1	
	Paramenthane hydroperoxide	Organic peroxide	UN2125	Organic peroxide	173.153	173.224	1 quart	1 quart	1,2	4	
	Paranitroaniline, solid. See Nitroaniline										
	Parathion and compressed gas mixture	Poison A	NA1967	Poison gas	None	173.334	Forbidden	Forbidden	1,3	5	
	Parathion, liquid	Poison B	NA2783	Poison	None	173.358	Forbidden	1 quart	1,3	1,3	
	Parathion mixture, dry	Poison B	NA2783	Poison	173.377	173.377	50 pounds	200 pounds	1,3	1,3	
	Parathion mixture, liquid	Poison B	NA2783	Poison	None	173.359	Forbidden	1 quart	1,3	1,3	
	Paris green, solid. See Copper acetoarsenite, solid										
	Pelargonyl peroxide, technically pure. See Organic peroxide, solid, n.o.s.										
AW	Pentaborane	Flammable liquid	UN1380	Flammable liquid and Poison	None	173.138	Forbidden	Forbidden	1	5	Segregation same as for flammable solids. Separate from flammable gases or liquids, oxidizing materials, or organic peroxides
	Pentaerythrite tetranitrate, desensitized, wet. See High explosive										
	Pentaerythrite tetranitrate (dry)	Forbidden									
	Pentane	Flammable liquid	UN1265	Flammable liquid	173.118	173.119	Forbidden	10 gallons	1,3	4	
	Pentanitroaniline (dry)	Forbidden									
	Pentolite, dry. See High explosive										
	Peracetic acid solution, not over 43% peracetic acid and not over 6% hydrogen peroxide	Organic peroxide	NA2131	Organic peroxide	173.223	173.223	1 pint	5 pints	1	4	Shade from radiant heat
	Perchlorate, n.o.s.	Oxidizer	NA1481	Oxidizer	173.153	173.154	25 pounds	100 pounds	1,3	1,3	Stow away from powdered metals

178

Chapter I—Research and Special Programs Administration §172.101

	Material	Class	ID No.	Label(s)	(a)	(b)	Packaging Exceptions	Packaging Specific Requirements	Max. Qty. Passenger	Max. Qty. Cargo	Cargo Loc.	Pass. Loc.	Other Requirements
	Perchloroethylene. See Tetrachloroethylene												
	Perchloric acid, exceeding 50% but not exceeding 72% strength	Oxidizer	NA 1897	Oxidizer			None	173.239	Forbidden	5 pints	1	5	Segregation same as for corrosive materials. Stow away from hydrazine.
	Perchloric acid exceeding 72% strength	Forbidden											
	Perchloric acid, not over 50% acid	Oxidizer	UN1873	Oxidizer			173.244	173.239	Forbidden	5 pints	1	1	Segregation same as for corrosive materials. Stow away from hydrazine.
+	Perchloromethyl mercaptan	Poison B	UN1802	Poison			173.345	173.260	Forbidden	10 pounds	1	4	
	Percussion cap	Class C explosive	UN1670	None			None	173.107	50 pounds	150 pounds	1,3	1,3	
	Percussion fuze	Class C explosive		Explosive C			None	173.105	50 pounds	150 pounds	1,3	1,3	
<	Perfluoro-2-butene	ORM-A	NA2422	None			173.505	173.606	10 gallons	55 gallons	1,2	1,2	Separate from ammonium compounds, hydrogen peroxide, and acids
	Permanganate, n.o.s.	Oxidizer	NA1482	Oxidizer			173.153	173.154	25 pounds	100 pounds			
	Permanganate of potash. See Potassium permanganate												
	Peroxide, organic. See Organic Peroxide												
	Peroxyacetic acid, not more than 43% and with not more than 6% hydrogen peroxide. See Peracetic acid solution, not over 43% peracetic acid and not over 6% hydrogen peroxide.		UN2131										
	Peroxyacetic acid, more than 43% and with more than 6% hydrogen peroxide	Forbidden											
W	Pesticide, water reactive, including but not limited to fungicides, and herbicides, etc., which contain manganese ethylenebisdithiocarbamate	ORM-C	NA2210	None			173.505	173.1040				2	Keep dry
W	Petroleum coke (uncalcined)	ORM-C		None			173.505	173.1045	No limit	No limit			Not permitted if temperature of material is at or above 130 deg F
	Petroleum distillate	Combustible liquid	UN1268	None			173.118a	173.510	No limit	No limit	1,2	1,2	
	Petroleum distillate	Flammable liquid	UN1268	Flammable liquid			173.118	173.119	1 quart	10 gallons	1,2	1,2	
	Petroleum ether	Flammable liquid	UN1271	Flammable liquid			173.118	173.119	1 quart	10 gallons	1,3	4	
	Petroleum gas, liquefied. See Liquefied petroleum gas												
	Petroleum naphtha	Combustible liquid	UN1255	None			173.118a	173.510	No limit	No limit	1,2	1,2	
	Petroleum naphtha	Flammable liquid	UN1255	Flammable liquid			173.118	173.119	1 quart	10 gallons	1,2	1	
	Petroleum oil, n.o.s. See Oil												
<+	Phencapton	ORM-A	NA2783	None			173.505	173.510	No limit	No limit	1,2	1,2	
	Phenol	Poison B	UN1671	Poison			173.364	173.369	50 pounds	250 pounds	1,2	1,2	
	Phenol, liquid or solution (liquid tar acid containing over 50% phenol)	Poison B	NA2821	Poison			173.345	173.349	1 quart	55 gallons			
	Phenoxy pesticide, liquid, n.o.s. (compounds and preparations)	Flammable liquid	UN2766	Flammable liquid			173.118	173.119	1 quart	10 gallons	1,2	1	
	Phenoxy pesticide, liquid, n.o.s. (compounds and preparations)	Poison B	UN2765	Poison			173.345	173.346	1 quart	55 gallons	1,2	1,2	
	Phenoxy pesticide, solid, n.o.s. (compounds and preparations)	Poison B	UN2765	Poison			173.364	173.365	50 pounds	200 pounds	1,2	1,2	
<	Phenylenediamine, meta or para, solid	ORM-A	UN1673	None			173.505	173.510	No limit	No limit	1	5	
	Phenyldichloroarsine	Poison B	NA1556	Poison			None	173.355	Forbidden	30 gallons			
	m-Phenylene diaminediperchlorate (dry)	Forbidden											

179

Chemical Hazard Communication Guidebook

§172.101   Title 49—Transportation

§172.101 Hazardous Materials Table

(1) +/A/W	(2) Hazardous materials descriptions and proper shipping names	(3) Hazard class	(3A) Identification number	(4) Label(s) required (if not excepted)	(5) Packaging (a) Exceptions	(5) Packaging (b) Specific requirements	(6) Maximum net quantity in one package (a) Passenger carrying aircraft or railcar	(6) (b) Cargo only aircraft	(7) Water shipments (a) Cargo vessel	(7) (b) Passenger vessel	(7) (c) Other requirements
	Phenyl mercaptan	Poison B	UN2337	Poison	173.345	173.346	Forbidden	10 gallons	1,2	1	
	Phenyltrichlorosilane	Corrosive material	UN1804	Corrosive	None	173.280	Forbidden	10 gallons	1	1	Keep dry
	Phenylurea pesticide, liquid, n.o.s. (compounds and preparations)	Flammable liquid	UN2768	Flammable liquid	173.118	173.119	1 quart	10 gallons	1,2	1	
	Phenylurea pesticide, liquid, n.o.s. (compounds and preparations)	Poison B	UN2767	Poison	173.345	173.346	1 quart	55 gallons	1,2	1,2	
	Phenylurea pesticide, solid, n.o.s. (compounds and preparations)	Poison B	UN2767	Poison	173.364	173.365	50 pounds	200 pounds	1,2	1,2	
	Phosgene (diphosgene)	Poison A	UN1076	Poison gas	None	173.333	Forbidden	Forbidden	1	5	Segregation same as for flammable gases
	Phosphine	Poison A	UN2199	Poison gas and Flammable gas	None	173.328	Forbidden	Forbidden	1	5	Segregation same as for flammable gases
	Phosphoric acid	Corrosive material	UN1805	Corrosive	173.244	173.245	1 quart	10 gallons	1,2	1,2	Glass carboys in hampers not permitted under deck
	Phosphoric acid triethyleneimine. See Tris-(1-aziridiyl) phosphine oxide										
	Phosphoric anhydride (phosphorus pentoxide)	Corrosive material	NA1807	Corrosive	None	173.188	Forbidden	100 pounds	1,2	1,2	Keep dry. Glass bottles not permitted under deck
	Phosphorus, amorphous, red	Flammable solid	UN1338	Flammable solid	None	173.189	Forbidden	11 pounds	1,2	1,2	
	Phosphorus bromide. See Phosphorus tribromide										
	Phosphorus chloride. See Phosphorus trichloride										
	Phosphorus heptasulfide	Flammable solid	UN1339	Flammable solid	None	173.225	Forbidden	10 pounds	1,2	1	Separate from oxidizing materials
	Phosphorus oxybromide	Corrosive material	UN1939	Corrosive	None	173.271	Forbidden	1 quart	1	1	Keep dry. Glass carboys not permitted on passenger vessels
	Phosphorus oxychloride	Corrosive material	UN1810	Corrosive	None	173.271	Forbidden	1 quart	1	1	Keep dry. Glass carboys not permitted on passenger vessels
	Phosphorus pentachloride, solid	Corrosive material	UN1806	Corrosive	None	173.191	Forbidden	5 pounds	1	1	Keep dry
	Phosphorus pentasulfide	Flammable solid	UN1340	Flammable solid and Dangerous when wet	None	173.225	Forbidden	11 pounds	1,2	1,2	Separate from oxidizing material

180

Chapter I—Research and Special Programs Administration   §172.101

Name	Class	UN/NA	Label	Packaging exceptions	Packaging requirements	Max qty passenger	Max qty cargo	Water shipments	Other
Phosphorus sesquisulfide	Flammable solid	UN1341	Flammable solid and Dangerous when wet	None	173.225	Forbidden	11 pounds	1,2	1	Separate from oxidizing materials
Phosphorus tribromide	Corrosive material	UN1808	Corrosive	None	173.270	Forbidden	1 quart	1	1	Keep dry. Glass carboys not permitted on passenger vessels
Phosphorus trichloride	Corrosive material	UN1809	Corrosive	None	173.271	Forbidden	1 quart	1	1	Keep dry. Glass carboys not permitted on passenger vessels
Phosphorus trisulfide	Flammable solid	UN1343	Flammable solid	None	173.225	Forbidden	10 pounds	1,2	1	Separate from oxidizing materials
Phosphorus, white or yellow, dry	Flammable solid	UN1381	Flammable solid and Poison	None	173.190	Forbidden	Forbidden	1,2	5	Separate from flammable gases or liquids, oxidizing materials, or organic peroxides
Phosphorus, white or yellow, in water	Flammable solid	UN1381	Flammable solid and Poison	None	173.190	Forbidden	25 pounds	1,2	5	Separate from flammable gases or liquids, oxidizing materials, or organic peroxides
Phosphorus (white or red) and a chlorate, mixtures of	Forbidden									
Phosphoryl chloride. See Phosphorus oxychloride										
Photographic film. See Film										
Photographic flash powder. See Fireworks, special or *Low explosive*										
Phthalimide derivative pesticide, liquid, n.o.s. *(compounds and preparations)*,	Flammable liquid	UN2774	Flammable liquid	173.118	173.119	1 quart	10 gallons	1,2	1	
Phthalimide derivative pesticide, liquid, n.o.s. *(compounds and preparations)*,	Poison B	UN2773	Poison	173.345	173.346	1 quart	55 gallons	1,2	1,2	
Phthalimide derivative pesticide, solid, n.o.s. *(compounds and preparations)*,	Poison B	UN2773	Poison	173.364	173.365	50 pounds	200 pounds	1,2	1,2	
Picrate, dry. See High explosive										
Picrate of ammonia. See High explosive										
Picric acid, wet, with not less than 10% water	Flammable solid	NA1344	Flammable solid	173.192	173.193	1 pound	25 pounds	1	5	Under deck stowage permitted on cargo vessels if wet with more than 30% water. Stow away from heavy metals and their compounds
Picric acid, dry. See High explosive										
Picric acid, wet with not less than 10% water, over 23 pounds. See High explosive										
Pinane hydroperoxide, *technically pure. See Organic peroxide, liquid or solution, n.o.s.*		UN2162								
Pinane hydroperoxide solution *not over 65% peroxide*	Organic peroxide Flammable liquid	UN2182	Organic peroxide Flammable liquid	173.153	173.224	1 quart	1 quart	1,2	4	
Pinene	Flammable liquid	UN2368	None	173.118	173.119	1 quart	10 gallons	1,2	1	
Pine oil	Combustible liquid	UN1272	None	173.118a	None	No limit	No limit	1,2	1,2	
Pinwheels. See Fireworks, common										
Pivaloyl chloride. See Trimethylacetyl chloride										
Poisonous liquid or gas, n.o.s	Poison A	NA 1953	Poison gas and Flammable gas	None	173.328	Forbidden	Forbidden	1	5	Segregation same as for flammable gas
Poisonous liquid, n.o.s. *or* Poison B, liquid, n.o.s.	Poison A Poison B	NA1955 UN2810	Poison gas Poison	None 173.345	173.328 173.346	Forbidden 1 quart	Forbidden 55 gallons	1 1,2	5 1	

181

Chemical Hazard Communication Guidebook

§172.101 **Title 49—Transportation**

§172.101 Hazardous Materials Table

(1) +/A/W	(2) Hazardous materials descriptions and proper shipping names	(3) Hazard class	(3A) Identification number	(4) Label(s) required (if not excepted)	(5) Packaging (a) Exceptions	(5) Packaging (b) Specific requirements	(6) Maximum net quantity in one package (a) Passenger carrying aircraft or railcar	(6) (b) Cargo only aircraft	(7) Water shipments (a) Cargo vessel	(7) (b) Passenger vessel	(7) (c) Other requirements
	Poisonous solid, corrosive, n.o.s.	Poison B	UN2928	Poison and Corrosive	173.364	173.365	25 pounds	100 pounds	1	4	
	Poisonous solid, n.o.s. or Poison B, solid, n.o.s.	Poison B	UN2811	Poison	173.364	173.365	50 pounds	200 pounds	1,2	1	
	Polymerizable material. See 173.21										
	Potassium arsenate, solid	Poison B	UN1677	Poison	173.364	173.365	50 pounds	200 pounds	1,2	1,2	
	Potassium arsenite, solid	Poison B	UN1678	Poison	173.364	173.365	50 pounds	200 pounds	1,2	1,2	
	Potassium bifluoride solution. *See Potassium hydrogen fluoride solution*										
	Potassium bromate	Oxidizer	UN1484	Oxidizer	173.153	173.154	25 pounds	100 pounds	1,2	1,2	
	Potassium carbonyl	Forbidden									
	Potassium chlorate (*potash chlorate*)	Oxidizer	UN1485	Oxidizer	173.153	173.163	25 pounds	100 pounds	1,2	1,2	Separate from ammonium compounds. Stow away from powdered metals.
	Potassium cyanide, solid	Poison B	UN1680	Poison	173.370	173.370	25 pounds	200 pounds	1,2	1,2	Separate from ammonium compounds. Stow away from powdered metals.
	Potassium cyanide solution	Poison B	UN1680	Poison	173.345	173.352	1 quart	55 gallons	1,2	1,2	Stow away from acids
	Potassium dichloro isocyanurate. See Potassium dichloro-s-triazinetrione										Stow away from acids
	Potassium dichloro-s-triazinetrione, *dry (containing more than 39% available chlorine)*	Oxidizer	NA2465	Oxidizer	173.153	173.217	50 pounds	100 pounds	1,2	1,2	
A	Potassium dichromate	ORM-A	NA1479	None	173.505	173.510	No limit	No limit	1,2	1,2	
A	Potassium fluoride	ORM-B	UN1812	None	173.505	173.510	No limit	No limit	1,2	1,2	
	Potassium fluoride solution	Corrosive material	UN1812	Corrosive	173.244	173.249	1 quart	5 gallons	1,2	1,2	
	Potassium hydrate. *See Potassium hydroxide*										
	Potassium hydrogen fluoride solution	Corrosive material	NA1811	Corrosive	173.244	173.249	1 quart	5 gallons	1,2	1,2	
A	Potassium hydrogen sulfate, solid	ORM-B	UN2509	None	173.505	173.800	25 pounds	100 pounds	1,2	1,2	Keep dry. Do not stow with metals or alloys such as brass, copper, tin, zinc, aluminum, solder, or lead
	Potassium hydroxide, dry, solid, flake, bead, or granular	Corrosive material	UN1813	Corrosive	173.244	173.245b	25 pounds	100 pounds	1,2	1,2	
	Potassium hydroxide, liquid *or* solution	Corrosive material	UN1814	Corrosive	173.244	173.249	1 quart	10 gallons	1,2	1,2	
	Potassium hypochlorite solution. *See Hypochlorite solutions containing more than 7% available chlorine by weight*										
A	Potassium metabisulfite	ORM-B	NA2693	None	173.505	173.510	No limit	No limit			

182

Chapter I—Research and Special Programs Administration §172.101

Name	Class	UN/NA	Label	Packaging (exceptions)	Packaging (specific)	Max qty passenger	Max qty cargo	Stowage	Other requirements
Potassium, metal or metallic	Flammable solid	UN2257	Flammable solid and Dangerous when wet	None	173.206	Forbidden	25 pounds	1,2	5	Segregation same as for flammable solids labeled Dangerous When Wet
Potassium, metal liquid alloy	Flammable solid	UN1420	Flammable solid and Dangerous when wet	None	173.202	Forbidden	1 pound	1,2	5	Segregation same as for flammable solids labeled Dangerous When Wet
Potassium nitrate	Oxidizer	UN1486	Oxidizer	173.153	173.182	25 pounds	100 pounds	1,2	1,2	
Potassium nitrate mixed (fused) with sodium nitrite. See Sodium nitrite mixed (fused) with potassium nitrate										
Potassium nitrite	Oxidizer	UN1488	Oxidizer	173.153	173.154	25 pounds	100 pounds	1,2	1,2	Separate from ammonium compounds and cyanides. Stow away from foodstuffs
Potassium perchlorate	Oxidizer	UN1489	Oxidizer	173.153	173.219	25 pounds	100 pounds	1,3	1,3	Stow away from powdered metals
Potassium permanganate	Oxidizer	UN1490	Oxidizer	173.153	173.154 173.194	25 pounds	100 pounds	1,2	1,2	Separate from ammonium compounds and hydrogen peroxide Keep dry
Potassium peroxide	Oxidizer	UN1491	Oxidizer	None	173.187	Forbidden	100 pounds	1,2	1,2	Separate from liquid acids, flammable gases or liquids, oxidizing materials or organic peroxides
Potassium persulfate	Oxidizer	UN1492	Oxidizer	173.153	173.154	50 pounds	200 pounds	1,2	1,2	
Potassium sulfide	Flammable solid	UN1382	Flammable solid	173.153	173.207	Forbidden	300 pounds	1,2	1,2	
Potassium superoxide	Oxidizer	UN2466	Oxidizer	None	173.187	Forbidden	100 pounds	1,2	1	Keep dry. Stow away from powdered metals, permanganates and combustible packagings and cargo
Pressurized product. See Compressed gas, n.o.s.										
Primer. See Cannon primer, Combination primer, or Small-arms primer										
Primer, detonating. See Detonating primers, Class A or Class C explosives										
Projectile, explosive. See Explosive projectile										
Projectile, gas, nonexplosive. See Chemical ammunition, nonexplosive (containing a Poison A, Poison B or irritating material, as appropriate)										
Projectile, gas, smoke, or incendiary, with burster or booster with or without detonating fuze. See Explosive projectile										
Projectile, illuminating, incendiary or smoke, with expelling charge but without bursting charge. See Fireworks, special										
Projectile, sand-loaded, empty or solid. See 173.55										
Propane or Liquefied petroleum gas. See Liquefied petroleum gas										
Propargyl alcohol	Flammable liquid	NA1986	Flammable liquid and Poison	None	173.119	Forbidden	1 quart	1,2	5	
Propellant explosive	Class A explosive			None	173.64	Forbidden	Forbidden	6	5	Magazine stowage authorized
Propellant explosive in water (smokeless powder)	Class B explosive		Explosive B	None	173.93	Forbidden	Forbidden	1,3	5	Magazine stowage authorized
Propellant explosive in water, unstable, condemned, or deteriorated (smokeless powder)	Class B explosive		Explosive B	None	173.93	Forbidden	Forbidden	1,3	5	

183

Chemical Hazard Communication Guidebook

§172.101 Title 49—Transportation

§172.101 Hazardous Materials Table

(1) +/A/W	(2) Hazardous materials descriptions and proper shipping names	(3) Hazard class	(3A) Identification number	(4) Label(s) required (if not excepted)	(5) Packaging (a) Exceptions	(5) Packaging (b) Specific requirements	(6) Maximum net quantity in one package (a) Passenger carrying aircraft or railcar	(6) Maximum net quantity in one package (b) Cargo only aircraft	(7) Water shipments (a) Cargo vessel	(7) Water shipments (b) Passenger vessel	(7) Water shipments (c) Other requirements
	Propellant explosive, liquid	Class B explosive		Explosive B	None	173.93	Forbidden	10 pounds	1,2	5	Magazine stowage authorized
	Propellant explosive, solid	Class B explosive		Explosive B	None	173.93	Forbidden	10 pounds	1,3	5	Magazine stowage authorized
	Propionaldehyde	Flammable liquid	UN1275	Flammable liquid	173.118	173.119	1 quart	10 gallons	1,2	1	
	Propionic acid	Corrosive material	UN1848	Corrosive	173.244	173.245	1 quart	5 gallons	1,2	1,2	Separated by a complete compartment or hold from organic peroxides
	Propionic acid solution	Corrosive material	UN1848	Corrosive	173.244	173.245	1 quart	10 gallons	1,2	1,2	Separated by a complete compartment or hold from organic peroxides
	Propionic anhydride	Corrosive material	UN2496	Corrosive	173.244	173.245	1 quart	1 quart	1,2	1	Keep dry
	Propionyl peroxide, not more than 28% in solution. See Organic peroxide, liquid or solution, n.o.s.	Forbidden	UN2132								
	Propionyl peroxide, more than 28% in solution										
	Propyl acetate	Flammable liquid	UN1276	Flammable liquid	173.118	173.119	1 quart	10 gallons	1,2	1	
	Propyl alcohol	Flammable liquid	UN1274	Flammable liquid	173.118	173.125	1 quart	10 gallons	1,2	1,2	
	Propylamine	Flammable liquid	UN1277	Flammable liquid	None	173.119	Forbidden	10 gallons	1,3	5	
	Propyl chloride	Flammable liquid	UN1278	Flammable liquid	None	173.119	Forbidden	10 gallons	1,3	5	
	Propylene or Liquefied petroleum gas. See Liquefied petroleum gas										
	Propylenediamine	Flammable liquid	UN2258	Flammable liquid	173.118	173.119	1 quart	10 gallons	1,2	1	
	Propylene dichloride	Flammable liquid	UN1279	Flammable liquid	173.118	173.119	1 quart	10 gallons	1,2	1	
	Propyleneimine, inhibited	Flammable liquid	UN1921	Flammable liquid	None	173.139	Forbidden	5 pints	1,2	1	
	Propylene oxide	Flammable liquid	UN1280	Flammable liquid	173.118	173.119	Forbidden	1 gallon	1,3	4	
	Propyl formate	Flammable liquid	UN1281	Flammable liquid	173.118	173.119	1 quart	10 gallons	1,2	1	
	Propyl mercaptan	Flammable liquid	NA2402	Flammable liquid	None	173.141	Forbidden	10 gallons	1,2	5	
	Propyl trichlorosilane	Corrosive material	UN1816	Corrosive	None	173.280	Forbidden	10 gallons	1	1	Keep dry

184

Appendix B: Regulations Reference Guide

Chapter I—Research and Special Programs Administration §172.101

Name	Class	ID No.	Col 1	Col 2	Pass. limit	Cargo limit	Label	Stow	Notes
Prussic acid. See Hydrocyanic acid, as appropriate									
Pyridine	Flammable liquid	UN1282	Flammable liquid	173.118	173.119	1 quart	10 gallons	1,2	1
Pyridine perchlorate	Forbidden								
Pyrophoric liquid, n.o.s. *or* Pyrofonic liquid, n.o.s.	Flammable liquid	UN2845	Flammable liquid	None	173.134	Forbidden	Forbidden	1	5	Shade from radiant heat. Separate from flammable gases or liquids, oxidizing materials, or organic peroxides
Pyrosulfuryl chloride	Corrosive material	UN1817	Corrosive	173.244	173.247	1 quart	1 quart	1	4	Keep dry. Glass carboys not permitted on passenger vessels
Pyroxylin plastic, scrap	Flammable solid	NA1325	Flammable solid	None	173.195	Forbidden	Forbidden	1	5	Shade from radiant heat
Pyroxylin plastic, rods, sheets, rolls, or tubes	Flammable solid	NA1325	Flammable solid	173.197	173.197	50 pounds	350 pounds	1,3	1
Pyrrolidine	Flammable liquid	UN1922	Flammable liquid	173.118	173.119	Forbidden	10 gallons	1,2	1
Quebrachitol pentanitrate	Forbidden								
Quicklime. See Calcium oxide									
Radioactive material, articles, manufactured from natural *or* depleted uranium *or* natural thorium	Radioactive material	UN2909	None	173.421–1 173.422	173.421– 173.424			1,2	1
Radioactive material, empty packages	Radioactive material	UN2908	Empty	173.421–1 173.427	173.421– 173.427			1,2	1,2
Radioactive material, fissile, n.o.s	Radioactive material	UN2918	Radioactive	173.453	173.417			1,2	1,2
Radioactive material, instruments and articles	Radioactive material	UN2911	None	173.421–1 173.422	173.421–			1,2	1,2
Radioactive material, limited quantity, n.o.s	Radioactive material	UN2910	None	173.421 173.421–1	173.422 173.421–			1,2	1,2
Radioactive material, low specific activity *or* LSA, n.o.s	Radioactive material	UN2912	Radioactive	173.421 173.422 173.424	173.425			1,2	1,2
Radioactive material, n.o.s	Radioactive material	UN2982	Radioactive	173.421 173.422 173.424	173.415 173.416			1,2	1,2
Radioactive material, special form, n.o.s	Radioactive material	UN2974	Radioactive	173.421 173.422	173.415 173.416			1,2	1,2
Railway Fusee. See Fusee									
Railway torpedo. See Torpedo, railway									
Refrigerant gas, n.o.s. *or* Dispersant gas n.o.s.	Nonflammable gas	UN1078	Nonflammable gas	173.306	173.304 173.314 173.315	150 pounds	300 pounds	1,2	1,2
Refrigerant gas, n.o.s. *or* Dispersant gas n.o.s.	Flammable gas	NA1954	Flammable gas	173.306	173.304 173.314 178.315	Forbidden	300 pounds	1,2	1,2
Refrigerating machine	Nonflammable gas	UN2857	Nonflammable gas	173.306 173.307 173.306		No limit	No limit	1,3	1,3
Refrigerating machine	Flammable gas	NA1954	Flammable gas			No limit	No limit	1,3	1,3
Refrigerating machine	Flammable liquid	NA1993	Flammable liquid	173.130 173.306		No limit	No limit	1,2	1
Resin solution *(resin compound, liquid)*	Flammable liquid	UN1866	Flammable liquid	173.118	173.119	1 quart	55 gallons	1,2	1

185

Chemical Hazard Communication Guidebook

§172.101 Title 49—Transportation

§172.101 Hazardous Materials Table

(1) +/A/W	(2) Hazardous materials descriptions and proper shipping names	(3) Hazard class	(3A) Identification number	(4) Label(s) required (if not excepted)	(5) Packaging (a) Exceptions	(5) Packaging (b) Specific requirements	(6) Maximum net quantity in one package (a) Passenger carrying aircraft or railcar	(6) (b) Cargo only aircraft	(7) Water shipments (a) Cargo vessel	(7) (b) Passenger vessel	(7) (c) Other requirements
	Rifle grenade. See Grenade, hand *or* rifle, explosive										
	Rifle powder. See Propellant explosive *or* Black powder										
	Road asphalt or tar, liquid See Asphalt, cut back										
	Road asphalt or tar (when heated to or above its flash point). See Asphalt										
	Rocket ammunition with empty, inert, or solid loaded projectile	Class A explosive		Explosive A	None	173.57	Forbidden	Forbidden	6	5	
	Rocket ammunition with empty projectile	Class B explosive		Explosive B	None	173.90	Forbidden	Forbidden	1,3	5	
	Rocket ammunition with explosive projectile	Class A explosive		Explosive A	None	173.57	Forbidden	Forbidden	6	5	
	Rocket ammunition with gas projectile	Class A explosive		Explosive A	None	173.57	Forbidden	Forbidden	6	5	
	Rocket ammunition with illuminating projectile	Class A explosive		Explosive A	None	173.57	Forbidden	Forbidden	6	5	
	Rocket ammunition with incendiary projectile	Class A explosive		Explosive B	None	173.90	Forbidden	Forbidden	6	5	
	Rocket ammunition with inert loaded projectile	Class A explosive		Explosive A	None	173.57	Forbidden	Forbidden	1,3	5	
	Rocket ammunition with smoke projectile	Class A explosive		Explosive A	None	173.57	Forbidden	Forbidden	6	5	
	Rocket ammunition with solid projectile	Class B explosive		Explosive B	None	173.90	Forbidden	Forbidden	1,3	5	
	Rocket body, with electric primer or electric squib. See 173.55										
	Rocket engine, liquid	Class B explosive		Explosive B	None	173.95	Forbidden	Forbidden	1,2	5	Magazine stowage authorized
	Rocket fireworks. See Fireworks, common										
	Rocket head. See Explosive projectile										
	Rocket motor	Class A explosive		Explosive A	None	173.79	Forbidden	Forbidden	6	5	
	Rocket motor	Class B explosive		Explosive B	None	173.92	Forbidden	550 pounds	1,3	5	
	Roman candle. See Fireworks, common										
	Rubidium metal	Flammable solid	UN1423	Flammable solid and Dangerous when wet	None	173.206	Forbidden	225 pounds	1,2	5	Segregation same as for flammable solid labeled Dangerous When Wet

186

Appendix B: Regulations Reference Guide

Chapter I—Research and Special Programs Administration   **§172.101**

Name	Hazard class	ID No.	Label(s)	Packaging Exceptions	Packaging Specific	Max qty passenger	Max qty cargo	Vessel stowage	Other	Segregation / Other requirements
Rubidium metal, in cartridges	Flammable solid	UN1423	Flammable solid and Dangerous when wet	173.206			25 pounds	1,2	4	Segregation same as for flammable solid labeled Dangerous When Wet
Safety fuse. *See* Fuse, safety										
Safety squib	Class C explosive		Explosive C	None	173.108	50 pounds	150 pounds	1,3	1,3	
Salute. See Fireworks common or special										
Samples. See 172.101(c)(12)										
Sand acid. *See* Hydrofluorosilicic acid										
Selenic acid, liquid	Corrosive material	UN1905	Corrosive	None	173.245	Forbidden	5 pints	1,2	1,2	Keep dry
Selenium nitride	Forbidden									
Selenium oxide	Poison B	NA2811	Poison	173.364	173.365	50 pounds	200 pounds	1,2	1,2	Stow away from foodstuffs
Self-lighting cigarette	Flammable solid	UN1867	Flammable solid	173.21		Forbidden	Forbidden	1,2	1,2	
Self propelled vehicle. *See* Motor vehicle										
Shaped charge, commercial. *See* High explosive (173.65(h))										
Shaped charges (commercial) containing more than 8 ounces of explosives	Forbidden									
Shell, fireworks. *See* Fireworks, common or special										
Ship, distress signal. *See* Fireworks, special										
Signal flare	Class C explosive		Explosive C	None	173.108	50 pounds	200 pounds	1,2	1,2	
Silicofluoric acid. *See* Hydrofluorosilicic acid										
Silicon chloride *or* Silicon tetrachloride	Corrosive material	UN1818	Corrosive	173.244	173.247	1 quart	1 gallon	1	1	Keep dry. Glass carboys not permitted on passenger vessels
Silicon chrome, exothermic. *See* Ferrochrome, exothermic										
Silicon tetrafluoride	Nonflammable gas	UN1859	Nonflammable gas	173.306	173.302	Forbidden	300 pounds	1	4	Stow away from acids
Silver acetylide (dry)	Forbidden									
Silver azide (dry)	Forbidden									
Silver chlorite (dry)	Forbidden									
Silver cyanide	Poison B	UN1684	Poison	173.370	173.370	25 pounds	200 pounds	1,2	1,2	Stow away from foodstuffs
Silver fulminate (dry)	Forbidden									
Silver nitrate	Oxidizer	UN1493	Oxidizer	173.153	173.182	25 pounds	100 pounds	1,2	1,2	Stow away from foodstuffs
Silver oxalate (dry)	Forbidden									
Silver picrate (dry)	Forbidden									
Small arms ammunition	Class C explosive ORM-D		None	173.101	173.1201	50 pounds	150 pounds	1,3	1,3	
Small arms ammunition, irritating (*tear gas*) cartridge	Class C explosive		None	178.101	178.101	65 pounds gross Forbidden	65 pounds gross 150 pounds	1,3	1,3	
Small arms primer	Class C explosive		Irritant	None	173.107	50 pounds	150 pounds	1,3	1,3	
Smoke candle	Class C explosive		Explosive C	None	173.108	50 pounds	200 pounds	1,3	1,3	
Smoke generator. See Chemical ammunition, nonexplosive (containing a Poison A, Poison B, or irritating material, as appropriate)										

187

Chemical Hazard Communication Guidebook

§172.101 Title 49—Transportation

§172.101 Hazardous Materials Table

(1) +/A/W	(2) Hazardous materials descriptions and proper shipping names	(3) Hazard class	(3A) Identification number	(4) Label(s) required (if not excepted)	(5) Packaging (a) Exceptions	(5) Packaging (b) Specific requirements	(6) Maximum net quantity in one package (a) Passenger carrying aircraft or railcar	(6) Maximum net quantity in one package (b) Cargo only aircraft	(7) Water shipments (a) Cargo vessel	(7) Water shipments (b) Passenger vessel	(7) Water shipments (c) Other requirements
	Smoke grenade	Class C explosive		Explosive C	None	173.108	50 pounds	150 pounds	1,3	1,3	
	Smokeless powder for cannon or small arms. See Propellant explosive, Class A or B, as appropriate										
	Smokeless powder for small arms (100 pounds or less)	Flammable solid	NA1325	Flammable solid	173.88	173.197a	Forbidden	Forbidden	1,3	1,3	Segregation same as for explosives
	Smoke pot	Class C explosive		Explosive C	None	173.108	50 pounds	200 pounds	1,3	1,3	
	Smoke projectile with bursting charge. See Explosive projectile										
	Smoke projectile with expelling charge but without bursting charge. See Fireworks, special										
	Smoke signal	Class C explosive		Explosive C	None	173.108	50 pounds	200 pounds	1,3	1,3	
	Soda amatol. See High explosive										
	Soda lime, solid	Corrosive material	UN1907	Corrosive	173.244	173.245b	25 pounds	100 pounds	1,2	1,2	Keep dry
	Sodium acid sulfate, solid or solution. See appropriate Sodium hydrogen sulfate entry										
A	Sodium aluminate, solid	ORM-B	UN2812	None	173.505	173.8(e)	25 pounds	100 pounds	1,2	1,2	
	Sodium aluminate solution	Corrosive material	UN1819	Corrosive	173.244	173.249	1 quart	5 gallons	1,2	5	
	Sodium aluminum hydride	Flammable solid	UN2835	Flammable solid and Dangerous when wet	None	173.206	Forbidden	25 pounds	1,2	5	Segregation same as for flammable solids labeled Dangerous When Wet
	Sodium amide	Flammable solid	UN1425	Flammable solid and Dangerous when wet	None	173.206	Forbidden	25 pounds	1,2	5	Segregation same as for flammable solids labeled Dangerous When Wet
	Sodium arsenate	Poison B	UN1685	Poison	173.364	173.365 173.368	50 pounds	200 pounds	1,2	1,2	
	Sodium arsenite, liquid, (solution)	Poison B	UN1686	Poison	173.345	173.346	1 quart	55 gallons	1,2	1,2	
+	Sodium azide	Poison B	UN1687	Poison	173.364	173.375	50 pounds	100 pounds	1,2	1,2	
	Sodium bifluoride, solid	Corrosive material	UN2439	Corrosive	173.244	173.245b	25 pounds	100 pounds	1,2	1,2	Stow away from heavy metals, especially lead and its compounds. Stow separate from acids

188

Chapter I—Research and Special Programs Administration §172.101

Sodium bifluoride, solution	Corrosive material	UN2439	Corrosive	173.244	173.245	1 quart	5 gallons	1,2	1,2
Sodium bisulfate, solid or solution. See Sodium hydrogen sulfate, solid or solution									
Sodium bisulfite, solid, or solution See Sodium hydrogen sulfite, solidorsolution									
Sodium bromate	Oxidizer	UN1494	Oxidizer	173.153	173.154	25 pounds	100 pounds	1,2	1,2 Stow separate from ammonium compounds. Stow away from powdered metals
Sodium chlorate (soda chlorate)	Oxidizer	UN1495	Oxidizer	173.153	173.153	25 pounds	100 pounds	1,2	1,2 Stow separate from ammonium compounds. Stow away from powdered metals
Sodium chlorite	Oxidizer	UN1496	Oxidizer	None	173.150	Forbidden	100 pounds	1,2	1,2 Stow separate from ammonium compounds. Stow away from powdered metals
Sodium chlorite solution (not exceeding 42% sodium chlorite)	Corrosive material	UN1908	Corrosive	173.244	173.263	1 quart	4 gallons	1,2	1 Glass carboys in hampers not permitted under deck
Sodium cyanide, solid	Poison B	UN1689	Poison	173.370	173.370	25 pounds	200 pounds	1,2	1,2 Stow away from acids
Sodium cyanide solution	Poison B	UN1689	Poison	173.345	173.352	1 quart	55 gallons	1,2	1,2 Stow away from acids
Sodium dichloroisocyanurate. See Sodium dichloro-s-triazinetrione									
Sodium dichloro-s-triazinetrione (dry, containing more than 39% available chlorine)	Oxidizer	UN2465	Oxidizer	173.153	173.217	50 pounds	100 pounds	1,2	1,2
< Sodium dichromate	ORM-A	NA1479	None	173.505	173.510	No limit	No limit	1,2	1,2
< Sodium fluoride, solid	ORM-B	UN1690	None	173.505	173.510	No limit	No limit	1,2	1,2
Sodium fluoride solution	Corrosive material	UN1690	Corrosive	173.244	173.245	1 quart	5 gallons	1,2	1,2 Stow away from acids
Sodium hydrate. See Sodium hydroxide									
Sodium hydride	Flammable solid	UN1427	Flammable solid and Dangerous when wet	None	173.198	Forbidden	25 pounds	1,2	5 Segregation same as for flammable solids labeled Dangerous When Wet
< Sodium hydrogen sulfate, solid	ORM-B	UN1821	None	173.505	173.800	25 pounds	100 pounds	1,2	1,2
Sodium hydrogen sulfate solution	Corrosive material	UN2837	Corrosive	173.244	173.245	1 quart	1 gallon	1,2	1,2
< Sodium hydrogen sulfite, solid	ORM-B	NA2693	None	173.505	173.800	25 pounds	100 pounds	1,2	1,2
Sodium hydrogen sulfite, solution	Corrosive material	NA2693	Corrosive	173.244	173.245	1 quart	5 gallons	1,2	1,2
Sodium hydrosulfide, solid (with less than 25% water of crystallization)	Flammable solid	UN2318	Flammable solid	173.153	173.154	25 pounds	100 pounds	1,2	1,2
Sodium hydrosulfide, solid (with not less than 25% water of crystallization)	Corrosive material	NA2923	Corrosive	173.244	173.245b	25 pounds	100 pounds	1,2	1,2
Sodium hydrosulfide, solution	Corrosive material	NA2922	Corrosive	173.244	173.245	1 quart	5 gallons	1,2	1,2
Sodium hydrosulfite (sodium dithionite)	Flammable solid	UN1384	Flammable solid	173.153	173.204	25 pounds	100 pounds	1,2	1,2 Keep dry. Below deck stowage in metal drums only. Separate from flammable gases, liquids, oxidizing materials, or organic peroxides
Sodium hydroxide, dry, solid, flake, bead, or granular	Corrosive material	UN1823	Corrosive	173.244	173.245b	25 pounds	200 pounds	1,2	1,2 Keep dry
Sodium hydroxide, liquid or solution	Corrosive material	UN1824	Corrosive	173.244	173.249	1 quart	5 gallons	1,2	1,2
Sodium hypochlorite. See Hypochlorite solution or Hypochlorite solution containing not more than 7% available chlorine									
< Sodium metabisulfite	ORM-B	NA2693	None	173.505	173.510	No limit	No limit		

189

Chemical Hazard Communication Guidebook

§172.101 — Title 49—Transportation

§172.101 Hazardous Materials Table

(1) +/A/W	(2) Hazardous materials descriptions and proper shipping names	(3) Hazard class	(3A) Identification number	(4) Label(s) required (if not excepted)	(5)(a) Packaging Exceptions	(5)(b) Specific requirements	(6)(a) Passenger carrying aircraft or railcar	(6)(b) Cargo only aircraft	(7)(a) Cargo vessel	(7)(b) Passenger vessel	(7)(c) Other requirements
	Sodium, metal or metallic	Flammable solid	UN1428	Flammable solid and Dangerous when wet	None	173.206	Forbidden	25 pounds	1,2	5	Segregation same as for flammable solids labeled Dangerous When Wet
	Sodium, metal dispersion in organic solvent	Flammable solid	UN1429	Flammable solid and Dangerous when wet	None	173.230	Forbidden	10 pounds	1,2	5	Segregation same as for flammable solids labeled Dangerous When Wet
	Sodium, metal liquid alloy	Flammable solid	NA1421	Flammable solid and Dangerous when wet	None	173.202	Forbidden	1 pound	1,2	5	Segregation same as for flammable solids labeled Dangerous When Wet
	Sodium methylate, alcohol mixture	Combustible liquid	NA1289	None	173.118a	None	No limit	No limit	1,2	1,2	
	Sodium methylate, alcohol mixture	Flammable liquid	NA1289	Flammable liquid	173.118	173.119	1 quart	10 gallons	1,2	1	
	Sodium methylate, alcohol mixture	Corrosive material	NA1289	Corrosive	173.244	173.245	1 quart	1 quart	1,2	1	
	Sodium methylate, dry	Flammable solid	UN1431	Flammable solid and Dangerous when wet	173.153	173.154	25 pounds	100 pounds	1,2	1	
	Sodium monoxide, solid	Corrosive material	UN1825	Corrosive	173.244	173.245b	25 pounds	100 pounds	1,2	1,2	Keep dry
	Sodium nitrate	Oxidizer	UN1498	Oxidizer	173.153	173.182	25 pounds	100 pounds	1,2	1,2	Stow separate from ammonium compounds and cyanides. Bagged material not permitted on passenger vessels
	Sodium nitrate bags. See Bags, sodium nitrate, empty and unwashed										
	Sodium nitrite	Oxidizer	UN1500	Oxidizer	173.153	173.234	25 pounds	100 pounds	1,2	1,2	Stow separate from ammonium compounds and cyanides
	Sodium nitrite mixed (fused) with potassium nitrate	Oxidizer	UN1487	Oxidizer	173.153	173.183	25 pounds	100 pounds	1,2	1,2	Stow separate from ammonium compounds and cyanides
	Sodium nitrite mixture (sodium nitrite, sodium nitrate, and potassium nitrate)	Oxidizer	NA1487	Oxidizer	173.153	173.234	25 pounds	100 pounds	1,2	1,2	Stow separate from ammonium compounds and cyanides
A	Sodium pentachlorophenate	ORM-A	UN2567	None	173.505	173.510	No limit	No limit	1,2	1,2	Stow away from powdered metals
	Sodium perchlorate	Oxidizer	UN1502	Oxidizer	173.153	173.154	25 pounds	100 pounds	1,3	1,3	Separate from ammonium compounds and hydrogen peroxide
	Sodium permanganate	Oxidizer	UN1503	Oxidizer	173.153	173.154	25 pounds	100 pounds	1,2	1,2	
	Sodium peroxide	Oxidizer	UN1504	Oxidizer	None	173.187	Forbidden	100 pounds	1,2	1	Keep dry. Stow away from powdered metals, permanganates, combustible packing of other cargo, and combustible foodstuffs

190

Chapter I—Research and Special Programs Administration §172.101

Appendix B: Regulations Reference Guide

Material	Classification	ID No.	Label(s) required	Packaging (Exceptions)	Packaging (Specific requirements)	Max qty per pkg (Passenger)	Max qty per pkg (Cargo)	Vessel stowage (Passenger)	Vessel stowage (Cargo)	Other stowage provisions
Sodium persulfate	Oxidizer	UN1505	Oxidizer	173.153	173.154	50 pounds	200 pounds	1,2	1,2	
Sodium phenolate, solid	Corrosive material	UN2497	Corrosive	173.244	173.245b	25 pounds	100 pounds	1,2	1,2	
Sodium phosphide	Flammable solid	UN1432	Flammable solid and Dangerous when wet	None	173.154	Forbidden	25 pounds	1	5	Stow away from heavy metals, especially lead, and its compounds
Sodium picramate, wet (with at least 20% water)	Flammable solid	UN1349	Flammable solid	None	173.205	Forbidden	25 pounds	1,2	5	Under deck stowage must be readily accessible. Segregation same as for flammable solid labeled Dangerous When Wet.
Sodium picryl peroxide	Forbidden									
Sodium potassium alloy (liquid)	Flammable solid	UN1422	Flammable solid and Dangerous when wet	None	173.202	Forbidden	1 pound	1,2	5	
Sodium potassium alloy (solid)	Flammable solid	UN1422	Flammable solid and Dangerous when wet	None	173.206	Forbidden	25 pounds	1,2	5	Under deck stowage must be readily accessible. Segregation same as for flammable solids labeled Dangerous When Wet
Sodium selenite	Poison B	UN2630	Poison	173.364	173.365	50 pounds	200 pounds	1,2	1,2	
Sodium sulfide, anhydrous or Sodium sulfide with less than 30% water of crystallization	Flammable solid	UN1385	Flammable solid	173.153	173.207	25 pounds	100 pounds	1,2	1,2	Stow separated from liquid acids
Sodium sulfide, hydrated with not less than 30% water	Corrosive material	UN1849	Corrosive	173.244	173.245b	25 pounds	100 pounds	1,2	1,2	Stow away from acids
Sodium superoxide	Oxidizer	UN2547	Oxidizer	None	173.187	Forbidden	100 pounds	1,2	5	Keep dry. Stow away from powdered metals, permanganates and combustible packagings and cargo
Sodium tetranitride	Forbidden									
Sparklers. See Fireworks, common										
Spent iron mass. See Iron mass, spent										
Spent iron sponge. See Iron sponge, spent										
Spent mixed acid. See Nitrating acid, spent										
Spent sulfuric acid. See Sulfuric acid, spent										
Spirits of nitroglycerin, (1 to 10%)	Flammable liquid	NA1204	Flammable liquid	None	173.133	Forbidden	6 quarts	1,2	5	Segregation same as for explosives
Spirits of nitroglycerin, not exceeding 1% nitroglycerin by weight	Flammable liquid	NA1204	Flammable liquid	173.118	173.133	1 quart	6 quarts	1,2	1	
Sporting powder. See Black powder or Propellant explosive, solid										
Spray starting fluid. See Engine starting fluid										
Spreader cartridge. See Fireworks, special										
Squib, electric or safety. See Electric squib or Safety squib										
Stannic phosphide	Flammable solid	UN1433	Flammable solid and Dangerous when wet	None	173.154	Forbidden	25 pounds	1	5	Segregation same as for flammable solid labeled Dangerous When Wet
ᴬ Stannous chloride, solid	ORM-B		None	173.505	173.510	No limit	No limit			
Starter cartridge	Class B explosive	NA1759	Explosive B	None	173.92	Forbidden	200 pounds	1,3	5	
Starter cartridge	Class C explosive		Explosive C	None	173.102	50 pounds	150 pounds	1,3	1,3	
Storage battery, wet. See Battery, electric storage, wet										
Strontium arsenite, solid	Poison B	UN1691	Poison	173.364	173.365	50 pounds	200 pounds	1,2	1,2	
Strontium chlorate	Oxidizer	UN1506	Oxidizer	173.153	173.168	25 pounds	100 pounds	1,2	1,2	Stow separate from ammonium compounds. Stow away from powdered metals

Chemical Hazard Communication Guidebook

§172.101   Title 49—Transportation

§172.101 Hazardous Materials Table

(1) +/A/W	(2) Hazardous materials descriptions and proper shipping names	(3) Hazard class	(3A) Identification number	(4) Label(s) required (if not excepted)	(5) Packaging (a) Exceptions	(5) Packaging (b) Specific requirements	(6) Maximum net quantity in one package (a) Passenger carrying aircraft or railcar	(6) (b) Cargo only aircraft	(7) Water shipments (a) Cargo vessel	(7) (b) Passenger vessel	(7) (c) Other requirements
	Strontium chlorate, wet	Oxidizer	UN1506	Oxidizer	173.153	173.163	25 pounds	200 pounds	1,2	1,2	Stow separate from ammonium compounds.
	Strontium nitrate	Oxidizer	UN1507	Oxidizer	173.153	173.182	25 pounds	100 pounds	1,2	1,2	Stow away from powdered metals
	Strontium peroxide	Oxidizer	UN1509	Oxidizer	173.153	173.154	25 pounds	100 pounds	1,2	1,2	Keep dry
	Strychnine salt, solid	Poison B	UN1692	Poison	173.364	173.365	50 pounds	200 pounds	1,2	1,2	
	Strychnine, solid	Poison B	UN1692	Poison	173.364	173.365	Forbidden	200 pounds	1,2	1,2	
	Styphnate of lead. See Initiating explosive										
	Styrene monomer, inhibited	Flammable liquid	UN2055	Flammable liquid	173.118	173.119	1 quart	10 gallons	1,2	1,2	
	Substituted nitrophenol pesticide, liquid, n.o.s. (compounds and preparations),	Flammable liquid	UN2780	Flammable liquid	173.118	173.119	1 quart	10 gallons	1,2	1	
	Substituted nitrophenol pesticide, liquid, n.o.s. (compounds and preparations),	Poison B	UN2779	Poison	173.345	173.346	1 quart	55 gallons	1,2	1,2	
	Substituted nitrophenol pesticide, solid, n.o.s. (compounds and preparations),	Poison B	UN2779	Poison	173.364	173.365	50 pounds	200 pounds	1,2	1,2	
	Succinic acid peroxide	Organic peroxide	UN2135	Organic peroxide	173.153	173.157 173.158	Forbidden	25 pounds	1	1	
	Succinic acid peroxide, technically pure. See Succinic acid peroxide.										
	Sucrose octanitrate (dry)	Forbidden									
	Sulfur and chlorate, loose mixtures of	Forbidden									
	Sulfur chloride (di)	Corrosive material	UN1828	Corrosive	None	173.247	Forbidden	1 gallon	1	1	Keep dry. Glass carboys not permitted on passenger vessels
	Sulfur chloride (mono)	Corrosive material	UN1828	Corrosive	None	173.247	Forbidden	1 gallon	1	1	Keep dry. Glass carboys not permitted on passenger vessels
	Sulfur dioxide	Nonflammable gas	UN1079	Nonflammable gas	173.306	173.304 173.314 173.315	Forbidden	300 pounds	1,2	4	Stow away from living quarters
	Sulfur flower. See Sulfur, solid										
	Sulfur hexafluoride	Nonflammable gas	UN1080	Nonflammable gas	173.306	173.304	150 pounds	300 pounds	1,2	1,2	
	Sulfur, molten	ORM-C	NA2448	None	173.505	173.1080	Forbidden	Forbidden	1	1	Stow away from oxidizers and living quarters.
	Sulfuric acid (*For fuming sulfuric acid, see Oleum*)	Corrosive material	UN1830	Corrosive	173.244	173.272	1 quart	1 gallon	1	1	Keep dry. Under deck stowage is permitted on cargo vessels only in metal drums
	Sulfuric acid, spent	Corrosive material	UN1832	Corrosive	None	173.248	Forbidden	1 quart	1	1	Under deck stowage is permitted on cargo vessels only in metal drums
	Sulfuric anhydride. See Sulfur trioxide										
	Sulfurous acid	Corrosive material	UN1833	Corrosive	173.244	173.245	2 gallons	2 gallons	1,2	1	Glass carboys in hampers not permitted under deck

192

Appendix B: Regulations Reference Guide

Chapter I—Research and Special Programs Administration   **§172.101**

	Material	Class	UN No.			Max qty passenger	Max qty cargo			Notes
W	Sulfur, solid	ORM-C	UN1350	None	173.506	173.1080	Forbidden	1 gallon	1,2	1,2	Protect from sparks and open flame. Stow separate from oxidizing materials. Segregation same as for flammable solids
	Sulfur trioxide	Corrosive material	UN1829	Corrosive	173.244	173.273	1 quart	1 quart	1,2	1,2	Keep dry. Glass bottles not permitted under deck
	Sulfuryl chloride	Corrosive material	UN1834	Corrosive	173.244	173.247			1	1	Keep dry. Glass carboys not permitted on passenger vessels
	Sulfuryl fluoride	Nonflammable gas	UN2191	Nonflammable gas	173.306	173.304 173.314	150 pounds	300 pounds	1,3	1	
	Sulphur. See Sulfur, solid										
	Supplementary charge (explosive)	Class A explosive		Explosive A	None	173.69	Forbidden	Forbidden	6	5	
▲	2,4,5-T. See 2,4,5-Trichlorophenoxyacetic acid										
	Tank car, containing residual phosphorus and filled with water or inert gas. See 173.190										
	Tank car, empty (previously used for a hazardous material). See 173.29										
	Tank car, empty (previously used for a Poison A material). See 172.510 and 173.29										
	Tank, portable, empty (previously used for a hazardous material). See 172.510, 172.514 and 173.29										
	Tank truck, empty. See 172.510, 172.514 and 173.29										
▲	TDE (1,1-Dichloro-2,2-bis(p-chlorophenyl) ethane)	ORM-A	NA2761	None	173.505	173.510	50 pounds	No limit	1,2	1,2	
	Tear gas ammunition. See Chemical ammunition, nonexplosive (containing an irritant material)										
	Tear gas candle	Irritating material	UN1700	Irritant	None	173.385	Forbidden	75 pounds	1	5	Stow away from living quarters
	Tear gas cartridge. See Small arms ammunition, irritating (tear gas) cartridge										
	Tear gas device	Irritating material	NA1693	Irritant	None	173.385	Forbidden	75 pounds	1	5	Stow away from living quarters
	Tear gas grenade. See Grenade, tear gas										
	Tertiary alcohol. See Alcohol, n.o.s.										
	Trinazido benzene quinone	Forbidden									
	Tetrachloroethane	ORM-A	UN1702	None	173.505	173.620	1 quart	10 gallons	1,2	1,2	
	Tetrachloroethylene or Perchloroethylene (dry)	ORM-A	UN1897	None	173.505	173.605	10 gallons	55 gallons	1,2	1,2	
AW	Tetraethylammonium perchlorate (dry)	Flammable solid	UN 1325	Flammable solid	173.153	173.154	25 pounds	25 pounds			
▲	Tetraethyl dithiopyrophosphate and compressed gas mixture	Poison A	UN1703	Poison gas	None	173.334	Forbidden	Forbidden	1	5	Shade from radiant heat. Stow away from living quarters. Segregation same as for nonflammable gases
	Tetraethyl dithiopyrophosphate, liquid	Poison B	UN1704	Poison	None	173.358	Forbidden	1 quart	1	5	
	Tetraethyl dithiopyrophosphate mixture, dry	Poison B	UN1704	Poison	None	173.377	Forbidden	200 pounds	1	5	
	Tetraethyl dithiopyrophosphate mixture, liquid	Poison B	UN1704	Poison	None	173.359	Forbidden	1 quart	1	5	
	Tetraethyl lead, liquid (including flash point for export shipment by water)	Poison B	NA1649	Poison	None	173.354	Forbidden	55 gallons	1	5	If flash point is 141 deg F. or less, segregation must be the same as for flammable liquids. Shade from radiant heat. Stow away from living quarters. Segregation same as for nonflammable gases
	Tetraethyl pyrophosphate and compressed gas mixture	Poison A	UN1705	Poison gas	None	173.334	Forbidden	Forbidden	1	5	
	Tetraethyl pyrophosphate, liquid	Poison B	NA2783	Poison	None	173.358	Forbidden	1 quart	1,2	5	

193

Chemical Hazard Communication Guidebook

§172.101   Title 49—Transportation

§172.101 Hazardous Materials Table

(1) +/A/W	(2) Hazardous materials descriptions and proper shipping names	(3) Hazard class	(3A) Identification number	(4) Label(s) required (if not excepted)	(5) Packaging (a) Exceptions	(5) Packaging (b) Specific requirements	(6) Maximum net quantity in one package (a) Passenger carrying aircraft or railcar	(6) Maximum net quantity in one package (b) Cargo only aircraft	(7) Water shipments (a) Cargo vessel	(7) Water shipments (b) Passenger vessel	(7) Water shipments (c) Other requirements
+	Tetraethyl pyrophosphate mixture, dry	Poison B	NA2783	Poison	None	173.377	Forbidden	200 pounds	1,2	5	Stow away from living quarters
	Tetraethyl pyrophosphate mixture, liquid	Poison B	NA2783	Poison	None	173.359	Forbidden	1 quart	1,2	5	
	Tetrafluoroethylene, inhibited	Flammable gas	UN1081	Flammable gas	173.306	173.304	Forbidden	300 pounds	1,2	1,2	
	1,2,3,6-Tetrahydrobenzaldehyde	Corrosive material	UN2498	Corrosive	173.244	173.245	1 quart	10 gallons	1,2	1,2	
	Tetrahydrofuran	Flammable liquid	UN2056	Flammable liquid	None	173.119	Forbidden	10 gallons	1,3	5	
	Tetralin hydroperoxide, technically pure. See Organic peroxide, solid, n.o.s.										
	Tetramethylammonium hydroxide, liquid	Corrosive material	UN1835	Corrosive	173.244	173.245	1 quart	10 gallons	1,2	1,2	
	1,1,3,3-Tetramethylbutyl hydroperoxide, technically pure. See Organic peroxide, liquid or solution, n.o.s.		UN2160								
	1,1,3,3-Tetramethylbutyl peroxy-2-ethylhexanoate, technically pure. See Organic peroxide, liquid or solution, n.o.s.		UN2161								
A	Tetramethylene diperoxide dicarbamide	Forbidden									
	Tetramethylmethylenediamine	ORM-A	NA9069	None	173.505	173.510	No limit	No limit	1		
	Tetranitro diglycerin	Forbidden									
	Tetranitromethane	Oxidizer	UN1510	Oxidizer	None	173.203	Forbidden	Forbidden		5	Shade from radiant heat. Stow away from foodstuffs
	2,3,4,6-Tetranitrophenol	Forbidden									
	2,3,4,6-Tetranitrophenyl methyl nitramine	Forbidden									
	2,3,4,6-Tetranitrophenylnitramine	Forbidden									
	Tetranitroresorcinol (dry)	Forbidden									
	2,3,5,6-Tetranitroso-1,4-dinitrobenzene	Forbidden									
	2,3,5,6-Tetranitroso nitrobenzene (dry)	Forbidden									
	Tetrazene (guanyl nitrosamino guanyltetrazene). See Initiating explosive										
	Tetrazine (dry)	Forbidden									
	Tetrazolyl azide (dry)	Forbidden									
	Teryl. See High explosive										
	Textile treating compound or mixture, liquid	Corrosive material	NA1760	Corrosive	173.244	173.249a	1 quart	10 gallons	1,2	1,2	
	Thallium salt, solid, n.o.s.	Poison B	NA1707	Poison	173.364	173.365	50 pounds	200 pounds	1,2	1,2	
	Thallium sulfate, solid	Poison B	NA1707	Poison	173.364	173.365	50 pounds	200 pounds	1,2	1,2	
	Thinner for rust prevention. See Paint related materials.										

194

Chapter I—Research and Special Programs Administration §172.101

Thiocarbonylchloride. See Thiophosgene									
Thioglycolic acid	Corrosive material	UN1940	Corrosive	173.244	173.245	1 quart	1 gallon	1,2	1,2	Glass carboys in hampers not permitted under deck
Thionyl chloride	Corrosive material	UN1836	Corrosive	None	173.247	Forbidden	1 gallon	1	1	Keep dry. Glass carboys not permitted on passenger vessels
Thiophenol. See Phenyl mercaptan									
Thiophosgene	Poison B	UN2474	Poison	None	173.356	Forbidden	1 gallon	1	5	Shade from radiant heat
Thiophosphoryl chloride	Corrosive material	UN1837	Corrosive	None	173.271	Forbidden	1 quart	1	1	Keep dry. Glass carboys not permitted on passenger vessels
+ Thiram	ORM-A	NA2771	None	173.505	173.510	No limit	No limit	1,2	1,2	
▲ Thorium metal, pyrophoric	Radioactive material	UN2975	Radioactive and Flammable Solid	None	173.418	Forbidden	Forbidden	1,2	1,2	
Thorium nitrate	Radioactive material	UN2976	Radioactive and Oxidizer	None	173.419	25 pounds	25 pounds	1,2	1,2	Separate longitudinally by a complete hold or compartment from explosives
Time fuze. See Fuze, time									
Tin chloride, fuming. See Tin tetrachloride, anhydrous									
Tin perchloride. See Tin tetrachloride, anhydrous									
Tin tetrachloride, anhydrous	Corrosive material	UN1827	Corrosive	173.244	173.247	1 quart	1 quart	1	1	Keep dry. Glass carboys not permitted on passenger vessels
Titanium metal powder, dry or wet with *less than 20% water*	Flammable solid	UN2546	Flammable solid	None	173.208	Forbidden	75 pounds	1,2	5	
Titanium metal powder, wet with 20% or more water	Flammable solid	UN1352	Flammable solid	None	173.208	Forbidden	150 pounds	1,2	5	
Titanium sulfate solution *containing not more than 45% sulfuric acid*	Corrosive material	NA1760	Corrosive	173.244	173.297	1 quart	1 gallon	1	4	Shade from radiant heat. Keep dry
Titanium tetrachloride	Corrosive material	UN1838	Corrosive	173.244	173.247	1 quart	10 gallons	1	1	Keep dry. Glass carboys not permitted on passenger vessels
Toluene (toluol)	Flammable liquid	UN1294	Flammable liquid	173.118	173.119	1 quart	10 gallons	1,2	1	
▲ Toluenediamine	ORM-A	NA1709	None	173.505	173.510	No limit	No limit	1,2	1,2	Shade from radiant heat
+ Toluene diisocyanate	Poison B	UN2078	Poison	173.345	173.346	Forbidden	55 gallons	1,3	1,3	
Toluene sulfonic acid, liquid	Corrosive material	UN2584	Corrosive	173.244	173.245	1 quart	10 gallons	1,2	1,2	
Torch. See Fireworks, common									
Torpedo, railway	Class B explosive		Explosive B	None	173.91	Forbidden	200 pounds	1,2	1,2	Passenger vessels in metal lockers only
▲ Toxaphene	ORM-A	NA2761	None	173.505	173.510	25 pounds	100 pounds	1,2	1,2	
Toy caps	Class C explosive		Explosive C	None	173.100 173.109	50 pounds	150 pounds	1,3	1,3	
Toy propellant device	Class C explosive		Explosive C	None	173.111	50 pounds	150 pounds	1,3	1,3	
Toy smoke device	Class C explosive		Explosive C	None	173.111	50 pounds	150 pounds	1,3	1,3	
Toy torpedo. See Fireworks, special									
2,4,5-TP. See 2,4,5-Trichlorophenoxypropionic acid									
Tracer	Class C explosive		Explosive C	None	173.106	50 pounds	150 pounds	1,3	1,3	
Tracer fuze	Class C explosive		Explosive C	None	173.106	50 pounds	150 pounds	1,3	1,3	
Tractor. See Motor vehicle									

195

Chemical Hazard Communication Guidebook

§172.101　　　　　　　　　　　　　　　　　　　　　　　　　Title 49—Transportation

§172.101 Hazardous Materials Table

(1)	(2)	(3)	(3A)	(4)	(5) Packaging		(6) Maximum net quantity in one package			(7) Water shipments	
+/A/W	Hazardous materials descriptions and proper shipping names	Hazard class	Identification number	Label(s) required (if not excepted)	(a) Exceptions	(b) Specific requirements	(a) Passenger carrying aircraft or railcar	(b) Cargo only aircraft	Cargo vessel (a)	Passenger vessel (b)	(c) Other requirements
	Trailer or truck body with refrigeration or heating equipment. *See Motor vehicle*										
	Triazine pesticide, liquid, n.o.s. *(compounds and preparations)*	Flammable liquid	UN2764	Flammable liquid	173.118	173.119	1 quart	10 gallons	1,2	1	
	Triazine pesticide, liquid, n.o.s. *(compounds and preparations)*, liquid	Poison B	UN2763	Poison	173.345	173.346	1 quart	55 gallons	1,2	1,2	
	Triazine pesticide, solid, n.o.s. *(compounds and preparations)*	Poison B	UN2763	Poison	173.364	173.365	50 pounds	200 pounds	1,2	1,2	
	Tri-(b-nitroxyethyl)ammonium nitrate	Forbidden									
A	Trichlorfon	ORM-A	NA2783	None	173.505	173.510	50 pounds	200 pounds	1,2	1,2	
	Trichloroacetic acid, solid	Corrosive material	UN1839	Corrosive	173.244	173.245b	25 pounds	100 pounds	1,2	1	
	Trichloroacetic acid solution	Corrosive material	UN2564	Corrosive	173.244	173.245	1 quart	1 quart	1,2	1,2	Glass carboys in hampers not permitted under deck
A	1,1,1-Trichloroethane	ORM-A	UN2831	None	173.505	173.605	10 gallons	55 gallons	1,2	1,2	
A	Trichloroethylene	ORM-A	UN1710	None	173.505	173.605	10 gallons	55 gallons	1,2	1,2	Shade from radiant heat. Keep dry, stow separated from nitrogen compounds
	Trichloroisocyanuric acid, dry	Oxidizer	UN2468	Oxidizer	173.153	173.217	10 pounds	50 pounds	1,2	1,2	
	Trichloromethyl perchlorate	Forbidden									
A	Trichlorophenol	ORM-A	NA2020	None	173.505	173.510	100 pounds	No limit	1,2	1,2	
A	2,4,5-Trichlorophenoxyacetic acid	ORM-A	NA2765	None	173.505	173.510	50 pounds	No limit	1,2	1,2	
A	2,4,5-Trichlorophenoxypropionic acid	ORM-A	NA2765	None	173.505	173.510	50 pounds	No limit	1,2	1,2	
	Trichlorosilane	Flammable liquid	UN1295	Flammable liquid	None	173.136	Forbidden	10 gallons	1	5	Segregation same as for flammable solids labeled Dangerous When Wet
	Trichloro-s-triazinetrione dry, containing over 39% available chlorine mono-(Trichloro) tetra-(monopotassium dichloro)-penta-s-triazinetrione, dry (containing over 39% available chlorine)	Oxidizer	NA2468	Oxidizer	173.153	173.217	50 pounds	100 pounds	1,2	1,2	Shade from radiant heat. Keep dry. Stow separated from nitrogen compounds
	Trick matches	Class C explosive		Explosive C	None	173.111	Forbidden	Forbidden	1,3	1,3	
	Trick noise maker, explosive	Class C explosive		Explosive C	None	173.111	50 pounds	150 pounds	1,3	1,3	
	Triethylamine	Flammable liquid	UN1296	Flammable liquid	173.118	173.119	1 quart	10 gallons	1,2	1	
	Trifluorochloroethylene	Flammable gas	UN1082	Flammable gas	173.306	173.304 173.314	Forbidden	10 gallons	1,2	1	
+	Trifluoromethane and chlorotrifluoromethane mixture *(constant boiling mixture) (R-503). See Refrigerant gas, n.o.s.*										

196

Chapter I—Research and Special Programs Administration   §172.101

Material		UN No.	Class						Notes
Triformoxime trinitrate	Forbidden								
1,3,5-Trimethyl-2,4,6-trinitrobenzene	Forbidden								
Trimethylacetyl chloride	Corrosive material	UN2438	Corrosive	173.244	173.247	1 quart	1 quart	1,2	1,2
Trimethylamine, anhydrous	Flammable gas	UN1083	Flammable gas	173.306	173.304 173.314 173.315	Forbidden	300 pounds	1	4
Trimethylamine, aqueous solution	Flammable liquid	UN1297	Flammable liquid	173.118	173.119	1 quart	10 gallons	1,2	1
Trimethylchlorosilane	Flammable liquid	UN1298	Flammable liquid	None	173.125	Forbidden	10 gallons	1,2	1
Trimethylene glycol diperchlorate	Forbidden								
Trimethylol nitromethane trinitrate	Forbidden								
2,4,6-Trinitro-1,3,5-triazido benzene (dry)	Forbidden								
2,4,6-Trinitro-1,3-diazobenzene	Forbidden								
Trinitroacetic acid	Forbidden								
Trinitroacetonitrile	Forbidden								
Trinitroamine cobalt	Forbidden								
Trinitrobenzene, dry. See High explosive									
Trinitrobenzene, wet containing at least 10% water	Flammable solid	UN1354	Flammable solid	173.212		1 pound	1 pound	1	4	Stow away from heavy metals and their compounds
Trinitrobenzene, wet, containing at least 10% water, over 16 ounces in one outside packaging. See High explosive									
Trinitrobenzoic acid, dry. See High explosive									
Trinitrobenzoic acid, wet, containing at least 10% water	Flammable solid	UN1355	Flammable solid	173.192	173.193	1 pound	25 pounds	1	5	Stow away from heavy metals and their compounds
Trinitrobenzoic acid, wet, containing at least 10% water, over 25 pounds in one outside packaging. See High explosives									
Trinitroethanol	Forbidden								
Trinitroethylnitrate	Forbidden								
Trinitromethane	Forbidden								
1,3,5-Trinitronaphthalene	Forbidden								
2,4,6-Trinitrophenyl guanidine (dry)	Forbidden								
2,4,6-Trinitrophenyl nitramine	Forbidden								
2,4,6-Trinitrophenyl trimethylol methyl nitramine trinitrate (dry)	Forbidden								
Trinitroresorcinol. See High explosive									
2,4,6-Trinitroso-3-methyl nitraminoanisole	Forbidden								
Trinitrotetramine cobalt nitrate	Forbidden								
Trinitrotoluene, dry. See High explosive									
Trinitrotoluene, wet containing at least 10% water	Flammable solid	UN1356	Flammable solid	173.212		1 pound	1 pound	1	4	Stow away from heavy metals and their compounds
Trinitrotoluene, wet, containing at least 10% water, over 16 ounces in one outside packaging. See High explosive									
Tris-(1-aziridinyl) phosphine oxide	Corrosive material	UN2501	Corrosive	173.244	173.299a	1 quart	1 gallon	1	1	Keep dry. Glass carboys not permitted on passenger vessels
Tris, bis(fluoroamino diethoxy propane (TVOPA)	Forbidden								
Tungsten hexafluoride	Corrosive material	UN2196	Corrosive	None	178.284	Forbidden	110 pounds	1	5	Segregation same as for nonflammable gases
Turpentine	Combustible liquid	UN1299	None	173.118a	None	No limit	No limit	1,2	1,2
Turpentine	Flammable liquid	UN1299	Flammable liquid	173.118	173.119	1 quart	10 gallons	1,2	1,2

197

Chemical Hazard Communication Guidebook

§172.101   Title 49—Transportation

§172.101 Hazardous Materials Table

(1)	(2)	(3)	(3A)	(4)	(5) Packaging		(6) Maximum net quantity in one package		(7) Water shipments		
+/A/W	Hazardous materials descriptions and proper shipping names	Hazard class	Identification number	Label(s) required (if not excepted)	(a) Exceptions	(b) Specific requirements	(a) Passenger carrying aircraft or railcar	(b) Cargo only aircraft	(a) Cargo vessel	(b) Passenger vessel	(c) Other requirements
	Uranium hexafluoride, fissile *(containing more than 1% U-235)*	Radioactive material	UN2977	Radioactive and Corrosive	173.453	173.417, 173.420			1,2	1,2	
	Uranium hexafluoride, low specific activity	Radioactive material	UN2978	Radioactive and Corrosive	173.421-2	173.420, 173.425			1,2	1,2	
	Uranium metal pyrophoric	Radioactive material	UN2979	Radioactive and Flammable solid	None	173.418	Forbidden	Forbidden	1,2	1,2	
	Uranyl acetate	Radioactive material	NA9180	Radioactive	173.421, 173.425	173.415, 173.416			1,2	1,2	
	Uranyl nitrate hexahydrate solution	Radioactive material	UN2980	Radioactive and Corrosive	173.421, 173.425	173.415, 173.416			1,2	1,2	
	Uranyl nitrate, solid	Radioactive material	UN2981	Radioactive and Oxidizer	None	173.417, 173.419	Forbidden	25 pounds	1,2	1,2	Separate rate longitudinally by an intervening hold or compartment from explosives
	Urea nitrate, dry. *See* High explosive Urea nitrate, wet with *10% or more water*.	Flammable solid	UN1357	Flammable solid	173.192	173.193	1 pound	25 pounds	1,2	1,2	
	Urea nitrate, wet with 10% or more water, over 25 pounds in one outside packaging. *See* High explosive										
	Urea peroxide	Organic peroxide	NA1511	Organic peroxide	173.153	173.227	2 pounds	25 pounds	1	4	Keep dry. Shade from radiant heat
	Valeric acid	Corrosive material	NA1760	Corrosive	173.244	173.245	1 quart	10 gallons	1,2	1,2	
	Valeryl chloride	Corrosive material	UN2502	Corrosive	173.244	173.245	1 quart	1 gallon	1,2	1,2	
	Vanadium oxytrichloride	Corrosive material	UN2443	Corrosive	173.244	173.247a	Forbidden	1 quart	1	4	Shade from radiant heat
	Vanadium oxytrichloride and titanium tetrachloride mixture	Corrosive material	NA2443	Corrosive	None	173.245, 173.245a	Forbidden	1 quart	1	4	Shade from radiant heat
	Vanadium tetrachloride	Corrosive material	UN2444	Corrosive	173.244	173.247a	Forbidden	1 quart	1	4	Shade from radiant heat
	Very signal cartridge	Class C explosive		Explosive C	None	173.108	50 pounds	200 pounds	1,3	1,3	
	Vinyl acetate	Flammable liquid	UN1301	Flammable liquid	173.118	173.119	1 quart	10 gallons	1,2	1	

198

Chapter I—Research and Special Programs Administration §172.101

+	Vinyl chloride	Flammable gas	UN1086	Flammable gas	173.306	173.304, 173.314, 173.315	Forbidden	300 pounds	1,2	4	Stow away from living quarters
	Vinyl ethyl ether, inhibited	Flammable liquid	UN1302	Flammable liquid	None	173.119	Forbidden	1 gallon	1,3	5	
	Vinyl fluoride, inhibited	Flammable gas	UN1860	Flammable gas	173.306	173.304, 173.314, 173.315	Forbidden	300 pounds	1	4	
	Vinylidene chloride, inhibited	Flammable liquid	UN1303	Flammable liquid	173.118	173.119	1 quart	10 gallons	1,3	4	
	Vinyl isobutyl ether	Flammable liquid	UN1304	Flammable liquid	173.118	173.119	1 quart	10 gallons	1,2	1	
	Vinyl methyl ether	Flammable gas	UN1087	Flammable gas	173.306	173.304, 173.314, 173.315	Forbidden	30 pounds	1,2	1	Stow away from living quarters.
	Vinyl nitrate polymer	Forbidden									
	Vinyl trichlorosilane	Flammable liquid	UN1305	Flammable liquid	None	173.135	Forbidden	10 gallons	1,2	1	
	Vitriol, oil of. See Sulfuric acid										
	War head. See Explosive projectile										
	Waste, hazardous. See Hazardous waste, liquid or solid, n.o.s.										
	Water pump system tank charged with compressed air or nitrogen	Nonflammable gas	NA1956	None	173.306		Forbidden	Forbidden		1	Segregation same as for flammable solids labeled Dangerous When Wet
	Water reactive solid, n.o.s.	Flammable solid	UN2813	Flammable solid and Dangerous when wet	173.153	173.154	25 pounds	25 pounds	1,2	4	
	Wheelchair, battery equipped. See Battery, electric storage, wet, with wheelchair										
	White acid (ammonium bifluoride and hydrofluoric acid mixture)	Corrosive material	NA1760	Corrosive	173.244	173.264	1 quart	1 gallon	1	1	
	Xenon	Nonflammable gas	UN2036	Nonflammable gas	173.306	173.302	150 pounds	300 pounds	1,2	1,2	
	X-ray film. See Film										
▲	Xylene (xylol)	Flammable liquid	UN1307	Flammable liquid	173.118	173.119	1 quart	10 gallons	1,2	1,	Stow away from living quarters
	Xylenol	ORM-A	UN2261	None	173.505	173.510	100 pounds	No limit	1,2	1,2	
	Xylyl bromide	Irritating material	UN1701	Irritant	None	173.382	Forbidden	75 pounds	1	5	
	p-Xylyl diazide	Forbidden									
	Zinc ammonium nitrite	Oxidizer	UN1512	Oxidizer	None	173.228	25 pounds	100 pounds	1,2	5	This material may be forbidden in water transportation by certain countries
	Zinc arsenate	Poison B	UN1712	Poison	173.364	173.365	50 pounds	200 pounds	1,2	1,2	Stow separate from ammonium compounds and away from powdered metals
	Zinc arsenite, solid	Poison B	UN1712	Poison	173.364	173.365	50 pounds	200 pounds	1,2	1,2	
	Zinc chlorate	Oxidizer	UN1513	Oxidizer	173.153	173.168	25 pounds	100 pounds	1,2	1,2	
	Zinc chloride solution	Corrosive material	UN1840	Corrosive	173.244	173.245	1 quart	1 quart	1,2	1,2	
	Zinc cyanide	Poison B	UN1713	Poison	173.370	173.370	25 pounds	200 pounds	1,2	1,2	Stow away from acids
	Zinc ethyl. See Pyrophoric liquid, n.o.s.										
	Zinc hydrosulfite	ORM-A	UN1931	None	173.505	173.510	50 pounds	100 pounds	1,2	1,2	Keep dry. Stow away from acids and oxidizers
	Zinc muriate solution. See Zinc chloride solution										
▲	Zinc nitrate	Oxidizer	UN1514	Oxidizer	173.153	173.182	25 pounds	100 pounds	1,2	1,2	Separate from ammonium compounds and hydrogen peroxide
	Zinc permanganate	Oxidizer	UN1515	Oxidizer	173.153	173.154	25 pounds	100 pounds	1,2	1,2	

199

Chemical Hazard Communication Guidebook

§172.101   Title 49—Transportation

§172.101 Hazardous Materials Table

(1)	(2)	(3)	(3A)	(4)	(5) Packaging		(6) Maximum net quantity in one package		(7) Water shipments		
+/ A/ W	Hazardous materials descriptions and proper shipping names	Hazard class	Identification number	Label(s) required (if not excepted)	(a) Exceptions	(b) Specific requirements	(a) Passenger carrying aircraft or railcar	(b) Cargo only aircraft	(a) Cargo vessel	(b) Passenger vessel	(c) Other requirements
	Zinc peroxide	Oxidizer	UN1516	Oxidizer	173.153	173.154	25 pounds	100 pounds	1,2	1,2	Keep dry
	Zinc phosphide	Poison B	UN1714	Poison	173.364	173.365	25 pounds	100 pounds	1,2	1,2	Stow away from acids and oxidizers
	Zirconium hydride	Flammable solid	UN1437	Flammable solid and Dangerous when wet	None	173.206	Forbidden	150 pounds	1,2	5	Segregation same as for flammable solids labeled Dangerous When Wet
	Zirconium metal, dry, chemically produced, finer than 20 mesh particle size	Flammable solid	UN2008	Flammable solid	None	173.214	Forbidden	75 pounds	1	5	Separate from flammable gases or liquids, oxidizing materials or organic peroxides
	Zirconium metal, dry, mechanically produced, finer than 270 mesh particle size	Flammable solid	UN2008	Flammable solid	None	173.214	Forbidden	75 pounds	1	5	Separate from flammable gases or liquids, oxidizing materials or organic peroxides
	Zirconium, metal, liquid, suspensions	Flammable liquid	UN1308	Flammable liquid	None	173.140	Forbidden	5 gallons	1	5	
	Zirconium metal, wet, chemically produced, finer than 20 mesh particle size	Flammable solid	UN1358	Flammable solid	None	173.214	Forbidden	150 pounds	1,2	5	
	Zirconium metal, wet, mechanically produced, finer than 270 mesh particle size	Flammable solid	UN1358	Flammable solid	None	173.214	Forbidden	150 pounds	1,2	5	
	Zirconium nitrate	Oxidizer	UN2728	Oxidizer	173.153	173.182	25 pounds	100 pounds	1,2	1,2	Stow away from heavy metals and their salts
	Zirconium picramate, wet with at least 20% of water	Flammable solid	UN1517	Flammable solid	None	173.216	Forbidden	25 pounds	1	1	
	Zirconium scrap (borings, clippings, shavings, sheets, or turnings)	Flammable solid	UN1932	Flammable solid	173.153	173.220	Forbidden	Forbidden	1	4	Separate from flammable gases or liquids, oxidizing materials, or organic peroxides
	Zirconium sulfate	ORM-B	NA9163	None	None	173.510	100 pounds	No limit	1,2	1,2	
A	Zirconium tetrachloride, solid	Corrosive material	UN2503	Corrosive	173.244	173.245b	25 pounds	100 pounds	1,2	1,2	

200

Appendix B: Regulations Reference Guide

Research and Special Programs Administration, DOT § 172.101, Appendix

EFFECTIVE DATE NOTE: At 54 FR 39501, Sept. 26, 1989, the introductory text to the appendix to § 172.101 was revised, effective October 31, 1989. For the convenience of the user, the superseded material follows.

APPENDIX TO § 172.101—LIST OF HAZARDOUS SUBSTANCES AND REPORTABLE QUANTITIES

1. This appendix lists materials and their corresponding reportable quantities (RQs) which are listed or designated as "hazardous substances" under section 101(14) of the Comprehensive Environmental Response, Compensation, and Liability Act (CERCLA; Pub. L. 96-510). A material in this list is regulated as a hazardous material and a hazardous substance under this subchapter if it meets the definition of a hazardous substance in § 171.8 of this subchapter.

2. Column 1 of the list, entitled "*Hazardous substances*", contains the names of hazardous substances. Elements and compounds are listed first in alphabetical sequence. Following the listing of elements and compounds is a listing of waste streams. These waste streams appear on the list in numerical sequence and are referenced by the appropriate "F" or "K" numbers. Column 2 of the list, entitled "*Synonyms*", contains the names of synonyms for certain elements and compounds listed in Column 1. No synonyms are listed for waste streams. Synonyms are useful in identifying hazardous substances and in identifying proper shipping names. Column 3 of the list, entitled "*Reportable quantity (RQ)*", contains the reportable quantity (RQ), in pounds and kilograms, for each hazardous substance listed in Column 1.

3. The procedure for selecting a proper shipping name for a hazardous substance is set forth in § 172.101(c)(9).

4. A series of notes is used throughout the list to provide additional information concerning certain hazardous substances. These notes are explained at the end of the list.

Chemical Hazard Communication Guidebook

Research and Special Programs Administration, DOT §172.101, Appendix, Note

LIST OF HAZARDOUS SUBSTANCES AND REPORTABLE QUANTITIES

Hazardous Substance	Synonyms	Reportable Quantity(RQ) Pounds(Kilograms)
Acenaphthene		100 (45.4)
Acenaphthylene		5000 (2270)
Acetaldehyde	Ethanal	1000 (454)
Acetaldehyde, chloro-	Chloroacetaldehyde	1000 (454)
Acetaldehyde, trichloro-	Chloral	1 (0.454)
Acetamide, N-(aminothioxomethyl)-	1-Acetyl-2-thiourea	1000 (454)
Acetamide, N-(4-ethoxyphenyl)-	Phenacetin	1 (0.454)
Acetamide, N-fluoren-2-yl-	2-Acetylaminofluorene	1 (0.454)
Acetamide, 2-fluoro-	Fluoroacetamide	100 (45.4)
Acetic acid *		5000 (2270)
Acetic acid, ethyl ester	Ethyl acetate	5000 (2270)
Acetic acid, fluoro-, sodium salt	Fluoroacetic acid, sodium salt	10 (4.54)
Acetic acid, lead salt	Lead acetate	5000 (2270)
Acetic acid, thallium(I) salt	Thallium(I) acetate	100 (45.4)
Acetic anhydride *		5000 (2270)
Acetimidic acid, N-[(methylcarbamoyl)oxy]thio-methyl ester	Methomyl	100 (45.4)
Acetone *	2-Propanone	5000 (2270)
Acetone cyanohydrin *	Propanenitrile, 2-hydroxy-2-methyl-	10 (4.54)
Acetonitrile *	2-Methyllactonitrile	5000 (2270)
	Ethanenitrile	
3-(alpha-Acetonylbenzyl)-4-hydroxycoumarin and salts	Warfarin	100 (45.4)
Acetophenone	Ethanone, 1-phenyl-	5000 (2270)
2-Acetylaminofluorene	Acetamide, N-fluoren-2-yl-	1 (0.454)
Acetyl bromide		5000 (2270)
Acetyl chloride *	Ethanoyl chloride	5000 (2270)
1-Acetyl-2-thiourea	Acetamide, N-(aminothioxomethyl)-	1000 (454)
Acrolein *	2-Propenal	1 (0.454)
Acrylamide	2-Propenamide	5000 (2270)
Acrylic acid *	2-Propenoic acid	5000 (2270)
Acrylonitrile *	2-Propenenitrile	100 (45.4)
Adipic acid		5000 (2270)
Alanine, 3-[p-bis(2-chloroethyl)amino]phenyl-, L-	Melphalan	1 (0.454)
Aldicarb	Propanal, 2-methyl-2-(methylthio)-, O-[(methylamino)carbonyl]oxime	1 (0.454)
Aldrin	1,2,3,4,10,10-Hexachloro-1,4,4a,5,8,8a-hexahydro-1,4:5,8-endo,exo-dimethanonaphthalene	1 (0.454)
Allyl alcohol *	2-Propen-1-ol	100 (45.4)
Allyl chloride *		1000 (454)
Aluminum phosphide *		100 (45.4)
Aluminum sulfate		5000 (2270)
2-Amino-1-methyl benzene	o-Toluidine	1 (0.454)
4-Amino-1-methyl benzene	p-Toluidine	1 (0.454)

243

§172.101, Appendix, Note 49 CFR Ch. I (10-1-89 Edition)

LIST OF HAZARDOUS SUBSTANCES AND REPORTABLE QUANTITIES—Continued

Hazardous Substance	Synonyms	Reportable Quantity(RQ) Pounds(Kilograms)
5-(Aminomethyl)-3-isoxazolol	3(2H)-Isoxazolone, 5-(aminomethyl)	1000 (454)
4-Aminopyridine	4-Pyridinamine	1000 (454)
Amitrole	1H-1,2,4-Triazol-3-amine	1 (0.454)
Ammonia *		100 (45.4)
Ammonium acetate		5000 (2270)
Ammonium benzoate		5000 (2270)
Ammonium bicarbonate		5000 (2270)
Ammonium bichromate	Ammonium dichromate @	1000 (454)
Ammonium bifluoride *		100 (45.4)
Ammonium bisulfite		5000 (2270)
Ammonium carbamate		5000 (2270)
Ammonium carbonate		5000 (2270)
Ammonium chloride		5000 (2270)
Ammonium chromate		1000 (454)
Ammonium citrate, dibasic		5000 (2270)
Ammonium dichromate @	Ammonium bichromate	1000 (454)
Ammonium fluoborate		5000 (2270)
Ammonium fluoride *		100 (45.4)
Ammonium hydroxide *		1000 (454)
Ammonium oxalate		5000 (2270)
Ammonium picrate	Phenol, 2,4,6-trinitro-, ammonium salt	10 (4.54)
Ammonium silicofluoride *		1000 (454)
Ammonium sulfamate		5000 (2270)
Ammonium sulfide *		100 (45.4)
Ammonium sulfite		5000 (2270)
Ammonium tartrate		5000 (2270)
Ammonium thiocyanate		5000 (2270)
Ammonium thiosulfate		5000 (2270)
Ammonium vanadate	Vanadic acid, ammonium salt	1000 (454)
Amyl acetate *		5000 (2270)
iso-Amyl acetate		
sec-Amyl acetate		
tert-Amyl acetate		
Aniline *	Benzenamine	5000 (2270)
Anthracene		5000 (2270)
Antimony ¢		5000 (2270)
Antimony pentachloride *		1000 (454)
Antimony potassium tartrate		100 (45.4)
Antimony tribromide		1000 (454)
Antimony trichloride		1000 (454)
Antimony trifluoride		1000 (454)
Antimony trioxide		1000 (454)

244

Chemical Hazard Communication Guidebook

Research and Special Programs Administration, DOT §172.101, Appendix, Note

Chemical	Value
Aroclor 1016	10 (4.54)
Aroclor 1221	10 (4.54)
Aroclor 1232	10 (4.54)
Aroclor 1242	10 (4.54)
Aroclor 1248	10 (4.54)
Aroclor 1254	10 (4.54)
Aroclor 1260	10 (4.54)
POLYCHLORINATED BIPHENYLS (PCBs)	
POLYCHLORINATED BIPHENYLS (PCBs)	
POLYCHLORINATED BIPHENYLS (PCBs)	
POLYCHLORINATED BIPHENYLS (PCBs)	
POLYCHLORINATED BIPHENYLS (PCBs)	
POLYCHLORINATED BIPHENYLS (PCBs)	
POLYCHLORINATED BIPHENYLS (PCBs)	
Arsenic * ¢	1 (0.454)
Arsenic acid *	1 (0.454)
Arsenic disulfide *	5000 (2270)
Arsenic(III) oxide	5000 (2270) — Arsenic trioxide *
Arsenic(V) oxide	5000 (2270) — Arsenic pentoxide *
Arsenic pentoxide *	5000 (2270)
Arsenic trichloride *	5000 (2270)
Arsenic trioxide *	5000 (2270) — Arsenic(III) oxide
Arsenic trisulfide *	5000 (2270)
Arsine, diethyl-	1 (0.454) — Diethylarsine
Asbestos * ¢¢	1 (0.454)
Auramine	10 (4.54) — Benzenamine, 4,4'-carbonimidoylbis (N,N-dimethyl-
Azaserine	1 (0.454) — L-Serine, diazoacetate (ester)
Aziridine	1 (0.454) — Ethylenimine
Azinphos methyl @	1 (0.454) — Guthion *
Azirino[2',3':3,4]pyrrolo(1,2-a)indole-4,7-dione,6-amino-8-[((aminocarbonyl)oxy)methyl]-1,1a,2,8,8a,8b-hexahydro-8a-methoxy-5-methyl-.	1 (0.454) — Mitomycin C
Barium cyanide *	10 (4.54)
Benz[j]aceanthrylene, 1,2-dihydro-3-methyl-	1 (0.454) — 3-Methylcholanthrene
Benz[c]acridine	1 (0.454)
3,4-Benzacridine	1 (0.454) — Benz[c]acridine
Benzal chloride *	5000 (2270) — Benzene, dichloromethyl-
Benz[a]anthracene	1 (0.454) — Benzo[a]anthracene
	1,2-Benzanthracene
1,2-Benzanthracene	1 (0.454) — Benz[a]anthracene
	Benzo[a]anthracene
1,2-Benzanthracene, 7,12-dimethyl-	7,12-Dimethylbenz[a]anthracene
Benzenamine	Aniline *
Benzenamine, 4,4'-carbonimidoylbis (N,N-dimethyl-	5000 (2270) — Auramine
Benzenamine, 4-chloro-	1000 (454) — p-Chloroaniline
Benzenamine, 4-chloro-2-methyl-, hydrochloride	1 (0.454) — 4-Chloro-o-toluidine, hydrochloride
Benzenamine, N,N-dimethyl-4-phenylazo-	1 (0.454) — Dimethylaminoazobenzene
Benzenamine, 4,4'-methylenebis(2-chloro-	1 (0.454) — 4,4'-Methylenebis(2-chloroaniline)
Benzenamine, 2-methyl-, hydrochloride	1 (0.454) — o-Toluidine hydrochloride
Benzenamine, 2-methyl-5-nitro-	5000 (2270) — 5-Nitro-o-toluidine
Benzenamine, 4-nitro-	1000 (454) — p-Nitroaniline
Benzene *	1000 (454)
Benzene, 1-bromo-4-phenoxy-	100 (45.4) — 4-Bromophenyl phenyl ether
Benzene, chloro-	100 (45.4) — Chlorobenzene *
Benzene, chloromethyl-	100 (45.4) — Benzyl chloride *

245

Appendix B: Regulations Reference Guide

§172.101, Appendix, Note   **49 CFR Ch. I (10-1-89 Edition)**

LIST OF HAZARDOUS SUBSTANCES AND REPORTABLE QUANTITIES—Continued

Hazardous Substance	Synonyms	Reportable Quantity(RQ) Pounds(Kilograms)
Benzene, 1,2-dichloro-	o-Dichlorobenzene *	100 (45.4)
Benzene, 1,3-dichloro-	1,2-Dichlorobenzene *	100 (45.4)
Benzene, 1,4-dichloro-	m-Dichlorobenzene *	100 (45.4)
	1,3-Dichlorobenzene *	
	p-Dichlorobenzene *	
	1,4-Dichlorobenzene *	
Benzene, dichloromethyl-	Benzal chloride	5000 (2270)
Benzene, 2,4-diisocyanatomethyl-	Toluene diisocyanate *	100 (45.4)
Benzene, dimethyl-	Xylene * (mixed)	1000 (454)
	m-	
	o-	
	p-	
Benzene, hexachloro-	Hexachlorobenzene *	1 (0.454)
Benzene, hexahydro-	Cyclohexane *	1000 (454)
Benzene, hydroxy-	Phenol *	1000 (454)
Benzene, methyl-	Toluene *	1000 (454)
Benzene, 1-methyl-2,4-dinitro-	2,4-Dinitrotoluene *	1000 (454)
Benzene, 1-methyl-2,6-dinitro-	2,6-Dinitrotoluene *	1000 (454)
Benzene, 1,2-methylenedioxy-4-allyl-	Safrole	1 (0.454)
Benzene, 1,2-methylenedioxy-4-propenyl-	Isosafrole	1 (0.454)
Benzene, 1,2-methylenedioxy-4-propyl-	Dihydrosafrole	1 (0.454)
Benzene, 1-methylethyl-	Cumene	5000 (2270)
Benzene, nitro-	Nitrobenzene *	1000 (454)
Benzene, pentachloro-	Pentachlorobenzene	10 (4.54)
Benzene, pentachloronitro-	Pentachloronitrobenzene	1 (0.454)
Benzene, 1,2,4,5-tetrachloro-	1,2,4,5-Tetrachlorobenzene	5000 (2270)
Benzene, trichloromethyl-	Benzotrichloride	1 (0.454)
Benzene, 1,3,5-trinitro-	sym-Trinitrobenzene *	10 (4.54)
Benzeneacetic acid, 4-chloro-alpha-(4-chlorophenyl)-alpha-hydroxy-, ethyl ester	Ethyl 4,4'-dichlorobenzilate	1 (0.454)
1,2-Benzenedicarboxylic acid anhydride	Phthalic anhydride	5000 (2270)
1,2-Benzenedicarboxylic acid, [bis(2-ethylhexyl)] ester	Bis(2-ethylhexyl)phthalate	1 (0.454)
1,2-Benzenedicarboxylic acid, dibutyl ester	Di-n-butyl phthalate	10 (4.54)
	Dibutyl phthalate	
	n-Butyl phthalate	
1,2-Benzenedicarboxylic acid, diethyl ester	Diethyl phthalate	1000 (454)
1,2-Benzenedicarboxylic acid, dimethyl ester	Dimethyl phthalate	5000 (2270)
1,2-Benzenedicarboxylic acid, di-n-octyl ester	Di-n-octyl phthalate	5000 (2270)
1,3-Benzenediol	Resorcinol	5000 (2270)
1,2-Benzenediol,4-(1-hydroxy-2-(methylamino)ethyl)-	Epinephrine	1000 (454)
Benzenesulfonic acid chloride	Benzenesulfonyl chloride	100 (45.4)
Benzenesulfonyl chloride	Benzenesulfonic acid chloride	100 (45.4)

246

Chemical Hazard Communication Guidebook

Research and Special Programs Administration, DOT §172.101, Appendix, Note

Chemical	Lbs (Kg)
Benzenethiol	
Phenyl mercaptan @	100 (45.4)
Benzidine *	
Thiophenol	
(1,1'-Biphenyl)-4,4'diamine	1 (0.454)
1,2-Benzisothiazolin-3-one,1,1-dioxide, and salts	
Saccharin and salts	1 (0.454)
Benzo[a]anthracene	
Benz[a]anthracene	
1,2-Benzanthracene	
Benzo[b]fluoranthene	1 (0.454)
Benzo[k]fluoranthene	1 (0.454)
Benzo[j,k]fluorene	
Fluoranthene	100 (45.4)
Benzoic acid	5000 (2270)
Benzonitrile	5000 (2270)
Benzo[g,h,i]perylene	5000 (2270)
3,4-Benzopyrene	
Benzo[a]pyrene	1 (0.454)
3,4-Benzopyrene	1 (0.454)
p-Benzoquinone	
1,4-Cyclohexadienedione	10 (4.54)
Benzotrichloride	
Benzene, trichloromethyl-	1 (0.454)
Benzoyl chloride	1000 (454)
1,2-Benzphenanthrene	
Chrysene	1 (0.454)
Benzyl chloride *	
Benzene, chloromethyl-	100 (45.4)
Beryllium ¢	
Beryllium dust ¢	1 (0.454)
Beryllium chloride *	5000 (2270)
Beryllium dust ¢	
Beryllium fluoride *	5000 (2270)
Beryllium nitrate *	5000 (2270)
alpha - BHC	1 (0.454)
beta - BHC	1 (0.454)
delta - BHC	1 (0.454)
gamma - BHC	1 (0.454)
Hexachlorocyclohexane (gamma isomer)	
Lindane *	
2,2'-Bioxirane	
1,2:3,4-Diepoxybutane	1 (0.454)
(1,1'-Biphenyl)-4,4'-diamine	
Benzidine *	1 (0.454)
(1,1'-Biphenyl)-4,4'-diamine,3,3'-dichloro-	
3,3'-Dichlorobenzidine	1 (0.454)
(1,1'-Biphenyl)-4,4'-diamine,3,3'-dimethoxy-	
3,3'-Dimethoxybenzidine	1 (0.454)
(1,1'-Biphenyl)-4,4'-diamine,3,3'-dimethyl-	
3,3'-Dimethylbenzidine	1 (0.454)
Bis(2-chloroethoxy) methane	
Ethane, 1,1'-[methylenebis(oxy)]bis(2-chloro-	1000 (454)
Bis(2-chloroethyl) ether	
Dichloroethyl ether	1 (0.454)
Ethane, 1,1'-oxybis(2-chloro-	
Bis(2-chloroisopropyl) ether	
Propane, 2,2'-oxybis(2-chloro-	1000 (454)
Bis(chloromethyl) ether	1 (0.454)
Methane, oxybis(chloro-	
Bis(dimethylthiocarbamoyl) disulfide	
Thiram	10 (4.54)
Bis(2-ethylhexyl)phthalate	
1,2-Benzenedicarboxylic acid, [bis(2-ethylhexyl)]ester	1000 (454)
Bromine cyanide	
Cyanogen bromide *	1000 (454)
Bromoacetone	
2-Propanone, 1-bromo-	1000 (454)
Bromoform	
Methane, tribromo-	100 (45.4)
4-Bromophenyl phenyl ether	
Benzene, 1-bromo-4-phenoxy-	100 (45.4)
Brucine	
Strychnidin-10-one, 2,3-dimethoxy-	100 (45.4)
1,3-Butadiene, 1,1,2,3,4,4-hexachloro-	
Hexachlorobutadiene *	1 (0.454)
1-Butanamine, N-butyl-N-nitroso-	
N-Nitrosodi-n-butylamine	1 (0.454)

247

Appendix B: Regulations Reference Guide

§172.101, Appendix, Note   **49 CFR Ch. I (10-1-89 Edition)**

LIST OF HAZARDOUS SUBSTANCES AND REPORTABLE QUANTITIES—Continued

Hazardous Substance	Synonyms	Reportable Quantity(RQ) Pounds(Kilograms)
Butanoic acid, 4-[bis(2-chloroethyl)amino]benzene-	Chlorambucil	1 (0.454)
1-Butanol	n-Butyl alcohol *	5000 (2270)
2-Butanone	Ethyl methyl ketone @	5000 (2270)
2-Butanone peroxide	Methyl ethyl ketone *	10 (4.54)
2-Butenal	Methyl ethyl ketone peroxide *	100 (45.4)
2-Butene, 1,4-dichloro-	Crotonaldehyde *	1 (0.454)
Butyl acetate *	1,4-Dichloro-2-butene	5000 (2270)
iso-Butyl acetate		
sec-Butyl acetate		
tert-Butyl acetate		
n-Butyl alcohol *	1-Butanol	5000 (2270)
Butylamine *		1000 (454)
iso-Butylamine		
sec-Butylamine		
tert-Butylamine		
Butyl benzyl phthalate		100 (45.4)
n-Butyl phthalate	Di-n-butyl phthalate	10 (4.54)
	Dibutyl phthalate	
	1,2-Benzenedicarboxylic acid, dibutyl ester	
Butyric acid		5000 (2270)
iso-Butyric acid		
Cacodylic acid	Hydroxydimethylarsine oxide	1 (0.454)
Cadmium ¢		1 (0.454)
Cadmium acetate		100 (45.4)
Cadmium bromide		100 (45.4)
Cadmium chloride		100 (45.4)
Calcium arsenate *		1000 (454)
Calcium arsenite *		1000 (454)
Calcium carbide *		10 (4.54)
Calcium chromate	Chromic acid, calcium salt	1000 (454)
Calcium cyanide *		10 (4.54)
Calcium dodecylbenzene sulfonate		1000 (454)
Calcium hypochlorite *		10 (4.54)
Camphene, octachloro-	Toxaphene *	10 (4.54)
Captan		10 (4.54)
Carbamic acid, ethyl ester	Ethyl carbamate (Urethan)	1 (0.454)
Carbamic acid, methylnitroso-, ethyl ester	N-Nitroso-N-methylurethane	1 (0.454)
Carbamide, N-ethyl-N-nitroso-	N-Nitroso-N-ethylurea	1 (0.454)
Carbamide, N-methyl-N-nitroso-	N-Nitroso-N-methylurea	1 (0.454)
Carbamide, thio-	Thiourea	1 (0.454)
Carbamimidoselenoic acid	Selenourea	1000 (454)

248

Chemical Hazard Communication Guidebook

Research and Special Programs Administration, DOT §172.101, Appendix, Note

Chemical	Synonym	Quantity
Carbamoyl chloride, dimethyl-	Dimethylcarbamoyl chloride	1 (0.454)
Carbaryl *		100 (45.4)
Carbofuran		10 (4.54)
Carbon bisulfide	Carbon disulfide *	100 (45.4)
Carbon bisulfide	Carbon disulfide *	100 (45.4)
Carbonic acid, dithallium (I) salt	Thallium(I) carbonate	100 (45.4)
Carbonochloridic acid, methyl ester	Methyl chlorocarbonate *	1000 (454)
	Methyl chloroformate @	1000 (454)
Carbon oxyfluoride	Carbonyl fluoride	1000 (454)
Carbon tetrachloride *	Methane, tetrachloro-	5000 (2270)
Carbonyl chloride	Phosgene	10 (4.54)
Carbonyl fluoride	Carbon oxyfluoride	1000 (454)
Chloral	Acetaldehyde, trichloro-	1 (0.454)
Chlorambucil	Butanoic acid, 4-[bis(2-chloroethyl)amino]benzene-	1 (0.454)
Chlordane	Chlordane, technical	1 (0.454)
	4,7-Methanoindan, 1,2,4,5,6,7,8,8-octachloro-3a,4,7,7a-tetrahydro-	
Chlordane, technical *	Chlordane	1 (0.454)
	4,7-Methanoindan, 1,2,4,5,6,7,8,8-octachloro-3a,4,7,7a-tetrahydro-	
Chlorine *	Cyanogen chloride	10 (4.54)
Chlorine cyanide		10 (4.54)
Chlornaphazine	2-Naphthylamine, N,N-bis(2-chloroethyl)-	1 (0.454)
Chloroacetaldehyde	Acetaldehyde, chloro-	1000 (454)
p-Chloroaniline	Benzenamine, 4-chloro-	1000 (454)
Chlorobenzene *	Benzene, chloro-	100 (45.4)
4-Chloro-m-cresol	p-Chloro-m-cresol	5000 (2270)
p-Chloro-m-cresol	Phenol, 4-chloro-3-methyl-	5000 (2270)
	Phenol, 4-chloro-3-methyl-	
	4-Chloro-m-cresol	
Chlorodibromomethane		100 (45.4)
1-Chloro-2,3-epoxypropane	Epichlorohydrin	1000 (454)
	Oxirane, 2-(chloromethyl)-	
Chloroethane	Ethyl chloride @	100 (45.4)
2-Chloroethyl vinyl ether	Ethene, 2-chloroethoxy-	1000 (454)
Chloroform *	Methane, trichloro-	5000 (2270)
Chloromethane	Methyl chloride *	1 (0.454)
Chloromethyl methyl ether	Methane, chloromethoxy-	1 (0.454)
	Methylchloromethyl ether @	
beta-Chloronaphthalene	Naphthalene, 2-chloro-	5000 (2270)
2-Chloronaphthalene	beta-Chloronaphthalene	5000 (2270)
	Naphthalene, 2-chloro-	
2-Chlorophenol	o-Chlorophenol	100 (45.4)
	Phenol, 2-chloro-	
o-Chlorophenol	Phenol, 2-chloro-	100 (45.4)
	2-Chlorophenol	
4-Chlorophenyl phenyl ether		5000 (2270)
1-(o-Chlorophenyl)thiourea	Thiourea, (2-chlorophenyl)-	100 (45.4)

249

§172.101, Appendix, Note   49 CFR Ch. I (10-1-89 Edition)

LIST OF HAZARDOUS SUBSTANCES AND REPORTABLE QUANTITIES—Continued

Hazardous Substance	Synonyms	Reportable Quantity(RQ) Pounds(Kilograms)
3-Chloropropionitrile	Propanenitrile, 3-chloro-	1000 (454)
Chlorosulfonic acid *		1000 (454)
4-Chloro-o-toluidine, hydrochloride	Benzenamine, 4-chloro-2-methyl-, hydrochloride	1 (0.454)
Chlorpyrifos *		1 (0.454)
Chromic acetate		1000 (454)
Chromic acid *		1000 (454)
Chromic acid, calcium salt	Calcium chromate	1000 (454)
Chromic sulfate		1000 (454)
Chromium ¢		1 (0.454)
Chromous chloride		1000 (454)
Chrysene	1,2-Benzphenanthrene	1 (0.454)
Cobaltous bromide		1000 (454)
Cobaltous formate		1000 (454)
Cobaltous sulfamate		1000 (454)
Coke Oven Emissions		1 (0.454)
Copper ¢		5000 (2270)
Copper cyanide *		10 (4.54)
Coumaphos *		10 (4.54)
Creosote		1 (0.454)
Cresols *	Cresylic acid	1000 (454)
m-Cresols	m-Cresylic acid	
o-Cresols	o-Cresylic acid	
p-Cresols	p-Cresylic acid	
Cresylic acid	Cresols *	1000 (454)
m-Cresols	m-Cresylic acid	
o-Cresols	o-Cresylic acid	
p-Cresols	p-Cresylic acid	
Crotonaldehyde *	2-Butenal	100 (45.4)
Cumene	Benzene, 1-methylethyl-	5000 (2270)
Cupric acetate		100 (45.4)
Cupric acetoarsenite *		100 (45.4)
Cupric chloride *		10 (4.54)
Cupric nitrate		100 (45.4)
Cupric oxalate		100 (45.4)
Cupric sulfate		10 (4.54)
Cupric sulfate ammoniated		100 (45.4)
Cupric tartrate		100 (45.4)
Cyanides (soluble cyanide salts), not elsewhere specified *		10 (4.54)
Cyanogen *		100 (45.4)
Cyanogen bromide *	Bromine cyanide	1000 (454)
Cyanogen chloride *	Chlorine cyanide	10 (4.54)
1,4-Cyclohexadienedione	p-Benzoquinone	10 (4.54)

250

Chemical Hazard Communication Guidebook

Research and Special Programs Administration, DOT §172.101, Appendix, Note

Chemical	Value
Cyclohexane	
Cyclohexanone	1000 (454)
1,3-Cyclopentadiene, 1,2,3,4,5,5-hexachloro-	5000 (2270)
Cyclophosphamide	1 (0.454)
2,4-D Acid	100 (45.4)
2,4-D Esters	100 (45.4)
2,4-D *, salts and esters	100 (45.4)
Daunomycin	1 (0.454)
DDD	1 (0.454)
4,4'-DDD	1 (0.454)
DDE	1 (0.454)
4,4'-DDE	1 (0.454)
DDT *	1 (0.454)
4,4'-DDT	1 (0.454)
Decachlorooctahydro-1,3,4-metheno-2H-cyclobuta[c,d]-pentalen-2-one	1 (0.454)
Diallate	1 (0.454)
Diamine	1 (0.454)
Diaminotoluene	1 (0.454)
Diazinon *	1 (0.454)
Dibenz[a,h]anthracene	1 (0.454)
1,2:5,6-Dibenzanthracene	1 (0.454)
Dibenzo[a,h]anthracene	1 (0.454)
1,2:7,8-Dibenzopyrene	1 (0.454)
Dibenz[a,i]pyrene	1 (0.454)
1,2-Dibromo-3-chloropropane	10 (0.454)
Dibutyl phthalate	10 (4.54)
Di-n-butyl phthalate	10 (4.54)
Dicamba	1000 (454)
Dichlobenil	100 (45.4)
Dichlone	1 (0.454)
S-(2,3-Dichloroallyl) diisopropylthiocarbamate	1 (0.454)
3,5-Dichloro-N-(1,1-dimethyl-2-propynyl)benzamide	5000 (2270)

251

§172.101, Appendix, Note   **49 CFR Ch. I (10-1-89 Edition)**

LIST OF HAZARDOUS SUBSTANCES AND REPORTABLE QUANTITIES—Continued

Hazardous Substance	Synonyms	Reportable Quantity(RQ) Pounds(Kilograms)
Dichlorobenzene (mixed)		100 (45.4)
1,2-Dichlorobenzene	Benzene, 1,2-dichloro- o-Dichlorobenzene *	100 (45.4)
1,3-Dichlorobenzene	Benzene, 1,3-dichloro- m-Dichlorobenzene	100 (45.4)@
1,4-Dichlorobenzene	Benzene, 1,4-dichloro- p-Dichlorobenzene *	100 (45.4)
m-Dichlorobenzene	Benzene, 1,3-dichloro- 1,3-Dichlorobenzene	100 (45.4)
o-Dichlorobenzene *	Benzene, 1,2-dichloro- 1,2-Dichlorobenzene	100 (45.4)
p-Dichlorobenzene *	Benzene, 1,4-dichloro- 1,4-Dichlorobenzene	100 (45.4)
3,3'-Dichlorobenzidine	(1,1'-Biphenyl)-4,4'-diamine,3,3'-dichloro-	1 (0.454)
Dichlorobromomethane		5000 (2270)
1,4-Dichloro-2-butene	2-Butene, 1,4-dichloro-	1 (0.454)
Dichlorodifluoromethane *	Methane, dichlorodifluoro-	5000 (2270)
Dichlorodiphenyl dichloroethane	DDD TDE * 4,4'-DDD	1 (0.454)
Dichlorodiphenyl trichloroethane *	DDT * 4,4'-DDT	1 (0.454)
1,1-Dichloroethane	Ethane, 1,1-dichloro- Ethylidene dichloride	1000 (454)
1,2-Dichloroethane	Ethane, 1,2-dichloro- Ethylene dichloride *	5000 (2270)
1,1-Dichloroethylene	Ethene, 1,1-dichloro- Vinylidene chloride *	5000 (2270)
1,2-trans-Dichloroethylene	Ethene, trans-1,2-dichloro-	1000 (454)
Dichloroethyl ether	Bis (2-chloroethyl) ether Ethane, 1,1'-oxybis(2-chloro-	1 (0.454)
2,4-Dichlorophenol	Phenol, 2,4-dichloro-	100 (45.4)
2,6-Dichlorophenol	Phenol, 2,6-dichloro-	100 (45.4)
2,4-Dichlorophenoxyacetic acid *, salts and esters	2,4-D Acid 2,4-D *, salts and esters	100 (45.4)
Dichlorophenylarsine	Phenyl dichloroarsine *	1 (0.454)
Dichloropropane		1000 (454)
1,1-Dichloropropane		
1,3-Dichloropropane		
1,2-Dichloropropane	Propylene dichloride *	1000 (454)
Dichloropropane - Dichloropropene (mixture)		100 (45.4)
Dichloropropene(s) *		100 (45.4)

252

Research and Special Programs Administration, DOT §172.101, Appendix, Note

2,3-Dichloropropene (isomer)	Propene, 1,3-dichloro-	100 (45.4)
1,3-Dichloropropene		5000 (2270)
2,2-Dichloropropionic acid *		10 (4.54)
Dichlorvos *		1 (0.454)
Dieldrin *	1,2,3,4,10,10-Hexachloro-6,7-epoxy-1,4,4a,5,6,7,8,8a-octahydro-endo,exo-1,4:5,8-dimethanonaphthalene.	1 (0.454)
1,2:3,4-Diepoxybutane	2,2'-Bioxirane	1 (0.454)
Diethylamine		100 (45.4)
Diethylarsine	Arsine, diethyl-	1 (0.454)
1,4-Diethylene dioxide	1,4-Dioxane	1 (0.454)
O,O-Diethyl S-[2-(ethylthio)ethyl] phosphorodithioate	Disulfoton *	1 (0.454)
N,N'-Diethylhydrazine	Hydrazine, 1,2-diethyl-	1 (0.454)
O,O-Diethyl S-methyl dithiophosphate	Phosphorodithioic acid, O,O-diethyl S-methyl ester	5000 (2270)
Diethyl-p-nitrophenyl phosphate	Phosphoric acid, diethyl p-nitrophenyl ester	100 (45.4)
Diethyl phthalate	1,2-Benzenedicarboxylic acid, diethyl ester	1000 (454)
O,O-Diethyl O-pyrazinyl phosphorothioate	Phosphorothioic acid, O,O-diethyl O-pyrazinyl ester	100 (45.4)
Diethylstilbestrol	4,4'-Stilbenediol, alpha,alpha'-diethyl-	1 (0.454)
1,2-Dihydro-3,6-pyridazinedione	Maleic hydrazide	5000 (2270)
Dihydrosafrole	Benzene, 1,2-methylenedioxy-4-propyl-	1 (0.454)
Diisopropyl fluorophosphate	Phosphorofluoridic acid, bis(1-methylethyl) ester	100 (45.4)
Dimethoate	Phosphorodithioic acid, O,O-dimethyl S-[2(methylamino)-2-oxoethyl] ester	10 (4.54)
3,3'-Dimethoxybenzidine	(1,1'-Biphenyl)-4,4'-diamine,3,3'-dimethoxy-	1 (0.454)
Dimethylamine *	Methanamine, N-methyl-	1000 (454)
Dimethylaminoazobenzene	Benzenamine, N,N-dimethyl-4-phenylazo-	1 (0.454)
7,12-Dimethylbenz[a]anthracene	1,2-Benzanthracene, 7,12-dimethyl-	1 (0.454)
3,3'-Dimethylbenzidine	(1,1'-Biphenyl)-4,4'-diamine,3,3'-dimethyl-	10 (4.54)
alpha,alpha-Dimethylbenzylhydroperoxide	Hydroperoxide, 1-methyl-1-phenylethyl-	10 (4.54)
3,3-Dimethyl-1-(methylthio)-2-butanone, O-[(methylamino)carbonyl] oxime	Thiofanox	100 (45.4)
Dimethylcarbamoyl chloride	Carbamoyl chloride, dimethyl-	1 (0.454)
Dimethylhydrazine, unsymmetrical @	1,1-Dimethylhydrazine	
1,1-Dimethylhydrazine	Dimethylhydrazine, unsymmetrical @	1 (0.454)
1,2-Dimethylhydrazine	Hydrazine, 1,1-dimethyl-	1 (0.454)
O,O-Dimethyl O-p-nitrophenyl phosphorothioate	Hydrazine, 1,2-dimethyl-	100 (45.4)
Dimethylnitrosamine	Methyl parathion *	1 (0.454)
alpha,alpha-Dimethylphenethylamine	N-Nitrosodimethylamine	5000 (2270)
2,4-Dimethylphenol	Ethanamine, 1,1-dimethyl-2-phenyl-	100 (45.4)
Dimethyl phthalate	Phenol, 2,4-dimethyl-	5000 (2270)
Dimethyl sulfate *	1,2-Benzenedicarboxylic acid, dimethyl ester	1 (0.454)
Dinitrobenzene * (mixed)	Sulfuric acid, dimethyl ester	100 (45.4)
m-Dinitrobenzene		
o-Dinitrobenzene		
p-Dinitrobenzene		
4,6-Dinitro-o-cresol and salts	Phenol, 2,4-dinitro-6-methyl-, and salts	10 (4.54)
4,6-Dinitro-o-cyclohexylphenol	Phenol, 2-cyclohexyl-4,6-dinitro-	100 (45.4)
Dinitrophenol		10 (4.54)
2,5-Dinitrophenol		

253

Appendix B: Regulations Reference Guide

§172.101, Appendix, Note   **49 CFR Ch. I (10-1-89 Edition)**

LIST OF HAZARDOUS SUBSTANCES AND REPORTABLE QUANTITIES—Continued

Hazardous Substance	Synonyms	Reportable Quantity(RQ) Pounds(Kilograms)
2,6-Dinitrophenol	Phenol, 2,4-dinitro-	10 (4.54)
Dinitrotoluene		1000 (454)
3,4-Dinitrotoluene		
2,4-Dinitrotoluene	Benzene, 1-methyl-2,4-dinitro-	1000 (454)
2,6-Dinitrotoluene	Benzene, 1-methyl-2,6-dinitro-	1000 (454)
Dinoseb	Phenol, 2,4-dinitro-6-(1-methylpropyl)-	1000 (454)
Di-n-octyl phthalate	1,2-Benzenedicarboxylic acid, di-n-octyl ester	5000 (2270)
1,4-Dioxane	1,4-Diethylene dioxide	1 (0.454)
1,2-Diphenylhydrazine	Hydrazine, 1,2-diphenyl-	1 (0.454)
Diphosphoramide, octamethyl-	Octamethylpyrophosphoramide	100 (45.4)
Dipropylamine	1-Propanamine, N-propyl-	5000 (2270)
Di-n-propylnitrosamine	N-Nitrosodi-n-propylamine	1 (0.454)
Diquat		1000 (454)
Disulfoton *	O,O-Diethyl S-[2-(ethylthio)ethyl]phosphorodithioate	1 (0.454)
2,4-Dithiobiuret	Thioimidodicarbonic diamide	100 (45.4)
Dithiopyrophosphoric acid, tetraethyl ester	Tetraethyldithiopyrophosphate	100 (45.4)
Diuron		1000 (454)
Dodecylbenzenesulfonic acid *		
Endosulfan *	5-Norbornene-2,3-dimethanol,1,4,5,6,7,7-hexachloro,cyclic sulfite	1 (0.454)
alpha - Endosulfan		1 (0.454)
beta - Endosulfan		1 (0.454)
Endosulfan sulfate		1 (0.454)
Endothall	7-Oxabicyclo[2.2.1]heptane-2,3-dicarboxylic acid	1000 (454)
Endrin	1,2,3,4,10,10-Hexachloro-6,7-epoxy-1,4,4a,5,6,7,8,8a-octahydro-endo,endo-1,4:5,8-dimethanonaphthalene.	1 (0.454)
Endrin aldehyde		1 (0.454)
Epichlorohydrin	1-Chloro-2,3-epoxypropane Oxirane, 2-(chloromethyl)-	1000 (454)
Epinephrine	1,2-Benzenediol,4-(1-hydroxy-2-(methylamino)ethyl].	1000 (454)
Ethanal	Acetaldehyde	1000 (454)
Ethanamine, 1,1-dimethyl-2-phenyl-	alpha,alpha-Dimethylphenethylamine	5000 (2270)
Ethanamine, N-ethyl-N-nitroso-	N-Nitrosodiethylamine	1 (0.454)
Ethane, 1,2-dibromo-	Ethylene dibromide	1000 (454)
Ethane, 1,1-dichloro-	Ethylidene dichloride	1000 (454)
Ethane, 1,2-dichloro-	1,1-Dichloroethane Ethylene dichloride * 1,2-Dichloroethane	5000 (2270)
Ethane, 1,1,1,2,2,2-hexachloro-	Hexachloroethane	1 (0.454)
Ethane, 1,1'-[methylenebis(oxy)]bis(2-chloro-	Bis(2-chloroethoxy)methane	1000 (454)
Ethane, 1,1'-oxybis-	Ethyl ether *	100 (45.4)

254

Chemical Hazard Communication Guidebook

Research and Special Programs Administration, DOT §172.101, Appendix, Note

Name	Synonym	Quantity
Ethane, 1,1'-oxybis(2-chloro-	Bis (2-chloroethyl) ether	1 (0.454)
	Dichloroethyl ether	1 (0.454)
Ethane, pentachloro-	Pentachloroethane	1 (0.454)
Ethane, 1,1,1,2-tetrachloro-	1,1,1,2-Tetrachloroethane	1 (0.454)
Ethane, 1,1,2,2-tetrachloro-	1,1,2,2-Tetrachloroethane	1 (0.454)
Ethane, 1,1,2-trichloro-	1,1,2-Trichloroethane	1 (0.454)
Ethane, 1,1,1-trichloro-2,2-bis(p-methoxyphenyl)-	Methoxychlor	1 (0.454)
1,2-Ethanediylbiscarbamodithioic acid	Ethylenebis(dithiocarbamic acid)	5000 (2270)
Ethanenitrile	Acetonitrile	5000 (2270)
Ethanethioamide	Thioacetamide	1 (0.454)
Ethanol, 2,2'-(nitrosoimino)bis-	N-Nitrosodiethanolamine	1 (0.454)
Ethanone, 1-phenyl-	Acetophenone	5000 (2270)
Ethanoyl chloride	Acetyl chloride *	5000 (2270)
Ethenamine, N-methyl-N-nitroso-	N-Nitrosomethylvinylamine	1 (0.454)
Ethene, chloro-	Vinyl chloride *	1 (0.454)
Ethene, 2-chloroethoxy-	2-Chloroethyl vinyl ether	1000 (454)
Ethene, 1,1-dichloro-	Vinylidene chloride *	5000 (2270)
	1,1-Dichloroethylene	
Ethene, trans-1,2-dichloro-	Perchloroethylene	1 (0.454)
	Tetrachloroethene	
	Tetrachloroethylene	
	1,2-trans-Dichloroethylene	
2-Ethoxyethanol	Ethylene glycol monoethyl ether *	1000 (454)
Ethion *		10 (4.54)
Ethyl acetate	Acetic acid, ethyl ester	5000 (2270)
Ethyl acrylate *	2-Propenoic acid, ethyl ester	1000 (454)
Ethylbenzene		1000 (454)
Ethyl carbamate (Urethan)	Carbamic acid, ethyl ester	1 (0.454)
Ethyl chloride @	Chloroethane	100 (45.4)
Ethyl cyanide	Propanenitrile	10 (4.54)
Ethylene 4,4'-dichlorobenzilate	Benzeneacetic acid, 4-chloro-alpha-(4-chlorophenyl)-alpha-hydroxy-, ethyl ester	1 (0.454)
Ethylene dibromide *	Ethane, 1,2-dibromo-	1000 (454)
Ethylene dichloride *	1,2-Dichloroethane	5000 (2270)
	Ethane, 1,2-dichloro-	
Ethylene glycol monoethyl ether *	2-Ethoxyethanol	1 (0.454)
Ethylene oxide *	Oxirane	1 (0.454)
Ethylenebis(dithiocarbamic acid)	1,2-Ethanediylbiscarbamodithioic acid	5000 (2270)
Ethylenediamine		5000 (2270)
Ethylenediamine tetraacetic acid (EDTA)		5000 (2270)
Ethylenethiourea	2-Imidazolidinethione	1 (0.454)
Ethylenimine	Aziridine	1 (0.454)
Ethyl ether *	Ethane, 1,1'-oxybis-	100 (45.4)
Ethylidene dichloride	Ethane, 1,1-dichloro-	1000 (454)
	1,1-Dichloroethane	
Ethyl methacrylate	2-Propenoic acid, 2-methyl-, ethyl ester	1000 (454)
Ethyl methanesulfonate	Methanesulfonic acid, ethyl ester	1 (0.454)
Ethyl methyl ketone @	2-Butanone	5000 (2270)
	Methyl ethyl ketone *	

255

Appendix B: Regulations Reference Guide

§172.101, Appendix, Note **49 CFR Ch. I (10-1-89 Edition)**

LIST OF HAZARDOUS SUBSTANCES AND REPORTABLE QUANTITIES—Continued

Hazardous Substance	Synonyms	Reportable Quantity (RQ) Pounds (Kilograms)
Famphur	Phosphorothioic acid, O,O-dimethyl O-[p-[(dimethylamino)-sulfonyl] phenyl] ester	1000 (454)
Ferric ammonium citrate		1000 (454)
Ferric ammonium oxalate		1000 (454)
Ferric chloride		1000 (454)
Ferric dextran	Iron dextran	5000 (2270)
Ferric fluoride		100 (45.4)
Ferric nitrate *		1000 (454)
Ferric sulfate *		1000 (454)
Ferrous ammonium sulfate		1000 (454)
Ferrous chloride *		100 (45.4)
Ferrous sulfate		1000 (454)
Fluoranthene	Benzo[j,k]fluorene	100 (45.4)
Fluorene		5000 (2270)
Fluorine *		10 (4.54)
Fluoroacetamide	Acetamide, 2-fluoro-	100 (45.4)
Fluoroacetic acid, sodium salt	Acetic acid, fluoro-, sodium salt	10 (4.54)
Formaldehyde *	Methylene oxide	1000 (454)
Formic acid *	Methanoic acid	5000 (2270)
Fulminic acid, mercury(II)salt	Mercury fulminate	10 (4.54)
Fumaric acid	2-Furancarboxaldehyde	5000 (2270)
Furan *	Furfuran	100 (45.4)
Furan, tetrahydro-	Tetrahydrofuran	1000 (454)
2-Furancarboxaldehyde	Furfural *	5000 (2270)
2,5-Furandione	Maleic anhydride	5000 (2270)
Furfural *	2-Furancarboxaldehyde	5000 (2270)
Furfuran	Furan *	100 (45.4)
D-Glucopyranose, 2-deoxy-2-(3-methyl-3-nitrosoureido)-	Streptozotocin	1 (0.454)
Glycidylaldehyde	1-Propanal, 2,3-epoxy-	1 (0.454)
Guanidine, N-nitroso-N-methyl-N'-nitro-	N-Methyl-N'-nitro-N-nitrosoguanidine	1 (0.454)
Guthion *	Azinphos methyl @	1 (0.454)
Heptachlor	4,7-Methano-1H-indene, 1,4,5,6,7,8,8-heptachloro-3a,4,7,7a-tetrahydro-	1 (0.454)
Heptachlor epoxide		1 (0.454)
Hexachlorobenzene	Benzene, hexachloro-	10 (4.54)
Hexachlorobutadiene *	1,3-Butadiene, 1,1,2,3,4,4-hexachloro-	1 (0.454)
Hexachlorocyclohexane (gamma isomer)	gamma - BHC Lindane	1 (0.454)
Hexachlorocyclopentadiene *	1,3-Cyclopentadiene, 1,2,3,4,5,5-hexachloro-	1 (0.454)
1,2,3,4,10,10-Hexachloro-6,7-epoxy-1,4,4a,5,6,7,8,8a-octahydro-endo,endo-1,4:5,8-dimethanonaphthalene.	Endrin *	1 (0.454)
1,2,3,4,10,10-Hexachloro-6,7-epoxy-1,4,4a,5,6,7,8,8a-octahydro-endo,exo-1,4:5,8-dimethanonaphthalene	Dieldrin *	1 (0.454)
Hexachloroethane *	Ethane, 1,1,1,2,2,2-hexachloro-	1 (0.454)

Chemical Hazard Communication Guidebook

Research and Special Programs Administration, DOT §172.101, Appendix, Note

Hexachlorohexahydro-endo,endo-dimethanonaphthalene	1,2,3,4,10,10-Hexachloro-1,4,4a,5,8,8a-hexahydro-1,4,5,8-endo,endo-dimethanonaphthalene	1 (0.454)
1,2,3,4,10,10-Hexachloro-1,4,4a,5,8,8a-hexahydro-1,4,5,8-endo,endo-dimethanonaphthalene	Hexachlorohexahydro-endo,endo-dimethanonaphthalene	1 (0.454)
1,2,3,4,10,10-Hexachloro-1,4,4a,5,8,8a-hexahydro-1,4:5,8-endo,exo-dimethanonaphthalene	Aldrin *	1 (0.454)
Hexachlorophene	2,2'-Methylenebis(3,4,6-trichlorophenol)	100 (45.4)
Hexachloropropene	1-Propene, 1,1,2,3,3,3-hexachloro-	1000 (454)
Hexaethyl tetraphosphate *	Tetraphosphoric acid, hexaethyl ester	100 (45.4)
Hydrazine *	Diamine	1 (0.454)
Hydrazine, 1,2-diethyl-	N,N'-Diethylhydrazine	1 (0.454)
Hydrazine, 1,1-dimethyl-	1,1-Dimethylhydrazine	1 (0.454)
	Dimethylhydrazine, unsymmetrical @	
Hydrazine, 1,2-dimethyl-	1,2-Dimethylhydrazine	1 (0.454)
Hydrazine, 1,2-diphenyl-	1,2-Diphenylhydrazine	1 (0.454)
Hydrazine, methyl-	Methyl hydrazine *	10 (4.54)
Hydrazinecarbothioamide	Thiosemicarbazide	100 (45.4)
Hydrochloric acid		5000 (2270)
Hydrocyanic acid *	Hydrogen cyanide	10 (4.54)
Hydrofluoric acid *	Hydrogen fluoride *	100 (45.4)
Hydrogen cyanide	Hydrocyanic acid *	10 (4.54)
Hydrogen fluoride *	Hydrofluoric acid *	100 (45.4)
Hydrogen phosphide	Phosphine *	100 (45.4)
Hydrogen sulfide *	Hydrosulfuric acid	100 (45.4)
	Sulfur hydride	
Hydroperoxide, 1-methyl-1-phenylethyl-	alpha,alpha-Dimethylbenzylhydroperoxide	10 (4.54)
Hydrosulfuric acid	Hydrogen sulfide *	100 (45.4)
	Sulfur hydride	
Hydroxydimethylarsine oxide	Cacodylic acid	1 (0.454)
2-Imidazolidinethione	Ethylenethiourea	1 (0.454)
Indeno(1,2,3-cd)pyrene	1,10-(1,2-Phenylene)pyrene	1 (0.454)
Iron dextran	Ferric dextran	5000 (2270)
Isobutyl alcohol	1-Propanol, 2-methyl-	5000 (2270)
Isocyanic acid, methyl ester	Methyl isocyanate *	1 (0.454)
Isophorone		100 (45.4)
Isoprene		1000 (454)
Isopropanolamine dodecylbenzene sulfonate		1000 (454)
Isosafrole	Benzene, 1,2-methylenedioxy-4-propenyl-	10 (4.54)
3(2H)-Isoxazolone, 5-(aminomethyl)-	5-(Aminomethyl)-3-isoxazolol	1 (0.454)
Kelthane	Decachlorooctahydro-1,3,4-metheno-2H-cyclobuta[c,d]-pentalen-2-one	1 (0.454)
Kepone		1 (0.454)
Lasiocarpine		1 (0.454)
Lead ¢		
Lead acetate	Acetic acid, lead salt	5000 (2270)
Lead arsenate *		5000 (2270)
Lead chloride		100 (45.4)
Lead fluoborate *		100 (45.4)
Lead fluoride *		100 (45.4)

257

§172.101, Appendix, Note   **49 CFR Ch. I (10-1-89 Edition)**

LIST OF HAZARDOUS SUBSTANCES AND REPORTABLE QUANTITIES—Continued

Hazardous Substance	Synonyms	Reportable Quantity(RQ) Pounds(Kilograms)
Lead iodide		100 (45.4)
Lead nitrate *		100 (45.4)
Lead phosphate	Phosphoric acid, lead salt	1 (0.454)
Lead stearate		5000 (2270)
Lead subacetate		1 (0.454)
Lead sulfate *		100 (45.4)
Lead sulfide		5000 (2270)
Lead thiocyanate		100 (45.4)
Lindane *	gamma - BHC. Hexachlorocyclohexane (gamma isomer).	1 (0.454)
Lithium chromate		1000 (454)
Malathion *		100 (45.4)
Maleic acid *		5000 (2270)
Maleic anhydride *	2,5-Furandione	5000 (2270)
Maleic hydrazide *	1,2-Dihydro-3,6-pyridazinedione	5000 (2270)
Malononitrile	Propanedinitrile.	1000 (454)
Melphalan	Alanine, 3-[p-bis(2-chloroethyl)amino]phenyl-, L-	1 (0.454)
Mercaptodimethur		10 (4.54)
Mercuric cyanide *		1 (0.454)
Mercuric nitrate *		10 (4.54)
Mercuric sulfate *		10 (4.54)
Mercuric thiocyanate *		10 (4.54)
Mercurous nitrate *		10 (4.54)
Mercury *		1 (0.454)
Mercury fulminate	Fulminic acid, mercury(II)salt.	10 (4.54)
Mercury, (acetato-O)phenyl-	Phenylmercuric acetate	100 (45.4)
Methacrylonitrile	2-Propenenitrile, 2-methyl-.	1000 (454)
Methanamine, N-methyl-	Dimethylamine.	1000 (454)
Methane, bromo-	Methyl bromide	1000 (454)
Methane, chloro-	Chloromethane. Methyl chloride.	1 (0.454)
Methane, chloromethoxy-	Chloromethyl methyl ether @ Methylchloromethyl ether @	1 (0.454)
Methane, dibromo-	Methylene bromide	1000 (454)
Methane, dichloro-	Methylene chloride	1000 (454)
Methane, dichlorodifluoro-	Dichlorodifluoromethane *	5000 (2270)
Methane, iodo-	Methyl iodide	1 (0.454)
Methane, oxybis(chloro-	Bis(chloromethyl) ether	1 (0.454)
Methane, tetrachloro-	Carbon tetrachloride *	5000 (2270)
Methane, tetranitro-	Tetranitromethane *	10 (4.54)
Methane, tribromo-	Bromoform	100 (45.4)
Methane, trichloro-	Chloroform *	5000 (2270)

258

Research and Special Programs Administration, DOT §172.101, Appendix, Note

Chemical	Synonym	Value
Methane, trichlorofluoro-	Trichloromonofluoromethane	5000 (2270)
Methanesulfenyl chloride, trichloro-	Perchloromethyl mercaptan @	100 (45.4)
Methanesulfonic acid, ethyl ester	Ethyl methanesulfonate	1 (0.454)
Methanethiol	Methyl mercaptan *	100 (45.4)
	Thiomethanol	
4,7-Methano-1H-indene, 1,4,5,6,7,8,8-heptachloro-a,4,7,7a-tetrahydro-	Heptachlor	1 (0.454)
Methanoic acid	Formic acid *	5000 (2270)
4,7-Methanoindan, 1,2,4,5,6,7,8,8-octachloro-3a,4,7,7a-tetrahydro-	Chlordane	1 (0.454)
	Chlordane, technical *	
Methanol *	Methyl alcohol *	5000 (2270)
Methapyrilene	Pyridine, 2-[(2-(dimethylamino)ethyl)-2-thenylamino]-	5000 (2270)
Methomyl	Acetimidic acid, N-[(methylcarbamoyl)oxy]thio-, methyl ester	100 (45.4)
Methoxychlor	Ethane, 1,1,1-trichloro-2,2-bis(p-methoxyphenyl)-	1 (0.454)
Methyl alcohol *	Methanol *	5000 (2270)
Methylamine @	Monomethylamine	100 (45.4)
2-Methylaziridine	1,2-Propylenimine	1 (0.454)
Methyl bromide *	Methane, bromo-	1000 (454)
1-Methylbutadiene	1,3-Pentadiene	100 (45.4)
Methyl chloride *	Chloromethane	1 (0.454)
	Methane, chloro-	
Methyl chlorocarbonate *	Carbonochloridic acid, methyl ester	1000 (454)
Methyl chloroform *	Methane, 1,1,1-trichloro-	1000 (454)
Methyl chloroformate @	Carbonochloridic acid, methyl ester	1000 (454)
	Methyl chlorocarbonate *	
Methylchloromethyl ether @	Chloromethyl methyl ether	1 (0.454)
	Methane, chloromethoxy-	
3-Methylcholanthrene	Benz[j]aceanthrylene, 1,2-dihydro-3-methyl-	1 (0.454)
4,4'-Methylenebis(2-chloroaniline)	Benzenamine, 4,4'-methylenebis(2-chloro-	1 (0.454)
2,2'-Methylenebis(3,4,6-trichlorophenol)	Hexachlorophene	100 (45.4)
Methylene bromide	Methane, dibromo-	1000 (454)
Methylene chloride *	Methane, dichloro-	1000 (454)
Methylene oxide *	Formaldehyde *	1000 (454)
Methyl ethyl ketone *	2-Butanone	5000 (2270)
	Ethyl methyl ketone @	
Methyl ethyl ketone peroxide *	2-Butanone peroxide	10 (4.54)
Methyl hydrazine	Hydrazine, methyl-	10 (4.54)
Methyl iodide	Methane, iodo-	1 (0.454)
Methyl isobutyl ketone	4-Methyl-2-pentanone	5000 (2270)
Methyl isocyanate *	Isocyanic acid, methyl ester	1 (0.454)
2-Methyllactonitrile	Acetone cyanohydrin *	10 (4.54)
	Propanenitrile, 2-hydroxy-2-methyl-	
Methyl mercaptan *	Methanethiol	100 (45.4)
	Thiomethanol	
Methyl methacrylate *	2-Propenoic acid, 2-methyl-, methyl ester	1000 (454)
N-Methyl-N'-nitro-N-nitrosoguanidine	Guanidine, N-nitroso-N-methyl-N'-nitro-	1 (0.454)
Methyl parathion	O,O-Dimethyl O-p-nitrophenyl phosphorothioate	100 (45.4)

259

Appendix B: Regulations Reference Guide

§172.101, Appendix, Note   **49 CFR Ch. I (10-1-89 Edition)**

LIST OF HAZARDOUS SUBSTANCES AND REPORTABLE QUANTITIES—Continued

Hazardous Substance	Synonyms	Reportable Quantity(RQ) Pounds(Kilograms)
4-Methyl-2-pentanone	Methyl isobutyl ketone	5000 (2270)
Methylthiouracil	4(1H)-Pyrimidinone, 2,3-dihydro-6-methyl-2-thioxo-	1 (0.454)
Mevinphos *		10 (4.54)
Mexacarbate *		1000 (454)
Mitomycin C	Azirino(2',3':3,4)pyrrolo(1,2-a)indole-4,7-dione,6-amino-8-(((aminocarbonyl)oxy) methyl)-1,1a,2,8,8a,8b-hexahydro-8a-methoxy-5-methyl-.	1 (0.454)
Monoethylamine *		100 (45.4)
Monomethylamine *	Methylamine @	100 (45.4)
Naled		10 (4.54)
5,12-Naphthacenedione, (8S-cis)-8-acetyl-10-[3-amino-2,3,6-trideoxy-alpha-L-lyxohexopyranosyl) oxy]-7,8,9,10-tetrahydro-6,8,11-trihydroxy-1-methoxy-.	Daunomycin	1 (0.454)
Naphthalene *		100 (45.4)
Naphthalene, 2-chloro-	beta-Chloronaphthalene 2-Chloronaphthalene	5000 (2270)
1,4-Naphthalenedione	1,4-Naphthoquinone	5000 (2270)
2,7-Naphthalenedisulfonic acid, 3,3'-[(3,3'-dimethyl-(1,1'-biphenyl)-4,4'-diyl)-bis(azo)]bis(5-amino-4-hydroxy)-tetrasodium salt.	Trypan blue	1 (0.454)
Naphthenic acid		100 (45.4)
1,4-Naphthoquinone	1,4-Naphthalenedione	5000 (2270)
alpha-Naphthylamine	1-Naphthylamine	1 (0.454)
beta-Naphthylamine	2-Naphthylamine	1 (0.454)
1-Naphthylamine	alpha-Naphthylamine	1 (0.454)
2-Naphthylamine	beta-Naphthylamine	1 (0.454)
2-Naphthylamine, N,N-bis(2-chloroethyl)-	Chlornaphazine	100 (45.4)
alpha-Naphthylthiourea	Thiourea, 1-naphthalenyl-	1 (0.454)
Nickel ¢		5000 (2270)
Nickel ammonium sulfate		1 (0.454)
Nickel carbonyl *		5000 (2270)
Nickel chloride		1 (0.454)
Nickel cyanide	Nickel(II) cyanide	1000 (454)
Nickel(II) cyanide	Nickel cyanide *	5000 (2270)
Nickel hydroxide		5000 (2270)
Nickel nitrate		5000 (2270)
Nickel sulfate		1 (0.454)
Nickel tetracarbonyl	Nickel carbonyl *	100 (45.4)
Nicotine * and salts	Pyridine, (S)-3-(1-methyl-2-pyrrolidinyl)-, and salts	1000 (454)
Nitric acid		10 (4.54)
Nitric oxide *		5000 (2270)
p-Nitroaniline	Benzenamine, 4-nitro-	1000 (454)
Nitrobenzene *	Benzene, nitro-	10 (4.54)
Nitrogen dioxide	Nitrogen(IV) oxide	10 (4.54)
Nitrogen(II) oxide	Nitric oxide *	

260

Chemical Hazard Communication Guidebook

Research and Special Programs Administration, DOT §172.101, Appendix, Note

Chemical	Quantity
Nitrogen(IV) oxide *	10 (4.54)
Nitroglycerine *	10 (4.54)
Nitrophenol (mixed)	100 (45.4)
m- 2-Nitrophenol	
o- 4-Nitrophenol	
p-	
o-Nitrophenol ... Phenol, 4-nitro-	100 (45.4)
p-Nitrophenol ... Phenol, 4-nitro-	100 (45.4)
2-Nitrophenol ... o-Nitrophenol	100 (45.4)
4-Nitrophenol ... p-Nitrophenol	100 (45.4)
2-Nitrophenol ... Phenol, 4-nitro-	100 (45.4)
2-Nitropropane ... Propane, 2-nitro-	1 (0.454)
N-Nitrosodi-n-butylamine ... 1-Butanamine, N-butyl-N-nitroso-	1 (0.454)
N-Nitrosodiethanolamine ... Ethanol, 2,2'-(nitrosoimino)bis-	1 (0.454)
N-Nitrosodiethylamine ... Ethanamine, N-ethyl-N-nitroso-	1 (0.454)
N-Nitrosodimethylamine ... Dimethylnitrosamine	1 (0.454)
N-Nitrosodiphenylamine	100 (45.4)
N-Nitrosodi-n-propylamine ... Di-n-propylnitrosamine	1 (0.454)
N-Nitroso-N-ethylurea ... Carbamide, N-ethyl-N-nitroso-	1 (0.454)
N-Nitroso-N-methylurea ... Carbamide, N-methyl-N-nitroso-	1 (0.454)
N-Nitroso-N-methylurethane ... Carbamic acid, methylnitroso-, ethyl ester	1 (0.454)
N-Nitrosomethylvinylamine ... Ethenamine, N-methyl-N-nitroso-	1 (0.454)
N-Nitrosopiperidine ... Pyridine, hexahydro-N-nitroso-	1 (0.454)
N-Nitrosopyrrolidine ... Pyrrole, tetrahydro-N-nitroso-	1 (0.454)
Nitrotoluene	1000 (454)
m-Nitrotoluene	
o-Nitrotoluene	
p-Nitrotoluene	
5-Nitro-o-toluidine ... Benzenamine, 2-methyl-5-nitro-	1 (0.454)
5-Norbornene-2,3-dimethanol,1,4,5,6,7,7-hexachloro,cyclic sulfite ... Endosulfan *	1 (0.454)
Octamethylpyrophosphoramide ... Diphosphoramide, octamethyl-	100 (45.4)
Osmium oxide ... Osmium tetroxide	1000 (454)
Osmium tetroxide ... Osmium oxide	1000 (454)
7-Oxabicyclo[2.2.1]heptane-2,3-dicarboxylic acid ... Endothall	1000 (454)
1,2-Oxathiolane, 2,2-dioxide ... 1,3-Propane sultone	1 (0.454)
2H-1,3,2-Oxazaphosphorine,2-[bis(2-chloroethyl) amino] tetrahydro-2-oxide ... Cyclophosphamide	1 (0.454)
Oxirane ... Ethylene oxide *	1 (0.454)
Oxirane, 2-(chloromethyl)- ... Epichlorohydrin *	1 (0.454)
1-Chloro-2,3-epoxypropane	1000 (454)
Paraformaldehyde *	1000 (454)
Paraldehyde *	1000 (454)
Parathion * ... Phosphorothioic acid, O,O-diethyl O-(p-nitrophenyl)ester	1 (0.454)
Pentachlorobenzene ... Benzene, pentachloro-	10 (4.54)
Pentachloroethane ... Ethane, pentachloro-	10 (4.54)
Pentachloronitrobenzene ... Benzene, pentachloronitro-	1 (0.454)
Pentachlorophenol ... Phenol, pentachloro-	10 (4.54)

261

Appendix B: Regulations Reference Guide

§172.101, Appendix, Note   49 CFR Ch. I (10-1-89 Edition)

LIST OF HAZARDOUS SUBSTANCES AND REPORTABLE QUANTITIES—Continued

Hazardous Substance	Synonym	Reportable Quantity(RQ) Pounds(Kilograms)
1,3-Pentadiene	1-Methylbutadiene	100 (45.4)
Perchloroethylene *	Ethene, 1,1,2,2-tetrachloro	1 (0.454)
	Tetrachloroethene	
	Tetrachloroethylene *	
Perchloromethyl mercaptan @	Methanesulfenyl chloride, trichloro-	100 (45.4)
	Trichloromethanesulfenyl chloride	
Phenacetin	Acetamide, N-(4-ethoxyphenyl)-	1 (0.454)
Phenanthrene		5000 (2270)
Phenol *	Benzene, hydroxy-	1000 (454)
Phenol, 2-chloro-	o-Chlorophenol	100 (45.4)
	2-Chlorophenol	
Phenol, 4-chloro-3-methyl-	p-Chloro-m-cresol	5000 (2270)
	4-Chloro-m-cresol	
Phenol, 2-cyclohexyl-4,6-dinitro-	4,6-Dinitro-o-cyclohexylphenol	100 (45.4)
Phenol, 2,4-dichloro-	2,4-Dichlorophenol	100 (45.4)
Phenol, 2,6-dichloro-	2,6-Dichlorophenol	100 (45.4)
Phenol, 2,4-dimethyl-	2,4-Dimethylphenol	100 (45.4)
Phenol, 2,4-dinitro-	2,4-Dinitrophenol	10 (4.54)
Phenol, 2,4-dinitro-6-(1-methylpropyl)-	Dinoseb	1000 (454)
Phenol, 2,4-dinitro-6-methyl-, and salts	4,6-Dinitro-o-cresol and salts	10 (4.54)
Phenol, 4-nitro-	p-Nitrophenol	100 (45.4)
	4-Nitrophenol	
Phenol, pentachloro-	Pentachlorophenol	10 (4.54)
Phenol, 2,3,4,6-tetrachloro-	2,3,4,6-Tetrachlorophenol	10 (4.54)
Phenol, 2,4,5-trichloro-	2,4,5-Trichlorophenol	10 (4.54)
Phenol, 2,4,6-trichloro-	2,4,6-Trichlorophenol	10 (4.54)
Phenol, 2,4,6-trinitro-, ammonium salt	Ammonium picrate	1 (0.454)
Phenyl dichloroarsine *	Dichlorophenylarsine	1 (0.454)
1,10-(1,2-Phenylene)pyrene	Indeno(1,2,3-cd)pyrene	100 (45.4)
Phenyl mercaptan @	Benzenethiol	
	Thiophenol	
Phenylmercuric acetate	Mercury, (acetato-O)phenyl-	100 (45.4)
N-Phenylthiourea	Thiourea, phenyl-	100 (45.4)
Phorate	Phosphorodithioic acid, O,O-diethyl S-(ethylthio), methylester	10 (4.54)
Phosgene *	Carbonyl chloride	10 (4.54)
Phosphine *	Hydrogen phosphide	100 (45.4)
Phosphoric acid		5000 (2270)
Phosphoric acid, diethyl p-nitrophenyl ester	Diethyl-p-nitrophenyl phosphate	100 (45.4)
Phosphoric acid, lead salt	Lead phosphate	1 (0.454)
Phosphorodithioic acid, O,O-diethyl S-(ethylthio), methyl ester	Phorate	10 (4.54)
Phosphorodithioic acid, O,O-diethyl S-methyl ester	O,O-Diethyl S-methyl dithiophosphate	5000 (2270)
Phosphorodithioic acid, O,O-dimethyl S-[2 (methylamino)-2-oxoethyl] ester	Dimethoate	10 (4.54)

Chemical Hazard Communication Guidebook

Research and Special Programs Administration, DOT §172.101, Appendix, Note

Chemical	Value
Diisopropyl fluorophosphate	100 (45.4)
Parathion *	1 (0.454)
O,O-Diethyl O-pyrazinyl phosphorothioate	100 (45.4)
Famphur	1000 (454)
Phosphorus *	1 (0.454)
Phosphorus sulfide	1000 (454)
Sulfur phosphide	1000 (454)
Phosphorus pentasulfide *	100 (45.4)
Sulfur phosphide	100 (45.4)
1,2-Benzenedicarboxylic acid anhydride	1000 (454)
Pyridine, 2-methyl-	5000 (2270)
Tetraethyl lead *	10 (4.54)
Aroclor 1016	10 (4.54)
Aroclor 1221	1000 (454)
Aroclor 1232	1000 (454)
Aroclor 1242	1000 (454)
Aroclor 1248	1000 (454)
Aroclor 1254	10 (4.54)
Aroclor 1260	1000 (454)
Potassium dichromate @	1000 (454)
Potassium bichromate	100 (45.4)
3,5-Dichloro-N-(1,1-dimethyl-2-propynyl)benzamide	5000 (2270)
Glycidylaldehyde	1 (0.454)
Aldicarb	5000 (2270)
n-Propylamine *	1 (0.454)
Dipropylamine	5000 (2270)
1,2-Dibromo-3-chloropropane	1000 (454)
2-Nitropropane	1 (0.454)
Bis(2-chloroisopropyl) ether	1000 (454)
1,2-Oxathiolane, 2,2-dioxide	1 (0.454)
Malononitrile	1000 (454)
Ethyl cyanide	10 (4.54)
3-Chloropropionitrile	1000 (454)
Acetone cyanohydrin *	10 (4.54)
2-Methyllactonitrile	10 (4.54)
Nitroglycerine *	10 (4.54)
Tris(2,3-dibromopropyl)phosphate	1 (0.454)
Isobutyl alcohol	5000 (2270)
Acetone *	5000 (2270)

Chemical
Phosphorofluoridic acid, bis(1-methylethyl) ester
Phosphorothioic acid, O,O-diethyl O-(p-nitrophenyl) ester
Phosphorothioic acid, O,O-diethyl O-pyrazinyl ester
Phosphorothioic acid, O,O-dimethyl O-[p-[(dimethylamino)-sulfonyl] phenyl] ester
Phosphorus *
Phosphorus oxychloride *
Phosphorus pentasulfide *
Phosphorus sulfide
Phosphorus trichloride *
Phthalic anhydride
2-Picoline
Plumbane, tetraethyl-
POLYCHLORINATED BIPHENYLS (PCBs)
Potassium arsenate *
Potassium arsenite *
Potassium bichromate
Potassium chromate
Potassium cyanide *
Potassium dichromate @
Potassium hydroxide *
Potassium permanganate *
Potassium silver cyanide
Pronamide
1-Propanal, 2,3-epoxy-
Propanal, 2-methyl-2-(methylthio)-,O-[(methylamino)carbonyl]oxime
1-Propanamine
1-Propanamine, N-propyl-
Propane, 1,2-dibromo-3-chloro-
Propane, 2-nitro-
Propane, 2,2'-oxybis(2-chloro-
1,3-Propane sultone
Propanedinitrile
Propanenitrile
Propanenitrile, 3-chloro-
Propanenitrile, 2-hydroxy-2-methyl-
1,2,3-Propanetriol, trinitrate
1-Propanol, 2,3-dibromo-, phosphate (3:1)
1-Propanol, 2-methyl-
2-Propanone

263

§172.101, Appendix, Note   **49 CFR Ch. I (10-1-89 Edition)**

LIST OF HAZARDOUS SUBSTANCES AND REPORTABLE QUANTITIES—Continued

Hazardous Substance	Synonyms	Reportable Quantity(RQ) Pounds(Kilograms)
2-Propanone, 1-bromo-	Bromoacetone *	1000 (454)
Propargite		10 (4.54)
Propargyl alcohol *	2-Propyn-1-ol	1000 (454)
2-Propenal	Acrolein *	1 (0.454)
2-Propenamide	Acrylamide	5000 (2270)
Propene, 1,3-dichloro-	1,3-Dichloropropene	100 (45.4)
1-Propene, 1,1,2,3,3,3-hexachloro-	Hexachloropropene	1000 (454)
2-Propenenitrile	Acrylonitrile *	100 (45.4)
2-Propenenitrile, 2-methyl-	Methacrylonitrile *	1000 (454)
2-Propenoic acid	Acrylic acid *	5000 (2270)
2-Propenoic acid, ethyl ester	Ethyl acrylate *	1000 (454)
2-Propenoic acid, 2-methyl-, ethyl ester	Ethyl methacrylate	1000 (454)
2-Propenoic acid, 2-methyl-, methyl ester	Methyl methacrylate *	1000 (454)
2-Propen-1-ol	Allyl alcohol *	100 (45.4)
Propionic acid *		5000 (2270)
Propionic acid, 2-(2,4,5-trichlorophenoxy)-	Silvex, 2,4,5-TP @, 2,4,5-TP acid	100 (45.4)
Propionic anhydride		5000 (2270)
n-Propylamine *	1-Propanamine	5000 (2270)
Propylene dichloride	1,2-Dichloropropane	1000 (454)
Propylene oxide *		100 (45.4)
1,2-Propylenimine *	2-Methylaziridine	1 (0.454)
2-Propyn-1-ol	Propargyl alcohol *	1000 (454)
Pyrene		5000 (2270)
Pyrethrins		1 (0.454)
4-Pyridinamine	4-Aminopyridine	1000 (454)
Pyridine *		1000 (454)
Pyridine, 2-[(2-(dimethylamino)ethyl)-2- thenylamino)]-	Methapyrilene	5000 (2270)
Pyridine, hexahydro-N-nitroso-	N-Nitrosopiperidine	1 (0.454)
Pyridine, 2-methyl-	2-Picoline	5000 (2270)
Pyridine, (S)-3-(1-methyl-2-pyrrolidinyl)-, and salts	Nicotine * and salts *	100 (45.4)
4(1H)-Pyrimidinone, 2,3-dihydro-6-methyl-2-thioxo-	Methylthiouracil	1 (0.454)
Pyrophosphoric acid, tetraethyl ester	Tetraethyl pyrophosphate *	10 (4.54)
Pyrrole, tetrahydro-N-nitroso-	N-Nitrosopyrrolidine	1 (0.454)
Quinoline		5000 (2270)
RADIONUCLIDES		See TABLE 2
Reserpine	Yohimban-16-carboxylic acid,11,17-dimethoxy-18-[(3,4,5-trimethoxybenzoyl)oxy]-, methyl ester.	5000 (2270)
Resorcinol	1,3-Benzenediol	5000 (2270)
Saccharin and salts	1,2-Benzisothiazolin-3-one,1,1-dioxide, and salts	1 (0.454)
Safrole	Benzene, 1,2-methylenedioxy-4-allyl-	1 (0.454)

264

Research and Special Programs Administration, DOT §172.101, Appendix, Note

Chemical	Amount lbs (kg)
Selenious acid	10 (4.54)
Selenium †	100 (45.4)
Selenium dioxide	10 (4.54)
Selenium disulfide	10 (4.54)
Selenium oxide *	10 (4.54)
Selenourea	1000 (454)
L-Serine, diazoacetate (ester)	1 (0.454)
Silver †	1000 (454)
Silver cyanide *	1 (0.454)
Silver nitrate *	1 (0.454)
Silvex	100 (45.4)
Sodium *	10 (4.54)
Sodium arsenate *	1000 (454)
Sodium arsenite *	1000 (454)
Sodium azide *	1000 (454)
Sodium bichromate *	1000 (454)
Sodium bifluoride *	100 (45.4)
Sodium bisulfite *	5000 (2270)
Sodium chromate *	1000 (454)
Sodium cyanide *	10 (4.54)
Sodium dichromate @	1000 (454)
Sodium dodecylbenzene sulfonate	1000 (454)
Sodium fluoride *	5000 (2270)
Sodium hydrosulfide *	1000 (454)
Sodium hydroxide *	100 (45.4)
Sodium hypochlorite *	100 (45.4)
Sodium methylate *	1000 (454)
Sodium nitrite *	100 (45.4)
Sodium phosphate, dibasic	5000 (2270)
Sodium phosphate, tribasic	5000 (2270)
Sodium selenite	100 (45.4)
4,4'-Stilbenediol, alpha,alpha'-diethyl-	
Streptozotocin	1 (0.454)
Strontium chromate	1000 (454)
Strontium sulfide	
Strychnidin-10-one, and salts *	10 (4.54)
Strychnidin-10-one, 2,3-dimethoxy-	100 (45.4)
Strychnine * and salts *	10 (4.54)
Styrene	1000 (454)
Sulfur hydride	100 (45.4)
Sulfur monochloride	1000 (454)
Sulfur phosphide	100 (45.4)
Sulfur selenide	1 (0.454)
Sulfuric acid	1000 (454)

Cross-references shown on page:
- Selenium oxide *
- Sulfur selenide
- Selenium dioxide
- Carbamimidoselenoic acid
- Azaserine
- Propionic acid, 2-(2,4,5-trichlorophenoxy)-
- 2,4,5-TP @
- 2,4,5-TP acid
- Sodium dichromate @
- Sodium bichromate
- Diethylstilbestrol
- D-Glucopyranose, 2-deoxy-2-(3-methyl-3-nitrosoureido)-
- Strychnine * and salts *
- Brucine
- Strychnidin-10-one, and salts
- Hydrogen sulfide *
- Hydrosulfuric acid
- Phosphorus pentasulfide *
- Phosphorus sulfide
- Selenium disulfide

265

§172.101, Appendix, Note

49 CFR Ch. I (10-1-89 Edition)

LIST OF HAZARDOUS SUBSTANCES AND REPORTABLE QUANTITIES—Continued

Hazardous Substance	Synonyms	Reportable Quantity(RQ) Pounds(Kilograms)
Sulfuric acid, dimethyl ester	Dimethyl sulfate *	1 (0.454)
Sulfuric acid, thallium(I) salt	Thallium(I) sulfate *	100 (45.4)
2,4,5-T *	2,4,5-T acid	1000 (454)
	2,4,5-Trichlorophenoxyacetic acid *	
2,4,5-T acid	2,4,5-T *	1000 (454)
	2,4,5-Trichlorophenoxyacetic acid *	
2,4,5-T amines		5000 (2270)
2,4,5-T esters		1000 (454)
2,4,5-T salts		1000 (454)
TDE	DDD	1(0.454)
	Dichlorodiphenyl dichloroethane	
	4,4'-DDD	
1,2,4,5-Tetrachlorobenzene	Benzene, 1,2,4,5-tetrachloro-	5000 (2270)
2,3,7,8-Tetrachlorodibenzo-p-dioxin (TCDD)		1 (0.454)
1,1,1,2-Tetrachloroethane	Ethane, 1,1,1,2-tetrachloro-	1 (0.454)
1,1,2,2-Tetrachloroethane	Ethane, 1,1,2,2-tetrachloro-	1 (0.454)
Tetrachloroethane	Ethane, 1,1,2,2-tetrachloro-	1 (0.454)
	Perchloroethylene *	
	Tetrachloroethylene *	
Tetrachloroethylene *	Ethene, 1,1,2,2-tetrachloro-	1 (0.454)
	Perchloroethylene *	
	Tetrachloroethene	
2,3,4,6-Tetrachlorophenol	Phenol, 2,3,4,6-tetrachloro-	10 (4.54)
Tetraethyl lead *	Plumbane, tetraethyl-	10 (4.54)
Tetraethyl pyrophosphate *	Pyrophosphoric acid, tetraethyl ester	10 (4.54)
Tetraethyldithiopyrophosphate	Dithiopyrophosphoric acid, tetraethyl ester	100 (45.4)
Tetrahydrofuran	Furan, tetrahydro-	1000 (454)
Tetranitromethane *	Methane, tetranitro-	10 (4.54)
Tetraphosphoric acid, hexaethyl ester	Hexaethyl tetraphosphate *	100 (45.4)
Thallic oxide	Thallium(III) oxide	100 (45.4)
Thallium t		100 (45.4)
Thallium(I) acetate	Acetic acid, thallium(I) salt	100(45.4)
Thallium(I) carbonate	Carbonic acid, dithallium (I) salt	100 (45.4)
Thallium(I) chloride		100 (45.4)
Thallium(I) nitrate		100 (45.4)
Thallium(III) oxide		100 (45.4)
Thallium(I) selenide		1000 (454)
Thallium(I) sulfate *	Sulfuric acid, thallium(I) salt	100 (45.4)
Thioacetamide	Ethanethioamide	1 (0.454)
Thiofanox	3,3-Dimethyl-1-(methylthio)-2-butanone, O-[(methylamino)carbonyl] oxime	100 (45.4)
Thiomidodicarbonic diamide	2,4-Dithiobiuret	100 (45.4)

266

Chemical Hazard Communication Guidebook

Research and Special Programs Administration, DOT §172.101, Appendix, Note

Thiomethanol	Methanethiol	100 (45.4)
Thiophenol *	Methyl mercaptan *	100 (45.4)
Thiosemicarbazide	Benzenethiol	100 (45.4)
Thiourea	Hydrazinecarbothioamide	1 (0.454)
Thiourea, (2-chlorophenyl)	Carbamide, thio-	100 (45.4)
Thiourea, 1-naphthalenyl-	1-(o-Chlorophenyl)thiourea	100 (45.4)
Thiourea, phenyl-	alpha-Naphthylthiourea	100 (45.4)
Thiram	N-Phenylthiourea	10 (4.54)
Toluene *	Bis(dimethylthiocarbamoyl) disulfide	1000 (454)
Toluenediamine	Benzene, methyl-	1 (0.454)
Toluene diisocyanate *	Diaminotoluene	100 (45.4)
o-Toluidine hydrochloride	Benzene, 2,4-diisocyanatomethyl-	1 (0.454)
o-Toluidine	Benzenamine, 2-methyl-, hydrochloride	1 (0.454)
p-Toluidine	2-Amino-1-methyl benzene	1 (0.454)
Toxaphene *	4-Amino-1-methyl benzene	1 (0.454)
2,4,5-TP @	Camphene, octachloro-	100 (45.4)
	Propionic acid, 2-(2,4,5-trichlorophenoxy)-	
	Silvex	
2,4,5-TP acid esters	2,4,5-TP acid	100 (45.4)
2,4,5-TP acid	Propionic acid, 2-(2,4,5-trichlorophenoxy)-	100 (45.4)
	Silvex	
1H-1,2,4-Triazol-3-amine	2,4,5-TP @	1 (0.454)
Trichlorfon	Amitrole	100 (45.4)
1,2,4-Trichlorobenzene		100 (45.4)
1,1,1-Trichloroethane *		1000 (454)
1,1,2-Trichloroethane	Methyl chloroform *	1 (0.454)
Trichloroethylene *	Ethane, 1,1,2-trichloro-	1000 (454)
Trichloroethene	Trichloroethylene *	1000 (454)
Trichloromethanesulfenyl chloride	Methanesulfenyl chloride, trichloro-	100 (45.4)
Trichloromonofluoromethane	Perchloromethyl mercaptan @	5000 (2270)
Trichlorophenol *	Methane, trichlorofluoro-	10 (4.54)
2,3,4-Trichlorophenol		
2,3,5-Trichlorophenol		
2,3,6-Trichlorophenol		
2,4,5-Trichlorophenol		
2,4,6-Trichlorophenol		
3,4,5-Trichlorophenol		
2,4,5-Trichlorophenol	Phenol, 2,4,5-trichloro-	10 (4.54)
2,4,6-Trichlorophenol	Phenol, 2,4,6-trichloro-	10 (4.54)
2,4,5-Trichlorophenoxyacetic acid *	2,4,5-T	1000 (454)
	2,4,5-T acid	
Triethanolamine dodecylbenzene sulfonate		1000 (454)
Triethylamine		5000 (2270)
Trimethylamine *		100 (45.4)

267

Appendix B: Regulations Reference Guide

§172.101, Appendix, Note   49 CFR Ch. I (10-1-89 Edition)

LIST OF HAZARDOUS SUBSTANCES AND REPORTABLE QUANTITIES—Continued

Hazardous Substance	Synonyms	Reportable Quantity(RQ) Pounds(Kilograms)
sym-Trinitrobenzene *	Benzene, 1,3,5-trinitro-	10 (4.54)
1,3,5-Trioxane, 2,4,6-trimethyl-	Paraldehyde	1000 (454)
Tris(2,3-dibromopropyl) phosphate	1-Propanol, 2,3-dibromo-, phosphate (3:1)	1 (0.454)
Trypan blue	2,7-Naphthalenedisulfonic acid, 3,3'-[(3,3'-dimethyl-(1,1'-biphenyl)-4,4'-diyl)- bis(azo)]bis(5-amino-4-hydroxy)-tetrasodium salt.	1 (0.454)
Unlisted Hazardous Wastes Characteristic of Corrosivity D002		100 (45.4)
Unlisted Hazardous Wastes Characteristic of EP Toxicity		
Arsenic D004		1 (0.454)
Barium D005		1000 (454)
Cadmium D006		1 (0.454)
Chromium D007		1 (0.454)
Lead D008		1 (0.454)
Mercury D009		1 (0.454)
Selenium D010		1 (0.454)
Silver D011		10 (4.54)
Endrin D012		1 (0.454)
Lindane D013		1 (0.454)
Methoxychlor D014		1 (0.454)
Toxaphene D015		1 (0.454)
2,4-D D016		100 (45.4)
2,4,5-TP D017		100 (45.4)
Unlisted Hazardous Wastes Characteristic of Ignitability D001		100 (45.4)
Unlisted Hazardous Wastes Characteristic of Reactivity D003		100 (45.4)
Uracil, 5-[bis(2-chloroethyl)amino]-	Uracil mustard	1 (0.454)
Uracil mustard	Uracil, 5-[bis(2-chloroethyl)amino]-	1 (0.454)
Uranyl acetate *		100 (45.4)
Uranyl nitrate *		100 (45.4)
Vanadic acid, ammonium salt	Ammonium vanadate	1000 (454)
Vanadium(V) oxide	Vanadium pentoxide	1000 (454)
Vanadium pentoxide	Vanadium(V) oxide	1000 (454)
Vanadyl sulfate		1000 (454)
Vinyl acetate *		5000 (2270)
Vinyl chloride *	Ethene, chloro-	1 (0.454)
Vinylidene chloride *	Ethene, 1,1-dichloro- 1,1-Dichloroethylene	5000 (2270)
Warfarin	3-(alpha-Acetonylbenzyl)-4-hydroxycoumarin and salts	100 (45.4)
Xylene * (mixed)	Benzene, dimethyl	1000 (454)
m-	m-	
o-	o-	
p-	p-	
Xylenol *		1000 (454)

268

Chemical Hazard Communication Guidebook

Research and Special Programs Administration, DOT §172.101, Appendix, Note

Yohimban-16-carboxylic acid,11,17-dimethoxy-18-[(3,4,5-trimethoxybenzoyl)oxy]-, methyl ester.	Reserpine	5000 (2270)
Zinc ¢		1000 (454)
Zinc acetate		1000 (454)
Zinc ammonium chloride		1000 (454)
Zinc borate		1000 (454)
Zinc bromide		1000 (454)
Zinc carbonate		1000 (454)
Zinc chloride		1000 (454)
Zinc cyanide *		10 (4.54)
Zinc fluoride		1000 (454)
Zinc formate		1000 (454)
Zinc hydrosulfite *		1000 (454)
Zinc nitrate *		1000 (454)
Zinc phenolsulfonate		5000 (2270)
Zinc phosphide *		100 (45.4)
Zinc silicofluoride		5000 (2270)
Zinc sulfate		1000 (454)
Zirconium nitrate *		5000 (2270)
Zirconium potassium fluoride		1000 (454)
Zirconium sulfate *		5000 (2270)
Zirconium tetrachloride *		5000 (2270)
F001		1 (0.454)
The following spent halogenated solvents used in degreasing and sludges from the recovery of these solvents in degreasing operations::		
(a) Tetrachloroethylene		1 (0.454)
(b) Trichloroethylene		1000 (454)
(c) Methylene chloride		1000 (454)
(d) 1,1,1-Trichloroethane		1000 (454)
(e) Carbon tetrachloride		5000 (2270)
(f) Chlorinated fluorocarbons		5000 (2270)
F002		1 (0.454)
The following spent halogenated solvents and the still bottoms from the recovery of these solvents::		
(a) Tetrachloroethylene		1 (0.454)
(b) Methylene chloride		1000 (454)
(c) Trichloroethylene		1000 (454)
(d) 1,1,1-Trichloroethane		1000 (454)
(e) Chlorobenzene		100 (45.4)
(f) 1,1,2-Trichloro-1,2,2-trifluoroethane		5000 (2270)
(g) o-Dichlorobenzene		100 (45.4)
(h) Trichlorofluoromethane		5000 (2270)
F003		100 (45.4)
The following spent non-halogenated solvents and solvents:		
(a) Xylene		1000 (454)
(b) Acetone		5000 (2270)
(c) Ethyl acetate		5000 (2270)
(d) Ethylbenzene		1000 (454)

269

§172.101, Appendix, Note   **49 CFR Ch. I (10-1-89 Edition)**

LIST OF HAZARDOUS SUBSTANCES AND REPORTABLE QUANTITIES—Continued

Hazardous Substance	Synonyms	Reportable Quantity(RQ) Pounds(Kilograms)
(e) Ethyl ether		100 (45.4)
(f) Methyl isobutyl ketone		5000 (2270)
(g) n-Butyl alcohol		5000 (2270)
(h) Cyclohexanone		5000 (2270)
(i) Methanol		5000 (2270)
F004 The following spent non-halogenated solvents and the stillbottoms from the recovery of these solvents:		1000 (454)
(a) Cresols/Cresylic acid		1000 (454)
(b) Nitrobenzene		1000 (454)
F005 The following spent non-halogenated solvents and the stillbottoms from the recovery of these solvents:		100 (45.4)
(a) Toluene		1000 (454)
(b) Methyl ethyl ketone		5000 (2270)
(c) Carbon disulfide		100 (45.4)
(d) Isobutanol		5000 (2270)
(e) Pyridine		1000 (454)
F006 Wastewater treatment sludges from electroplating operations except from the following processes: (1) Sulfuric acid anodizing of aluminum,(2) tin plating on carbon steel, (3) zinc plating (segregated basis) on carbonsteel, (4) aluminum or zinc-aluminum plating on carbon steel, (5) cleaning/stripping associated with tin, zinc and aluminum plating on carbon steel, and (6) chemical etching and milling of aluminum.		1 (0.454)
F007 Spent cyanide plating bath solutions from electroplating operations.		10 (4.54)
F008 Plating bath sludges from the bottom of plating baths fromelectroplating operations where cyanides are used in the process (except for precious metals electroplating plating bath sludges).		10 (4.54)
F009 Spent stripping and cleaning bath solutions from electroplating operations where cyanides are used in the process (except for precious metals electroplating spent stripping and cleaning bath solutions).		10 (4.54)
F010 Quenching bath sludge from oil baths from metal heat treating operations where cyanides are used in the process (except for precious metals heat-treating quenching bath sludges).		10 (4.54)

270

Chemical Hazard Communication Guidebook

Research and Special Programs Administration, DOT §172.101, Appendix, Note

Waste No.	Description	Qty lbs (kg)
F011	Spent cyanide solutions from salt bath pot cleaning from metal heat treating operations (except for precious metals heat treating spent cyanide solutions from salt bath pot cleaning).	10 (4.54)
F012	Quenching wastewater treatment sludges from metal heat treating operations where cyanides are used in the process (except for precious metals heat treating quenching wastewater treatment sludges).	10 (4.54)
F019	Wastewater treatment sludges from the chemical conversion coating of aluminum.	1 (0.454)
F020	Wastes (except wastewater and spent carbon from hydrogen chloride purification) from the production or manufacturing use (as a reactant, chemical intermediate, or component in a formulating process) of tri- or tetrachlorophenol, or of intermediates used to produce their pesticide derivatives. (This listing does not include wastes from the production of hexachlorophene from highly purified 2,4,5-trichlorophenol.).	1 (0.454)
F021	Wastes (except wastewater and spent carbon from hydrogen chloride purification) from the production or manufacturing use (as a reactant, chemical intermediate, or component in a formulating process) of pentachlorophenol, or of intermediates used to produce its derivatives.	1 (0.454)
F022	Wastes (except wastewater and spent carbon from hydrogen chloride purification) from the manufacturing use (as a reactant, chemical intermediate, or component in a formulating process) of tetra-, penta-, or hexachlorobenzenes under alkaline conditions.	1 (0.454)
F023	Wastes (except wastewater and spent carbon from hydrogen chloride purification) from the production of materials on equipment previously used for the production or manufacturing use (as a reactant, chemical intermediate, or component in a formulating process) of tri- and tetrachlorophenols. (This listing does not include wastes from equipment used only for the productionor use of hexachlorophene from highly purified 2,4,5-trichlorophenol.).	1 (0.454)
F024	Wastes, including but not limited to distillation residues, heavy ends, tars, and reactor cleanout wastes, from the production of chlorinated aliphatic hydrocarbons, having carbon content from one to five, utilizing free radical catalyzed processes. (This listing does not include light ends, spent filters and filter aids, spent dessicants(sic), wastewater, wastewater treatment sludges, spent catalysts, and wastes listed in 40 CFR 261.32.).	1 (0.454)
F026	Wastes (except wastewater and spent carbon from hydrogen chloride purification) from the production of materials on equipment previously used for the manufacturing use (as a reactant, chemical intermediate, or component in a formulating process) of tetra-, penta-, or hexachlorobenzene under alkaline conditions.	1 (0.454)

271

§172.101, Appendix, Note   **49 CFR Ch. I (10-1-89 Edition)**

LIST OF HAZARDOUS SUBSTANCES AND REPORTABLE QUANTITIES—Continued

Hazardous Substance	Synonyms	Reportable Quantity(RQ) Pounds(Kilograms)
F027. Discarded unused formulations containing tri-, tetra-, or pentachlorophenol or discarded unused formulations containing compounds derived from these chlorophenols. (This listing does not include formulations containing hexachlorophene synthesized from prepurified 2,4,5-trichlorophenol as the sole component.).		1 (0.454)
F028. Residues resulting from the incineration or thermal treatment of soil contaminated with EPA Hazardous Waste Nos. F020, F021, F022, F023, F026, and F027.		1 (0.454)
K001. Bottom sediment sludge from the treatment of wastewaters from wood preserving processes that use creosote and/or pentachlorophenol.		1 (0.454)
K002. Wastewater treatment sludge from the production of chrome yellow and orange pigments.		1 (0.454)
K003. Wastewater treatment sludge from the production of molybdate orange pigments		1 (0.454)
K004. Wastewater treatment sludge from the production of zinc yellow pigments.		1 (0.454)
K005. Wastewater treatment sludge from the production of chrome green pigments		1 (0.454)
K006. Wastewater treatment sludge from the production of chrome oxide green pigments (anhydrous and hydrated).		1 (0.454)
K007. Wastewater treatment sludge from the production of iron blue pigments		1 (0.454)
K008. Oven residue from the production of chrome oxide green pigments.		1 (0.454)
K009. Distillation bottoms from the production of acetaldehyde from ethylene		1 (0.454)
K010. Distillation side cuts from the production of acetaldehyde from ethylene		1 (0.454)
K011. Bottom stream from the wastewater stripper in the production of acrylonitrile		1 (0.454)
K013. Bottom stream from the acetonitrile column in the production of acrylonitrile		1 (0.454)
K014. Bottoms from the acetonitrile purification column in the production of acrylonitrile		5000 (2270)
K015. Still bottoms from the distillation of benzyl chloride		1 (0.454)
K016. Heavy ends or distillation residues from the production of carbon tetrachloride		1 (0.454)

272

Chemical Hazard Communication Guidebook

Research and Special Programs Administration, DOT §172.101, Appendix, Note

Waste	Quantity
K017 Heavy ends (still bottoms) from the purification column in the production of epichlorohydrin.	1 (0.454)
K018 Heavy ends from the fractionation column in ethyl chloride production.	1 (0.454)
K019 Heavy ends from the distillation of ethylene dichloride in ethylene dichloride production.	1 (0.454)
K020 Heavy ends from the distillation of vinyl chloride in vinyl chloride monomer production.	1 (0.454)
K021 Aqueous spent antimony catalyst waste from fluoromethanes production.	1 (0.454)
K022 Distillation bottom tars from the production of phenol/acetone from cumene.	5000 (2270)
K023 Distillation light ends from the production of phthalic anhydride from naphthalene.	5000 (2270)
K024 Distillation bottoms from the production of phthalic anhydride from naphthalene.	1 (0.454)
K025 Distillation bottoms from the production of nitrobenzene by the nitration of benzene.	1000 (454)
K026 Stripping still tails from the production of methyl ethyl pyridines.	1 (0.454)
K027 Centrifuge and distillation residues from toluene diisocyanate production.	1 (0.454)
K028 Spent catalyst from the hydrochlorinator reactor in the production of 1,1,1-trichloroethane.	1 (0.454)
K029 Waste from the product steam stripper in the production of 1,1,1-trichloroethane.	1 (0.454)
K030 Column bottoms or heavy ends from the combined production of trichloroethylene and perchloroethylene.	1 (0.454)
K031 By-product salts generated in the production of MSMA and cacodylic acid.	1 (0.454)
K032 Wastewater treatment sludge from the production of chlordane.	1 (0.454)
K033 Wastewater and scrub water from the chlorination of cyclopentadiene in the production of chlordane.	1 (0.454)
K034 Filter solids from the filtration of hexachlorocyclopentadiene in the production of chlordane.	1 (0.454)
K035 Wastewater treatment sludges generated in the production of creosote.	1 (0.454)
K036 Still bottoms from toluene reclamation distillation in the production of disulfoton.	1 (0.454)

273

§172.101, Appendix, Note	49 CFR Ch. I (10-1-89 Edition)

LIST OF HAZARDOUS SUBSTANCES AND REPORTABLE QUANTITIES—Continued

Hazardous Substance	Synonyms	Reportable Quantity(RQ) Pounds(Kilograms)
K037 Wastewater treatment sludges from the production of disulfoton.		1 (0.454)
K038 Wastewater from the washing and stripping of phorate production.		1 (0.454)
K039 Filter cake from the filtration of diethylphosphorodithioic acid in the production of phorate.		10 (4.54)
K040 Wastewater treatment sludge from the production of phorate.		1 (0.454)
K041 Wastewater treatment sludge from the production of toxaphene.		1 (0.454)
K042 Heavy ends or distillation residues from the distillation of tetrachlorobenzene in the production of 2,4,5-T.		1 (0.454)
K043 2,6-dichlorophenol waste from the production of 2,4-D.		1 (0.454)
K044 Wastewater treatment sludges from the manufacturing and processing of explosives.		10 (4.54)
K045 Spent carbon from the treatment of wastewater containing explosives.		10 (4.54)
K046 Wastewater treatment sludges from the manufacturing, formulation and loading of lead-based initiating compounds.		100 (45.4)
K047 Pink/red water from TNT operations.		10 (4.54)
K048 Dissolved air flotation (DAF) float from the petroleum refining industry.		1 (0.454)
K049 Slop oil emulsion solids from the petroleum refining industry.		1 (0.454)
K050 Heat exchanger bundle cleaning sludge from the petroleum refining industry.		1 (0.454)
K051 API separator sludge from the petroleum refining industry.		1 (0.454)
K052 Tank bottoms (leaded) from the petroleum refining industry.		10 (4.54)
K060 Ammonia still lime sludge from coking operations.		1 (0.454)
K061 Emission control dust/sludge from the primary production of steel in electric furnaces.		1 (0.454)

274

Chemical Hazard Communication Guidebook

Research and Special Programs Administration, DOT §172.101, Appendix, Note

Waste code	Description	Value
K062	Spent pickle liquor from steel finishing operations	1 (0.454)
K069	Emission control dust/sludge from secondary lead smelting	1 (0.454)
K071	Brine purification muds from the mercury cell process in chlorine production, where separately prepurified brine is not used	1 (0.454)
K073	Chlorinated hydrocarbon waste from the purification step of the diaphragm cell process using graphite anodes in chlorine production	1 (0.454)
K083	Distillation bottoms from aniline extraction	100 (45.4)
K084	Wastewater treatment sludges generated during the production of veterinary pharmaceuticals from arsenic or organo-arsenic compounds	1 (0.454)
K085	Distillation or fractionation column bottoms from the production of chlorobenzenes	1 (0.454)
K086	Solvent washes and sludges, caustic washes and sludges, orwater washes and sludges from cleaning tubs and equipment used in the formulation of ink from pigments, driers, soaps, and stabilizers containing chromium and lead	1 (0.454)
K087	Decanter tank tar sludge from coking operations	100 (45.4)
K093	Distillation light ends from the production of phthalic anhydride from ortho-xylene	5000 (2270)
K094	Distillation bottoms from the production of phthalic anhydride from ortho-xylene	5000 (2270)
K095	Distillation bottoms from the production of 1,1,1-trichloroethane	1 (0.454)
K096	Heavy ends from the heavy ends column from the production of 1,1,1-trichloroethane	1 (0.454)
K097	Vacuum stripper discharge from the chlordane chlorinator in the production of chlordane	1 (0.454)
K098	Untreated process wastewater from the production of toxaphene	1 (0.454)
K099	Untreated wastewater from the production of 2,4-D	1 (0.454)
K100	Waste leaching solution from acid leaching of emission control dust/sludge from secondary lead smelting	1 (0.454)
K101	Distillation tar residues from the distillation of aniline-based compounds in the production of veterinary pharmaceuticals from arsenic or organo-arsenic compounds	1 (0.454)

275

Appendix B: Regulations Reference Guide

§172.101, Appendix, Note   **49 CFR Ch. I (10-1-89 Edition)**

LIST OF HAZARDOUS SUBSTANCES AND REPORTABLE QUANTITIES—Continued

Hazardous Substance	Synonyms	Reportable Quantity(RQ) Pounds(Kilograms)
K102 Residue from the use of activated carbon for decolorization in the production of veterinary pharmaceuticals from arsenic or organo-arsenic compounds.		1 (0.454)
K103 Process residues from aniline extraction from the production of aniline.		100 (45.4)
K104 Combined wastewater streams generated from nitrobenzene/aniline chlorobenzenes.		1 (0.454)
K105 Separated aqueous stream from the reactor product washing step in the production of chlorobenzenes.		1 (0.454)
K106 Wastewater treatment sludge from the mercury cell process in chlorine production.		1 (0.454)
K111 Product washwaters from the production of dinitrotoluene via nitration of toluene.		1 (0.454)
K112 Reaction by-product water from the drying column in the production of toluenediamine via hydrogenation of dinitrotoluene.		1 (0.454)
K113 Condensed liquid light ends from the purification of toluenediamine in the production of toluenediamine via hydrogenation of dinitrotoluene.		1 (0.454)
K114 Vicinals from the purification of toluenediamine in the production of toluenediamine via hydrogenation of dinitrotoluene.		1 (0.454)
K115 Heavy ends from the purification of toluenediamine in the production of toluenediamine via hydrogenation of dinitrotoluene.		1 (0.454)
K116 Organic condensate from the solvent recovery column in the production of toluene diisocyanate via phosgenation of toluenediamine.		1 (0.454)
K117 Wastewater from the reaction vent gas scrubber in the production of ethylene bromide via bromination of ethene.		1 (0.454)
K118 Spent absorbent solids from purification of ethylene dibromide in the production of ethylene dibromide.		1 (0.454)
K136 Still bottoms from the purification of ethylene dibromide in the production of ethylene dibromide via bromination of ethene.		1 (0.454)

Footnotes:
¢ - the RQ for these hazardous substances is limited to those pieces of the metal having a diameter smaller than 100 micrometers (0.004 inches)
¢¢ - the RQ for asbestos is limited to friable forms only
* - indicates that this material appears by name in the Hazardous Materials Table
@ - indicates that the name was added by RSPA because (1) the name is a synonym for a specific hazardous substance and (2) the name appears in the Hazardous Materials Table as a proper shipping name.

[Amdt. 172-108, 52 FR 4825, Feb. 17, 1987]

276

APPENDIX C

DEFINITIONS OF HAZARDS

This appendix provides the definitions of hazard-related terms used in the OSHA Standard, the Emergency Planning and Community Right-To-Know Act, and the Department of Transportation's hazardous materials transportation regulations. It is intended to assist the reader in comparing and contrasting the definitions in each type of regulation.

Each of these requirements defines the types of hazard covered, which may be, in some cases, the same term but defined differently. For example, corrosivity is defined in the OSHA Standard by its effect on skin; however, corrosive materials in DOT's regulations are defined not only by their effect on living tissue, but also by the effect on steel as defined by a particular test. In other cases, the definition in one regulation may refer to the definitions in another regulation. In EPA's definition of immediate and delayed health hazards, specific reference is made to OSHA's definitions of hazards which are subsumed within the EPA definition.

To clarify how these terms relate to each other, Table C-1 provides an approximate comparison. The table relates the type of hazard in the regulations within several major groups. At the top of the table indicates the chemicals covered in each area, such as DOT's regulation of hazardous materials and SARA's definition of extremely hazardous substances. The use in SARA of the term hazardous chemical is by reference to OSHA's definition. Below these are the terms in each area of regulation that relate to each other. Combustible liquids under DOT and OSHA have similar definitions, for example.

Chemical Hazard Communication Guidebook

Table C-1
A COMPARISION OF HAZARD-RELATED TERMS USED IN THE REGULATIONS

TYPE OF HAZARD	SARA RIGHT-TO-KNOW	OSHA STANDARD	TRANSPORTATION
Covered Chemicals	Extremely Haz. Substance Hazardous Chemical Toxic Chemicals	Hazardous Chemical	Hazardous Materials
Health Effects	Immediate (Acute) health hazard	Sensitizer Target organ effects Corrosive Highly toxic Irritant Toxic	Corrosive material Poisonous materials -Irritating materials -Poison A -Poison B
	Delayed (chronic) health hazard	Carcinogen Target organ effects	
Physical Effects	Fire hazard	Combustible liquid Flammable -aerosol -gas -liquid -solid Oxidizer Pyrophoric	Combustible liquid Flammable liquid Flammable solid Compressed gas (flammable) Pyrophoric liquid
	Reactive hazard	Organic peroxide Unstable (reactive) Water-reactive	Organic peroxide Oxidizer
	Sudden release of pressure	Compressed gas	Compressed gas -liquefied -non-liquefied -refrigerant gas or dispersant gas -in solution Cryogenic liquid
Other		Explosive	Explosives -Class A -Class B -Class C Etiologic agents ORM-A ORM-B ORM-C ORM-D ORM-E Radioactive material -Fissile material

Appendix C: Definitions of Hazards

OSHA Hazard Communication Standard

Acute effects

Appendix A of the Standard describes acute effects, as follows:

"Acute" effects usually occur rapidly as a result of short-term exposures, and are of short duration. .... The acute effects referred to most frequently are those defined by the American National Standards Institute (ANSI) standard for Precautionary Labeling of Hazardous Industrial Chemicals (Z129.1-1982)--irritation, corrosivity, sensitization and lethal dose. Although these are important health effects, they do not adequately cover the considerable range of acute effects which may occur as a result of occupational exposure, such as, for example, narcosis.

Refer To: 52 FR 31884

Carcinogen

A chemical is considered to be a carcinogen if:

(a) It has been evaluated by the International Agency for Research on Cancer (IARC) and found to be a carcinogen or potential carcinogen; or

(b) It is listed as a carcinogen or potential carcinogen in the Annual Report on Carcinogens published by the National Toxicology Program (NTP) (latest edition); or

(c) It is regulated by OSHA as a carcinogen.

Refer To: 52 FR 31884

OSHA Hazard Communication Standard (cont.)

Chronic effects

Appendix A to the Standard describes chronic effects, as follows:

"Chronic" effects generally occur as a result of long-term exposure, and are of long duration. .... The term chronic effect is often used to cover only carcinogenicity, teratogenicity, and mutagenicity. These effects are obviously a concern in the workplace, but ... do not adequately cover the area of chronic effects, excluding, for example, blood dyscrasias (such as enemia), chronic bronchitis and liver atrophy.
Refer To: 52 FR 31884

Combustible liquid

"Combustible liquid" means any liquid having a flashpoint at or above 100 degrees F but below 200 degrees F except any mixture having components with flashpoints of 200 degrees F or higher, the total volume of which make up 99% or more of the total volume of the mixture.
Refer To: 52 FR 31878

Compressed gas

"Compressed gas" means:

(i) a gas or mixture of gases having, in a container, an absolute pressure exceeding 40 psi at 70 degrees F; or

Appendix C: Definitions of Hazards

OSHA Hazard Communication Standard (cont.)

(ii) a gas or mixture of gases having, in a container, an absolute pressure exceeding 104 psi at 130 degrees F, regardless of the pressure at 70 degrees F; or

(iii) a liquid having a vapor pressure exceeding 40 psi at 100 degrees F as determined by ASTM D-323-72.
Refer To: 52 FR 31878

Corrosive

"Corrosive" refers to a chemical that causes visible destruction of, or irreversible alterations in, living tissue by chemical action at the site of contact. For example, a chemical is considered to be corrosive if, when tested on the intact skin of albino rabbits by the method described by the U.S. Department of Transportation in Appendix A to 49 CFR Part 173, it destroys or changes irreversibly the structure of the tissue at the site of contact following an exposure period of four hours. This term does not refer to action on inanimate surfaces.
Refer To: 52 FR 31884

Explosive

"Explosive" means a chemical that causes a sudden, almost instantaneous release of pressure, gas, and heat when subjected to sudden shock, pressure, or high temperature.
Refer To: 52 FR 31878

Chemical Hazard Communication Guidebook

OSHA Hazard Communication Standard (cont.)

Flammable

"Flammable" means a chemical that falls into one of the following categories:

a. Aerosol, flammable

b. Gas, flammable

c. Liquid, flammable

d. Solid, flammable
Refer To: 52 FR 31878

Flammable aerosol

"Aerosol, flammable" means an aerosol that, when tested by the method described in 16 CFR 1500.45 yields a flame projection exceeding 18 inches at full valve opening, or a flashback (a flame extending back to the valve) at any degree of valve opening.
Refer To: 52 FR 31878

Flammable gas

"Gas, flammable" means:

(A) a gas that, at ambient temperature and pressure, forms a flammable mixture with air at a concentration of 13% by volume or less; or

(B) a gas that, at ambient temperature and pressure, forms a range of flammable mixtures with air wider than 12% by volume, regardless of the lower limit.
Refer To: 52 FR 31878

Appendix C: Definitions of Hazards

OSHA Hazard Communication Standard (cont.)

Flammable liquid

"Liquid, flammable" means any liquid having a flashpoint below 100 degrees F except any mixture having components with flashpoints of 100 degrees F or higher, the total of which make up 99 percent or more of the total volume of the mixture.
Refer To: 52 FR 31878

Flammable solid

"Solid, flammable" means a solid other than a blasting agent or explosive as defined in section 190.109(a) that is liable to cause fire through friction, absorption of moisture, spontaneous chemical change, or retained heat from manufacturing or processing, or which can be ignited readily and when ignited burns so vigorously and persistently as to create a serious hazard. A chemical must be considered to be a flammable solid if, when tested by the method described in 16 CFR 1500.44, it ignites and burns with a self-sustained flame at a rate greater than one-tenth of an inch per second along its major axis.
Refer To: 52 FR 31878

Hazard determination

See also "Hazard Evaluation"

The Standard requires chemical manufacturers, importers, and employers evaluating chemicals to perform a hazard determination in order to determine if

Chemical Hazard Communication Guidebook

OSHA Hazard Communication Standard (cont.)

the chemical is hazardous. Appendix B to the Standard describes the process of hazard determination.

Hazard evaluation

As described in Appendix B to the Standard:

Hazard evaluation is a process which relies heavily on the professional judgment of the evaluator, particularly in the area of chronic hazards. The performance-orientation of the hazard determination does not diminish the duty of the chemical manufacturer, importer or employer to conduct a thorough evaluation, examining all relevant data and producing a scientifically defensible evaluation. For purposes of this standard, the following criteria must be used in making hazard determinations that meet the requirements of the HC standard.

1. Carcinogenicity: As described in paragraph (d)(4) and Appendix A of the HCS, a determination by the National Toxicology Program, the International Agency for Research on Cancer, or OSHA that a chemical is a carcinogen or potential carcinogen will be considered conclusive evidence for purposes of the HCS.

2. Human data: Where available epidemiological studies and case reports of adverse health effects must be considered in the evaluation.

3. Animal data: Human evidence of health effects in exposed populations is generally not available for the majority of chemicals produced or used in the

Appendix C: Definitions of Hazards

OSHA Hazard Communication Standard (cont.)

workplace. Therefore, the available results of toxicological testing in animal populations must be used to predict the health effects that may be experienced by exposed workers....

4. Adequacy and reporting of data. The results of any studies which are designed and conducted according to established scientific principles, and which report statistically significant conclusions regarding the health effects of a chemical, must be a sufficient basis for a hazard determination and reported on any material safety data sheet. The chemical manufacturer, importer, or employer may also report the results of other scientifically valid studies which tend to refute the findings of hazard.
Refer To: 52 FR 31885

Hazardous Chemical

"Hazardous chemical" means any chemical which is a physical hazard or a health hazard. (at 31879)

Note: the following are OSHA's exemptions under the HC standard:

(i) Any hazardous waste as such term is defined by the Solid Waste Disposal Act, as amended by the Resource Conservation and Recovery Act of 1976, as amended (42 U.S.C. 6901 et seq.), when subject to regulations issued under that Act by the Environmental Protection Agency;
(ii) Tobacco or tobacco products;
(iii) Wood or wood products;
(iv) Articles;
(v) Food, drugs, cosmetics, or alcoholic beverages in a retail establishment which are packaged for sale to consumers;

OSHA Hazard Communication Standard (cont.)

(vi) Foods, drugs, or cosmetics intended for personal consumption by employees while in the workplace;
(vii) Any consumer product or hazardous substance, as those terms are defined in the Consumer Product Safety Act (15 U.S.C. 2051 et seq.) and Federal Hazardous Substances Act (15 U.S.C. 1261 et seq.) respectively; and
(viii) Any drug as that term is defined in the Federal Food, Drug, and Cosmetic Act (21 U.S.C. 301 et seq.).
Refer To: 52 FR 31879

Health hazard

"Health hazard" means a chemical for which there is statistically significant evidence based on at least one study conducted in accordance with established scientific principles that acute or chronic health effects may occur in exposed employees.

Although safety hazards related to the physical characteristics of a chemical can be objectively defined in terms of testing requirements (e.g. flammability), health hazard definitions are less precise and more subjective. Health hazards may cause measurable changes in the body--such as decreased pulmonary function. These changes are generally indicated by the occurrence of signs and symptoms in the exposed employees--such as shortness of breath, a non-measurable, subjective feeling. Employees exposed to such hazards must be apprised of both the change in body function and the signs and symptoms that may occur to signal that change.
Refer To: 52 FR 31879, 31883

Appendix C: Definitions of Hazards

OSHA Hazard Communication Standard (cont.)

Irritant

"Irritant" refers to a chemical, which is not corrosive, but which causes a reversible inflammatory effect on living tissue by chemical action at the site of contact. A chemical is a skin irritant if, when tested on the intact skin of albino rabbits by the methods of 16 CFR 1500.41 for four hours exposure or by other appropriate techniques, it results in an empirical score of five or more. A chemical is an eye irritant if so determined under the procedure listed in 16 CFR 1500.42 or other appropriate techniques.

Organic peroxide

Refer To: 52 FR 31879

"Organic peroxide" means an organic compound that contains the bivalent -O-O- structure and which may be considered to be a structural derivative of hydrogen peroxide where one or both of the hydrogen atoms has been replaced by an organic radical.
Refer To: 52 FR 31884

Oxidizer

"Oxidizer" means a chemical other than a blasting agent or explosive as defined in section 1910.109(a) that initiates or promotes combustion in other materials, thereby causing fire either of itself or through the release of oxygen or other gases.
Refer To: 52 FR 31879

Chemical Hazard Communication Guidebook

OSHA Hazard Communication Standard (cont.)

Pyrophoric

"Pyrophoric" means a chemical that will ignite spontaneously in air at a temperature of 130 degrees F or below.
Refer To: 52 FR 31879

Sensitizer

"Sensitizer" refers to a chemical that causes a substantial proportion of exposed people or animals to develop an allergic reaction in normal tissue after repeated exposure to the chemical.
Refer To: 52 FR 31884

Target organ effects

The following is a target organ categorization of effects which may occur, including examples of signs and symptoms and chemicals which have been found to cause such effects. These examples are presented to illustrate the range and diversity of effects and hazards found in the workplace, and the broad scope employers must consider in this area, but are not intended to be all-inclusive:

(a) Hepatotoxins: chemicals which produce liver damage
Signs & Symptoms: Jaundice; liver enlargement
Chemicals: Carbon tetrachloride; nitrosamines

(b) Nephrotoxins: chemicals which produce kidney damage
Signs & Symptoms: Edema; proteinuria

Appendix C: Definitions of Hazards

OSHA Hazard Communication Standard (cont.)

Chemicals: Halogenated hydrocarbons; uranium

(c) Neurotoxins: chemicals which produce their primary toxic effects on the nervous system
Signs & Symptoms: Narcosis; behavioral changes; decrease in motor functions
Chemicals: Mercury; carbon disulfide

(d) Agents which act on the blood or hematopoietic system: decrease hemoglobin function; deprive the body tissues of oxygen
Signs & Symptoms: Cyanosis; loss of consciousness
Chemicals: Carbon monoxide; cyanides

(e) Agents which damage the lung: chemicals which irritate or damage the pulmonary tissue
Signs & Symptoms: Cough; tightness in chest; shortness of breath
Chemicals: Silica; asbestos

(f) Reproductive toxins: chemicals which affect the reproductive capabilities including chromosomal damage (mutations) and effects on fetuses (teratogenesis)
Signs & Symptoms: Birth defects; sterility
Chemicals: Lead; DBCP

(g) Cutaneous hazards: chemicals which effect the dermal layer of the body
Signs & Symptoms: Defatting of the skin; rashes; irritation
Chemicals: Ketones; chlorinated compounds

(h) Eye hazards: chemicals which affect the eye or visual capacity
Signs & Symptoms: Conjunctivitis; corneal damage
Chemicals: Organic solvents; acids
Refer To: 52 FR 31884

Chemical Hazard Communication Guidebook

OSHA Hazard Communication Standard (cont.)

Toxic

"Toxic" refers to a chemical falling within any of the following categories:

(a) A chemical that has a median lethal dose of more than 50 milligrams per kilogram but not more than 500 milligrams per kilogram of body weight when administered orally to albino rats weighing between 200 and 300 grams each.

(b) A chemical that has a median lethal dose of more than 200 milligrams per kilogram but not more than 1,000 milligrams per kilogram of body weight when administered by continuous contact for 24 hours (or less if death occurs within 24 hours) with the bare skin of albino rabbits weighing between two and three kilograms each.

(c) A chemical that has a median lethal concentration in air of more than 200 parts per million but not more than 2,000 parts per million by volume of gas or vapor, or more than two milligrams per liter but not more than 20 milligrams per liter of mist, fume, or dust, when administered by continuous inhalation for one hour (or less if death occurs within one hour) to albino rats weighing between 200 and 300 grams each.
Refer To: 52 FR 31884

Toxic (Highly)

"Highly toxic" refers to a chemical falling within any of the following categories: (a) A chemical that has a median lethal dose of 50 milligrams or less per kilogram of body weight when

Appendix C: Definitions of Hazards

OSHA Hazard Communication Standard (cont.)

administered orally to albino rats weighing between 200 and 300 grams each. (b) A chemical that has a median lethal dose of 200 milligrams or less per kilogram of body weight when administered by continuous contact for 24 hours (or less if death occurs within 24 hours) with the bare skin of albino rabbits weighing between two and three kilograms each. (c) A chemical that has a median lethal concentration in air of 200 parts per million by volume or less of gas or vapor, or 2 milligrams per liter or less of mist, fume, or dust, when administered by continuous inhalation for one hour (or less if death occurs within one hour) to albino rats weighing between 200 and 300 grams each.
Refer To: 52 FR 31884

Unstable (reactive)

"Unstable (reactive)" means a chemical which in the pure state, or as produced or transported, will vigorously polymerize, decompose, condense, or will become self-reactive under conditions of shocks, pressure, or temperature.
Refer To: 52 FR 31879

Water-reactive

"Water-reactive" means a chemical that reacts with water to release a gas that is either flammable or presents a health hazard.
Refer To: 52 FR 31879

Chemical Hazard Communication Guidebook

SARA Right-To-Know Reporting

Delayed (chronic) health hazard

"Delayed (chronic) health hazard" includes "carcinogens" as defined under section 1910.1200 of Title 29 of the Code of Federal Regulations) and other hazardous chemicals that cause an adverse effect to a target organ and which effect generally occurs as a result of long term exposure and is of long duration.
Refer To: 52 FR 38364

Extremely Hazardous Substance

"Extremely hazardous substance" means a substance listed in the Appendices to 40 CFR Part 344, Emergency Planning and Notification.
"Extremely hazardous substances" are substances subject to the emergency planning provisions of Title III and are defined in section 302 of SARA.
Refer To: 52 FR 38364, 38347

Hazard Category - Fire hazard

"Fire hazard" includes "flammable," "combustible liquid," "pyrophoric," and "oxidizer" (as defined under section 1910.1200 or Title 29 of the CFR).
Refer To: 52 FR 38364

Appendix C: Definitions of Hazards

SARA Right-To-Know Reporting (cont.)

Hazardous Chemical

"Hazardous chemical" means any hazardous chemical as defined under section 1910.1200 of Title 29 of the Code of Federal Regulations, except that such term does not include the substances exempted from the regulation. Title III uses several different terms to describe related groups of substances. .... "Hazardous chemical" comprises the group of substances subject to sections 311 and 312 and is defined as all "hazardous chemicals" as defined under OSHA and its implementing regulations, but with five additional exclusions under section 311(e) of Title III.
Refer To: 52 FR 38364, 38347

Hazardous Chemical - Agriculture Exemption

Any substance to the extent it is used in routine agricultural operations or is a fertilizer held for sale by a retailer to the ultimate customer. Section 311(e)(5) is a two-part exemption that excludes retailers of fertilizer from reporting requirements for the fertilizer and also excludes any substance when used in routine agricultural operations. EPA has interpreted this exemption to eliminate reporting of fertilizers, pesticides, and other chemical substances when applied, administered, or otherwise used as part of routine agricultural activities. Fertilizers handled by retailers, even though not directly utilized by such persons for agricultural purposes, are also excluded.

SARA Right-To-Know Reporting (cont.)

Hazardous Chemical - Food and Drug Exemption

Refer To: 52 FR 38364

Any food, food additive, color additive, drug, or cosmetic regulated by the Food and Drug Administration.
Refer To: 52 FR 38349

Hazardous Chemical - Household Product Exemption

Any substance to the extent it is used for personal, family, or household purposes, or is present in the same form and concentration as a product packaged for distribution and use by the general public.

"Present in the same form and concentration as a product packaged for distribution and use by the general public" means a substance packaged in a similar manner and present in the same concentration as the substance when packaged for use by the general public, whether or not it is intended for distribution to the general public or used for the same purpose as when it is packaged for use by the general public. EPA interprets this exclusion to apply to household or consumer products, either in use by the general public or in commercial or industrial use when in the same form and concentration as the product intended for use by the public.

This exemption is for general household and domestic products, and thus the clearest example of its application is ordinary household products stored in a home or located on a retailer's shelf.

Appendix C: Definitions of Hazards

SARA Right-To-Know Reporting (cont.)

However, EPA has interpreted this exemption to apply to such products prior to distribution to the consumer when in the same form and concentration, and to such products when not intended for use by the general public. Thus, the exemption also applies to any substance packaged in the same form and concentration as a consumer product whether or not it is used for the same purpose as the consumer product. In addition, the exemption applies to such products when purchased in larger quantities by industrial facilities if packaged in substantially the same form as the consumer product and present in the same concentration. The exemption would not apply to substances present in different concentrations from the consumer products even if the substance is only used in small quantities.
Refer To: 52 FR 38364, 38348

Hazardous Chemical - Research Laboratory Exemption

Any substance to the extent it is used in a research laboratory or a hospital or other medical facility under the direct supervision of a technically qualified individual.
Section 311(e)(4) of SARA and section 370.2 of the regulations exclude from the definition of "hazardous chemical" any substance to the extent it is used in a research laboratory or a hospital or other medical facility under the direct supervision of a technically qualified individual. EPA has interpreted this exclusion to exempt facilities where small amounts of many types of chemicals are used, or stored for short periods, that are not hazardous to the general public when administered or used under appropriate supervision. In addition, it is

Chemical Hazard Communication Guidebook

SARA Right-To-Know Reporting (cont.)

important to recognize that the exemption applies to the substance used, rather than to the entire facility. Thus, research and medical facilities are not exempted from reporting requirements under sections 311 and 312; rather, they would not need to provide information on many of their chemicals.

With respect to research laboratories, EPA interprets the exclusion to apply to research facilities as well as quality control laboratory operations located within manufacturing facilities. However, laboratories that produce chemical specialty products or full-scale pilot plant operations are considered to be part of manufacturing rather than research operations and would not be a "research laboratory."
Refer To: 52 FR

Hazardous Chemical - Solids in Manufactured Item Exemption

Any substance present as a solid in any manufactured item to the extent exposure to the substance does not occur under normal conditions of use. Steel and other similar non-reactive solids are generally exempt from MSDS requirements under OSHA (and thus from sections 311 and 312) when they are articles shaped during manufacture whose end use depends upon that shape. (section 1900.1200(b)). Even if subject to the OSHA MSDS requirements, steel and other manufactured solids are excluded from sections 311 and 312 reporting under section 311(e)(2).
Refer To: 52 FR 38364, 38349

Appendix C: Definitions of Hazards

SARA Right-To-Know Reporting (cont.)

Immediate (acute) health hazard

"Immediate (acute) health hazard" includes "highly toxic," "toxic," "irritant," "sensitizer," "corrosive," (as defined under section 1900.1200 of Title 29 of CFR) and other hazardous chemicals that cause an adverse effect to a target organ and which effect usually occurs rapidly as a result of short term exposure and is of short duration.
Refer To: 52 FR 38364

Reactive hazard

"Reactive" includes "unstable reactive," "organic peroxide," and "water reactive" (as defined under section 1900.1200 of Title 29 of the CFR).
Refer To: 52 FR 38364

Sudden release of pressure

"Sudden release of pressure" includes "explosive" and "compressed gas" (as defined under section 1900.1200 of Title 29 of the CFR).
Refer To: 52 FR 38364

Hazardous Materials Transportation

Class A Explosive

There are nine types of Class A explosives including solid or liquid explosives, and ammunition, which can be detonated under conditions specified by DOT. Regulations provide specific descriptions of and tests for the different types of Class A explosives.
Refer To: 49 CFR 173.53

Class B Explosives

Explosives, class B, are defined as those explosives which in general function by rapid combustion rather than detonation and include some explosive devices such as special fireworks, flash powders, some pyrotechnic signal devices and liquid or solid propellant explosives which include some smokeless powders. Regulations provide specific descriptions of and tests for class B explosives.
Refer To: 49 CFR 173.88

Class C Explosives

Explosives, class C, are defined as certain types of manufactured articles which contain class A, or class B explosives, or both as components but in restricted quantities, and certain types of fireworks. Regulations include specific descriptions of, prescribed uses for, and tests for class C explosives.
Refer To: 49 CFR 173.100

Appendix C: Definitions of Hazards

Hazardous Materials Transportation (cont)

Combustible liquid

A combustible liquid is defined as any liquid that does not meet the definition of any other classification specified in Subchapter C and has a flash point at or above 100 degrees F. and below 200 degrees F. except any mixture having one component or more with a flash point at 200 degrees F. or higher, that makes up at least 99 percent of the total volume of the mixture.
Refer To: 49 CFR 173.115

Compressed gas

The term "compressed gas" designates any material or mixture having in the container an absolute pressure exceeding 40 p.s.i. at 70 degrees F. or regardless of the pressure at 70 degrees F., having an absolute pressure exceeding 104 p.s.i. at 130 degrees F.; or any liquid flammable material having a vapor pressure exceeding 40 p.s.i. absolute at 100 degrees F. as determined by ASTM Test D-323. The regulation includes additional descriptions of and specifications for tests for compressed gases.
Refer To: 49 CFR 173.300

Compressed gas (flammable)

Any compressed gas is designated a flammable compressed gas if:

a. Either a mixture of 13 percent or less (by volume) with air forms a flammable mixture or the flammable range with air is wider than 12 percent regardless of the lower limit;

Chemical Hazard Communication Guidebook

Hazardous Materials Transportation (cont)

b. Using the Bureau of Explosives': (i) Flame Projection Apparatus, the flame projects more than 18 inches beyond the ignition source with valve opened fully, or, the flame flashes back and burns at the valve with any degree of valve opening; (ii) Open Drum Apparatus, there is any significant propagation of flame away from the ignition source; (iii) Closed Drum Apparatus, there is any explosion of the vapor-air mixture in the drum.
Refer To: 49 CFR 173.300

Compressed gas (liquefied)

A liquefied compressed gas is a gas which, under the charged pressure, is partially liquid at a temperature of 70 degrees F.
Refer To: 49 CFR 173.300

Compressed gas (non-liquefied)

A non-liquefied compressed gas is a gas other than gas in solution which under the charged pressure is entirely gaseous at a temperature of 70 degrees F.
Refer To: 49 CFR 173.300

Compressed gas (Refrigerant gas or Dispersant gas)

The term "refrigerant gas" or "dispersant gas" applies to all flammable or nonflammable, nonpoisonous refrigerant gases, dispersant gases (fluorocarbons) referred to in 49 CFR 173.300 (i) and mixtures thereof, or any other compressed gas meeting one of the following:

Hazardous Materials Transportation (cont)

o a nonflammable mixture containing not less than 50% fluorocarbon content, having a vapor pressure not exceeding 260 psig at 130 degrees F.

o a flammable mixture containing not less than 50% fluorocarbon content, not over 40% by weight of a flammable component, having a vapor pressure not exceeding 260 psig at 130 F.
Refer To: 49 CFR 173.300

Compressed gas in solution

A compressed gas in solution is a non-liquefied compressed gas which is dissolved in a solvent.
Refer To: 49 CFR 173.300

Corrosive material

A corrosive material is a liquid or solid that cause visible destruction or irreversible alteration in human skin tissue at the site of contact, or in the case of leakage from its packaging, a liquid that has a severe corrosion rate on steel. Regulations include specific tests for and descriptions of corrosive capacity.
Refer To: 49 CFR 173.240

Cryogenic liquid

A cryogenic liquid is a refrigerated liquefied gas having a boiling point colder than -130 degrees F. at one atmosphere, absolute. A material meeting this definition is subject to the requirements for compressed gases

Chemical Hazard Communication Guidebook

Hazardous Materials Transportation (cont)

without regard to whether it meets the standard definition of a compressed gas.
Refer To: 49 CFR 173.300

Etiologic agents

An etiologic agent is a viable microorganism, or its toxin, which causes or may cause human disease, and is limited to those agents listed by the Department of Health, Education, and Welfare in 42 CFR 72.3.
Refer To: 49 CFR 173.386

Explosive

Any chemical compound, mixture, or device, the primary or common purpose of which is to function by explosion, i.e., with substantially instantaneous release of gas and heat, unless such compound, mixture, or device is otherwise specifically classified in Parts 170-189 of Subchapter C. Specific tests are indicated for different classes of explosives.
Refer To: 49 CFR 173.50

Flammable liquid

A flammable liquid means any liquid having a flash point below 100 degrees F, except for:

a. Any liquid meeting one of the definitions of a compressed gas; and

b. Any mixture having one component or more with a flash point of 100 degrees F or higher that makes up at least 99 percent of the total volume of the mixture.

Appendix C: Definitions of Hazards

Hazardous Materials Transportation (cont)

A distilled spirit of 140 proof or lower is considered to have a flash point no lower than 73 degrees F.
Refer To: 49 CFR 173.115

Flammable solid

A flammable solid is any solid material other than one classed as an explosive, which, under conditions normally incident to transportation is liable to cause fires through friction, retained heat from manufacturing or processing, or which can be ignited readily and when ignited burns so vigorously and persistently as to create a serious transportation hazard. Included in this class are spontaneously combustible and water-reactive materials.
Refer To: 49 CFR 173.150

Organic peroxide

An organic compound containing the bivalent -O-O- structure and which may be considered a derivative of hydrogen peroxide where one or more of the hydrogen atoms have been replace by organic radicals must be classed as an organic peroxide unless:

a. The material meets the definition of an explosive A or explosive B, in which case it must be classed as such;

b. The material is forbidden to be offered for transportation according to sections 172.101 or 173.21 of subchapter C;

Hazardous Materials Transportation (cont)

c. It is determined that the predominant hazard of the material containing an organic peroxide is other than that of an organic peroxide;

d. According to data on file with the Research and Special Programs Administration, it has been determined that the material does not present a hazard in transportation
Refer To: 49 CFR 173.151a

ORM-A

An ORM-A material is a material which has an anesthetic, irritating, noxious, toxic, or other similar property and which can cause extreme annoyance or discomfort to passengers and crew in the event of leakage during transportation. A list of chemicals specified as ORM-A materials is established in the regulation.
Refer To: 49 CFR 173.500

ORM-B

An ORM-B material is a material (including a solid when wet with water) capable of causing significant damage to a transport vehicle from leakage during transportation. Materials meeting one or both of the following criteria are ORM-B materials:

a. A liquid substance that has a corrosion rate exceeding 0.250 inch per year on aluminum at a test temperature of 130 degrees F; and

Appendix C: Definitions of Hazards

Hazardous Materials Transportation (cont)

b. Specifically designated by name in the Hazardous Materials Table.

A list of chemicals specified as ORM-B materials is established in the regulation.
Refer To: 49 CFR 173.500

ORM-C

An ORM-C material is a material which has other inherent characteristics not described as an ORM-A or ORM-B but which make it unsuitable for shipment unless properly identified and prepared for transportation. Each ORM-C material is specifically named in the Hazardous Materials Table.
Refer To: 49 CFR 173.500

ORM-D

An ORM-D material is a material such as a consumer commodity, which though otherwise subject to the regulations of this subchapter, presents a limited hazard during transportation due to its form, quantity and packaging. They must be materials for which exceptions are provided in 49 CFR 172.101. A shipping description applicable to each ORM-D material or category is found in the Hazardous Materials Table. In order to be transported under the proper shipping name of "consumer commodity," a material must meet that definition. It may be reclassed and offered for shipment as ORM-D material provided that an ORM-D exception is authorized in specific sections applicable to the material.
Refer To: 49 CFR 173.500

Chemical Hazard Communication Guidebook

Hazardous Materials Transportation (cont)

ORM-E

An ORM-E is a material that is not included in any other hazard class, but is subject to the requirements of Title 49, Subchapter C. ORM-E materials include hazardous waste and hazardous substances as defined in 49 CFR 171.8.
Refer To: 49 CFR 173.500

Other Regulated Material (ORM)

An Other Regulated Material (ORM) is a material that:

a. May pose an unreasonable risk to health and safety or property when transported in commerce; and

b. Does not meet any of the definitions of the other hazard classes specified in subchapter C; or

c. Has been reclassed an ORM (specifically or permissively).
Refer To: 49 CFR 173.500

Oxidizer

An oxidizer is a substance such as a chlorate, permanganate, inorganic peroxide, or a nitrate, that yields oxygen readily to stimulate the combustion of organic matter.
Refer To: 49 CFR 173.151

Poisonous materials

Poisonous materials are divided into three groups according to the degree of hazard in transportation:

Appendix C: Definitions of Hazards

Hazardous Materials Transportation (cont)

a. Poison A;

b. Poison B; and

c. Irritating material
Refer To: 49 CFR 173.325

Poisonous materials (Irritating materials)

An irritating material is a liquid or solid substance which upon contact with fire or when exposed to air gives off dangerous or intensely irritating fumes, such as brombenzylcyanide, chloracetophenone, diphenylaminechlorarsine, and diphenylchlorarsine, but not including any Class A poisonous material.
Refer To: 49 CFR 173.381

Poisonous materials (Poison A)

Extremely dangerous poisons, class A, are poisonous gases or liquids of such nature that a very small amount of the gas, or vapor of the liquid, mixed with air is dangerous to life. A partial list of such chemicals is included in the specifications for Poison A materials.
Refer To: 49 CFR 173.326

Poisonous materials (Poison B)

Class B poisons are those substances, liquid or solid (including pastes and semisolids), other than Class A poisons or Irritating materials, which are known to be so toxic to man as to afford a hazard to health during transportation; or which, in the absence of adequate data on human toxicity, are presumed to

Chemical Hazard Communication Guidebook

Hazardous Materials Transportation (cont)

be toxic to man because they fall within the categories, in tests specified by the regulation, for oral toxicity, toxicity on inhalation, or toxicity by skin absorption when tested on laboratory animals.
Refer To: 49 CFR 173.343

Pyrophoric liquid

A pyrophoric liquid is any liquid that ignites spontaneously in dry or moist air at or below 130 degrees F.
Refer To: 49 CFR 173.115

Radioactive material

"Radioactive material" means any material having a specific activity greater than 0.002 microcuries per gram (uCi/g). Specifications and additional descriptions are established in the regulation.
Refer To: 49 CFR 173.403

Radioactive material (Fissile material)

"Fissile material" means any material consisting of or containing one or more fissile radionuclides. Fissile radionuclides are plutonium-238, plutonium-239, plutonium-241, uranium-233, and uranium-235. Neither natural nor depleted uranium are fissile material. Fissile materials are classified according to the controls needed to provide nuclear criticality safety during transportation in section 173.455.
Refer To: 49 CFR 173.403

APPENDIX D

EXTREMELY HAZARDOUS SUBSTANCES
(Alphabetical Order)

This appendix provides the list of Extremely Hazardous Substances subject to emergency planning in SARA. The threshold planning quantities are listed under the column "TPQ." Where a reportable quantity has been established under CERCLA or a statutory reportable quantity has been set pursuant to SARA, it is listed under the column "RQ." These quantities would apply to an emergency release of the substance in determining whether a report would have to be submitted. A guide to the notes is included at the end.

Chemical Hazard Communication Guidebook

CAS	CHEMICAL NAME	RQ (LBS)	TPQ (LBS)	NOTES
75-86-5	ACETONE CYANOHYDRIN	10	1000	
1752-30-3	ACETONE THIOSEMICARBAZIDE	1	1000/10000	e
107-02-8	ACROLEIN	1	500	
79-06-1	ACRYLAMIDE	5000	1000/10000	d,l
107-13-1	ACRYLONITRILE	100	10000	d,l
814-68-6	ACRYLYL CHLORIDE	1	100	e,h
111-69-3	ADIPONITRILE	1	1000	e,l
116-06-3	ALDICARB	1	100/10000	c
309-00-2	ALDRIN	1	500/10000	d
309-00-2	ALDRIN [1,4:5,8-DIMETHANONAPHTHALENE,1,2,3,4,10,10-HEXACHLORO-1,4,4A,5	1	500/10000	d
107-18-6	ALLYL ALCOHOL	100	1000	
107-11-9	ALLYL AMINE	1	500	e
20859-73-8	ALUMINUM PHOSPHIDE	100	500	b
54-62-6	AMINOPTERIN	1	500/10000	e
78-53-5	AMITON	1	500	e
3734-97-2	AMITON OXALATE	1	100/10000	e
7664-41-7	AMMONIA	100	500	l
300-62-9	AMPHETAMINE	1	1000	e
62-53-3	ANILINE	5000	1000	d,l
88-05-1	ANILINE, 2,4,6-TRIMETHYL-	1	500	e
7783-70-2	ANTIMONY PENTAFLUORIDE	1	500	e
1397-94-0	ANTIMYCIN A	1	1000/10000	c,e
86-88-4	ANTU	100	500/10000	
1303-28-2	ARSENIC PENTOXIDE	1	100/10000	d
1327-53-3	ARSENOUS OXIDE		100/10000	d,h
7784-34-1	ARSENOUS TRICHLORIDE	1	500	d
7784-42-1	ARSINE	1	100	e
2642-71-9	AZINPHOS-ETHYL	1	100/10000	e
86-50-0	AZINPHOS-METHYL	1	10/10000	
98-87-3	BENZAL CHLORIDE	5000	500	d
98-16-8	BENZENAMINE, 3-(TRIFLUOROMETHYL)-	1	500	e
100-14-1	BENZENE, 1-(CHLOROMETHYL)-4-NITRO-	1	500/10000	e
98-05-5	BENZENEARSONIC ACID	1	10/10000	e
3615-21-2	BENZIMIDAZOLE, 4,5-DICHLORO-2-(TRIFLUOROMETHYL)-	1	500/10000	e,g
98-07-7	BENZOIC TRICHLORIDE (BENZOTRICHLORIDE)	10	100	d
98-07-7	BENZOTRICHLORIDE	10	100	d
100-44-7	BENZYL CHLORIDE	100	500	d
140-29-4	BENZYL CYANIDE	1	500	e,h
15271-41-7	BICYCLO(2.2.1)HEPTANE-2-CARBONITRILE, 5-CHLORO-6-((((METHYLAMINO)CARBO	1	500/10000	e
534-07-6	BIS(CHLOROMETHYL) KETONE	1	10/10000	e
4044-65-9	BITOSCANATE	1	500/10000	e
10294-34-5	BORON TRICHLORIDE	1	500	e
7637-07-2	BORON TRIFLUORIDE	1	500	e

Appendix D: Extremely Hazardous Substances (Alphabetical Order)

CAS	CHEMICAL NAME	RQ (LBS)	TPQ (LBS)	NOTES
353-42-4	BORON TRIFLUORIDE COMPOUND WITH METHYL ETHER (1:1)	1	1000	e
28772-56-7	BROMADIOLONE	1	100/10000	e
7726-95-6	BROMINE	1	500	e,l
1306-19-0	CADMIUM OXIDE	1	100/10000	e
2223-93-0	CADMIUM STEARATE	1	1000/10000	c,e
7778-44-1	CALCIUM ARSENATE	1	500/10000	d
8001-35-2	CAMPHECHLOR	1	500/10000	d
56-25-7	CANTHARIDIN	1	100/10000	e
51-83-2	CARBACHOL CHLORIDE	1	500/10000	e
26419-73-8	CARBAMIC ACID, METHYL-, O-(((2,4-DIMETHYL-1,3-DITHIOLAN-2-YL)METHYLENE	1	100/10000	e
1563-66-2	CARBOFURAN	10	10/10000	
75-15-0	CARBON DISULFIDE	100	10000	l
786-19-6	CARBOPHENOTHION	1	500	e
57-74-9	CHLORDANE	1	1000	d
57-74-9	CHLORDANE [4,7-METHANOINDAN,1,2,4,5,6,7,8,8-OCTACHLORO-2,3,3A,4,7,7A-H	1	1000	d
470-90-6	CHLORFENVINFOS	1	500	e
7782-50-5	CHLORINE	10	100	
24934-91-6	CHLORMEPHOS	1	500	e
999-81-5	CHLORMEQUAT CHLORIDE	1	100/10000	e,h
79-11-8	CHLOROACETIC ACID	1	100/10000	e
107-07-3	CHLOROETHANOL	1	500	e
627-11-2	CHLOROETHYL CHLOROFORMATE	1	1000	e
67-66-3	CHLOROFORM	10	10000	d,l
542-88-1	CHLOROMETHYL ETHER	10	100	d,h
107-30-2	CHLOROMETHYL METHYL ETHER	10	100	c,d
3691-35-8	CHLOROPHACINONE	1	100/10000	e
1982-47-4	CHLOROXURON	1	500/10000	e
21923-23-9	CHLORTHIOPHOS	1	500	e,h
10025-73-7	CHROMIC CHLORIDE	1	1/10000	e
10210-68-1	COBALT CARBONYL	1	10/10000	e,h
62207-76-5	COBALT, ((2,2'-(1,2-ETHANEDIYLBIS(NITRILOMETHYLIDENE)BIS(PHENALOTO))(	1	100/10000	e
64-86-8	COLCHICINE	1	10/10000	e,h
56-72-4	COUMAPHOS	10	100/10000	
5836-29-3	COUMATETRALYL	1	500/10000	e
95-48-7	O-CRESOL	1000	1000/10000	d
535-89-7	CRIMIDINE	1	100/10000	e
4170-30-3	CROTONALDEHYDE	100	1000	
123-73-9	CROTONALDEHYDE (E)-	100	1000	
506-68-3	CYANOGEN BROMIDE	1000	500/10000	
506-78-5	CYANOGEN IODIDE	1	1000/10000	e
2636-26-2	CYANOPHOS	1	1000	e
675-14-9	CYANURIC FLUORIDE	1	100	e
66-81-9	CYCLOHEXIMIDE	1	100/10000	e

Chemical Hazard Communication Guidebook

CAS	CHEMICAL NAME	RQ (LBS)	TPQ (LBS)	NOTES
108-91-8	CYCLOHEXYLAMINE	1	10000	e,l
17702-41-9	DECABORANE	1	500/10000	e
8065-48-3	DEMETON	1	500	e
919-86-8	DEMETON-S-METHYL	1	500	e
10311-84-9	DIALIFOS	1	100/10000	e
19287-45-7	DIBORANE	1	100	e
110-57-6	TRANS-1,4-DICHLOROBUTENE	1	500	e
111-44-4	DICHLOROETHYL ETHER	10	10000	d
149-74-6	DICHLOROMETHYLPHENYLSILANE	1	1000	e
62-73-7	DICHLORVOS	10	1000	
62-73-7	DICHLORVOS [PHOSPHORIC ACID, 2,2-DICHLOROETHENYL DIMETHYL ESTER]	10	1000	
141-66-2	DICROTOPHOS	1	100	e
1464-53-5	DIEPOXYBUTANE	10	500	d
814-49-3	DIETHYL CHLOROPHOSPHATE	1	500	e,h
1642-54-2	DIETHYLCARBAMAZINE CITRATE	1	100/10000	e
71-63-6	DIGITOXIN	1	100/10000	c,e
2238-07-5	DIGLYCIDYL ETHER	1	1000	e
20830-75-5	DIGOXIN	1	10/10000	e,h
115-26-4	DIMEFOX	1	500	e
60-51-5	DIMETHOATE	10	500/10000	
2524-03-0	DIMETHYL PHOSPHOROCHLORIDOTHIOATE	1	500	e
77-78-1	DIMETHYL SULFATE	100	500	d
99-98-9	DIMETHYL-P-PHENYLENEDIAMINE	1	10/10000	e
75-78-5	DIMETHYLDICHLOROSILANE	1	500	e,h
57-14-7	DIMETHYLHYDRAZINE	10	1000	d
644-64-4	DIMETILAN	1	500/10000	e
534-52-1	DINITROCRESOL	10	10/10000	
88-85-7	DINOSEB	1000	100/10000	
1420-07-1	DINOTERB	1	500/10000	e
78-34-2	DIOXATHION	1	500	e
82-66-6	DIPHACINONE	1	10/10000	e
152-16-9	DIPHOSPHORAMIDE, OCTAMETHYL-	100	100	
298-04-4	DISULFOTON	1	500	
514-73-8	DITHIAZANINE IODIDE	1	500/10000	e
541-53-7	DITHIOBIURET	100	100/10000	
316-42-7	EMETINE, DIHYDROCHLORIDE	1	1/10000	e,h
115-29-7	ENDOSULFAN	1	10/10000	
2778-04-3	ENDOTHION	1	500/10000	e
72-20-8	ENDRIN	1	500/10000	
106-89-8	EPICHLOROHYDRIN	100	1000	d,l
2104-64-5	EPN	1	100/10000	e
50-14-6	ERGOCALCIFEROL	1	1000/10000	c,e
379-79-3	ERGOTAMINE TARTRATE	1	500/10000	e
1622-32-8	ETHANESULFONYL CHLORIDE, 2-CHLORO-	1	500	e
10140-87-1	ETHANOL, 1,2-DICHLORO-, ACETATE	1	1000	e
563-12-2	ETHION	10	1000	

Appendix D: Extremely Hazardous Substances (Alphabetical Order)

CAS	CHEMICAL NAME	RQ (LBS)	TPQ (LBS)	NOTES
13194-48-4	ETHOPROPHOS	1	1000	e
542-90-5	ETHYL THIOCYANATE	1	10000	e
538-07-8	ETHYLBIS(2-CHLOROETHYL)AMINE	1	500	e,h
107-15-3	ETHYLENE DIAMINE	5000	10000	
371-62-0	ETHYLENE FLUOROHYDRIN	1	10	ceh
75-21-8	ETHYLENE OXIDE	10	1000	d,l
151-56-4	ETHYLENEIMINE	1	500	d
22224-92-6	FENAMIPHOS	1	10/10000	e
122-14-5	FENITROTHION	1	500	e
115-90-2	FENSULFOTHION	1	500	e,h
4301-50-2	FLUENETIL	1	100/10000	e
7782-41-4	FLUORINE	10	500	k
640-19-7	FLUOROACETAMIDE	100	100/10000	j
144-49-0	FLUOROACETIC ACID	1	10/10000	e
359-06-8	FLUOROACETYL CHLORIDE	1	10	c,e
51-21-8	FLUOROURACIL	1	500/10000	e
944-22-9	FONOFOS	1	500	e
50-00-0	FORMALDEHYDE	100	500	d,l
107-16-4	FORMALDEHYDE CYANOHYDRIN	1	1000	e,h
23422-53-9	FORMETANATE HYDROCHLORIDE	1	500/10000	e,h
2540-82-1	FORMOTHION	1	100	e
17702-57-7	FORMPARANATE	1	100/10000	e
21548-32-3	FOSTHIETAN	1	500	e
3878-19-1	FUBERIDAZOLE	1	100/10000	e
110-00-9	FURAN	100	500	
13450-90-3	GALLIUM TRICHLORIDE	1	500/10000	e
77-47-4	HEXACHLOROCYCLOPENTADIENE	10	100	d,h
4835-11-4	HEXAMETHYLENEDIAMINE, N,N'-DIBUTYL-	1	500	e
302-01-2	HYDRAZINE	1	1000	d
74-90-8	HYDROCYANIC ACID	10	100	
7647-01-0	HYDROGEN CHLORIDE	5000	500	elm
7647-01-0	HYDROGEN CHLORIDE GAS (ONLY)	5000	500	elm
7664-39-3	HYDROGEN FLUORIDE	100	100	
7722-84-1	HYDROGEN PEROXIDE	1	1000	e,l
7783-07-5	HYDROGEN SELENIDE	1	10	e
7783-06-4	HYDROGEN SULFIDE	100	500	l
123-31-9	HYDROQUINONE	1	500/10000	l
13463-40-6	IRON PENTACARBONYL	1	100	e
297-78-9	ISOBENZAN	1	100/10000	e
78-82-0	ISOBUTYRONITRILE	1	1000	e,h
102-36-3	ISOCYANIC ACID, 3,4-DICHLOROPHENYL ESTER	1	500/10000	e
465-73-6	ISODRIN	1	100/10000	
55-91-4	ISOFLUORPHATE	100	100	c
4098-71-9	ISOPHORONE DIISOCYANATE	1	100	b,e
108-23-6	ISOPROPYL CHLOROFORMATE	1	1000	e
119-38-0	ISOPROPYLMETHYLPYRAZOLYL DIMETHYLCARBAMATE	1	500	e

Chemical Hazard Communication Guidebook

CAS	CHEMICAL NAME	RQ (LBS)	TPQ (LBS)	NOTES
78-97-7	LACTONITRILE	1	1000	e
21609-90-5	LEPTOPHOS	1	500/10000	e
541-25-3	LEWISITE	1	10	ceh
58-89-9	LINDANE	1	1000/10000	d
58-89-9	LINDANE [CYCLOHEXANE,1,2,3,4,5,6-HEXACHLORO-,(1.ALPHA.,2.ALPHA.,3.BETA	1	1000/10000	d
7580-67-8	LITHIUM HYDRIDE	1	100	b,e
109-77-3	MALONONITRILE	1000	500/10000	
12108-13-3	MANGANESE, TRICARBONYL METHYLCYCLOPENTADIENYL	1	100	e,h
51-75-2	MECHLORETHAMINE	1	10	c,e
950-10-7	MEPHOSFOLAN	1	500	e
1600-27-7	MERCURIC ACETATE	1	500/10000	e
7487-94-7	MERCURIC CHLORIDE	1	500/10000	e
21908-53-2	MERCURIC OXIDE	1	500/10000	e
10476-95-6	METHACROLEIN DIACETATE	1	1000	e
760-93-0	METHACRYLIC ANHYDRIDE	1	500	e
126-98-7	METHACRYLONITRILE	1000	500	h
920-46-7	METHACRYLOYL CHLORIDE	1	100	e
30674-80-7	METHACRYLOYLOXYETHYL ISOCYANATE	1	100	e,h
10265-92-6	METHAMIDOPHOS	1	100/10000	e
558-25-8	METHANESULFONYL FLUORIDE	1	1000	e
950-37-8	METHIDATHION	1	500/10000	e
2032-65-7	METHIOCARB	10	500/10000	
16752-77-5	METHOMYL	100	500/10000	h
151-38-2	METHOXYETHYLMERCURIC ACETATE	1	500/10000	e
80-63-7	METHYL 2-CHLOROACRYLATE	1	500	e
74-83-9	METHYL BROMIDE	1000	1000	l
79-22-1	METHYL CHLOROFORMATE	1000	500	d,h
624-83-9	METHYL ISOCYANATE	1	500	f
556-61-6	METHYL ISOTHIOCYANATE	1	500	b,e
74-93-1	METHYL MERCAPTAN	100	500	
3735-23-7	METHYL PHENKAPTON	1	500	e
676-97-1	METHYL PHOSPHONIC DICHLORIDE	1	100	b,e
556-64-9	METHYL THIOCYANATE	1	10000	e
75-79-6	METHYL TRICHLOROSILANE	1	500	e,h
78-94-4	METHYL VINYL KETONE	1	10	e
60-34-4	METHYLHYDRAZINE	10	500	
502-39-6	METHYLMERCURIC DICYANAMIDE	1	500/10000	e
1129-41-5	METOLCARB	1	100/10000	e
7786-34-7	MEVINPHOS	10	500	
315-18-4	MEXACARBATE	1000	500/10000	
50-07-7	MITOMYCIN C	10	500/10000	d
6923-22-4	MONOCROTOPHOS	1	10/10000	e
2763-96-4	MUSCIMOL	1000	10000	a,h
505-60-2	MUSTARD GAS	1	500	e,h
13463-39-3	NICKEL CARBONYL	10	1	d

Appendix D: Extremely Hazardous Substances (Alphabetical Order)

CAS	CHEMICAL NAME	RQ (LBS)	TPQ (LBS)	NOTES
54-11-5	NICOTINE	100	100	c
65-30-5	NICOTINE SULFATE	1	100/10000	e
7697-37-2	NITRIC ACID	1000	1000	
10102-43-9	NITRIC OXIDE	10	100	c
98-95-3	NITROBENZENE	1000	10000	l
1122-60-7	NITROCYCLOHEXANE	1	500	e
10102-44-0	NITROGEN DIOXIDE	10	100	
991-42-4	NORBORMIDE	1	100/10000	e
PMN82147	ORGANORHODIUM COMPLEX	1	10/10000	e
23135-22-0	OXAMYL	1	100/10000	e
78-71-7	OXETANE, 3,3-BIS(CHLOROMETHYL)	1	500	e
2497-07-6	OXYDISULFOTON	1	500	e,h
10028-15-6	OZONE	1	100	e
1910-42-5	PARAQUAT	1	10/10000	e
2074-50-2	PARAQUAT METHOSULFATE	1	10/10000	e
56-38-2	PARATHION	10	100	c,d
56-38-2	PARATHION [PHOSPHOROTHIOIC ACID, 0,0-DI-ETHYL-0-(4-NITROPHENYL) ESTER]	10	100	c,d
298-00-0	PARATHION-METHYL	100	100/10000	c
12002-03-8	PARIS GREEN	1	500/10000	d
19624-22-7	PENTABORANE	1	500	e
2570-26-5	PENTADECYLAMINE	1	100/10000	e
79-21-0	PERACETIC ACID	1	500	e
594-42-3	PERCHLOROMETHYL MERCAPTAN	100	500	
108-95-2	PHENOL	1000	500/10000	
4418-66-0	PHENOL, 2,2'-THIOBIS(4-CHLORO-6-METHYL-	1	100/10000	e
64-00-6	PHENOL, 3-(1-METHYLETHYL)-, METHYLCARBAMATE	1	500/10000	e
58-36-6	PHENOXARSINE, 10, 10'-OXIDI	1	500/10000	e
696-28-6	PHENYL DICHLOROARSINE	1	500	d,h
59-88-1	PHENYLHYDRAZINE HYDROCHLORIDE	1	1000/10000	e
62-38-4	PHENYLMERCURY ACETATE	100	500/10000	
2097-19-0	PHENYLSILATRANE	1	100/10000	e,h
103-85-5	PHENYLTHIOUREA	100	100/10000	
298-02-2	PHORATE	10	10	
4104-14-7	PHOSACETIM	1	100/10000	e
947-02-4	PHOSFOLAN	1	100/10000	e
75-44-5	PHOSGENE	10	10	l
732-11-6	PHOSMET	1	10/10000	e
13171-21-6	PHOSPHAMIDON	1	100	e
7803-51-2	PHOSPHINE	100	500	
2665-30-7	PHOSPHONOTHIOIC ACID, METHYL-, O-(4-NITROPHENYL) O-PHENYL ESTER	1	500	e
2703-13-1	PHOSPHONOTHIOIC ACID, METHYL-, O-ETHYL O-[4-(METHYLTHIO)PHENYL] ESTER	1	500	e
50782-69-9	PHOSPHONOTHIOIC ACID, METHYL-, S-(2-(BIS(1-METHYLETHYL)AMINO)ETHYL)O-E	1	100	e
3254-63-5	PHOSPHORIC ACID, DIMETHYL 4-(METHYLTHIO)PHENYL ESTER	1	500	e

Chemical Hazard Communication Guidebook

CAS	CHEMICAL NAME	RQ (LBS)	TPQ (LBS)	NOTES
2587-90-8	PHOSPHOROTHIOIC ACID, O,O-DIMETHYL-S-2-METHYLTHIO) ETHYL ESTER	1	500	ceg
7723-14-0	PHOSPHORUS	1	100	b,h
10025-87-3	PHOSPHORUS OXYCHLORIDE	1000	500	d
10026-13-8	PHOSPHORUS PENTACHLORIDE	1	500	b,e
1314-56-3	PHOSPHORUS PENTOXIDE	1	10	b,e
7719-12-2	PHOSPHORUS TRICHLORIDE	1000	1000	
57-47-6	PHYSOSTIGMINE	1	100/10000	e
57-64-7	PHYSOSTIGMINE, SALICYLATE (1:1)	1	100/10000	e
124-87-8	PICROTOXIN	1	500/10000	e
110-89-4	PIPERIDINE	1	1000	e
23505-41-1	PIRIMIFOS-ETHYL	1	1000	e
10124-50-2	POTASSIUM ARSENITE	1	500/10000	d
151-50-8	POTASSIUM CYANIDE	10	100	b
506-61-6	POTASSIUM SILVER CYANIDE	1	500	b
2631-37-0	PROMECARB	1	500/10000	e,h
106-96-7	PROPARGYL BROMIDE	1	10	e
57-57-8	BETA-PROPIOLACTONE	1	500	e
107-12-0	PROPIONITRILE	10	500	
70-69-9	PROPIOPHENONE, 4-AMINO	1	100/10000	e,g
109-61-5	PROPYL CHLOROFORMATE	1	500	e
75-56-9	PROPYLENE OXIDE	100	10000	l
75-55-8	PROPYLENEIMINE	1	10000	d
2275-18-5	PROTHOATE	1	100/10000	e
129-00-0	PYRENE	5000	1000/10000	c
140-76-1	PYRIDINE, 2-METHYL-5-VINYL-	1	500	e
504-24-5	PYRIDINE, 4-AMINO-	1000	500/10000	h
1124-33-0	PYRIDINE, 4-NITRO-, 1-OXIDE	1	500/10000	e
53558-25-1	PYRIMINIL	1	100/10000	e,h
630-60-4	QUABAIN	1	100/10000	c,e
14167-18-1	SALCOMINE	1	500/10000	e
107-44-8	SARIN	1	10	e,h
7783-00-8	SELENIOUS ACID	10	1000/10000	
7791-23-3	SELENIUM OXYCHLORIDE	1	500	e
563-41-7	SEMICARBAZIDE HYDROCHLORIDE	1	1000/10000	e
3037-72-7	SILANE, (4-AMINOBUTYL)DIETHOXYMETHYL-	1	1000	e
7631-89-2	SODIUM ARSENATE	1	1000/10000	d
7784-46-5	SODIUM ARSENITE	1	500/10000	d
26628-22-8	SODIUM AZIDE	1000	500	b
124-65-2	SODIUM CACODYLATE	1	100/10000	e
143-33-9	SODIUM CYANIDE	10	100	b
62-74-8	SODIUM FLUOROACETATE	10	10/10000	
13410-01-0	SODIUM SELENATE	1	100/10000	e
10102-18-8	SODIUM SELENITE	100	100/10000	h
10102-20-2	SODIUM TELLURITE	1	500/10000	e
900-95-8	STANNANE, ACETOXYTRIPHENYL-	1	500/10000	e,g
57-24-9	STRYCHNINE	10	100/10000	c

Appendix D: Extremely Hazardous Substances (Alphabetical Order)

CAS	CHEMICAL NAME	RQ (LBS)	TPQ (LBS)	NOTES
60-41-3	STRYCHNINE, SULFATE	1	100/10000	e
3689-24-5	SULFOTEP	100	500	
3569-57-1	SULFOXIDE, 3-CHLOROPROPYL OCTYL	1	500	e
7446-09-5	SULFUR DIOXIDE	1	500	e,l
7783-60-0	SULFUR TETRAFLUORIDE	1	100	e
7446-11-9	SULFUR TRIOXIDE	1	100	b,e
7664-93-9	SULFURIC ACID	1000	1000	
77-81-6	TABUN	1	10	ceh
13494-80-9	TELLURIUM	1	500/10000	e
7783-80-4	TELLURIUM HEXAFLUORIDE	1	100	e,k
107-49-3	TEPP	10	100	
13071-79-9	TERBUFOS	1	100	e,h
78-00-2	TETRAETHYL LEAD	10	100	c,d
597-64-8	TETRAETHYLTIN	1	100	c,e
75-74-1	TETRAMETHYL LEAD	1	100	cel
509-14-8	TETRANITROMETHANE	10	500	
10031-59-1	THALLIUM SULFATE	100	100/10000	h
6533-73-9	THALLOUS CARBONATE	100	100/10000	c,h
7791-12-0	THALLOUS CHLORIDE	100	100/10000	c,h
2757-18-8	THALLOUS MALONATE	1	100/10000	ceh
7446-18-6	THALLOUS SULFATE	100	100/10000	
2231-57-4	THIOCARBAZIDE	1	1000/10000	e
39196-18-4	THIOFANOX	100	100/10000	
297-97-2	THIONAZIN	100	500	
108-98-5	THIOPHENOL	100	500	
79-19-6	THIOSEMICARBAZIDE	100	100/10000	
5344-82-1	THIOUREA, (2-CHLOROPHENYL)-	100	100/10000	
614-78-8	THIOUREA, (2-METHYLPHENYL)-	1	500/10000	e
7550-45-0	TITANIUM TETRACHLORIDE	1	100	e
584-84-9	TOLUENE-2,4-DIISOCYANATE	100	500	
91-08-7	TOLUENE-2,6-DIISOCYANATE	100	100	
1031-47-6	TRIAMIPHOS	1	500/10000	e
24017-47-8	TRIAZOFOS	1	500	e
1558-25-4	TRICHLORO(CHLOROMETHYL)SILANE	1	100	e
27137-85-5	TRICHLORO(DICHLOROPHENYL)SILANE	1	500	e
76-02-8	TRICHLOROACETYL CHLORIDE	1	500	e
115-21-9	TRICHLOROETHYLSILANE	1	500	e,h
327-98-0	TRICHLORONATE	1	500	e,k
98-13-5	TRICHLOROPHENYLSILANE	1	500	e,h
998-30-1	TRIETHOXYSILANE	1	500	e
75-77-4	TRIMETHYLCHLOROSILANE	1	1000	e
824-11-3	TRIMETHYLOLPROPANE PHOSPHITE	1	100/10000	e,h
1066-45-1	TRIMETHYLTIN CHLORIDE	1	500/10000	e
639-58-7	TRIPHENYLTIN CHLORIDE	1	500/10000	e
555-77-1	TRIS(2-CHLOROETHYL)AMINE	1	100	e,h
2001-95-8	VALINOMYCIN	1	1000/10000	c,e
1314-62-1	VANADIUM PENTOXIDE	1000	100/10000	

Chemical Hazard Communication Guidebook

CAS	CHEMICAL NAME	RQ (LBS)	TPQ (LBS)	NOTES
108-05-4	VINYL ACETATE MONOMER	5000	1000	d,l
81-81-2	WARFARIN	100	500/10000	
129-06-6	WARFARIN SODIUM	1	100/10000	e,h
28347-13-9	XYLYLENE DICHLORIDE	1	100/10000	e
1314-84-7	ZINC PHOSPHIDE	100	500	b
58270-08-9	ZINC, DICHLORO(4,4-DIMETHYL-5-((((METHYLAMINO)CARBONYL)OXY)IMINO)PENTA	1	100/10000	e

NOTES

a. This chemical does not meet acute toxicity criteria. Its TPQ is set at 10,000 pounds.

b. This material is a reactive solid. The Threshold Planning Quantity does not default to 10,000 pounds for non-powder, non-molten, non-solution form.

c. EPA changed the calculated Threshold Planning Quantity and the reader is referred to the *Federal Register* of April 22, 1987 for further details.

d. EPA has indicated that the reportable quantity is likely to change when the assessment of potential carcinogenicity and chronic toxicity is completed.

e. Statutory reportable quantity for purposes of emergency notification under Section 304(a)(2) of the Emergency Planning and Community Right-to-Know Act.

f. EPA has indicated that the statutory one pound reportable quantity for methyl isocyanate may be adjusted in a future rulemaking.

g. New chemicals added that were not part of the original list of 402 substances.

h. Revised TPQ based on new or re-evaluated toxicity data.

j. TPQ is revised to its calculated value and does not change due to technical review as in proposed rule.

k. The TPQ was revised after proposal due to calculation error.

l. Chemicals on the original list that do not meet the toxicity criteria but because of their high production volume and recognized toxicity are considered chemicals of concern.

m. Hydrogen chloride is regulated as an extremely hazardous substance for the gas only.

APPENDIX E

EXTREMELY HAZARDOUS SUBSTANCES
(CAS Order)

This appendix provides the list of Extremely Hazardous Substances subject to emergency planning in SARA. The threshold planning quantities are listed under the column "TPQ." Where a reportable quantity has been established under CERCLA or a statutory reportable quantity has been set pursuant to SARA, it is listed under the column "RQ." These quantities would apply to an emergency release of the substance in determining whether a report would have to be submitted. A guide to the notes is included at the end.

Chemical Hazard Communication Guidebook

CAS	CHEMICAL NAME	RQ (LBS)	TPQ (LBS)	NOTES
50-00-0	FORMALDEHYDE	100	500	d,l
50-07-7	MITOMYCIN C	10	500/10000	d
50-14-6	ERGOCALCIFEROL	1	1000/10000	c,e
51-21-8	FLUOROURACIL	1	500/10000	e
51-75-2	MECHLORETHAMINE	1	10	c,e
51-83-2	CARBACHOL CHLORIDE	1	500/10000	e
54-11-5	NICOTINE	100	100	c
54-62-6	AMINOPTERIN	1	500/10000	e
55-91-4	ISOFLUORPHATE	100	100	c
56-25-7	CANTHARIDIN	1	100/10000	e
56-38-2	PARATHION	10	100	c,d
56-38-2	PARATHION [PHOSPHOROTHIOIC ACID, O,O-DI-ETHYL-O-(4-NITROPHENYL) ESTER]	10	100	c,d
56-72-4	COUMAPHOS	10	100/10000	
57-14-7	DIMETHYLHYDRAZINE	10	1000	d
57-24-9	STRYCHNINE	10	100/10000	c
57-47-6	PHYSOSTIGMINE	1	100/10000	e
57-57-8	BETA-PROPIOLACTONE	1	500	e
57-64-7	PHYSOSTIGMINE, SALICYLATE (1:1)	1	100/10000	e
57-74-9	CHLORDANE	1	1000	d
57-74-9	CHLORDANE [4,7-METHANOINDAN,1,2,4,5,6,7,8,8-OC-TACHLORO-2,3,3A,4,7,7A-H	1	1000	d
58-36-6	PHENOXARSINE, 10, 10'-OXIDI	1	500/10000	e
58-89-9	LINDANE	1	1000/10000	d
58-89-9	LINDANE [CYCLOHEXANE,1,2,3,4,5,6-HEXACHLORO-,(1.ALPHA.,2.ALPHA.,3.BETA	1	1000/10000	d
59-88-1	PHENYLHYDRAZINE HYDROCHLORIDE	1	1000/10000	e
60-34-4	METHYLHYDRAZINE	10	500	
60-41-3	STRYCHNINE, SULFATE	1	100/10000	e
60-51-5	DIMETHOATE	10	500/10000	
62-38-4	PHENYLMERCURY ACETATE	100	500/10000	
62-53-3	ANILINE	5000	1000	d,l
62-73-7	DICHLORVOS	10	1000	
62-73-7	DICHLORVOS [PHOSPHORIC ACID, 2,2-DICHLOROETHENYL DIMETHYL ESTER]	10	1000	
62-74-8	SODIUM FLUOROACETATE	10	10/10000	
64-00-6	PHENOL, 3-(1-METHYLETHYL)-, METHYLCARBAMATE	1	500/10000	e
64-86-8	COLCHICINE	1	10/10000	e,h
65-30-5	NICOTINE SULFATE	1	100/10000	e
66-81-9	CYCLOHEXIMIDE	1	100/10000	e
67-66-3	CHLOROFORM	10	10000	d,l
70-69-9	PROPIOPHENONE, 4-AMINO	1	100/10000	e,g
71-63-6	DIGITOXIN	1	100/10000	c,e
72-20-8	ENDRIN	1	500/10000	
74-83-9	METHYL BROMIDE	1000	1000	l
74-90-8	HYDROCYANIC ACID	10	100	
74-93-1	METHYL MERCAPTAN	100	500	
75-15-0	CARBON DISULFIDE	100	10000	l

Appendix E: Extremely Hazardous Substances (CAS Order)

CAS	CHEMICAL NAME	RQ (LBS)	TPQ (LBS)	NOTES
75-21-8	ETHYLENE OXIDE	10	1000	d,l
75-44-5	PHOSGENE	10	10	l
75-55-8	PROPYLENEIMINE	1	10000	d
75-56-9	PROPYLENE OXIDE	100	10000	l
75-74-1	TETRAMETHYL LEAD	1	100	cel
75-77-4	TRIMETHYLCHLOROSILANE	1	1000	e
75-78-5	DIMETHYLDICHLOROSILANE	1	500	e,h
75-79-6	METHYL TRICHLOROSILANE	1	500	e,h
75-86-5	ACETONE CYANOHYDRIN	10	1000	
76-02-8	TRICHLOROACETYL CHLORIDE	1	500	e
77-47-4	HEXACHLOROCYCLOPENTADIENE	10	100	d,h
77-78-1	DIMETHYL SULFATE	100	500	d
77-81-6	TABUN	1	10	ceh
78-00-2	TETRAETHYL LEAD	10	100	c,d
78-34-2	DIOXATHION	1	500	e
78-53-5	AMITON	1	500	e
78-71-7	OXETANE, 3,3-BIS(CHLOROMETHYL)	1	500	e
78-82-0	ISOBUTYRONITRILE	1	1000	e,h
78-94-4	METHYL VINYL KETONE	1	10	e
78-97-7	LACTONITRILE	1	1000	e
79-06-1	ACRYLAMIDE	5000	1000/10000	d,l
79-11-8	CHLOROACETIC ACID	1	100/10000	e
79-19-6	THIOSEMICARBAZIDE	100	100/10000	
79-21-0	PERACETIC ACID	1	500	e
79-22-1	METHYL CHLOROFORMATE	1000	500	d,h
80-63-7	METHYL 2-CHLOROACRYLATE	1	500	e
81-81-2	WARFARIN	100	500/10000	
82-66-6	DIPHACINONE	1	10/10000	e
86-50-0	AZINPHOS-METHYL	1	10/10000	
86-88-4	ANTU	100	500/10000	
88-05-1	ANILINE, 2,4,6-TRIMETHYL-	1	500	e
88-85-7	DINOSEB	1000	100/10000	
91-08-7	TOLUENE-2,6-DIISOCYANATE	100	100	
95-48-7	O-CRESOL	1000	1000/10000	d
98-05-5	BENZENEARSONIC ACID	1	10/10000	e
98-07-7	BENZOIC TRICHLORIDE (BENZOTRICHLORIDE)	10	100	d
98-07-7	BENZOTRICHLORIDE	10	100	d
98-13-5	TRICHLOROPHENYLSILANE	1	500	e,h
98-16-8	BENZENAMINE, 3-(TRIFLUOROMETHYL)-	1	500	e
98-87-3	BENZAL CHLORIDE	5000	500	d
98-95-3	NITROBENZENE	1000	10000	l
99-98-9	DIMETHYL-P-PHENYLENEDIAMINE	1	10/10000	e
100-14-1	BENZENE, 1-(CHLOROMETHYL)-4-NITRO-	1	500/10000	e
100-44-7	BENZYL CHLORIDE	100	500	d
102-36-3	ISOCYANIC ACID, 3,4-DICHLOROPHENYL ESTER	1	500/10000	e
103-85-5	PHENYLTHIOUREA	100	100/10000	
106-89-8	EPICHLOROHYDRIN	100	1000	d,l

Chemical Hazard Communication Guidebook

CAS	CHEMICAL NAME	RQ (LBS)	TPQ (LBS)	NOTES
106-96-7	PROPARGYL BROMIDE	1	10	e
107-02-8	ACROLEIN	1	500	
107-07-3	CHLOROETHANOL	1	500	e
107-11-0	ALLYL AMINE	1	500	e
107-12-0	PROPIONITRILE	10	500	
107-13-1	ACRYLONITRILE	100	10000	d,l
107-15-3	ETHYLENE DIAMINE	5000	10000	
107-16-4	FORMALDEHYDE CYANOHYDRIN	1	1000	e,h
107-18-6	ALLYL ALCOHOL	100	1000	
107-30-2	CHLOROMETHYL METHYL ETHER	10	100	c,d
107-44-8	SARIN	1	10	e,h
107-49-3	TEPP	10	100	
108-05-4	VINYL ACETATE MONOMER	5000	1000	d,l
108-23-6	ISOPROPYL CHLOROFORMATE	1	1000	e
108-91-8	CYCLOHEXYLAMINE	1	10000	e,l
108-95-2	PHENOL	1000	500/10000	
108-98-5	THIOPHENOL	100	500	
109-61-5	PROPYL CHLOROFORMATE	1	500	e
109-77-3	MALONONITRILE	1000	500/10000	
110-00-9	FURAN	100	500	
110-57-6	TRANS-1,4-DICHLOROBUTENE	1	500	e
110-89-4	PIPERIDINE	1	1000	e
111-44-4	DICHLOROETHYL ETHER	10	10000	d
111-69-3	ADIPONITRILE	1	1000	e,l
115-21-9	TRICHLOROETHYLSILANE	1	500	e,h
115-26-4	DIMEFOX	1	500	e
115-29-7	ENDOSULFAN	1	10/10000	
115-90-2	FENSULFOTHION	1	500	e,h
116-06-3	ALDICARB	1	100/10000	c
119-38-0	ISOPROPYLMETHYLPYRAZOLYL DIMETHYLCARBAMATE	1	500	e
122-14-5	FENITROTHION	1	500	e
123-31-9	HYDROQUINONE	1	500/10000	l
123-73-9	CROTONALDEHYDE (E)-	100	1000	
124-65-2	SODIUM CACODYLATE	1	100/10000	e
124-87-8	PICROTOXIN	1	500/10000	e
126-98-7	METHACRYLONITRILE	1000	500	h
129-00-0	PYRENE	5000	1000/10000	c
129-06-6	WARFARIN SODIUM	1	100/10000	e,h
140-29-4	BENZYL CYANIDE	1	500	e,h
140-76-1	PYRIDINE, 2-METHYL-5-VINYL-	1	500	e
141-66-2	DICROTOPHOS	1	100	e
143-33-9	SODIUM CYANIDE	10	100	b
144-49-0	FLUOROACETIC ACID	1	10/10000	e
149-74-6	DICHLOROMETHYLPHENYLSILANE	1	1000	e
151-38-2	METHOXYETHYLMERCURIC ACETATE	1	500/10000	e
151-50-8	POTASSIUM CYANIDE	10	100	b

Appendix E: Extremely Hazardous Substances (CAS Order)

CAS	CHEMICAL NAME	RQ (LBS)	TPQ (LBS)	NOTES
151-56-4	ETHYLENEIMINE	1	500	d
152-16-9	DIPHOSPHORAMIDE, OCTAMETHYL-	100	100	
297-78-9	ISOBENZAN	1	100/10000	e
297-97-2	THIONAZIN	100	500	
298-00-0	PARATHION-METHYL	100	100/10000	c
298-02-2	PHORATE	10	10	
298-04-4	DISULFOTON	1	500	
300-62-9	AMPHETAMINE	1	1000	e
302-01-2	HYDRAZINE	1	1000	d
309-00-2	ALDRIN	1	500/10000	d
309-00-2	ALDRIN [1,4:5,8-DIMETHANONAPHTHALENE,1,2,3,4,10,10-HEXACHLORO-1,4,4A,5	1	500/10000	d
315-18-4	MEXACARBATE	1000	500/10000	
316-42-7	EMETINE, DIHYDROCHLORIDE	1	1/10000	e,h
327-98-0	TRICHLORONATE	1	500	e,k
353-42-4	BORON TRIFLUORIDE COMPOUND WITH METHYL ETHER (1:1)	1	1000	e
359-06-8	FLUOROACETYL CHLORIDE	1	10	c,e
371-62-0	ETHYLENE FLUOROHYDRIN	1	10	ceh
379-79-3	ERGOTAMINE TARTRATE	1	500/10000	e
465-73-6	ISODRIN	1	100/10000	
470-90-6	CHLORFENVINFOS	1	500	e
502-39-6	METHYLMERCURIC DICYANAMIDE	1	500/10000	e
504-24-5	PYRIDINE, 4-AMINO-	1000	500/10000	h
505-60-2	MUSTARD GAS	1	500	e,h
506-61-6	POTASSIUM SILVER CYANIDE	1	500	b
506-68-3	CYANOGEN BROMIDE	1000	500/10000	
506-78-5	CYANOGEN IODIDE	1	1000/10000	e
509-14-8	TETRANITROMETHANE	10	500	
514-73-8	DITHIAZANINE IODIDE	1	500/10000	e
534-07-6	BIS(CHLOROMETHYL) KETONE	1	10/10000	e
534-52-1	DINITROCRESOL	10	10/10000	
535-89-7	CRIMIDINE	1	100/10000	e
538-07-8	ETHYLBIS(2-CHLOROETHYL)AMINE	1	500	e,h
541-25-3	LEWISITE	1	10	ceh
541-53-7	DITHIOBIURET	100	100/10000	
542-88-1	CHLOROMETHYL ETHER	10	100	d,h
542-90-5	ETHYL THIOCYANATE	1	10000	e
555-77-1	TRIS(2-CHLOROETHYL)AMINE	1	100	e,h
556-61-6	METHYL ISOTHIOCYANATE	1	500	b,e
556-64-9	METHYL THIOCYANATE	1	10000	e
558-25-8	METHANESULFONYL FLUORIDE	1	1000	e
563-12-2	ETHION	10	1000	
563-41-7	SEMICARBAZIDE HYDROCHLORIDE	1	1000/10000	e
584-84-9	TOLUENE-2,4-DIISOCYANATE	100	500	
594-42-3	PERCHLOROMETHYL MERCAPTAN	100	500	
597-64-8	TETRAETHYLTIN	1	100	c,e

Chemical Hazard Communication Guidebook

CAS	CHEMICAL NAME	RQ (LBS)	TPQ (LBS)	NOTES
614-78-8	THIOUREA, (2-METHYLPHENYL)-	1	500/10000	e
624-83-9	METHYL ISOCYANATE	1	500	f
627-11-2	CHLOROETHYL CHLOROFORMATE	1	1000	e
630-60-4	QUABAIN	1	100/10000	c,e
639-58-7	TRIPHENYLTIN CHLORIDE	1	500/10000	e
640-19-7	FLUOROACETAMIDE	100	100/10000	j
644-64-4	DIMETILAN	1	500/10000	e
675-14-9	CYANURIC FLUORIDE	1	100	e
676-97-1	METHYL PHOSPHONIC DICHLORIDE	1	100	b,e
696-28-6	PHENYL DICHLOROARSINE	1	500	d,h
732-11-6	PHOSMET	1	10/10000	e
760-93-0	METHACRYLIC ANHYDRIDE	1	500	e
786-19-6	CARBOPHENOTHION	1	500	e
814-49-3	DIETHYL CHLOROPHOSPHATE	1	500	e,h
814-68-6	ACRYLYL CHLORIDE	1	100	e,h
824-11-3	TRIMETHYLOLPROPANE PHOSPHITE	1	100/10000	e,h
900-95-8	STANNANE, ACETOXYTRIPHENYL-	1	500/10000	e,g
919-86-8	DEMETON-S-METHYL	1	500	e
920-46-7	METHACRYLOYL CHLORIDE	1	100	e
944-22-9	FONOFOS	1	500	e
947-02-4	PHOSFOLAN	1	100/10000	e
950-10-7	MEPHOSFOLAN	1	500	e
950-37-8	METHIDATHION	1	500/10000	e
991-42-4	NORBORMIDE	1	100/10000	e
998-30-1	TRIETHOXYSILANE	1	500	e
999-81-5	CHLORMEQUAT CHLORIDE	1	100/10000	e,h
1031-47-6	TRIAMIPHOS	1	500/10000	e
1066-45-1	TRIMETHYLTIN CHLORIDE	1	500/10000	e
1122-60-7	NITROCYCLOHEXANE	1	500	e
1124-33-0	PYRIDINE, 4-NITRO-, 1-OXIDE	1	500/10000	e
1129-41-5	METOLCARB	1	100/10000	e
1303-28-2	ARSENIC PENTOXIDE	1	100/10000	d
1306-19-0	CADMIUM OXIDE	1	100/10000	e
1314-56-3	PHOSPHORUS PENTOXIDE	1	10	b,e
1314-62-1	VANADIUM PENTOXIDE	1000	100/10000	
1314-84-7	ZINC PHOSPHIDE	100	500	b
1327-53-3	ARSENOUS OXIDE		100/10000	d,h
1397-94-0	ANTIMYCIN A	1	1000/10000	c,e
1420-07-1	DINOTERB	1	500/10000	e
1464-53-5	DIEPOXYBUTANE	10	500	d
1558-25-4	TRICHLORO(CHLOROMETHYL)SILANE	1	100	e
1563-66-2	CARBOFURAN	10	10/10000	
1600-27-7	MERCURIC ACETATE	1	500/10000	e
1622-32-8	ETHANESULFONYL CHLORIDE, 2-CHLORO-	1	500	e
1642-54-2	DIETHYLCARBAMAZINE CITRATE	1	100/10000	e
1752-30-3	ACETONE THIOSEMICARBAZIDE	1	1000/10000	e
1910-42-5	PARAQUAT	1	10/10000	e

Appendix E: Extremely Hazardous Substances (CAS Order)

CAS	CHEMICAL NAME	RQ (LBS)	TPQ (LBS)	NOTES
1982-47-4	CHLOROXURON	1	500/10000	e
2001-95-8	VALINOMYCIN	1	1000/10000	c,e
2032-65-7	METHIOCARB	10	500/10000	
2074-50-2	PARAQUAT METHOSULFATE	1	10/10000	e
2097-19-0	PHENYLSILATRANE	1	100/10000	e,h
2104-64-5	EPN	1	100/10000	e
2223-93-0	CADMIUM STEARATE	1	1000/10000	c,e
2231-57-4	THIOCARBAZIDE	1	1000/10000	e
2238-07-5	DIGLYCIDYL ETHER	1	1000	e
2275-18-5	PROTHOATE	1	100/10000	e
2497-07-6	OXYDISULFOTON	1	500	e,h
2524-03-0	DIMETHYL PHOSPHOROCHLORIDOTHIOATE	1	500	e
2540-82-1	FORMOTHION	1	100	e
2570-26-5	PENTADECYLAMINE	1	100/10000	e
2587-90-8	PHOSPHOROTHIOIC ACID, O,O-DIMETHYL-S-2-METHYLTHIO) ETHYL ESTER	1	500	ceg
2631-37-0	PROMECARB	1	500/10000	e,h
2636-26-2	CYANOPHOS	1	1000	e
2642-71-9	AZINPHOS-ETHYL	1	100/10000	e
2665-30-7	PHOSPHONOTHIOIC ACID, METHYL-, 0-(4-NITROPHENYL) O-PHENYL ESTER	1	500	e
2703-13-1	PHOSPHONOTHIOIC ACID, METHYL-, 0-ETHYL 0-[4-(METHYLTHIO)PHENYL) ESTER	1	500	e
2757-18-8	THALLOUS MALONATE	1	100/10000	ceh
2763-96-4	MUSCIMOL	1000	10000	a,h
2778-04-3	ENDOTHION	1	500/10000	e
3037-72-7	SILANE, (4-AMINOBUTYL)DIETHOXYMETHYL-	1	1000	e
3254-63-5	PHOSPHORIC ACID, DIMETHYL 4-(METHYLTHIO)PHENYL ESTER	1	500	e
3569-57-1	SULFOXIDE, 3-CHLOROPROPYL OCTYL	1	500	e
3615-21-2	BENZIMIDAZOLE, 4,5-DICHLORO-2-(TRIFLUOROMETHYL)-	1	500/10000	e,g
3689-24-5	SULFOTEP	100	500	
3691-35-8	CHLOROPHACINONE	1	100/10000	e
3734-97-2	AMITON OXALATE	1	100/10000	e
3735-23-7	METHYL PHENKAPTON	1	500	e
3878-19-1	FUBERIDAZOLE	1	100/10000	e
4044-65-9	BITOSCANATE	1	500/10000	e
4098-71-9	ISOPHORONE DIISOCYANATE	1	100	b,e
4104-14-7	PHOSACETIM	1	100/10000	e
4170-30-3	CROTONALDEHYDE	100	1000	
4301-50-2	FLUENETIL	1	100/10000	e
4418-66-0	PHENOL, 2,2'-THIOBIS(4-CHLORO-6-METHYL-	1	100/10000	e
4835-11-4	HEXAMETHYLENEDIAMINE, N,N'-DIBUTYL-	1	500	e
5344-82-1	THIOUREA, (2-CHLOROPHENYL)-	100	100/10000	
5836-29-3	COUMATETRALYL	1	500/10000	e
6533-73-9	THALLOUS CARBONATE	100	100/10000	c,h
6923-22-4	MONOCROTOPHOS	1	10/10000	e

Chemical Hazard Communication Guidebook

CAS	CHEMICAL NAME	RQ (LBS)	TPQ (LBS)	NOTES
7446-09-5	SULFUR DIOXIDE	1	500	e,l
7446-11-9	SULFUR TRIOXIDE	1	100	b,e
7446-18-6	THALLOUS SULFATE	100	100/10000	
7487-94-7	MERCURIC CHLORIDE	1	500/10000	e
7550-45-0	TITANIUM TETRACHLORIDE	1	100	e
7580-67-8	LITHIUM HYDRIDE	1	100	b,e
7631-89-2	SODIUM ARSENATE	1	1000/10000	d
7637-07-2	BORON TRIFLUORIDE	1	500	e
7647-01-0	HYDROGEN CHLORIDE	5000	500	e l m
7647-01-0	HYDROGEN CHLORIDE GAS (ONLY)	5000	500	e l m
7664-39-3	HYDROGEN FLUORIDE	100	100	
7664-41-7	AMMONIA	100	500	l
7664-93-9	SULFURIC ACID	1000	1000	
7697-37-2	NITRIC ACID	1000	1000	
7719-12-2	PHOSPHORUS TRICHLORIDE	1000	1000	
7722-84-1	HYDROGEN PEROXIDE	1	1000	e,l
7723-14-0	PHOSPHORUS	1	100	b,h
7726-95-6	BROMINE	1	500	e,l
7778-44-1	CALCIUM ARSENATE	1	500/10000	d
7782-41-4	FLUORINE	10	500	k
7782-50-5	CHLORINE	10	100	
7783-00-8	SELENIOUS ACID	10	1000/10000	
7783-06-4	HYDROGEN SULFIDE	100	500	l
7783-07-5	HYDROGEN SELENIDE	1	10	e
7783-60-0	SULFUR TETRAFLUORIDE	1	100	e
7783-70-2	ANTIMONY PENTAFLUORIDE	1	500	e
7783-80-4	TELLURIUM HEXAFLUORIDE	1	100	e,k
7784-34-1	ARSENOUS TRICHLORIDE	1	500	d
7784-42-1	ARSINE	1	100	e
7784-46-5	SODIUM ARSENITE	1	500/10000	d
7786-34-7	MEVINPHOS	10	500	
7791-12-0	THALLOUS CHLORIDE	100	100/10000	c,h
7791-23-3	SELENIUM OXYCHLORIDE	1	500	e
7803-51-2	PHOSPHINE	100	500	
8001-35-2	CAMPHECHLOR	1	500/10000	d
8065-48-3	DEMETON	1	500	e
10025-73-7	CHROMIC CHLORIDE	1	1/10000	e
10025-87-3	PHOSPHORUS OXYCHLORIDE	1000	500	d
10026-13-8	PHOSPHORUS PENTACHLORIDE	1	500	b,e
10028-15-6	OZONE	1	100	e
10031-59-1	THALLIUM SULFATE	100	100/10000	h
10102-18-8	SODIUM SELENITE	100	100/10000	h
10102-20-2	SODIUM TELLURITE	1	500/10000	e
10102-43-9	NITRIC OXIDE	10	100	c
10102-44-0	NITROGEN DIOXIDE	10	100	
10124-50-2	POTASSIUM ARSENITE	1	500/10000	d
10140-87-1	ETHANOL, 1,2-DICHLORO-, ACETATE	1	1000	e

Appendix E: Extremely Hazardous Substances (CAS Order)

CAS	CHEMICAL NAME	RQ (LBS)	TPQ (LBS)	NOTES
10210-68-1	COBALT CARBONYL	1	10/10000	e,h
10265-92-6	METHAMIDOPHOS	1	100/10000	e
10294-34-5	BORON TRICHLORIDE	1	500	e
10311-84-9	DIALIFOS	1	100/10000	e
10476-95-6	METHACROLEIN DIACETATE	1	1000	e
12002-03-8	PARIS GREEN	1	500/10000	d
12108-13-3	MANGANESE, TRICARBONYL METHYLCYCLOPENTADIENYL	1	100	e,h
13071-79-9	TERBUFOS	1	100	e,h
13171-21-6	PHOSPHAMIDON	1	100	e
13194-48-4	ETHOPROPHOS	1	1000	e
13410-01-0	SODIUM SELENATE	1	100/10000	e
13450-90-3	GALLIUM TRICHLORIDE	1	500/10000	e
13463-39-3	NICKEL CARBONYL	10	1	d
13463-40-6	IRON PENTACARBONYL	1	100	e
13494-80-9	TELLURIUM	1	500/10000	e
14167-18-1	SALCOMINE	1	500/10000	e
15271-41-7	BICYCLO(2.2.1)HEPTANE-2-CARBONITRILE, 5-CHLORO-6-((((METHYLAMINO)CARBO	1	500/10000	e
16752-77-5	METHOMYL	100	500/10000	h
17702-41-9	DECABORANE	1	500/10000	e
17702-57-7	FORMPARANATE	1	100/10000	e
19287-45-7	DIBORANE	1	100	e
19624-22-7	PENTABORANE	1	500	e
20830-75-5	DIGOXIN	1	10/10000	e,h
20859-73-8	ALUMINUM PHOSPHIDE	100	500	b
21548-32-3	FOSTHIETAN	1	500	e
21609-90-5	LEPTOPHOS	1	500/10000	e
21908-53-2	MERCURIC OXIDE	1	500/10000	e
21923-23-9	CHLORTHIOPHOS	1	500	e,h
22224-92-6	FENAMIPHOS	1	10/10000	e
23135-22-0	OXAMYL	1	100/10000	e
23422-53-9	FORMETANATE HYDROCHLORIDE	1	500/10000	e,h
23505-41-1	PIRIMIFOS-ETHYL	1	1000	e
24017-47-8	TRIAZOFOS	1	500	e
24934-91-6	CHLORMEPHOS	1	500	e
26419-73-8	CARBAMIC ACID, METHYL-, O-(((2,4-DIMETHYL-1,3-DITHIOLAN-2-YL)METHYLENE	1	100/10000	e
26628-22-8	SODIUM AZIDE	1000	500	b
27137-85-5	TRICHLORO(DICHLOROPHENYL)SILANE	1	500	e
28347-13-9	XYLYLENE DICHLORIDE	1	100/10000	e
28772-56-7	BROMADIOLONE	1	100/10000	e
30674-80-7	METHACRYLOYLOXYETHYL ISOCYANATE	1	100	e,h
39196-18-4	THIOFANOX	100	100/10000	
50782-69-9	PHOSPHONOTHIOIC ACID, METHYL-, S-(2-(BIS(1-METHYLETHYL)AMINO)ETHYL)O-E	1	100	e
53558-25-1	PYRIMINIL	1	100/10000	e,h

Chemical Hazard Communication Guidebook

CAS	CHEMICAL NAME	RQ (LBS)	TPQ (LBS)	NOTES
58270-08-9	ZINC, DICHLORO(4,4-DIMETHYL-5-((((METHYLAMINO)CARBONYL)OXY)IMINO)PENTA	1	100/10000	e
62207-76-5	COBALT, ((2,2'-(1,2-ETHANEDIYLBIS(NITRILOMETHYLIDENE)BIS(PHENALOTO))(	1	100/10000	e
PMN82147	ORGANORHODIUM COMPLEX	1	10/10000	e

NOTES

a. This chemical does not meet acute toxicity criteria. Its TPQ is set at 10,000 pounds.

b. This material is a reactive solid. The Threshold Planning Quantity does not default to 10,000 pounds for non-powder, non-molten, non-solution form.

c. EPA changed the calculated Threshold Planning Quantity and the reader is referred to the *Federal Register* of April 22, 1987 for further details.

d. EPA has indicated that the reportable quantity is likely to change when the assessment of potential carcinogenicity and chronic toxicity is completed.

e. Statutory reportable quantity for purposes of emergency notification under Section 304(a)(2) of the Emergency Planning and Community Right-to-Know Act.

f. EPA has indicated that the statutory one pound reportable quantity for methyl isocyanate may be adjusted in a future rulemaking.

g. New chemicals added that were not part of the original list of 402 substances.

h. Revised TPQ based on new or re-evaluated toxicity data.

j. TPQ is revised to its calculated value and does not change due to technical review as in proposed rule.

k. The TPQ was revised after proposal due to calculation error.

Appendix E: Extremely Hazardous Substances (CAS Order)

l. Chemicals on the original list that do not meet the toxicity criteria but because of their high production volume and recognized toxicity are considered chemicals of concern.

m. Hydrogen chloride is regulated as an extremely hazardous substance for the gas only.

APPENDIX F

HAZARDOUS SUBSTANCES
(Alphabetical Order)

This appendix provides the list of Hazardous Substances subject to spill reporting in CERCLA/SARA. Where a reportable quantity has been established under CERCLA or a statutory reportable quantity has been set pursuant to SARA, it is listed under the column "RQ." As an aid to the reader, if the CERCLA Hazardous Substance has also been listed as an extremely hazardous substance subject to SARA's emergency planning provisions, the threshold planning quantities are provided under the column, "TPQ." A guide to the notes is included at the end.

Chemical Hazard Communication Guidebook

CAS	CHEMICAL NAME	RQ (LBS)	TPQ (LBS)	NOTES
K043	*2,6-DICHLOROPHENOL WASTE FROM 2,4-D PROD.	10		
K064	*ACID PLANT BLOWDOWN/SLURRY/SLUDGE RESULTING FROM ...PRIMARY COPPER	1		
K060	*AMMONIA STILL LIME SLUDGE FROM COKING OPERATIONS	1		
K051	*API SEPARATOR SLUDGE FR. PETROLEUM REFINING INDUSTRY	1		
K021	*AQUEOUS SPENT ANTIMONY CATALYST WASTE FR. FLUOROMETHANES PROD.	10		
K126	*BAGHOUSE DUST, SWEEP. IN MILL., PACK.,PRD ETHYLENEBISDITHIOCARBAMICAC	10		
K011	*BOTTOM STREAM FROM WASTEWATER STRIPPER IN PRODUCTION OF ACRYLONITRILE	10		
K013	*BOTTOM STREAM FROM ACETONITRILE COLUMN IN PRODUCTION OF ACRYLONITRILE	10		
K014	*BOTTOMS FROM ACETONITRILE PURIF. COLUMN IN PROD. OF ACRYLONITRILE	5000		
K071	*BRINE PURIF. MUDS FR. MERCURY CELL PROCESS IN CHLORINE PROD. EXCEPT..	1		
K031	*BY-PRODUCT SALTS GENERATED IN PROD. OF MSMA AND CACODYLIC ACID	1		
K027	*CENTRIFUGE AND DISTILLATION RESIDUES FR. TOLUENE DIISOCYANATE PROD.	10		
K073	*CHLORINATED HYDROCARBON WASTE FR. PURIF. STEP OF ... IN CHLORINE PROD	10		
K107	*COLUMN BOTTOMS FROM PRODUCT SEPERATION FR... CARBOXYLIC ACID HYDAZINE	10		
K030	*COLUMN BOTTOMS/HEAVY ENDS FR. COMBO PROD. TRI- AND PER-CHLOROETHYLENE	1		
K104	*COMBINED WASTEWATER STREAMS GEN. NITROBENZENE/ANILINE CHLOROBENZINES	10		
K108	*CONDENSED COLUMN OVERHEADS FROM PRODUCT S... CARBOXYLIC ACID HYDRAZIN	10		
K110	*CONDENSED COLUMN OVERHEADS FROM INTERMEDI... CARBOXYLIC ACID HYDRAZIN	10		
F025	*CONDENSED LIGHT ENDS,SPENT FILTERS...WASTES FROM CHLOR. ALIPHATIC HYD	1		
K113	*CONDENSED LIQUID LIGHT ENDS FR. PURIF. TOLUENEDIAMINE IN PROD. VIA...	10		
K087	*DECANTER TANK TAR SLUDGE FROM COKING OPERATIONS	100		
F027	*DISCARDED,UNUSED FORMU.W/ TRI,TETRA,PENTA-CHLOROPHENOLS OR DERIVATIVES	1		
K048	*DISSOLVED AIR FLOTATION (DAF) FLOAT FR. PETROLEUM REFINING INDUSTRY	1		
K009	*DISTILLATION BOTTOMS FROM PRODUCTION OF ACETALDEHYDE FROM ETHYLENE	10		
K022	*DISTILLATION BOTTOM TARS FR. PROD. OF PHENOL/ACETONE FR. CUMENE	1		
K024	*DISTILLATION BOTTOMS FR. PROD. PHTHALIC ANHYDRIDE FR. NAPHTHALENE	5000		
K025	*DISTILLATION BOTTOMS FR. PROD. NITROBENZENE BY NITRATION OF BENZENE	10		

Appendix F: Hazardous Substances (Alphabetical Order)

CAS	CHEMICAL NAME	RQ (LBS)	TPQ (LBS)	NOTES
K083	*DISTILLATION BOTTOMS FROM ANILINE EXTRACTION	100		
K094	*DISTILLATION BOTTOMS FR. PROD. PHTHALIC ANHYDRIDE FR. ORTHO-XYLENE	5000		
K095	*DISTILLATION BOTTOMS FR. PROD. 1,1,1-TRICHLOROETHANE	100		
K023	*DISTILLATION LIGHT ENDS FR. PROD. PHTHALIC ANHYDRIDE FR. NAPHTHALENE	5000		
K093	*DISTILLATION LIGHT ENDS FR. PROD. PHTHALIC ANHYDRIDE FR. ORTHO-XYLENE	5000		
K085	*DISTILLATION OR FRACTIONATION COLUMN BOTTOMS FROM CHLOROBENZENE PROD.	10		
K010	*DISTILLATION SIDE CUTS FROM PRODUCTION OF ACETALDEHYDE FROM ETHYLENE	10		
K101	*DISTILLATION TAR RESIDUES FR. ANILINE-BASED COMPOUNDS VET. PHARMACEUT	1		
F008	*ELECTROPLATING BATH SLUDGES FROM BOTTOMS USING CYANIDES	10		
K061	*EMISSION CONTROL DUST/SLUDGE FR. PRIM. PROD. STEEL IN ELEC. FURNACES	1		
K069	*EMISSION CONTROL DUST/SLUDGE FR. SECONDARY LEAD SMELTING	1		
K090	*EMISSION CONTROL DUST OR SLUDGE FROM FERROCHROMIUM SILICON PRODUCTION	1		
K091	*EMISSION CONTROL DUST OR SLUDGE FROM FERROCHROMIUM PRODUCTION	1		
K039	*FILTER CAKE FR. FILTR. DIETHYLPHOSPHORODITHIOIC ACID IN PHORATE PROD.	10		
K034	*FILTER SOLIDS FR. FILTR. HEXACHLOROCYCLOPENTADIENE IN CHLORDANE PROD.	10		
K125	*FILTRATION, EVAP.,CENTRIFUG., PROD. ETHEYLENEBISDITHIOCARBAMIC ACIDS	10		
K050	*HEAT EXCHANGER BUNDLE CLEANING SLUDGE FR. PETROLEUM REFINING INDUSTRY	10		
K096	*HEAVY ENDS FR. HEAVY ENDS COLUMN FR. PROD. 1,1,1-TRICHLOROETHANE	100		
K019	*HEAVY ENDS FROM DISTILLATION IN PRODUCTION OF ETHYLENE DICHLORIDE	1		
K020	*HEAVY ENDS FROM DISTILLATION IN PRODUCTION OF VINYL CHLORIDE MONOMERS	1		
K018	*HEAVY ENDS FROM FRACTIONATION COLUMN IN PRODUCTION OF ETHYL CHLORIDE	1		
K016	*HEAVY ENDS OR DISTILLATION RESIDUES FROM PROD.OF CARBON TETRACHLORIDE	1		
K115	*HEAVY ENDS PURIF. TOLUENEDIAMINE IN PROD. VIA HYDROG. DINITROTOLUENE	10		
K017	*HEAVY ENDS(STILL BOTTOMS) FROM PURIF. COLUMN IN PROD. EPICHLOROHYDRIN	10		
K042	*HEAVY ENDS/DIST. RESIDUES FR. DIST. TETRACHLOROBENZENE IN 2,4,5-T PR.	10		
F039	*LEACHATE RESULTING FROM THE TREATMENT, STORAGE, OR DISPOSAL OF WASTES	1		
F039	*MULTI SOURCE LEACHATE	1		

Chemical Hazard Communication Guidebook

CAS	CHEMICAL NAME	RQ (LBS)	TPQ (LBS)	NOTES
K116	*ORGANIC CONDENS. FR. SOLVENT RECOVERY COLUMN TOLUENE DIISOCYANATE VIA	10		
K008	*OVEN RESIDUE FROM THE PRODUCTION OF CHROME OXIDE GREEN PIGMENTS	10		
F037	*PETROLEUM REFINERY PRIMARY OIL/WATER/SOLIDS SEPARATION SLUDGE--ANY SL	1		
F038	*PETROLEUM REFINERY SECONDARY (EMULSIFIED) OIL/WATER/SOLIDS SEPARATION	1		
K047	*PINK/RED WATER FROM TNT OPERATIONS	10		
K103	*PROCESS RESIDUES FROM ANILINE EXTRACTION FROM ANILINE PROD.	100		
F024	*PROCESS WASTES INCLUDING PROD. CHLORINATED ALIPHATIC HYDROCARBONS	1		
K123	*PROCESS WASTEWATER ...FROM PROD. ETHYLENEBISDITHIOCARBAMIC ACID/SALTS	10		
K111	*PRODUCT WASHWATERS FROM PROD. DINITROTOLUENE VIA NITRATION OF TOLUENE	10		
F010	*QUENCHING BATH SLUDGE (OIL BATH) (METAL HEAT TREATING) USING CYANIDES	10		
F012	*QUENCHING WASTEWATER TREAT SLUDGES FR METAL HEAT TREAT USING CYANIDES	10		
K112	*REACT. BY-PROD. WATER FR. DRYING COLUMN PROD. TOLUENEDIAMINE VIA...	10		
K124	*REACTOR VENT SCRUBBER WTER PRD. ETHYLENEBISDITHIOACARBAMIC ACID/SALTS	10		
K102	*RESIDUE FR. ACTIVATED CARBON FOR DECOLORIZATION PROD. VET. PHARMACEUT	1		
F028	*RESIDUE FR. INCIN./THERMAL TREAT. SOIL CONTAMINATED W/SPECIFIED WASTE	1		
K105	*SEPARATED AQUEOUS STREAM FR. REACTOR PROD.WASHING STEP IN CHLOROBENZ.	10		
K049	*SLOP OIL EMULSION SOLIDS FROM THE PETROLEUM REFINING INDUSTRY	1		
K066	*SLUDGE FROM TREATMENT OF PROCESS WASTEWATER AND/OR ACID..PRIMARY ZINC	1		
K001	*SLUDGE OF WOOD PRESERVING PROCESSES USING CREOSOTE/PENTACHLOROPHENOL	1		
K086	*SOLVENT,WATER,CAUSTIC WASHES & SLUDGES CLEANING EQUIP. FOR INK FORMU.	1		
F001	*SPECIFIED SPENT HALOGENATED SOLVENTS USED DEGREASING & SLUDGES FR REC	10		
F002	*SPECIFIED SPENT HALOGENATED SOLVENTS AND STILL BOTTOMS FR. RECOVERY	10		
F003	*SPECIFIED SPENT NON-HALOGENATED SOLVENTS & STILL BOTTOMS FR. RECOVERY	100		
K132	*SPENT ABSORBENT AND WASTEWATER SOLIDS FROM PROD. METHYL BROMIDE	1000		
K118	*SPENT ABSORBENT SOLIDS FR. PURIF. ETHYLENE DIBROMIDE IN PROD. OF IT	1		
K045	*SPENT CARBON FROM TREAT. OF WASTEWATER CONTAINING EXPLOSIVES	10		
K028	*SPENT CATALYST FR. HYDROCHLORINATOR REACTOR IN 1,1,1-TRICHLOROETHANE	1		

Appendix F: Hazardous Substances (Alphabetical Order)

CAS	CHEMICAL NAME	RQ (LBS)	TPQ (LBS)	NOTES
F007	*SPENT CYANIDE ELECTROPLATING BATH SOLUTIONS W/ SPECIFIED EXCEPTIONS	10		
F011	*SPENT CYANIDE SOLUTIONS FR SALT BATH POT CLEANING (METAL HEAT TREAT)	10		
K109	*SPENT FILTER CARTRIDGES FROM PRODUCT PURI... CARBOXYLIC ACID HYDRAZIN	10		
K062	*SPENT PICKLE LIQUOR FR. STEEL FINISHING OPERATIONS	1		
K088	*SPENT POTLINERS FROM PRIMARY ALUMINUM REDUCTION	1		
F009	*SPENT STRIPPING & CLEANING SOLUTIONS (ELECTROPLATING) USING CYANIDES	10		
K136	*STILL BOTTOMS FR. PURIF. ETHYLENE DIBROMIDE IN PROD. VIA BROMINATION	1		
K036	*STILL BOTTOMS FR. TOLUENE RECLAMATION DISTIL. IN DISULFOTON PROD.	1		
K015	*STILL BOTTOMS FROM THE DISTILLATION OF BENZYL CHLORIDE	10		
K136	*STILL BOTTOMS FROM PURIF. ETHYLENE DIBROMIDE IN PROD. VIA BROMIN. ETH	1		
K026	*STRIPPING STILL TAILS FR. PROD. METHYL ETHYL PYRIDINES	1000		
K065	*SURFACE IMPOUNDMENT SOLIDS CONTAINED IN AND DEGRADED...LEAD SMELTING	1		
K052	*TANK BOTTOMS (LEADED) FR. PETROLEUM REFINING INDUSTRY	10		
F004	*THE FOLLOWING SPENT NON-HALOGENATED SOLVENTS...CRESOLS...NITROBENZENE	1000		
F005	*THE FOLLOWING SPENT NON-HALOGENATED SOLVENTS...TOLUENE...METHYL ETHYL	100		
D004	*UNLISTED HAZ. WASTE CHARACTERISTIC OF EP TOXICITY - ARSENIC	1		
D005	*UNLISTED HAZ. WASTE CHARACTERISTIC OF EP TOXICITY - BARIUM	1000		
D006	*UNLISTED HAZ. WASTE CHARACTERISTIC OF EP TOXICITY - CADMIUM	10		
D007	*UNLISTED HAZ. WASTE CHARACTERISTIC OF EP TOXICITY - CHROMIUM	10		
D008	*UNLISTED HAZ. WASTE CHARACTERISTIC OF EP TOXICITY - LEAD	1		
D009	*UNLISTED HAZ. WASTE CHARACTERISTIC OF EP TOXICITY - MERCURY	1		
D010	*UNLISTED HAZ. WASTE CHARACTERISTIC OF EP TOXICITY - SELENIUM	10		
D011	*UNLISTED HAZ. WASTE CHARACTERISTIC OF EP TOXICITY - SILVER	1		
D012	*UNLISTED HAZ. WASTE CHARACTERISTIC OF EP TOXICITY - ENDRIN	1		
D013	*UNLISTED HAZ. WASTE CHARACTERISTIC OF EP TOXICITY - LINDANE	1		
D014	*UNLISTED HAZ. WASTE CHARACTERISTIC OF EP TOXICITY - METHOXYCHLOR	1		
D015	*UNLISTED HAZ. WASTE CHARACTERISTIC OF EP TOXICITY - TOXAPHENE	1		

Chemical Hazard Communication Guidebook

CAS	CHEMICAL NAME	RQ (LBS)	TPQ (LBS)	NOTES
D016	*UNLISTED HAZ. WASTE CHARACTERISTIC OF EP TOXICITY - 2,4-D	100		
D017	*UNLISTED HAZ. WASTE CHARACTERISTIC OF EP TOXICITY - 2,4,5-TP	100		
D---	*UNLISTED HAZARDOUS WASTES - CHARACTERISTIC OF EP TOXICITY			
D001	*UNLISTED HAZARDOUS WASTES - CHARACTERISTIC OF IGNITABILITY	100		
D002	*UNLISTED HAZARDOUS WASTES - CHARACTERISTIC OF CORROSIVITY	100		
D003	*UNLISTED HAZARDOUS WASTES - CHARACTERISTIC OF REACTIVITY	100		
D007	*UNLISTED HAZARDOUS WASTES CHARACTERISTIC OF EP TOXICITY CHROMIUM(VI)	10		
K098	*UNTREATED PROCESS WASTEWATER FR. TOXAPHENE PROD.	1		
K099	*UNTREATED WASTEWATER FR. 2,4-D PROD.	10		
K097	*VACUUM STRIPPER DISCHARGE FR. CHLORDANE CHLORINATOR IN CHLORDANE PROD	1		
K114	*VICINALS FR. PURIF. TOLUENEDIAMINE IN PROD. VIA HYDROG. DINITROTOLUEN	10		
K029	*WASTE FR. PRODUCT STEAM STRIPPER IN 1,1,1-TRICHLOROETHANE PROD.	1		
K100	*WASTE LEACHING SOLUTION ... (COMPONENTS IDENTICAL WITH THOSE OF K069)	1		
K033	*WASTE- & SCRUBWATER FR. CHLORIN. CYCLOPENTADIENE IN CHLORDANE PROD.	10		
F022	*WASTES FR. MANU. USE OF TETRA-,PENTA-,OR HEXACHLOROBENZENE (ALKALINE)	1		
F026	*WASTES FR. PROD. MATERIALS ON EQUIP. FOR TETRA,PENTA,HEXACHLOROBENZEN	1		
F023	*WASTES FR. PROD. OF MATERIALS ON EQUIP. FOR TRI- & TETRACHLOROPHENOLS	1		
F020	*WASTES IN PROD. OR MANU. OF TRI- OR TETRACHLOROPHENOL, OR PEST. DERIV	1		
F021	*WASTES IN PROD. OR MANU. OF PENTACHLOROPHENOL/INTERMED. TO PROD DERIV	1		
K117	*WASTEWATER FR. REACTION VENT GAS SCRUBBER PROD. ETHYLENE BROMIDE VIA	1		
K038	*WASTEWATER FR. WASHING AND STRIPPING OF PHORATE PROD.	10		
K131	*WASTEWATER FROM REACTOR,SPENT SULFURIC ACID IN PROD. METHYL BROMIDE	100		
K002	*WASTEWATER SLUDGE FROM PRODUCTION OF CHROME YELLOW & ORANGE PIGMENTS	1		
K003	*WASTEWATER SLUDGE FROM PRODUCTION OF MOLYBDATE ORANGE PIGMENTS	1		
K006	*WASTEWATER SLUDGE FR PROD. CHROME OXIDE GREEN PIGMENTS (ANHY. & HYD.)	10		
K046	*WASTEWATER TR. SLUDGE FROM MANU,FORMU,LOADING OF LEAD-BASED INITIATOR	100		
K032	*WASTEWATER TREAT. SLUDGE FROM PROD. OF CHLORDANE	10		

Appendix F: Hazardous Substances (Alphabetical Order)

CAS	CHEMICAL NAME	RQ (LBS)	TPQ (LBS)	NOTES
K035	*WASTEWATER TREAT. SLUDGES GENERATED IN CREOSOTE PROD.	1		
K037	*WASTEWATER TREAT. SLUDGES FROM DISULFOTON PROD.	1		
K040	*WASTEWATER TREAT. SLUDGE FROM PHORATE PROD.	10		
K041	*WASTEWATER TREAT. SLUDGE FROM TOXAPHENE PROD.	1		
K044	*WASTEWATER TREAT. SLUDGES FROM MANU. & PROCESSING OF EXPLOSIVES	10		
K084	*WASTEWATER TREAT. SLUDGES IN PROD. OF VETERINARY PHARMACEUT (ARSENIC)	1		
F006	*WASTEWATER TREATMENT SLUDGES (ELECTROPLATING) W/ SPECIFIED EXCEPTIONS	10		
F019	*WASTEWATER TREATMENT SLUDGES- CHEMICAL CONVERSION COATING OF ALUMINUM	10		
K004	*WASTEWATER TREATMENT SLUDGE FROM PRODUCTION OF ZINC YELLOW PIGMENTS	10		
K005	*WASTEWATER TREATMENT SLUDGE FROM PRODUCTION OF CHROME GREEN PIGMENTS	1		
K007	*WASTEWATER TREATMENT SLUDGE FROM PRODUCTION OF IRON BLUE PIGMENTS	10		
K106	*WASTEWATER TREATMENT SLUDGE FR. MERCURY CELL PROCESS IN CHLORINE PROD	1		
83-32-9	ACENAPHTHENE	100		
208-96-8	ACENAPHTHYLENE	5000		
75-07-0	ACETALDEHYDE	1000		
107-20-0	ACETALDEHYDE, CHLORO-	1000		
75-87-6	ACETALDEHYDE, TRICHLORO-	5000		
640-19-7	ACETAMIDE, 2-FLUORO-	100		
62-44-2	ACETAMIDE, N-(4-ETHOXYPHENYL)-	100		
591-08-2	ACETAMIDE, N-(AMINOTHIOXOMETHYL)-	1000		
53-96-3	ACETAMIDE, N-9H-FLUOREN-2-YL-	1		
64-19-7	ACETIC ACID	5000		
141-78-6	ACETIC ACID, ETHYL ESTER	5000		
62-74-8	ACETIC ACID, FLUORO-, SODIUM SALT	10		
301-04-2	ACETIC ACID, LEAD SALT	5000		
563-68-8	ACETIC ACID, THALLIUM (1+) SALT	100		
108-24-7	ACETIC ANHYDRIDE	5000		
16752-77-5	ACETIMIDIC ACID, N-(METHYLCARBAMOYL)OXY-LTHIO-, METHYL ESTER	100		
67-64-1	ACETONE	5000		
75-86-5	ACETONE CYANOHYDRIN	10	1000	
75-05-8	ACETONITRILE	5000		
81-81-2	3-(ALPHA-ACETONYLBENZYL)-4-HYDROXYCOUMARIN AND SALTS	100		
98-86-2	ACETOPHENONE	5000		
506-96-7	ACETYL BROMIDE	5000		
75-36-5	ACETYL CHLORIDE	5000		
591-08-2	1-ACETYL-2-THIOUREA	1000		
53-96-3	2-ACETYLAMINOFLUORENE	1		

Chemical Hazard Communication Guidebook

CAS	CHEMICAL NAME	RQ (LBS)	TPQ (LBS)	NOTES
107-02-8	ACROLEIN	1	500	
79-06-1	ACRYLAMIDE	5000	1000/10000	d,l
79-10-7	ACRYLIC ACID	5000		
107-13-1	ACRYLONITRILE	100	10000	d,l
124-04-9	ADIPIC ACID	5000		
148-82-3	ALANINE, 3-[P-BIS(2-CHLOROETHYL)AMINO]PHENYL-, L-	1		
116-06-3	ALDICARB	1	100/10000	c
309-00-2	ALDRIN	1	500/10000	d
309-00-2	ALDRIN [1,4:5,8-DIMETHANONAPHTHALENE,1,2,3,4,10,10-HEXACHLORO-1,4,4A,5	1	500/10000	d
107-18-6	ALLYL ALCOHOL	100	1000	
107-05-1	ALLYL CHLORIDE	1000		
20859-73-8	ALUMINUM PHOSPHIDE	100	500	b
10043-01-3	ALUMINUM SULFATE	5000		
95-53-4	2-AMINO-1-METHYL BENZENE	100		
106-49-0	4-AMINO-1-METHYL BENZENE	100		
2763-96-4	5-(AMINOMETHYL)-3-ISOXAZOLOL	1000		
504-24-5	4-AMINOPYRIDINE	1000		
61-82-5	AMITROLE	10		
7664-41-7	AMMONIA	100	500	l
631-61-8	AMMONIUM ACETATE	5000		
1863-63-4	AMMONIUM BENZOATE	5000		
1066-33-7	AMMONIUM BICARBONATE	5000		
7789-09-5	AMMONIUM BICHROMATE	10		
1341-49-7	AMMONIUM BIFLUORIDE	100		
10192-30-0	AMMONIUM BISULFITE	5000		
1111-78-0	AMMONIUM CARBAMATE	5000		
506-87-6	AMMONIUM CARBONATE	5000		
12125-02-9	AMMONIUM CHLORIDE	5000		
7788-98-9	AMMONIUM CHROMATE	10		
3012-65-5	AMMONIUM CITRATE DIBASIC	5000		
13826-83-0	AMMONIUM FLUOBORATE	5000		
12125-01-8	AMMONIUM FLUORIDE	100		
1336-21-6	AMMONIUM HYDROXIDE	1000		
5972-73-6	AMMONIUM OXALATE	5000		
6009-70-7	AMMONIUM OXALATE	5000		
14258-49-2	AMMONIUM OXALATE	5000		
131-74-8	AMMONIUM PICRATE	10		
16919-19-0	AMMONIUM SILICOFLUORIDE	1000		
7773-06-0	AMMONIUM SULFAMATE	5000		
12135-76-1	AMMONIUM SULFIDE	100		
10196-04-0	AMMONIUM SULFITE	5000		
3164-29-2	AMMONIUM TARTRATE	5000		
14307-43-8	AMMONIUM TARTRATE	5000		
1762-95-4	AMMONIUM THIOCYANATE	5000		
7803-55-6	AMMONIUM VANADATE	1000		

Appendix F: Hazardous Substances (Alphabetical Order)

CAS	CHEMICAL NAME	RQ (LBS)	TPQ (LBS)	NOTES
628-63-7	AMYL ACETATE	5000		
123-92-2	ISO-AMYL ACETATE	5000		
626-38-0	SEC-AMYL ACETATE	5000		
625-16-1	TERT-AMYL ACETATE	5000		
62-53-3	ANILINE	5000	1000	d,l
120-12-7	ANTHRACENE	5000		
7440-36-0	ANTIMONY	5000		
--	ANTIMONY COMPOUNDS			
7647-18-9	ANTIMONY PENTACHLORIDE	1000		
28300-74-5	ANTIMONY POTASSIUM TARTRATE	100		
7789-61-9	ANTIMONY TRIBROMIDE	1000		
10025-91-9	ANTIMONY TRICHLORIDE	1000		
7783-56-4	ANTIMONY TRIFLUORIDE	1000		
1309-64-4	ANTIMONY TRIOXIDE	1000		
12674-11-2	AROCLOR 1016	1		
11104-28-2	AROCLOR 1221	1		
11141-16-5	AROCLOR 1232	1		
53469-21-9	AROCLOR 1242	1		
12672-29-6	AROCLOR 1248	1		
11097-69-1	AROCLOR 1254	1		
11096-82-5	AROCLOR 1260	1		
7440-38-2	ARSENIC	1		
1327-52-2	ARSENIC ACID	1		
7778-39-4	ARSENIC ACID	1		
--	ARSENIC COMPOUNDS			
1303-32-8	ARSENIC DISULFIDE	1		
1303-28-2	ARSENIC PENTOXIDE	1	100/10000	d
7784-34-1	ARSENIC TRICHLORIDE	1		
1327-53-3	ARSENIC TRIOXIDE	1		
1303-33-9	ARSENIC TRISULFIDE	1		
1327-53-3	ARSENIC(III) OXIDE	1		
1303-28-2	ARSENIC(V) OXIDE	1		
7784-34-1	ARSENOUS TRICHLORIDE	1	500	d
692-42-2	ARSINE, DIETHYL	1		
1332-21-4	ASBESTOS	1		
492-80-8	AURAMINE	100		
115-02-6	AZASERINE	1		
151-56-4	AZIRIDINE	1		
50-07-7	AZIRINO[2',3':3,4]PYRROLO[1,2-A]INDOLE-4,7-DIONE, 6-AMINO-8-[[(AMINOCA	10		
542-62-1	BARIUM CYANIDE	10		
225-51-4	3,4-BENZACRIDINE	100		
98-87-3	BENZAL CHLORIDE	5000	500	d
56-55-3	1,2-BENZANTHRACENE	10		
57-97-6	1,2-BENZANTHRACENE, 7,12-DIMETHYL-	1		
62-53-3	BENZENAMINE	5000		
95-53-4	BENZENAMINE, 2-METHYL-	100		

Chemical Hazard Communication Guidebook

CAS	CHEMICAL NAME	RQ (LBS)	TPQ (LBS)	NOTES
99-55-8	BENZENAMINE, 2-METHYL-5-NITRO	100		
636-21-5	BENZENAMINE, 2-METHYL-, HYDROCHLORIDE	100		
492-80-8	BENZENAMINE, 4,4'-CARBONIMIDOYLBIS(N,N-DIMETHYL-	100		
101-14-4	BENZENAMINE, 4,4'-METHYLENEBIS(2-CHLORO-	10		
106-47-8	BENZENAMINE, 4-CHLORO	1000		
3165-93-3	BENZENAMINE, 4-CHLORO-2-METHYL-, HYDROCHLORIDE	100		
100-01-6	BENZENAMINE, 4-NITRO	5000		
60-11-7	BENZENAMINE, N,N-DIMETHYL-4-PHENYLAZO-	10		
D018	BENZENE	10		
71-43-2	BENZENE	10		
94-59-7	BENZENE 1,2-METHYLENEDIOXY-4-ALLYL-	100		
108-90-7	BENZENE CHLORO-	100		
98-09-9	BENZENE SULFONYL CHLORIDE	100		
95-94-3	BENZENE, 1,2,4,5-TETRACHLORO-	5000		
95-50-1	BENZENE, 1,2-DICHLORO-	100		
94-58-6	BENZENE, 1,2-METHYLENEDIOXY-4-PROPYL-	10		
120-58-1	BENZENE, 1,2-METHYLENEDIOXY-4-PROPENYL-	100		
99-35-4	BENZENE, 1,3,5-TRINITRO-	10		
541-73-1	BENZENE, 1,3-DICHLORO-	100		
106-46-7	BENZENE, 1,4-DICHLORO-	100		
101-55-3	BENZENE, 1-BROMO-4-PHENOXY-	100		
121-14-2	BENZENE, 1-METHYL-2,4-DINITRO-	10		
606-20-2	BENZENE, 1-METHYL-2,6-DINITRO-	100		
98-82-8	BENZENE, 1-METHYLETHYL-	5000		
91-08-7	BENZENE, 2,4-DIISOCYANATOMETHYL-	100		
584-84-9	BENZENE, 2,4-DIISOCYANATOMETHYL-	100		
26471-62-5	BENZENE, 2,4-DIISOCYANATOMETHYL-	100		
100-44-7	BENZENE, CHLOROMETHYL-	100		
98-87-3	BENZENE, DICHLOROMETHYL-	5000		
108-38-3	M-BENZENE, DIMETHYL	1000		
95-47-6	O-BENZENE, DIMETHYL	1000		
106-42-3	P-BENZENE, DIMETHYL	1000		
1330-20-7	BENZENE, DIMETHYL-	1000		
118-74-1	BENZENE, HEXACHLORO-	10		
110-82-7	BENZENE, HEXAHYDRO-	1000		
108-95-2	BENZENE, HYDROXY-	1000		
108-88-3	BENZENE, METHYL-	1000		
98-95-3	BENZENE, NITRO	1000		
82-68-8	BENZENE, PENTACHLORONITRO-	100		
608-93-5	BENZENE, PENTACHLORO-	10		
98-07-7	BENZENE, TRICHLOROMETHYL	10		
510-15-6	BENZENEACETIC ACID, 4-CHLORO-ALPHA- (4-CHLOROPHENYL)-ALPHA-HYDROXY-, E	10		
84-66-2	1,2-BENZENEDICARBOXYLIC ACID, DIETHYL ESTER	1000		
84-74-2	1,2-BENZENEDICARBOXYLIC ACID, DIBUTYL ESTER	10		
85-44-9	1,2-BENZENEDICARBOXYLIC ACID ANHYDRIDE	5000		

Appendix F: Hazardous Substances (Alphabetical Order)

CAS	CHEMICAL NAME	RQ (LBS)	TPQ (LBS)	NOTES
117-81-7	1,2-BENZENEDICARBOXYLIC ACID, [BIS(2-ETHYLHEXYL)] ESTER	100		
117-84-0	1,2-BENZENEDICARBOXYLIC ACID, DI-N-OCTYL ESTER	5000		
131-11-3	1,2-BENZENEDICARBOXYLIC ACID, DIMETHYL ESTER	5000		
108-46-3	1,3-BENZENEDIOL	5000		
51-43-4	1,2-BENZENEDIOL, 4-(1-HYDROXY-2-(METHYLAMINO)ETHYL)-	1000		
98-09-9	BENZENESULFONIC ACID CHLORIDE	100		
108-98-5	BENZENETHIOL	100		
92-87-5	BENZIDINE	1		
81-07-2	1,2-BENZISOTHIAZOLIN-3-ONE,1,1-DIOXIDE, AND SALTS	100		
120-58-1	1,3-BENZODIOXOLE, 5-(1-PROPENYL)-	100		
94-58-6	1,3-BENZODIOXOLE, 5-PROPYL-	10		
65-85-0	BENZOIC ACID	5000		
98-07-7	BENZOIC TRICHLORIDE (BENZOTRICHLORIDE)	10	100	d
100-47-0	BENZONITRILE	5000		
50-32-8	3,4-BENZOPYRENE	1		
106-51-4	P-BENZOQUINONE	10		
98-07-7	BENZOTRICHLORIDE	10	100	d
98-88-4	BENZOYL CHLORIDE	1000		
56-55-3	BENZO[A]ANTHRACENE	10		
50-32-8	BENZO[A]PYRENE	1		
205-99-2	BENZO[B]FLUORANTHENE	1		
191-24-2	BENZO[GHI]PERYLENE	5000		
206-44-0	BENZO[J.K.]FLUORENE	100		
207-08-9	BENZO[K]FLUORANTHENE	5000		
189-55-9	BENZO[RST]PENTAPHENE	10		
218-01-9	1,2-BENZPHENANTHRENE	100		
100-44-7	BENZYL CHLORIDE	100	500	d
56-55-3	BENZ[A]ANTHRACENE	10		
225-51-4	BENZ[C]ACRIDINE	100		
56-49-5	BENZ[J]ACEANTHRYLENE, 1,2-DIHYDRO-3-METHYL	10		
7440-41-7	BERYLIUM DUST	10		
7440-41-7	BERYLLIUM	10		
7787-47-5	BERYLLIUM CHLORIDE	1		
--	BERYLLIUM COMPOUNDS			
7440-41-7	BERYLLIUM DUST	10		
7787-49-7	BERYLLIUM FLUORIDE	1		
7787-55-5	BERYLLIUM NITRATE	1		
13597-99-4	BERYLLIUM NITRATE	1		
319-84-6	ALPHA-BHC	10		
319-85-7	BETA-BHC	1		
319-86-8	DELTA-BHC	1		
58-89-9	GAMMA-BHC	1		
1464-53-5	2,2'-BIOXIRANE	10		
92-87-5	(1,1'-BIPHENYL)-4,4'DIAMINE	1		

Chemical Hazard Communication Guidebook

CAS	CHEMICAL NAME	RQ (LBS)	TPQ (LBS)	NOTES
119-90-4	(1,1'-BIPHENYL)-4,4'DIAMINE,3,3'DIMETHOXY-	100		
119-93-7	(1,1'-BIPHENYL)-4,4'DIAMINE,3,3'DIMETHYL-	10		
91-94-1	[1,1'-BIPHENYL]-4,4'-DIAMINE, 3,3'-DICHLORO-	1		
111-91-1	BIS(2-CHLOROETHOXY) METHANE	1000		
111-44-4	BIS(2-CHLOROETHYL) ETHER	10		
108-60-1	BIS(2-CHLOROISOPROPYL) ETHER	1000		
117-81-7	BIS(2-ETHYLHEXYL) PHTHALATE	100		
542-88-1	BIS(CHLOROMETHYL) ETHER	10		
137-26-8	BIS(DIMETHYLTHIOCARBAMOYL) DISULFIDE	10		
506-68-3	BROMINE CYANIDE	1000		
598-31-2	BROMOACETONE	1000		
75-25-2	BROMOFORM	100		
101-55-3	4-BROMOPHENYL PHENYL ETHER	100		
357-57-3	BRUCINE	100		
87-68-3	1,3-BUTADIENE, 1,1,2,3,4,4-HEXACHLORO-	1		
924-16-3	1-BUTANAMINE, N-BUTYL-N-NITROSO-	10		
305-03-3	BUTANOIC ACID,4-[BIS(2-CHLOROETHYL)-AMINO]BENZENE-	10		
71-36-3	1-BUTANOL	5000		
78-93-3	2-BUTANONE	5000		
1338-23-4	2-BUTANONE PEROXIDE	10		
123-73-9	2-BUTENAL	100		
4170-30-3	2-BUTENAL	100		
764-41-0	2-BUTENE, 1,4-DICHLORO-	1		
303-34-4	2-BUTENOIC ACID, 2-METHYL-, 7-[[2,3-DIHYDROXY-2-(1-METHOXYETHYL)-3-METHY	10		
123-86-4	BUTYL ACETATE	5000		
110-19-0	ISO-BUTYL ACETATE	5000		
105-46-4	SEC-BUTYL ACETATE	5000		
540-88-5	TERT-BUTYL ACETATE	5000		
71-36-3	N-BUTYL ALCOHOL	5000		
85-68-7	BUTYL BENZYL PHTHALATE	100		
84-74-2	N-BUTYL PHTHALATE	10		
109-73-9	BUTYLAMINE	1000		
78-81-9	ISO-BUTYLAMINE	1000		
513-49-5	SEC-BUTYLAMINE	1000		
13952-84-6	SEC-BUTYLAMINE	1000		
75-64-9	TERT-BUTYLAMINE	1000		
107-92-6	BUTYRIC ACID	5000		
79-31-2	ISO-BUTYRIC ACID	5000		
75-60-5	CACODYLIC ACID	1		
7440-43-9	CADMIUM	10		
543-90-8	CADMIUM ACETATE	10		
7789-42-6	CADMIUM BROMIDE	10		
10108-64-2	CADMIUM CHLORIDE	10		
--	CADMIUM COMPOUNDS			
7778-44-1	CALCIUM ARSENATE	1	500/10000	d

Appendix F: Hazardous Substances (Alphabetical Order)

CAS	CHEMICAL NAME	RQ (LBS)	TPQ (LBS)	NOTES
75-20-7	CALCIUM CARBIDE	10		
13765-19-0	CALCIUM CHROMATE	10		
592-01-8	CALCIUM CYANIDE	10		
26264-C0-2	CALCIUM DODECYLBENZENE SULFONATE	1000		
7778-54-3	CALCIUM HYPOCHLORITE	10		
8001-35-2	CAMPHENE, OCTACHLORO-	1		
133-06-2	CAPTAN	10		
133-06-2	CAPTAN [1H-ISOINDOLE-1,3(2H)-DIONE,3A,4,7,7A-TETRAHYDRO-2-[(TRICHLOROM	10		
51-79-6	CARBAMIC ACID, ETHYL ESTER	100		
615-53-2	CARBAMIC ACID, METHYLNITROSO-, ETHYL ESTER	1		
759-73-9	CARBAMIDE, N-ETHYL-N-NITROSO-	1		
684-93-5	CARBAMIDE, N-METHYL-N-NITROSO-	1		
62-56-6	CARBAMIDE, THIO-	10		
630-10-4	CARBAMIMIDOSELENOIC ACID	1000		
79-44-7	CARBAMOYL CHLORIDE, DIMETHYL-	1		
63-25-2	CARBARYL	100		
63-25-2	CARBARYL [1-NAPHTHALENOL, METHYLCARBAMATE]	100		
1563-66-2	CARBOFURAN	10	10/10000	
75-15-0	CARBON BISULFIDE	100		
75-15-0	CARBON DISULFIDE	100	10000	l
353-50-4	CARBON OXYFLUORIDE	1000		
D019	CARBON TETRACHLORIDE	10		
56-23-5	CARBON TETRACHLORIDE	10		
6533-73-9	CARBONIC ACID DITHALLIUM (I) SALT	100	100/10000	c,h
79-22-1	CARBONOCHLORIDIC ACID, METHYL ESTER	1000		
75-44-5	CARBONYL CHLORIDE	10		
353-50-4	CARBONYL FLUORIDE	1000		
75-87-6	CHLORAL	5000		
305-03-3	CHLORAMBUCIL	10		
D020	CHLORDANE	1		
57-74-9	CHLORDANE	1	1000	d
57-74-9	CHLORDANE (ALPHA AND GAMMA ISOMERS)	1		
--	CHLORDANE (TECHNICAL MIXTURE AND METABOLITES)			
57-74-9	CHLORDANE TECHNICAL	1		
57-74-9	CHLORDANE [4,7-METHANOINDAN,1,2,4,5,6,7,8,8-OCTACHLORO-2,3,3A,4,7,7A-H	1	1000	d
--	CHLORINATED BENZENES			
--	CHLORINATED ETHANES			
--	CHLORINATED NAPHTHALENE			
--	CHLORINATED PHENOLS			
7782-50-5	CHLORINE	10	100	
506-77-4	CHLORINE CYANIDE	10		
494-03-1	CHLORNAPHAZINE	100		
106-89-8	1-CHLORO-2,3-EPOXYPROPANE	100		
59-50-7	P-CHLORO-M-CRESOL	5000		

Chemical Hazard Communication Guidebook

CAS	CHEMICAL NAME	RQ (LBS)	TPQ (LBS)	NOTES
3165-93-3	4-CHLORO-O-TOLUIDINE HYDROCHLORIDE	100		
107-20-0	CHLOROACETALDEHYDE	1000		
--	CHLOROALKYL ETHERS			
106-47-8	P-CHLOROANILINE	1000		
D021	CHLOROBENZENE	100		
108-90-7	CHLOROBENZENE	100		
124-48-1	CHLORODIBROMOMETHANE	100		
75-00-3	CHLOROETHANE	100		
110-75-8	2-CHLOROETHYL VINYL ETHER	1000		
D022	CHLOROFORM	10		
67-66-3	CHLOROFORM	10	10000	d,l
107-30-2	CHLOROMETHYL METHYL ETHER	10	100	c,d
91-58-7	BETA-CHLORONAPHTHALENE	5000		
95-57-8	2-CHLOROPHENOL	100		
7005-72-3	4-CHLOROPHENYL PHENYL ETHER	5000		
5344-82-1	1-(O-CHLOROPHENYL)THIOUREA	100		
542-76-7	3-CHLOROPROPIONITRILE	1000		
7790-94-5	CHLOROSULFONIC ACID	1000		
2921-88-2	CHLORPYRIFOS	1		
1066-30-4	CHROMIC ACETATE	1000		
7738-94-5	CHROMIC ACID	10		
11115-74-5	CHROMIC ACID	10		
13765-19-0	CHROMIC ACID, AS /H2CRO4/, CALCIUM SALT (1:1)	10		
13765-19-0	CHROMIC ACID, CALCIUM SALT	10		
10101-53-8	CHROMIC SULFATE	1000		
7440-47-3	CHROMIUM	5000		
--	CHROMIUM COMPOUNDS			
10049-05-5	CHROMOUS CHLORIDE	1000		
218-01-9	CHRYSENE	100		
7789-43-7	COBALTOUS BROMIDE	1000		
544-18-3	COBALTOUS FORMATE	1000		
14017-41-5	COBALTOUS SULFAMATE	1000		
--	COKE OVEN EMISSIONS	1		
7440-50-8	COPPER	5000		
--	COPPER COMPOUNDS			
544-92-3	COPPER CYANIDE	10		
56-72-4	COUMAPHOS	10	100/10000	
8001-58-9	CREOSOTE	1		
D026	CRESOL	1000		
D024	M-CRESOL	1000		
108-39-4	M-CRESOL	1000		
D023	O-CRESOL	1000		
95-48-7	O-CRESOL	1000	1000/10000	d
D025	P-CRESOL	1000		
106-44-5	P-CRESOL	1000		
1319-77-3	CRESOL(S)	1000		
1319-77-3	CRESYLIC ACID	1000		

Appendix F: Hazardous Substances (Alphabetical Order)

CAS	CHEMICAL NAME	RQ (LBS)	TPQ (LBS)	NOTES
108-39-4	M-CRESYLIC ACID	1000		
95-48-7	O-CRESYLIC ACID	1000		
106-44-5	P-CRESYLIC ACID	1000		
4170-30-3	CROTONALDEHYDE	100	1000	
123-73-9	CROTONALDEHYDE (E)-	100	1000	
98-82-8	CUMENE	5000		
142-71-2	CUPRIC ACETATE	100		
12002-03-8	CUPRIC ACETOARSENITE	1		
7447-39-4	CUPRIC CHLORIDE	10		
3251-23-8	CUPRIC NITRATE	100		
5893-66-3	CUPRIC OXALATE	100		
7758-98-7	CUPRIC SULFATE	10		
10380-29-7	CUPRIC SULFATE AMMONIATED	100		
815-82-7	CUPRIC TARTRATE	100		
--	CYANIDES			
151-50-8	CYANIDES			
57-12-5	CYANIDES (SOLUBLE CYANIDE SALTS), NOT ELSE-WHERE SPECIFIED	10		
57-12-5	CYANIDES, AS /CN/			
460-19-5	CYANOGEN	100		
506-68-3	CYANOGEN BROMIDE	1000	500/10000	
506-77-4	CYANOGEN CHLORIDE	10		
106-51-4	1,4-CYCLOHEXADIENEDIONE	10		
110-82-7	CYCLOHEXANE	1000		
108-94-1	CYCLOHEXANONE	5000		
77-47-4	1,3-CYCLOPENTADIENE, 1,2,3,4,5,5-HEXACHLORO-	10		
50-18-0	CYCLOPHOSPHAMIDE	10		
94-75-7	2,4-D ACID	100		
94-11-1	2,4-D ESTERS	100		
94-79-1	2,4-D ESTERS	100		
94-80-4	2,4-D ESTERS	100		
1320-18-9	2,4-D ESTERS	100		
1928-38-7	2,4-D ESTERS	100		
1928-61-6	2,4-D ESTERS	100		
1929-73-3	2,4-D ESTERS	100		
2971-38-2	2,4-D ESTERS	100		
25168-26-7	2,4-D ESTERS	100		
53467-11-1	2,4-D ESTERS	100		
94-75-7	2,4-D SALTS AND ESTERS	100		
20830-81-3	DAUNOMYCIN	10		
72-54-8	DDD	1		
72-54-8	4,4'-DDD	1		
72-55-9	4,4'-DDE	1		
50-29-3	DDT	1		
--	DDT AND METABOLITES			
143-50-0	DECACHLOROOCTAHYDRO-1,3,4-METHENO-2H-CYCLOBUTA[C,D]-PENTALEN-2-ONE	1		

Chemical Hazard Communication Guidebook

CAS	CHEMICAL NAME	RQ (LBS)	TPQ (LBS)	NOTES
84-74-2	DI-N-BUTYL PHTHALATE	10		
117-84-0	DI-N-OCTYL PHTHALATE	5000		
621-64-7	DI-N-PROPYLNITROSAMINE	10		
2303-16-4	DIALLATE	100		
302-01-2	DIAMINE	1		
496-72-0	DIAMINOTOLUENE	10		
823-40-5	DIAMINOTOLUENE	10		
25376-45-8	DIAMINOTOLUENE	10		
5333-41-5	DIAZINON	1		
53-70-3	1,2:5,6-DIBENZANTHRACENE	1		
189-55-9	1,2:7,8-DIBENZOPYRENE	10		
53-70-3	DIBENZO[A,H]ANTHRACENE	1		
189-55-9	DIBENZO[A,I]PYRENE	10		
53-70-3	DIBENZ[A,H]ANTHRACENE	1		
189-55-9	DIBENZ[A,I]PYRENE	10		
96-12-8	1,2-DIBROMO-3-CHLOROPROPANE	1		
106-93-4	1,2-DIBROMOETHANE	1		
84-74-2	DIBUTYL PHTHALATE	10		
1918-00-9	DICAMBA	1000		
1194-65-6	DICHLOBENIL	100		
117-80-6	DICHLONE	1		
50-29-3	DICHLORO DIPHENYL TRICHLOROETHANE	1		
764-41-0	1,4-DICHLORO-2-BUTENE	1		
23950-58-5	3,5-DICHLORO-N-(1,1-DIMETHYL-2-PRO-PYNYL)BENZAMIDE	5000		
2303-16-4	S-(2,3-DICHLOROALLYL) DIISOPROPYLTHIOCARBA-MATE	100		
95-50-1	1,2-DICHLOROBENZENE	100		
541-73-1	1,3-DICHLOROBENZENE	100		
D027	1,4-DICHLOROBENZENE	100		
106-46-7	P-DICHLOROBENZENE	100		
25321-22-6	DICHLOROBENZENE (MIXED)	100		
91-94-1	3,3'-DICHLOROBENZIDENE	1		
--	DICHLOROBENZIDINE			
75-27-4	DICHLOROBROMOMETHANE	5000		
75-71-8	DICHLORODIFLUOROMETHANE	5000		
72-54-8	DICHLORODIPHENYL DICHLOROETHANE	1		
75-34-3	1,1-DICHLOROETHANE	1000		
D028	1,2-DICHLOROETHANE	100		
107-06-2	1,2-DICHLOROETHANE	100		
111-44-4	DICHLOROETHYL ETHER	10	10000	d
D029	1,1-DICHLOROETHYLENE	100		
75-35-4	1,1-DICHLOROETHYLENE	100		
156-60-5	1,2-TRANS-DICHLOROETHYLENE	1000		
75-09-2	DICHLOROMETHANE	1000		
120-83-2	2,4-DICHLOROPHENOL	100		
87-65-0	2,6-DICHLOROPHENOL	100		

Appendix F: Hazardous Substances (Alphabetical Order)

CAS	CHEMICAL NAME	RQ (LBS)	TPQ (LBS)	NOTES
94-75-7	2,4-DICHLOROPHENOXYACETIC ACID, SALTS AND ESTERS	100		
696-28-6	DICHLOROPHENYLARSINE	1		
26638-19-7	DICHLOROPROPANE	1000		
78-99-9	1,1-DICHLOROPROPANE	1000		
78-87-5	1,2-DICHLOROPROPANE	1000		
142-28-9	1,3-DICHLOROPROPANE	1000		
8003-19-8	DICHLOROPROPANE - DICHLOROPROPENE (MIXTURE)	100		
26952-23-8	DICHLOROPROPENE	100		
78-88-6	2,3-DICHLOROPROPENE (ISOMER)	100		
26952-23-8	DICHLOROPROPENE, N.O.S.	100		
75-99-0	2,2-DICHLOROPROPIONIC ACID	5000		
62-73-7	DICHLORVOS	10	1000	
62-73-7	DICHLORVOS [PHOSPHORIC ACID, 2,2-DICHLOROETHENYL DIMETHYL ESTER]	10	1000	
115-32-2	DICOFOL	10		
60-57-1	DIELDRIN	1		
1464-53-5	DIEPOXYBUTANE	10	500	d
1464-53-5	1,2:3,4-DIEPOXYBUTANE	10		
297-97-2	O,O-DIETHYL O-PYRAZINYL PHOSPHOROTHIOATE	100		
84-66-2	DIETHYL PHTHALATE	1000		
3288-58-2	O,O-DIETHYL S-METHYL DITHIOPHOSPHATE	5000		
298-04-4	O,O-DIETHYL S-[2-(ETHYLTHIO)ETHYL]PHOSPHORODITHIOATE	1		
311-45-5	DIETHYL-P-NITROPHENYL PHOSPHATE	100		
109-89-7	DIETHYLAMINE	100		
692-42-2	DIETHYLARSINE	1		
123-91-1	1,4-DIETHYLENE DIOXIDE	100		
1615-80-1	N,N'-DIETHYLHYDRAZINE	10		
56-53-1	DIETHYLSTILBESTEROL	1		
123-33-1	1,2-DIHYDRO-3,6-PYRIDAZINEDIONE	5000		
94-58-6	DIHYDROSAFROLE	10		
55-91-4	DIISOPROPYL FLUOROPHOSPHATE	100		
72-20-8	2,7:3,6-DIMETHANONAPHTH[2,3-B]OXIRENE, 3,4,5,6,9,9-HEXACHLORO-1A,2,2A,3,6,6A,7	1		
60-51-5	DIMETHOATE	10	500/10000	
119-90-4	3,3'-DIMETHOXYBENZIDINE	100		
39196-18-4	3,3-DIMETHYL-1-(METHYLTHIO)-2-BUTANONE,O-[(METHYLAMINO)CARBONYL] OXIME	100		
298-00-0	O,O-DIMETHYL O-P-NITROPHENYL PHOSPHOROTHIOATE	100		
122-09-8	ALPHA,ALPHA-DIMETHYL PHENETHYLAMINE	5000		
131-11-3	DIMETHYL PHTHALATE	5000		
77-78-1	DIMETHYL SULFATE	100	500	d
124-40-3	DIMETHYLAMINE	1000		
119-93-7	3,3'-DIMETHYLBENZIDINE	10		
80-15-9	ALPHA,ALPHA-DIMETHYLBENZYLHYDROPEROXIDE	10		
57-97-6	7,12-DIMETHYLBENZ[A]ANTHRACENE	1		

Chemical Hazard Communication Guidebook

CAS	CHEMICAL NAME	RQ (LBS)	TPQ (LBS)	NOTES
79-44-7	DIMETHYLCARBOMOYL CHLORIDE	1		
57-14-7	DIMETHYLHYDRAZINE	10	1000	d
57-14-7	1,1-DIMETHYLHYDRAZINE	10		
540-73-8	1,2-DIMETHYLHYDRAZINE	1		
62-75-9	DIMETHYLNITROSAMINE	10		
105-67-9	2,4-DIMETHYLPHENOL	100		
534-52-1	4,6-DINITRO-O-CRESOL AND SALTS	10		
131-89-5	4,6-DINITRO-O-CYCLOHEXYLPHENOL	100		
99-65-0	M-DINITROBENZENE	100		
528-29-0	O-DINITROBENZENE	100		
100-25-4	P-DINITROBENZENE	100		
25154-54-5	DINITROBENZENE (MIXED)	100		
25550-58-7	DINITROPHENOL	10		
51-28-5	2,4-DINITROPHENOL	10		
329-71-5	2,5-DINITROPHENOL	10		
573-56-8	2,6-DINITROPHENOL	10		
25321-14-6	DINITROTOLUENE	10		
D030	2,4-DINITROTOLUENE	10		
121-14-2	2,4-DINITROTOLUENE	10		
606-20-2	2,6-DINITROTOLUENE	100		
610-39-9	3,4-DINITROTOLUENE	10		
88-85-7	DINOSEB	1000	100/10000	
123-91-1	1,4-DIOXANE	100		
122-66-7	1,2-DIPHENYL HYDRAZINE (HYDRAZOBENZENE)	10		
--	DIPHENYLHYDRAZINE			
152-16-9	DIPHOSPHORAMIDE, OCTAMETHYL-	100	100	
142-84-7	DIPROPYLAMINE	5000		
85-00-7	DIQUAT	1000		
2764-72-9	DIQUAT	1000		
298-04-4	DISULFOTON	1	500	
3689-24-5	DITHIOPYROPHOSPHORIC ACID, TETRAETHYL ESTER	100		
330-54-1	DIURON	100		
27176-87-0	DODECYLBENZENE-SULFONIC ACID	1000		
115-29-7	ENDOSULFAN	1	10/10000	
959-98-8	ALPHA-ENDOSULFAN	1		
33213-65-9	BETA-ENDOSULFAN	1		
--	ENDOSULFAN AND METABOLITES			
1031-07-8	ENDOSULFAN SULFATE	1		
145-73-3	ENDOTHALL	1000		
72-20-8	ENDRIN	1	500/10000	
7421-93-4	ENDRIN ALDEHYDE	1		
--	ENDRIN AND METABOLITES			
72-20-8	ENDRIN AND METABOLITES	1		
106-89-8	EPICHLOROHYDRIN	100	1000	d,l
51-43-4	EPINEPHRINE	1000		
75-07-0	ETHANAL	1000		

Appendix F: Hazardous Substances (Alphabetical Order)

CAS	CHEMICAL NAME	RQ (LBS)	TPQ (LBS)	NOTES
122-09-8	ETHANAMINE, 1,1-DIMETHYL-2-PHENYL	5000		
55-18-5	ETHANAMINE, N-ETHYL-N-NITROSO	1		
630-20-6	ETHANE 1,1,1,2-TETRACHLORO-	100		
72-43-5	ETHANE 1,1,1-TRICHLORO-2,2-BIS-(P-METHOXYPHENYL)-	1		
75-34-3	ETHANE 1,1-DICHLORO-	1000		
76-01-7	ETHANE PENTACHLORO-	10		
111-44-4	ETHANE, 1,1'-OXYBIS(2-CHLORO-	10		
60-29-7	ETHANE, 1,1'-OXYBIS-	100		
111-91-1	ETHANE, 1,1'-[METHYLENEBIS(OXY)]BIS(2-CHLORO-	1000		
67-72-1	ETHANE, 1,1,1,2,2,2-HEXACHLORO-	100		
79-34-5	ETHANE, 1,1,2,2-TETRACHLORO-	100		
79-00-5	ETHANE, 1,1,2-TRICHLORO-	100		
106-93-4	ETHANE, 1,2-DIBROMO	1		
107-06-2	ETHANE, 1,2-DICHLORO-	100		
111-54-6	1,2-ETHANEDIYLBISCARBAMODITHIOIC ACID	5000		
75-05-8	ETHANENITRILE	5000		
62-55-5	ETHANETHIOAMIDE	10		
1116-54-7	ETHANOL, 2,2'-(NITROSOIMINO)BIS	1		
98-86-2	ETHANONE, 1-PHENYL-	5000		
75-36-5	ETHANOYL CHLORIDE	5000		
4549-40-0	ETHENAMINE, N-METHYL-N-NITROSO-	10		
75-01-4	ETHENE CHLORO-	1		
75-35-4	ETHENE, 1,1 DICHLORO-	100		
127-18-4	ETHENE, 1,1,2,2-TETRACHLORO-	100		
110-75-8	ETHENE, 2-CHLOROETHOXY-	1000		
156-60-5	ETHENE, TRANS-1,2,DICHLORO-	1000		
563-12-2	ETHION	10	1000	
110-80-5	2-ETHOXYETHANOL	1000		
510-15-6	ETHYL 4,4'-DICHLOROBENZILATE	10		
141-78-6	ETHYL ACETATE	5000		
140-88-5	ETHYL ACRYLATE	1000		
100-41-4	ETHYL BENZENE	1000		
51-79-6	ETHYL CARBAMATE	100		
107-12-0	ETHYL CYANIDE	10		
60-29-7	ETHYL ETHER	100		
97-63-2	ETHYL METHACRYLATE	1000		
62-50-0	ETHYL METHANESULFONATE	1		
107-15-3	ETHYLENE DIAMINE	5000	10000	
106-93-4	ETHYLENE DIBROMIDE	1		
107-06-2	ETHYLENE DICHLORIDE	100		
110-80-5	ETHYLENE GLYCOL MONOETHYL ETHER	1000		
75-21-8	ETHYLENE OXIDE	10	1000	d,l
96-45-7	ETHYLENE THIOUREA	10		
111-54-6	ETHYLENEBIS(DITHIOCARBAMIC ACID)	5000		
60-00-4	ETHYLENEDIAMINE TETRAACETIC ACID	5000		
151-56-4	ETHYLENIMINE	1		

Chemical Hazard Communication Guidebook

CAS	CHEMICAL NAME	RQ (LBS)	TPQ (LBS)	NOTES
75-34-3	ETHYLIDENE DICHLORIDE	1000		
52-85-7	FAMPHUR	1000		
1185-57-5	FERRIC AMMONIUM CITRATE	1000		
2944-67-4	FERRIC AMMONIUM OXALATE	1000		
55488-87-4	FERRIC AMMONIUM OXALATE	1000		
7705-08-0	FERRIC CHLORIDE	1000		
9004-66-4	FERRIC DEXTRAN	5000		
7783-50-8	FERRIC FLUORIDE	100		
10421-48-4	FERRIC NITRATE	1000		
10028-22-5	FERRIC SULFATE	1000		
10045-89-3	FERROUS AMMONIUM SULFATE	1000		
7758-94-3	FERROUS CHLORIDE	100		
7720-78-7	FERROUS SULFATE	1000		
7782-63-0	FERROUS SULFATE	1000		
206-44-0	FLUORANTHENE	100		
86-73-7	FLUORENE	5000		
7782-41-4	FLUORINE	10	500	k
640-19-7	FLUOROACETAMIDE	100	100/10000	j
62-74-8	FLUOROACETIC ACID, SODIUM SALT	10		
50-00-0	FORMALDEHYDE	100	500	d,l
64-18-6	FORMIC ACID	5000		
628-86-4	FULMINIC ACID, MERCURY(II) SALT	10		
110-17-8	FUMARIC ACID	5000		
110-00-9	FURAN	100	500	
109-99-9	FURAN, TETRAHYDRO-	1000		
98-01-1	2-FURANCARBOXALDEHYDE	5000		
108-31-6	2,5-FURANDIONE	5000		
98-01-1	FURFURAL	5000		
110-00-9	FURFURAN	100		
18883-66-4	D-GLUCOPYRANOSE, 2-DEOXY-2-(3-METHYL-3-NITROSOUREIDO)-	1		
765-34-4	GLYCIDYLALDEHYDE	10		
70-25-7	GUANIDINE, N-NITROSO-N-METHYL-N'-NITRO-	10		
86-50-0	GUTHION	1		
--	HALOETHERS			
--	HALOMETHANES			
76-44-8	HEPTACHLOR	1		
D031	HEPTACHLOR (AND ITS HYDROXIDE)	1		
--	HEPTACHLOR AND METABOLITES			
1024-57-3	HEPTACHLOR EPOXIDE	1		
76-44-8	HEPTACHLOR [1,4,5,6,7,8,8-HEPTACHLORO-3A,4,7,7A-TETRAHYDRO-4,7-METHANO	1		
60-57-1	1,2,3,4,10,10-HEXACHLORO- 6,7-EPOXY-1,4,4A,5,6,7,8,8A-OCTAHYDRO-ENDO,EXO-1,4:5, 8-D	1		
309-00-2	1,2,3,4,10,10-HEXACHLORO-1,4,4A,5,8,8A-HEXAHYDRO-1,4,5,8-ENDO,EXO-DIMETHANONAPHTH.	1		

Appendix F: Hazardous Substances (Alphabetical Order)

CAS	CHEMICAL NAME	RQ (LBS)	TPQ (LBS)	NOTES
465-73-6	1,2,3,4,10,10-HEXACHLORO-1,4,4A,5,8,8A-HEXAHYDRO-1,4,5,8-ENDO,ENDO-DIMETHANONAPHT ...	1		
D032	HEXACHLOROBENZENE	10		
118-74-1	HEXACHLOROBENZENE	10		
D033	HEXACHLOROBUTADIENE	1		
87-68-3	HEXACHLOROBUTADIENE	1		
58-89-9	HEXACHLOROCYCLOHEXANE (GAMMA ISOMER)	1		
608-73-1	HEXACHLOROCYCLOHEXANE (ALL ISOMERS)			
77-47-4	HEXACHLOROCYCLOPENTADIENE	10	100	d,h
D034	HEXACHLOROETHANE	100		
67-72-1	HEXACHLOROETHANE	100		
465-73-6	HEXACHLOROHEXAHYDRO-ENDO,ENDO-DIMETHANONAPHTHALENE	1		
70-30-4	HEXACHLOROPHENE	100		
1888-71-7	HEXACHLOROPROPENE	1000		
757-58-4	HEXAETHYL TETRAPHOSPHATE	100		
302-01-2	HYDRAZINE	1	1000	d
1615-80-1	HYDRAZINE 1,2-DIETHYL-	10		
57-14-7	HYDRAZINE, 1,1-DIMETHYL	10		
540-73-8	HYDRAZINE, 1,2-DIMETHYL	1		
122-66-7	HYDRAZINE, 1,2-DIPHENYL	10		
60-34-4	HYDRAZINE, METHYL	10		
79-19-6	HYDRAZINECARBOTHIOAMIDE	100		
7647-01-0	HYDROCHLORIC ACID	5000		
74-90-8	HYDROCYANIC ACID	10	100	
7664-39-3	HYDROFLUORIC ACID	100		
74-90-8	HYDROGEN CYANIDE	10		
7664-39-3	HYDROGEN FLUORIDE	100	100	
7803-51-2	HYDROGEN PHOSPHIDE	100		
7783-06-4	HYDROGEN SULFIDE	100	500	l
80-15-9	HYDROPEROXIDE, 1-METHYL-1-PHENYLETHYL-	10		
7783-06-4	HYDROSULFURIC ACID	100		
75-60-5	HYDROXYDIMETHYLARSINE OXIDE	1		
96-45-7	2-IMIDAZOLIDINETHIONE	10		
193-39-5	INDENO [1,2,3-CD]PYRENE	100		
78-83-1	ISOBUTYL ALCOHOL	5000		
624-83-9	ISOCYANIC ACID, METHYL ESTER	1		
78-59-1	ISOPHORONE	5000		
78-79-5	ISOPRENE	100		
42504-46-1	ISOPROPANOLAMINE DODECYLBENZENESULFONATE	1000		
120-58-1	ISOSAFROLE	100		
2763-96-4	3(2H)-ISOXAZOLONE, 5-(AMINOMETHYL)-	1000		
143-50-0	KEPONE	1		
303-34-4	LASIOCARPINE	10		
7439-92-1	LEAD	1		
301-04-2	LEAD ACETATE	5000		

Chemical Hazard Communication Guidebook

CAS	CHEMICAL NAME	RQ (LBS)	TPQ (LBS)	NOTES
7645-25-2	LEAD ARSENATE	1		
7784-40-9	LEAD ARSENATE	1		
10102-48-4	LEAD ARSENATE	1		
7758-95-4	LEAD CHLORIDE	100		
--	LEAD COMPOUNDS			
13814-96-5	LEAD FLUOBORATE	100		
7783-46-2	LEAD FLUORIDE	100		
10101-63-0	LEAD IODIDE	100		
10099-74-8	LEAD NITRATE	100		
7446-27-7	LEAD PHOSPHATE	1		
1072-35-1	LEAD STEARATE	5000		
7428-48-0	LEAD STEARATE	5000		
52652-59-2	LEAD STEARATE	5000		
56189-09-4	LEAD STEARATE	5000		
1335-32-6	LEAD SUBACETATE	100		
7446-14-2	LEAD SULFATE	100		
15739-80-7	LEAD SULFATE	100		
1314-87-0	LEAD SULFIDE	5000		
592-87-0	LEAD THIOCYANATE	100		
1335-32-6	LEAD, BIS(ACETATO-O)TETRAHYDROXYTRI-	100		
58-89-9	LINDANE	1	1000/10000	d
58-89-9	LINDANE [CYCLOHEXANE,1,2,3,4,5,6-HEXACHLORO-,(1.ALPHA.,2.ALPHA.,3.BETA	1	1000/10000	d
14307-35-8	LITHIUM CHROMATE	10		
121-75-5	MALATHION	100		
110-16-7	MALEIC ACID	5000		
108-31-6	MALEIC ANHYDRIDE	5000		
123-33-1	MALEIC HYDRAZIDE	5000		
109-77-3	MALONONITRILE	1000	500/10000	
148-82-3	MELPHALAN	1		
2032-65-7	MERCAPTODIMETHUR	10		
592-04-1	MERCURIC CYANIDE	1		
10045-94-0	MERCURIC NITRATE	10		
7783-35-9	MERCURIC SULFATE	10		
592-85-8	MERCURIC THIOCYANATE	10		
7782-86-7	MERCUROUS NITRATE	10		
10415-75-5	MERCUROUS NITRATE	10		
7439-97-6	MERCURY	1		
--	MERCURY COMPOUNDS			
628-86-4	MERCURY FULMINATE	10		
62-38-4	MERCURY, (ACETATO-O)PHENYL-	100		
126-98-7	METHACRYLONITRILE	1000	500	h
62-75-9	METHAMINE, N-METHYL-N-NITROSO-	10		
124-40-3	METHANAMINE, N-METHYL-	1000		
74-87-3	METHANE CHLORO-	100		
74-95-3	METHANE DIBROMO-	1000		
75-09-2	METHANE DICHLORO-	1000		

Appendix F: Hazardous Substances (Alphabetical Order)

CAS	CHEMICAL NAME	RQ (LBS)	TPQ (LBS)	NOTES
74-88-4	METHANE IODO-	100		
56-23-5	METHANE TETRACHLORO-	10		
509-14-8	METHANE TETRANITRO -	10		
75-25-2	METHANE TRIBROMO-	100		
67-66-3	METHANE TRICHLORO-	10		
75-69-4	METHANE TRICHLOROFLUORO-	5000		
74-83-9	METHANE, BROMO-	1000		
107-30-2	METHANE, CHLOROMETHOXY-	10		
75-71-8	METHANE, DICHLORODIFLUORO-	5000		
542-88-1	METHANE, OXYBIS(CHLORO-	10		
594-42-3	METHANESULFENYL CHLORIDE, TRICHLORO-	100		
62-50-0	METHANESULFONIC ACID, ETHYL ESTER	1		
74-93-1	METHANETHIOL	100		
76-44-8	4,7-METHANO-1H-INDENE, 1,4,5,6,7,8,8-HEPTACHLORO- 3A,3,7,7A-TETRAHYDRO-	1		
64-18-6	METHANOIC ACID	5000		
57-74-9	4,7-METHANOINDAN, 1,2,4,5,6,7,8,8-OCTACHLORO-3A,4,7,7A-TETRAHYDRO-	1		
67-56-1	METHANOL	5000		
91-80-5	METHAPYRILENE	5000		
16752-77-5	METHOMYL	100	500/10000	h
72-43-5	METHOXYCHLOR	1		
72-43-5	METHOXYCHLOR [BENZENE, 1,1'-(2,2,2-TRICHLOROETHYLIDENE)BIS[4-METHOXY-	1		
67-56-1	METHYL ALCOHOL	5000		
74-83-9	METHYL BROMIDE	1000	1000	l
504-60-9	1-METHYL BUTADIENE	100		
74-87-3	METHYL CHLORIDE	100		
79-22-1	METHYL CHLOROCARBONATE	1000		
71-55-6	METHYL CHLOROFORM	1000		
D035	METHYL ETHYL KETONE	5000		
78-93-3	METHYL ETHYL KETONE	5000		
1338-23-4	METHYL ETHYL KETONE PEROXIDE	10		
74-88-4	METHYL IODIDE	100		
108-10-1	METHYL ISOBUTYL KETONE	5000		
624-83-9	METHYL ISOCYANATE	1	500	f
74-93-1	METHYL MERCAPTAN	100	500	
80-62-6	METHYL METHACRYLATE	1000		
298-00-0	METHYL PARATHION	100		
108-10-1	4-METHYL-2-PENTANONE	5000		
70-25-7	N-METHYL-N-NITRO-N-NITROSOGUANIDINE	10		
75-55-8	2-METHYLAZIRIDINE	1		
56-49-5	3-METHYLCHOLANTHRENE	10		
101-14-4	4,4'-METHYLENE BIS(2-CHLOROANILINE) (MOCA)	10		
74-95-3	METHYLENE BROMIDE	1000		
75-09-2	METHYLENE CHLORIDE	1000		
50-00-0	METHYLENE OXIDE	100		
70-30-4	2,2'-METHYLENEBIS(3,4,6-TRICHLOROPHENOL)	100		

Chemical Hazard Communication Guidebook

CAS	CHEMICAL NAME	RQ (LBS)	TPQ (LBS)	NOTES
60-34-4	METHYLHYDRAZINE	10	500	
75-86-5	2-METHYLLACTONITRILE	10		
56-04-2	METHYLTHIOURACIL	10		
7786-34-7	MEVINPHOS	10	500	
315-18-4	MEXACARBATE	1000	500/10000	
50-07-7	MITOMYCIN C	10	500/10000	d
75-04-7	MONOETHYLAMINE	100		
74-89-5	MONOMETHYLAMINE	100		
300-76-5	NALED	10		
20830-81-3	5,12-NAPHTHACENEDIONE, (8S-CIS)- 8-ACETYL-10-(3-AMINO-2,3,6 TRIDEOXY-ALPHA-	10		
91-59-8	2-NAPHTHALENAMINE	10		
91-20-3	NAPHTHALENE	100		
91-58-7	NAPHTHALENE, 2-CHLORO-	5000		
130-15-4	1,4-NAPHTHALENEDIONE	5000		
72-57-1	2,7-NAPHTHALENEDISULFONIC ACID, 3,3'-[(3,3'-DIMETHYL-(1,1'-BIPHENYL)-4,4'-	10		
1338-24-5	NAPHTHENIC ACID	100		
130-15-4	1,4-NAPHTHOQUINONE	5000		
86-88-4	ALPHA-NAPHTHYL THIOUREA	100		
134-32-7	1-NAPHTHYLAMINE	100		
134-32-7	ALPHA-NAPHTHYLAMINE	100		
91-59-8	BETA-NAPHTHYLAMINE	10		
494-03-1	2-NAPHTHYLAMINE, N,N-BIS(2-CHLOROETHYL)-	100		
7440-02-0	NICKEL	100		d
15699-18-0	NICKEL AMMONIUM SULFATE	100		
13463-39-3	NICKEL CARBONYL	10	1	d
7718-54-9	NICKEL CHLORIDE	100		
37211-05-5	NICKEL CHLORIDE	100		
--	NICKEL COMPOUNDS			
557-19-7	NICKEL CYANIDE	10		
12054-48-7	NICKEL HYDROXIDE	10		
14216-75-2	NICKEL NITRATE	100		
7786-81-4	NICKEL SULFATE	100		
13463-39-3	NICKEL TETRACARBONYL	10		
557-19-7	NICKEL(II) CYANIDE	10		
54-11-5	NICOTINE AND SALTS	100		
7697-37-2	NITRIC ACID	1000	1000	
10102-43-9	NITRIC OXIDE	10	100	c
99-55-8	5-NITRO-O-TOLUIDINE	100		
100-01-6	P-NITROANILINE	5000		
D036	NITROBENZENE	1000		
98-95-3	NITROBENZENE	1000	10000	l
10102-43-9	NITROGEN (II) OXIDE	10		
10102-44-0	NITROGEN (IV) OXIDE	10		
10544-72-6	NITROGEN (IV) OXIDE	10		
10102-44-0	NITROGEN DIOXIDE	10	100	

Appendix F: Hazardous Substances (Alphabetical Order)

CAS	CHEMICAL NAME	RQ (LBS)	TPQ (LBS)	NOTES
10544-72-6	NITROGEN DIOXIDE	10		
55-63-0	NITROGLYCERINE	10		
88-75-5	2-NITROPHENOL	100		
100-02-7	4-NITROPHENOL	100		
25154-55-6	NITROPHENOL (MIXED)	100		
554-84-7	M-NITROPHENOL (MIXED)	100		
88-75-5	O-NITROPHENOL (MIXED)	100		
100-02-7	P-NITROPHENOL (MIXED)	100		
--	NITROPHENOLS			
79-46-9	2-NITROPROPANE	10		
--	NITROSAMINES			
759-73-9	N-NITROSO-N-ETHYLUREA	1		
684-93-5	N-NITROSO-N-METHYLUREA	1		
615-53-2	N-NITROSO-N-METHYLURETHANE	1		
924-16-3	N-NITROSODI-N-BUTYLAMINE	10		
621-64-7	N-NITROSODI-N-PROPYLAMINE	10		
1116-54-7	N-NITROSODIETHANOLAMINE	1		
55-18-5	N-NITROSODIETHYLAMINE	1		
62-75-9	N-NITROSODIMETHYLAMINE	10		
86-30-6	N-NITROSODIPHENYLAMINE	100		
4549-40-0	N-NITROSOMETHYLVINYLAMINE	10		
100-75-4	N-NITROSOPIPERIDINE	10		
930-55-2	N-NITROSOPYRROLIDINE	1		
1321-12-6	NITROTOLUENE	1000		
99-08-1	M-NITROTOLUENE	1000		
88-72-2	O-NITROTOLUENE	1000		
99-99-0	P-NITROTOLUENE	1000		
115-29-7	5-NORBORNENE-2,3-DIMETHANOL, 1,4,5,6,7,7-HEXACHLORO, CYCLIC SULFITE	1		
152-16-9	OCTAMETHYLPYROPHOSPHORAMIDE	100		
20816-12-0	OSMIUM OXIDE	1000		
20816-12-0	OSMIUM TETROXIDE	1000		
145-73-3	7-OXABICYCLO[2.2.1]HEPTANE-2,3-DICARBOXYLIC ACID	1000		
1120-71-4	1,2-OXATHIOLANE 2,2-DIOXIDE	10		
50-18-0	2H-1,3,2-OXAZAPHOSPHORINE,2-[BIS(2-CHLOROETHYL)AMINO] TETRAHYDRO-2-OXIDE	10		
75-21-8	OXIRANE	10		
106-89-8	OXIRANE, 2-(CHLOROMETHYL)-	100		
30525-89-4	PARAFORMALDEHYDE	1000		
123-63-7	PARALDEHYDE	1000		
56-38-2	PARATHION	10	100	c,d
56-38-2	PARATHION [PHOSPHOROTHIOIC ACID, 0,0-DI-ETHYL-0-(4-NITROPHENYL) ESTER]	10	100	c,d
608-93-5	PENTACHLOROBENZENE	10		
76-01-7	PENTACHLOROETHANE	10		d
82-68-8	PENTACHLORONITROBENZENE	100		
D037	PENTACHLOROPHENOL	10		

Chemical Hazard Communication Guidebook

CAS	CHEMICAL NAME	RQ (LBS)	TPQ (LBS)	NOTES
87-86-5	PENTACHLOROPHENOL	10		d
504-60-9	1,3-PENTADIENE	100		
62-44-2	PHENACETIN	100		
85-01-8	PHENANTHRENE	5000		
108-95-2	PHENOL	1000	500/10000	
131-89-5	PHENOL 2-CYCLOHEXYL-4-6-DINITRO-	100		
696-28-6	PHENOL DICHLOROARSINE	1		
58-90-2	PHENOL, 2,3,4,6-TETRACHLORO-	10		
95-95-4	PHENOL, 2,4,5-TRICHLORO-	10		
88-06-2	PHENOL, 2,4,6-TRICHLORO-	10		
131-74-8	PHENOL, 2,4,6-TRINITRO- AMMONIUM SALT	10		
120-83-2	PHENOL, 2,4-DICHLORO-	100		
105-67-9	PHENOL, 2,4-DIMETHYL-	100		
51-28-5	PHENOL, 2,4-DINITRO-	10		
88-85-7	PHENOL, 2,4-DINITRO-6-(1-METHYLPROPYL)	1000		
534-52-1	PHENOL, 2,4-DINITRO-6-METHYL- AND SALTS	10		
87-65-0	PHENOL, 2,6-DICHLORO-	100		
95-57-8	PHENOL, 2-CHLORO-	100		
59-50-7	PHENOL, 4-CHLORO-3-METHYL-	5000		
100-02-7	PHENOL, 4-NITRO	100		
87-86-5	PHENOL, PENTACHLORO-	10		
696-28-6	PHENYL DICHLOROARSINE	1	500	d,h
193-39-5	1,10-(1,2-PHENYLENE)PYRENE	100		
62-38-4	PHENYLMERCURIC ACETATE	100		
298-02-2	PHORATE	10	10	
75-44-5	PHOSGENE	10	10	l
7803-51-2	PHOSPHINE	100	500	
7664-38-2	PHOSPHORIC ACID	5000		
311-45-5	PHOSPHORIC ACID, DIETHYL P-NITROPHENYL ESTER	100		
7446-27-7	PHOSPHORIC ACID, LEAD SALT	1		
60-51-5	PHOSPHORODITHIOIC ACID,O,O-DIMETHYL S-[2(METHYLAMINO)-2-OXOETHYL]ESTER	10		
298-02-2	PHOSPHORODITHIOIC ACID,O,O-DIETHYL S-(ETHYLTHIO) METHYL ESTER	10		
3288-58-2	PHOSPHORODITHIOIC ACID,O,O-DIETHYL S-METHYLESTER	5000		
55-91-4	PHOSPHOROFLUORIDIC ACID, BIS(1-METHYLETHYL) ESTER	100		
52-85-7	PHOSPHOROTHIOIC ACID, O,O-DIMETHYL O-[P-[(DIMETHYLAMINO)-SULFONYL]PHEN	1000		
56-38-2	PHOSPHOROTHIOIC ACID, O,O-DIETHYL O-(P-NITROPHENYL) ESTER	10		
297-97-2	PHOSPHOROTHIOIC ACID, O,O-DIETHYL O-PYRAZINYL ESTER	100		
7723-14-0	PHOSPHORUS	1	100	b,h
10025-87-3	PHOSPHORUS OXYCHLORIDE	1000	500	d
1314-80-3	PHOSPHORUS PENTASULFIDE	100		
1314-80-3	PHOSPHORUS SULFIDE	100		

Appendix F: Hazardous Substances (Alphabetical Order)

CAS	CHEMICAL NAME	RQ (LBS)	TPQ (LBS)	NOTES
7719-12-2	PHOSPHORUS TRICHLORIDE	1000	1000	
--	PHTHALATE ESTERS			
85-44-9	PHTHALIC ANHYDRIDE	5000		
109-06-8	2-PICOLINE	5000		
78-00-2	PLUMBANE TETRAETHYL-	10		
1336-36-3	POLYCHLORINATED BIPHENYLS	1		
11096-82-5	POLYCHLORINATED BIPHENYLS	1		
11097-69-1	POLYCHLORINATED BIPHENYLS	1		
11104-28-2	POLYCHLORINATED BIPHENYLS	1		
11141-16-5	POLYCHLORINATED BIPHENYLS	1		
12672-29-6	POLYCHLORINATED BIPHENYLS	1		
12674-11-2	POLYCHLORINATED BIPHENYLS	1		
53469-21-9	POLYCHLORINATED BIPHENYLS	1		
--	POLYNUCLEAR AROMATIC HYDROCARBONS			
7784-41-0	POTASSIUM ARSENATE	1		
10124-50-2	POTASSIUM ARSENITE	1	500/10000	d
7778-50-9	POTASSIUM BICHROMATE	10		
7789-00-6	POTASSIUM CHROMATE	10		
151-50-8	POTASSIUM CYANIDE	10	100	b
1310-58-3	POTASSIUM HYDROXIDE	1000		
7722-64-7	POTASSIUM PERMANGANATE	100		
506-61-6	POTASSIUM SILVER CYANIDE	1	500	b
23950-58-5	PRONAMIDE	5000		
765-34-4	1-PROPANAL, 2,3-EPOXY-	10		
116-06-3	PROPANAL, 2-METHYL-2-(METHYLTHIO)-, O-[(METHYLAMINO)CARBONYL]-OXIME	1		
107-10-8	1-PROPANAMINE	5000		
142-84-7	1-PROPANAMINE, N-PROPYL	5000		
1120-71-4	1,3-PROPANE SULTONE	10		
96-12-8	PROPANE, 1,2-DIBROMO-3-CHLORO	1		
108-60-1	PROPANE, 2,2'-OXYBIS(2-CHLORO-	1000		
79-46-9	PROPANE, 2-NITRO-	10		
109-77-3	PROPANEDINITRILE	1000		
107-12-0	PROPANENITRILE	10		
75-86-5	PROPANENITRILE, 2-HYDROXY-2-METHYL-	10		
542-76-7	PROPANENITRILE, 3-CHLORO-	1000		
55-63-0	1,2,3-PROPANETRIOL, TRINITRATE-	10		
93-72-1	PROPANOIC ACID, 2(2,4,5-TRICHLOROPHENOXY)-	100		
126-72-7	1-PROPANOL, 2,3-DIBROMO-, PHOSPHATE (3:1)	10		
765-34-4	1-PROPANOL, 2,3-EPOXY-	10		
78-83-1	1-PROPANOL, 2-METHYL-	5000		
67-64-1	2-PROPANONE	5000		
598-31-2	2-PROPANONE, 1-BROMO-	1000		
2312-35-8	PROPARGITE	10		
107-19-7	PROPARGYL ALCOHOL	1000		
107-18-6	2-PROPEN-1-OL	100		
107-02-8	2-PROPENAL	1		

Chemical Hazard Communication Guidebook

CAS	CHEMICAL NAME	RQ (LBS)	TPQ (LBS)	NOTES
79-06-1	2-PROPENAMIDE	5000		
1888-71-7	1-PROPENE, 1,1,2,3,3,3-HEXACHLORO-	1000		
542-75-6	PROPENE, 1,3-DICHLORO	100		
107-13-1	2-PROPENENITRILE	100		
126-98-7	2-PROPENENITRILE, 2-METHYL	1000		
79-10-7	2-PROPENOIC ACID	5000		
80-62-6	2-PROPENOIC ACID, 2-METHYL-, METHYL ESTER	1000		
97-63-2	2-PROPENOIC ACID, 2-METHYL-, ETHYL ESTER	1000		
140-88-5	2-PROPENOIC ACID, ETHYL ESTER	1000		
79-09-4	PROPIONIC ACID	5000		
107-10-8	N-PROPYLAMINE	5000		
78-87-5	PROPYLENE DICHLORIDE	1000		
75-56-9	PROPYLENE OXIDE	100	10000	I
75-55-8	1,2-PROPYLENIMINE	1		
107-19-7	2-PROPYN-1-OL	1000		
129-00-0	PYRENE	5000	1000/10000	c
121-21-1	PYRETHRINS	1		
121-29-9	PYRETHRINS	1		
8003-34-7	PYRETHRINS	1		
504-24-5	4-PYRIDINAMINE	1000		
D038	PYRIDINE	1000		
110-86-1	PYRIDINE	1000		
54-11-5	PYRIDINE, (S)-3-(1-METHYL-2-PYRROLIDINYL)-, AND SALTS	100		
109-06-8	PYRIDINE, 2-METHYL-	5000		
91-80-5	PYRIDINE, 2-[(2-(DIMETHYLAMINO)ETHYL)-2-THE-NYLA...	5000		
100-75-4	PYRIDINE, HEXAHYDRO-N-NITROSO-	10		
66-75-1	2,4(1H,3H)-PYRIMIDINEDIONE, 5-[BIS(2-CHLOROETHYL)AMINO]-	10		
56-04-2	4(1H)-PYRIMIDINONE, 2,3-DIHYDRO-6-METHYL-2-THIOXO-	10		
107-49-3	PYROPHOSPHORIC ACID, TETRAETHYL ESTER	10		
930-55-2	PYRROLE, TETRAHYDRO-N-NITROSO-	1		
91-22-5	QUINOLINE	5000		
82-68-8	QUINTOZENE [PENTACHLORONITROBENZENE]	100		
--	RADIONUCLIDES	1		
50-55-5	RESERPINE	5000		
108-46-3	RESORCINOL	5000		
81-07-2	SACCHARIN AND SALTS	100		
94-59-7	SAFROLE	100		
7783-00-8	SELENIOUS ACID	10	1000/10000	
7782-49-2	SELENIUM	100		
--	SELENIUM COMPOUNDS			
7446-08-4	SELENIUM DIOXIDE	10		
7446-08-4	SELENIUM DIOXIDE, AS /SEO2/	10		
7488-56-4	SELENIUM DISULFIDE	10		
7488-56-4	SELENIUM SULFIDE	10		

Appendix F: Hazardous Substances (Alphabetical Order)

CAS	CHEMICAL NAME	RQ (LBS)	TPQ (LBS)	NOTES
630-10-4	SELENOUREA	1000		
115-02-6	L-SERINE, DIAZOACETATE (ESTER)	1		
7440-22-4	SILVER	1000		
--	SILVER COMPOUNDS			
506-64-9	SILVER CYANIDE	1		
7761-88-8	SILVER NITRATE	1		
93-72-1	SILVEX	100		
7440-23-5	SODIUM	10		
7631-89-2	SODIUM ARSENATE	1	1000/10000	d
7784-46-5	SODIUM ARSENITE	1	500/10000	d
26628-22-8	SODIUM AZIDE	1000	500	b
10588-01-9	SODIUM BICHROMATE	10		
1333-83-1	SODIUM BIFLUORIDE	100		
7631-90-5	SODIUM BISULFITE	5000		
7775-11-3	SODIUM CHROMATE	10		
143-33-9	SODIUM CYANIDE	10	100	b
25155-30-0	SODIUM DODECYLBENZENE SULFONATE	1000		
7681-49-4	SODIUM FLUORIDE	1000		
16721-80-5	SODIUM HYDROSULFIDE	5000		
1310-73-2	SODIUM HYDROXIDE	1000		
7681-52-9	SODIUM HYPOCHLORITE	100		
10022-70-5	SODIUM HYPOCHLORITE	100		
124-41-4	SODIUM METHYLATE	1000		
7632-00-0	SODIUM NITRITE	100		
7558-79-4	SODIUM PHOSPHATE DIBASIC	5000		
10039-32-4	SODIUM PHOSPHATE DIBASIC	5000		
7601-54-9	SODIUM PHOSPHATE TRIBASIC	5000		
7758-29-4	SODIUM PHOSPHATE TRIBASIC	5000		
7785-84-4	SODIUM PHOSPHATE TRIBASIC	5000		
10101-89-0	SODIUM PHOSPHATE TRIBASIC	5000		
10361-89-4	SODIUM PHOSPHATE TRIBASIC	5000		
10140-65-5	SODIUM PHOSPHATE, DIBASIC	5000		
10124-56-8	SODIUM PHOSPHATE, TRIBASIC	5000		
7782-82-3	SODIUM SELENITE	1000		
10102-18-8	SODIUM SELENITE	100	100/10000	h
56-53-1	4,4'-STILBENEDIOL, ALPHA,ALPHA'DIETHYL-	1		
18883-66-4	STREPTOZOTOCIN	1		
7789-06-2	STRONTIUM CHROMATE	10		
57-24-9	STRYCHNIDIN-10-ONE, AND SALTS	10		
357-57-3	STRYCHNIDIN-10-ONE, 2,3-DIMETHOXY-	100		
57-24-9	STRYCHNINE AND SALTS	10		
100-42-5	STYRENE	1000		
7783-06-4	SULFUR HYDRIDE	100		
12771-08-3	SULFUR MONOCHLORIDE	1000		
1314-80-3	SULFUR PHOSPHIDE	100		
7488-56-4	SULFUR SELENIDE	10		
7664-93-9	SULFURIC ACID	1000	1000	

Chemical Hazard Communication Guidebook

CAS	CHEMICAL NAME	RQ (LBS)	TPQ (LBS)	NOTES
8014-95-7	SULFURIC ACID	1000		
77-78-1	SULFURIC ACID, DIMETHYL ESTER	100		
7446-18-6	SULFURIC ACID, THALLIUM(I) SALT	100		
10031-59-1	SULFURIC ACID, THALLIUM(I) SALT	100		
93-76-5	2,4,5-T	1000		
93-76-5	2,4,5-T ACID	1000		
1319-72-8	2,4,5-T AMINES	5000		
2008-46-0	2,4,5-T AMINES	5000		
3813-14-7	2,4,5-T AMINES	5000		
6369-96-6	2,4,5-T AMINES	5000		
6369-97-7	2,4,5-T AMINES	5000		
93-79-8	2,4,5-T ESTERS	1000		
1928-47-8	2,4,5-T ESTERS	1000		
2545-59-7	2,4,5-T ESTERS	1000		
25168-15-4	2,4,5-T ESTERS	1000		
61792-07-2	2,4,5-T ESTERS	1000		
13560-99-1	2,4,5-T SALTS	1000		
72-54-8	TDE	1		
95-94-3	1,2,4,5-TETRACHLOROBENZENE	5000		
1746-01-6	2,3,7,8-TETRACHLORODIBENZO-P-DIOXIN	1		
630-20-6	1,1,1,2-TETRACHLOROETHANE	100		
79-34-5	1,1,2,2-TETRACHLOROETHANE	100		
D039	TETRACHLOROETHYLENE	100		
127-18-4	TETRACHLOROETHYLENE	100		
58-90-2	2,3,4,6-TETRACHLOROPHENOL	10		
78-00-2	TETRAETHYL LEAD	10	100	c,d
107-49-3	TETRAETHYL PYROPHOSPHATE	10		
3689-24-5	TETRAETHYLDITHIOPYROPHOSPHATE	100		
109-99-9	TETRAHYDROFURAN	1000		
509-14-8	TETRANITROMETHANE	10	500	
757-58-4	TETRAPHOSPHORIC ACID, HEXAETHYL ESTER	100		
1314-32-5	THALLIC OXIDE	100		
7440-28-0	THALLIUM	1000		
563-68-8	THALLIUM (I) ACETATE	100		
--	THALLIUM COMPOUNDS			
6533-73-9	THALLIUM(I) CARBONATE	100		
7791-12-0	THALLIUM(I) CHLORIDE	100		
10102-45-1	THALLIUM(I) NITRATE	100		
12039-52-0	THALLIUM(I) SELENIDE	1000		
7446-18-6	THALLIUM(I) SULFATE	100		
10031-59-1	THALLIUM(I) SULFATE	100		
1314-32-5	THALLIUM(III) OXIDE	100		
62-55-5	THIOACETAMIDE	10		
39196-18-4	THIOFANOX	100	100/10000	
541-53-7	THIOIMIDODICARBONIC DIAMIDE	100		
74-93-1	THIOMETHANOL	100		
108-98-5	THIOPHENOL	100	500	

Appendix F: Hazardous Substances (Alphabetical Order)

CAS	CHEMICAL NAME	RQ (LBS)	TPQ (LBS)	NOTES
79-19-6	THIOSEMICARBAZIDE	100	100/10000	
62-56-6	THIOUREA	10		
5344-82-1	THIOUREA, (2-CHLOROPHENYL)-	100	100/10000	
86-88-4	THIOUREA, 1-NAPHTHALENYL-	100		
103-85-5	THIOUREA, PHENYL-	100		
137-26-8	THIRAM	10		
108-88-3	TOLUENE	1000		
91-08-7	TOLUENE DIISOCYANATE	100		
584-84-9	TOLUENE DIISOCYANATE	100		
26471-62-5	TOLUENE DIISOCYANATE	100		
95-80-7	TOLUENEDIAMINE	10		
496-72-0	TOLUENEDIAMINE	10		
823-40-5	TOLUENEDIAMINE	10		
25376-45-8	TOLUENEDIAMINE	10		
95-53-4	O-TOLUIDINE	100		
106-49-0	P-TOLUIDINE	100		
636-21-5	O-TOLUIDINE HYDROCHLORIDE	100		
8001-35-2	TOXAPHENE	1		
93-72-1	2,4,5-TP ACID	100		
32534-95-5	2,4,5-TP ACID ESTERS	100		
61-82-5	1H-1,2,4-TRIAZOL-3-AMINE	10		
52-68-6	TRICHLORFON	100		
52-68-6	TRICHLORFON [PHOSPHONIC ACID,(2,2,2-TRICHLORO-1-HYDROXYETHYL)-,DIMETHY	100		
120-82-1	1,2,4-TRICHLOROBENZENE	100		
71-55-6	1,1,1-TRICHLOROETHANE	1000		
79-00-5	1,1,2-TRICHLOROETHANE	100		
79-01-6	TRICHLOROETHENE	100		
D040	TRICHLOROETHYLENE	100		
79-01-6	TRICHLOROETHYLENE	100		
594-42-3	TRICHLOROMETHANESULFENYL CHLORIDE	100		
75-69-4	TRICHLOROMONOFLUOROMETHANE	5000		
25167-82-2	TRICHLOROPHENOL	10		
15950-66-0	2,3,4-TRICHLOROPHENOL	10		
933-78-8	2,3,5-TRICHLOROPHENOL	10		
933-75-5	2,3,6-TRICHLOROPHENOL	10		
D041	2,4,5-TRICHLOROPHENOL	10		
95-95-4	2,4,5-TRICHLOROPHENOL	10		
D042	2,4,6-TRICHLOROPHENOL	10		
88-06-2	2,4,6-TRICHLOROPHENOL	10		
609-19-8	3,4,5-TRICHLOROPHENOL	10		
93-76-5	2,4,5-(TRICHLOROPHENOXY) ACETIC ACID	1000		
27323-41-7	TRIETHANOLAMINE DODECYLBENZENESULFONATE	1000		
121-44-8	TRIETHYLAMINE	5000		
75-50-3	TRIMETHYLAMINE	100		
99-35-4	SYM-TRINITROBENZENE	10		
123-63-7	1,3,5-TRIOXANE, 2,4,6-TRIMETHYL-	1000		

Chemical Hazard Communication Guidebook

CAS	CHEMICAL NAME	RQ (LBS)	TPQ (LBS)	NOTES
126-72-7	TRIS(2,3-DIBROMOPROPYL)PHOSPHATE	10		
72-57-1	TRYPAN BLUE	10		
66-75-1	URACIL 5[BIS(2-CHLOROETHYL) AMINO]-	10		
66-75-1	URACIL MUSTARD	10		
541-09-3	URANYL ACETATE	100		
10102-06-4	URANYL NITRATE	100		
36478-76-9	URANYL NITRATE	100		
51-79-6	URETHANE	100		
7803-55-6	VANADIC ACID, AMMONIUM SALT	1000		
1314-62-1	VANADIUM (V) OXIDE	1000		
1314-62-1	VANADIUM PENTOXIDE	1000	100/10000	
27774-13-6	VANADYL SULFATE	1000		
108-05-4	VINYL ACETATE	5000		
D043	VINYL CHLORIDE	1		
75-01-4	VINYL CHLORIDE	1		
75-35-4	VINYLIDENE CHLORIDE	100		
81-81-2	WARFARIN	100	500/10000	
108-38-3	M-XYLENE	1000		
95-47-6	O-XYLENE	1000		
106-42-3	P-XYLENE	1000		
1330-20-7	XYLENE (MIXED ISOMERS)	1000		
1300-71-6	XYLENOL	1000		
50-55-5	YOHIMBAN-16-CARBOXYLIC ACID, 11,17-DIMETHOXY-18-[(3,4,5- TRI METHOXYB	5000		
7440-66-6	ZINC	1000		
557-34-6	ZINC ACETATE	1000		
14639-97-5	ZINC AMMONIUM CHLORIDE	5000		
14639-98-6	ZINC AMMONIUM CHLORIDE	5000		
52628-25-8	ZINC AMMONIUM CHLORIDE	1000		
1332-07-6	ZINC BORATE	1000		
7699-45-8	ZINC BROMIDE	1000		
3486-35-9	ZINC CARBONATE	1000		
7646-85-7	ZINC CHLORIDE	1000		
--	ZINC COMPOUNDS			
557-21-1	ZINC CYANIDE	10		
7783-49-5	ZINC FLUORIDE	1000		
557-41-5	ZINC FORMATE	1000		
7779-86-4	ZINC HYDROSULFITE	1000		
7779-88-6	ZINC NITRATE	1000		
127-82-2	ZINC PHENOLSULFONATE	5000		
1314-84-7	ZINC PHOSPHIDE	100	500	b
16871-71-9	ZINC SILICOFLUORIDE	5000		
7733-02-0	ZINC SULFATE	1000		
13746-89-9	ZIRCONIUM NITRATE	5000		
16923-95-8	ZIRCONIUM POTASSIUM FLUORIDE	1000		
14644-61-2	ZIRCONIUM SULFATE	5000		
10026-11-6	ZIRCONIUM TETRACHLORIDE	5000		

Appendix F: Hazardous Substances (Alphabetical Order)

NOTES

a. This chemical does not meet acute toxicity criteria. Its TPQ is set at 10,000 pounds.

b. This material is a reactive solid. The Threshold Planning Quantity does not default to 10,000 pounds for non-powder, non-molten, non-solution form.

c. EPA changed the calculated Threshold Planning Quantity and the reader is referred to the *Federal Register* of April 22, 1987 for further details.

d. EPA has indicated that the reportable quantity is likely to change when the assessment of potential carcinogenicity and chronic toxicity is completed.

e. Statutory reportable quantity for purposes of emergency notification under Section 304(a)(2) of the Emergency Planning and Community Right-to-Know Act.

f. EPA has indicated that the statutory one pound reportable quantity for methyl isocyanate may be adjusted in a future rulemaking.

g. New chemicals added that were not part of the original list of 402 substances.

h. Revised TPQ based on new or re-evaluated toxicity data.

j. TPQ is revised to its calculated value and does not change due to technical review as in proposed rule.

k. The TPQ was revised after proposal due to calculation error.

l. Chemicals on the original list that do not meet the toxicity criteria but because of their high production volume and recognized toxicity are considered chemicals of concern.

m. Hydrogen chloride is regulated as an extremely hazardous substance for the gas only.

APPENDIX G

HAZARDOUS SUBSTANCES
(CAS Order)

This appendix provides the list of Hazardous Substances subject to spill reporting in CERCLA/SARA. Where a reportable quantity has been established under CERCLA or a statutory reportable quantity has been set pursuant to SARA, it is listed under the column "RQ." As an aid to the reader, if the CERCLA Hazardous Substance has also been listed as an extremely hazardous substance subject to SARA's emergency planning provisions, the threshold planning quantities are provided under the column "TPQ." A guide to the notes is included at the end.

Chemical Hazard Communication Guidebook

CAS	CHEMICAL NAME	RQ (LBS)	TPQ (LBS)	NOTES
--	ANTIMONY COMPOUNDS			
--	ARSENIC COMPOUNDS			
--	BERYLLIUM COMPOUNDS			
--	CADMIUM COMPOUNDS			
--	CHLORDANE (TECHNICAL MIXTURE AND METABOLITES)			
--	CHLORINATED BENZENES			
--	CHLORINATED ETHANES			
--	CHLORINATED NAPHTHALENE			
--	CHLORINATED PHENOLS			
--	CHLOROALKYL ETHERS			
--	CHROMIUM COMPOUNDS			
--	COKE OVEN EMISSIONS	1		
--	COPPER COMPOUNDS			
--	CYANIDES			
--	DDT AND METABOLITES			
--	DICHLOROBENZIDINE			
--	DIPHENYLHYDRAZINE			
--	ENDOSULFAN AND METABOLITES			
--	ENDRIN AND METABOLITES			
--	HALOETHERS			
--	HALOMETHANES			
--	HEPTACHLOR AND METABOLITES			
--	LEAD COMPOUNDS			
--	MERCURY COMPOUNDS			
--	NICKEL COMPOUNDS			
--	NITROPHENOLS			
--	NITROSAMINES			
--	PHTHALATE ESTERS			
--	POLYNUCLEAR AROMATIC HYDROCARBONS			
--	RADIONUCLIDES	1		
--	SELENIUM COMPOUNDS			
--	SILVER COMPOUNDS			
--	THALLIUM COMPOUNDS			
--	ZINC COMPOUNDS			
D--	*UNLISTED HAZARDOUS WASTES - CHARACTERISTIC OF EP TOXICITY			
D001	*UNLISTED HAZARDOUS WASTES - CHARACTERISTIC OF IGNITABILITY	100		
D002	*UNLISTED HAZARDOUS WASTES - CHARACTERISTIC OF CORROSIVITY	100		
D003	*UNLISTED HAZARDOUS WASTES - CHARACTERISTIC OF REACTIVITY	100		
D004	*UNLISTED HAZ. WASTE CHARACTERISTIC OF EP TOXICITY - ARSENIC	1		
D005	*UNLISTED HAZ. WASTE CHARACTERISTIC OF EP TOXICITY - BARIUM	1000		
D006	*UNLISTED HAZ. WASTE CHARACTERISTIC OF EP TOXICITY - CADMIUM	10		

Appendix G: Hazardous Substances (CAS Order)

CAS	CHEMICAL NAME	RQ (LBS)	TPQ (LBS)	NOTES
D007	*UNLISTED HAZ. WASTE CHARACTERISTIC OF EP TOXICITY - CHROMIUM	10		
D007	*UNLISTED HAZARDOUS WASTES CHARACTERISTIC OF EP TOXICITY CHROMIUM(VI)	10		
D008	*UNLISTED HAZ. WASTE CHARACTERISTIC OF EP TOXICITY - LEAD	1		
D009	*UNLISTED HAZ. WASTE CHARACTERISTIC OF EP TOXICITY - MERCURY	1		
D010	*UNLISTED HAZ. WASTE CHARACTERISTIC OF EP TOXICITY - SELENIUM	10		
D011	*UNLISTED HAZ. WASTE CHARACTERISTIC OF EP TOXICITY - SILVER	1		
D012	*UNLISTED HAZ. WASTE CHARACTERISTIC OF EP TOXICITY - ENDRIN	1		
D013	*UNLISTED HAZ. WASTE CHARACTERISTIC OF EP TOXICITY - LINDANE	1		
D014	*UNLISTED HAZ. WASTE CHARACTERISTIC OF EP TOXICITY - METHOXYCHLOR	1		
D015	*UNLISTED HAZ. WASTE CHARACTERISTIC OF EP TOXICITY - TOXAPHENE	1		
D016	*UNLISTED HAZ. WASTE CHARACTERISTIC OF EP TOXICITY - 2,4-D	100		
D017	*UNLISTED HAZ. WASTE CHARACTERISTIC OF EP TOXICITY - 2,4,5-TP	100		
D018	BENZENE	10		
D019	CARBON TETRACHLORIDE	10		
D020	CHLORDANE	1		
D021	CHLOROBENZENE	100		
D022	CHLOROFORM	10		
D023	O-CRESOL	1000		
D024	M-CRESOL	1000		
D025	P-CRESOL	1000		
D026	CRESOL	1000		
D027	1,4-DICHLOROBENZENE	100		
D028	1,2-DICHLOROETHANE	100		
D029	1,1-DICHLOROETHYLENE	100		
D030	2,4-DINITROTOLUENE	10		
D031	HEPTACHLOR (AND ITS HYDROXIDE)	1		
D032	HEXACHLOROBENZENE	10		
D033	HEXACHLOROBUTADIENE	1		
D034	HEXACHLOROETHANE	100		
D035	METHYL ETHYL KETONE	5000		
D036	NITROBENZENE	1000		
D037	PENTACHLOROPHENOL	10		
D038	PYRIDINE	1000		
D039	TETRACHLOROETHYLENE	100		
D040	TRICHLOROETHYLENE	100		
D041	2,4,5-TRICHLOROPHENOL	10		
D042	2,4,6-TRICHLOROPHENOL	10		
D043	VINYL CHLORIDE	1		

Chemical Hazard Communication Guidebook

CAS	CHEMICAL NAME	RQ (LBS)	TPQ (LBS)	NOTES
F001	*SPECIFIED SPENT HALOGENATED SOLVENTS USED DEGREASING & SLUDGES FR REC	10		
F002	*SPECIFIED SPENT HALOGENATED SOLVENTS AND STILL BOTTOMS FR. RECOVERY	10		
F003	*SPECIFIED SPENT NON-HALOGENATED SOLVENTS & STILL BOTTOMS FR. RECOVERY	100		
F004	*THE FOLLOWING SPENT NON-HALOGENATED SOLVENTS...CRESOLS...NITROBENZENE	1000		
F005	*THE FOLLOWING SPENT NON-HALOGENATED SOLVENTS...TOLUENE...METHYL ETHYL	100		
F006	*WASTEWATER TREATMENT SLUDGES (ELECTROPLATING) W/ SPECIFIED EXCEPTIONS	10		
F007	*SPENT CYANIDE ELECTROPLATING BATH SOLUTIONS W/ SPECIFIED EXCEPTIONS	10		
F008	*ELECTROPLATING BATH SLUDGES FROM BOTTOMS USING CYANIDES	10		
F009	*SPENT STRIPPING & CLEANING SOLUTIONS (ELECTROPLATING) USING CYANIDES	10		
F010	*QUENCHING BATH SLUDGE (OIL BATH) (METAL HEAT TREATING) USING CYANIDES	10		
F011	*SPENT CYANIDE SOLUTIONS FR SALT BATH POT CLEANING (METAL HEAT TREAT)	10		
F012	*QUENCHING WASTEWATER TREAT SLUDGES FR METAL HEAT TREAT USING CYANIDES	10		
F019	*WASTEWATER TREATMENT SLUDGES- CHEMICAL CONVERSION COATING OF ALUMINUM	10		
F020	*WASTES IN PROD. OR MANU. OF TRI- OR TETRACHLOROPHENOL, OR PEST. DERIV	1		
F021	*WASTES IN PROD. OR MANU. OF PENTACHLOROPHENOL/INTERMED. TO PROD DERIV	1		
F022	*WASTES FR. MANU. USE OF TETRA-,PENTA-,OR HEXACHLOROBENZENE (ALKALINE)	1		
F023	*WASTES FR. PROD. OF MATERIALS ON EQUIP. FOR TRI- & TETRACHLOROPHENOLS	1		
F024	*PROCESS WASTES INCLUDING PROD. CHLORINATED ALIPHATIC HYDROCARBONS	1		
F025	*CONDENSED LIGHT ENDS,SPENT FILTERS...WASTES FROM CHLOR. ALIPHATIC HYD	1		
F026	*WASTES FR. PROD. MATERIALS ON EQUIP. FOR TETRA,PENTA,HEXACHLOROBENZEN	1		
F027	*DISCARDED,UNUSED FORMU.W/ TRI,TETRA,PENTA-CHLOROPHENOLS OR DERIVATIVES	1		
F028	*RESIDUE FR. INCIN./THERMAL TREAT. SOIL CONTAMINATED W/SPECIFIED WASTE	1		
F037	*PETROLEUM REFINERY PRIMARY OIL/WATER/SOLIDS SEPARATION SLUDGE--ANY SL	1		
F038	*PETROLEUM REFINERY SECONDARY (EMULSIFIED) OIL/WATER/SOLIDS SEPARATION	1		
F039	*LEACHATE RESULTING FROM THE TREATMENT, STORAGE, OR DISPOSAL OF WASTES	1		
F039	*MULTI SOURCE LEACHATE	1		
K001	*SLUDGE OF WOOD PRESERVING PROCESSES USING CREOSOTE/PENTACHLOROPHENOL	1		

Appendix G: Hazardous Substances (CAS Order)

CAS	CHEMICAL NAME	RQ (LBS)	TPQ (LBS)	NOTES
K002	*WASTEWATER SLUDGE FROM PRODUCTION OF CHROME YELLOW & ORANGE PIGMENTS	1		
K003	*WASTEWATER SLUDGE FROM PRODUCTION OF MOLYBDATE ORANGE PIGMENTS	1		
K004	*WASTEWATER TREATMENT SLUDGE FROM PRODUCTION OF ZINC YELLOW PIGMENTS	10		
K005	*WASTEWATER TREATMENT SLUDGE FROM PRODUCTION OF CHROME GREEN PIGMENTS	1		
K006	*WASTEWATER SLUDGE FR PROD. CHROME OXIDE GREEN PIGMENTS (ANHY. & HYD.)	10		
K007	*WASTEWATER TREATMENT SLUDGE FROM PRODUCTION OF IRON BLUE PIGMENTS	10		
K008	*OVEN RESIDUE FROM THE PRODUCTION OF CHROME OXIDE GREEN PIGMENTS	10		
K009	*DISTILLATION BOTTOMS FROM PRODUCTION OF ACETALDEHYDE FROM ETHYLENE	10		
K010	*DISTILLATION SIDE CUTS FROM PRODUCTION OF ACETALDEHYDE FROM ETHYLENE	10		
K011	*BOTTOM STREAM FROM WASTEWATER STRIPPER IN PRODUCTION OF ACRYLONITRILE	10		
K013	*BOTTOM STREAM FROM ACETONITRILE COLUMN IN PRODUCTION OF ACRYLONITRILE	10		
K014	*BOTTOMS FROM ACETONITRILE PURIF. COLUMN IN PROD. OF ACRYLONITRILE	5000		
K015	*STILL BOTTOMS FROM THE DISTILLATION OF BENZYL CHLORIDE	10		
K016	*HEAVY ENDS OR DISTILLATION RESIDUES FROM PROD. OF CARBON TETRACHLORIDE	1		
K017	*HEAVY ENDS(STILL BOTTOMS) FROM PURIF. COLUMN IN PROD. EPICHLOROHYDRIN	10		
K018	*HEAVY ENDS FROM FRACTIONATION COLUMN IN PRODUCTION OF ETHYL CHLORIDE	1		
K019	*HEAVY ENDS FROM DISTILLATION IN PRODUCTION OF ETHYLENE DICHLORIDE	1		
K020	*HEAVY ENDS FROM DISTILLATION IN PRODUCTION OF VINYL CHLORIDE MONOMERS	1		
K021	*AQUEOUS SPENT ANTIMONY CATALYST WASTE FR. FLUOROMETHANES PROD.	10		
K022	*DISTILLATION BOTTOM TARS FR. PROD. OF PHENOL/ACETONE FR. CUMENE	1		
K023	*DISTILLATION LIGHT ENDS FR. PROD. PHTHALIC ANHYDRIDE FR. NAPHTHALENE	5000		
K024	*DISTILLATION BOTTOMS FR. PROD. PHTHALIC ANHYDRIDE FR. NAPHTHALENE	5000		
K025	*DISTILLATION BOTTOMS FR. PROD. NITROBENZENE BY NITRATION OF BENZENE	10		
K026	*STRIPPING STILL TAILS FR. PROD. METHYL ETHYL PYRIDINES	1000		
K027	*CENTRIFUGE AND DISTILLATION RESIDUES FR. TOLUENE DIISOCYANATE PROD.	10		
K028	*SPENT CATALYST FR. HYDROCHLORINATOR REACTOR IN 1,1,1-TRICHLOROETHANE	1		
K029	*WASTE FR. PRODUCT STEAM STRIPPER IN 1,1,1-TRICHLOROETHANE PRCD.	1		

Chemical Hazard Communication Guidebook

CAS	CHEMICAL NAME	RQ (LBS)	TPQ (LBS)	NOTES
K030	*COLUMN BOTTOMS/HEAVY ENDS FR. COMBO PROD. TRI- AND PER-CHLOROETHYLENE	1		
K031	*BY-PRODUCT SALTS GENERATED IN PROD. OF MSMA AND CACODYLIC ACID	1		
K032	*WASTEWATER TREAT. SLUDGE FROM PROD. OF CHLORDANE	10		
K033	*WASTE- & SCRUBWATER FR. CHLORIN. CYCLOPENTADIENE IN CHLORDANE PROD.	10		
K034	*FILTER SOLIDS FR. FILTR. HEXACHLOROCYCLOPENTADIENE IN CHLORDANE PROD.	10		
K035	*WASTEWATER TREAT. SLUDGES GENERATED IN CREOSOTE PROD.	1		
K036	*STILL BOTTOMS FR. TOLUENE RECLAMATION DISTIL. IN DISULFOTON PROD.	1		
K037	*WASTEWATER TREAT. SLUDGES FROM DISULFOTON PROD.	1		
K038	*WASTEWATER FR. WASHING AND STRIPPING OF PHORATE PROD.	10		
K039	*FILTER CAKE FR. FILTR. DIETHYLPHOSPHORODITHIOIC ACID IN PHORATE PROD.	10		
K040	*WASTEWATER TREAT. SLUDGE FROM PHORATE PROD.	10		
K041	*WASTEWATER TREAT. SLUDGE FROM TOXAPHENE PROD.	1		
K042	*HEAVY ENDS/DIST. RESIDUES FR. DIST. TETRACHLOROBENZENE IN 2,4,5-T PR.	10		
K043	*2,6-DICHLOROPHENOL WASTE FROM 2,4-D PROD.	10		
K044	*WASTEWATER TREAT. SLUDGES FROM MANU. & PROCESSING OF EXPLOSIVES	10		
K045	*SPENT CARBON FROM TREAT. OF WASTEWATER CONTAINING EXPLOSIVES	10		
K046	*WASTEWATER TR. SLUDGE FROM MANU,FORMU,LOADING OF LEAD-BASED INITIATOR	100		
K047	*PINK/RED WATER FROM TNT OPERATIONS	10		
K048	*DISSOLVED AIR FLOTATION (DAF) FLOAT FR. PETROLEUM REFINING INDUSTRY	1		
K049	*SLOP OIL EMULSION SOLIDS FROM THE PETROLEUM REFINING INDUSTRY	1		
K050	*HEAT EXCHANGER BUNDLE CLEANING SLUDGE FR. PETROLEUM REFINING INDUSTRY	10		
K051	*API SEPARATOR SLUDGE FR. PETROLEUM REFINING INDUSTRY	1		
K052	*TANK BOTTOMS (LEADED) FR. PETROLEUM REFINING INDUSTRY	10		
K060	*AMMONIA STILL LIME SLUDGE FROM COKING OPERATIONS	1		
K061	*EMISSION CONTROL DUST/SLUDGE FR. PRIM. PROD. STEEL IN ELEC. FURNACES	1		
K062	*SPENT PICKLE LIQUOR FR. STEEL FINISHING OPERATIONS	1		
K064	*ACID PLANT BLOWDOWN/SLURRY/SLUDGE RESULTING FROM ...PRIMARY COPPER	1		

Appendix G: Hazardous Substances (CAS Order)

CAS	CHEMICAL NAME	RQ (LBS)	TPQ (LBS)	NOTES
K065	*SURFACE IMPOUNDMENT SOLIDS CONTAINED IN AND DEGRADED...LEAD SMELTING	1		
K066	*SLUDGE FROM TREATMENT OF PROCESS WASTEWATER AND/OR ACID..PRIMARY ZINC	1		
K069	*EMISSION CONTROL DUST/SLUDGE FR. SECONDARY LEAD SMELTING	1		
K071	*BRINE PURIF. MUDS FR. MERCURY CELL PROCESS IN CHLORINE PROD. EXCEPT..	1		
K073	*CHLORINATED HYDROCARBON WASTE FR. PURIF. STEP OF ... IN CHLORINE PROD	10		
K083	*DISTILLATION BOTTOMS FROM ANILINE EXTRACTION	100		
K084	*WASTEWATER TREAT. SLUDGES IN PROD. OF VETERINARY PHARMACEUT (ARSENIC)	1		
K085	*DISTILLATION OR FRACTIONATION COLUMN BOTTOMS FROM CHLOROBENZENE PROD.	10		
K086	*SOLVENT,WATER,CAUSTIC WASHES & SLUDGES CLEANING EQUIP. FOR INK FORMU.	1		
K087	*DECANTER TANK TAR SLUDGE FROM COKING OPERATIONS	100		
K088	*SPENT POTLINERS FROM PRIMARY ALUMINUM REDUCTION	1		
K090	*EMISSION CONTROL DUST OR SLUDGE FROM FERROCHROMIUM SILICON PRODUCTION	1		
K091	*EMISSION CONTROL DUST OR SLUDGE FROM FERROCHROMIUM PRODUCTION	1		
K093	*DISTILLATION LIGHT ENDS FR. PROD. PHTHALIC ANHYDRIDE FR. ORTHO-XYLENE	5000		
K094	*DISTILLATION BOTTOMS FR. PROD. PHTHALIC ANHYDRIDE FR. ORTHO-XYLENE	5000		
K095	*DISTILLATION BOTTOMS FR. PROD. 1,1,1-TRICHLOROETHANE	100		
K096	*HEAVY ENDS FR. HEAVY ENDS COLUMN FR. PROD. 1,1,1-TRICHLOROETHANE	100		
K097	*VACUUM STRIPPER DISCHARGE FR. CHLORDANE CHLORINATOR IN CHLORDANE PROD	1		
K098	*UNTREATED PROCESS WASTEWATER FR. TOXAPHENE PROD.	1		
K099	*UNTREATED WASTEWATER FR. 2,4-D PROD.	10		
K100	*WASTE LEACHING SOLUTION ... (COMPONENTS IDENTICAL WITH THOSE OF K069)	1		
K101	*DISTILLATION TAR RESIDUES FR. ANILINE-BASED COMPOUNDS VET. PHARMACEUT	1		
K102	*RESIDUE FR. ACTIVATED CARBON FOR DECOLORIZATION PROD. VET. PHARMACEUT	1		
K103	*PROCESS RESIDUES FROM ANILINE EXTRACTION FROM ANILINE PROD.	100		
K104	*COMBINED WASTEWATER STREAMS GEN. NITROBENZENE/ANILINE CHLOROBENZINES	10		
K105	*SEPARATED AQUEOUS STREAM FR. REACTOR PROD.WASHING STEP IN CHLOROBENZ.	10		
K106	*WASTEWATER TREATMENT SLUDGE FR. MERCURY CELL PROCESS IN CHLORINE PROD	1		

Chemical Hazard Communication Guidebook

CAS	CHEMICAL NAME	RQ (LBS)	TPQ (LBS)	NOTES
K107	*COLUMN BOTTOMS FROM PRODUCT SEPERATION FR... CARBOXYLIC ACID HYDAZINE	10		
K108	*CONDENSED COLUMN OVERHEADS FROM PRODUCT S... CARBOXYLIC ACID HYDRAZIN	10		
K109	*SPENT FILTER CARTRIDGES FROM PRODUCT PURI... CARBOXYLIC ACID HYDRAZIN	10		
K110	*CONDENSED COLUMN OVERHEADS FROM INTERMEDI... CARBOXYLIC ACID HYDRAZIN	10		
K111	*PRODUCT WASHWATERS FROM PROD. DINITROTOLUENE VIA NITRATION OF TOLUENE	10		
K112	*REACT. BY-PROD. WATER FR. DRYING COLUMN PROD. TOLUENEDIAMINE VIA...	10		
K113	*CONDENSED LIQUID LIGHT ENDS FR. PURIF. TOLUENEDIAMINE IN PROD. VIA...	10		
K114	*VICINALS FR. PURIF. TOLUENEDIAMINE IN PROD. VIA HYDROG. DINITROTOLUEN	10		
K115	*HEAVY ENDS PURIF. TOLUENEDIAMINE IN PROD. VIA HYDROG. DINITROTOLUENE	10		
K116	*ORGANIC CONDENS. FR. SOLVENT RECOVERY COLUMN TOLUENE DIISOCYANATE VIA	10		
K117	*WASTEWATER FR. REACTION VENT GAS SCRUBBER PROD. ETHYLENE BROMIDE VIA	1		
K118	*SPENT ABSORBENT SOLIDS FR. PURIF. ETHYLENE DIBROMIDE IN PROD. OF IT	1		
K123	*PROCESS WASTEWATER ...FROM PROD. ETHYLENEBISDITHIOCARBAMIC ACID/SALTS	10		
K124	*REACTOR VENT SCRUBBER WTER PRD. ETHYLENEBISDITHIOACARBAMIC ACID/SALTS	10		
K125	*FILTRATION, EVAP.,CENTRIFUG., PROD. ETHEYLENEBISDITHIOCARBAMIC ACIDS	10		
K126	*BAGHOUSE DUST, SWEEP. IN MILL., PACK.,PRD ETHYLENEBISDITHIOCARBAMICAC	10		
K131	*WASTEWATER FROM REACTOR,SPENT SULFURIC ACID IN PROD. METHYL BROMIDE	100		
K132	*SPENT ABSORBENT AND WASTEWATER SOLIDS FROM PROD. METHYL BROMIDE	1000		
K136	*STILL BOTTOMS FR. PURIF. ETHYLENE DIBROMIDE IN PROD. VIA BROMINATION	1		
K136	*STILL BOTTOMS FROM PURIF. ETHYLENE DIBROMIDE IN PROD. VIA BROMIN. ETH	1		
50-00-0	FORMALDEHYDE	100	500	d,l
50-00-0	METHYLENE OXIDE	100		
50-07-7	AZIRINO[2',3':3,4]PYRROLO[1,2-A]INDOLE-4,7-DIONE, 6-AMINO-8-[[(AMINOCA	10		
50-07-7	MITOMYCIN C	10	500/10000	d
50-18-0	CYCLOPHOSPHAMIDE	10		
50-18-0	2H-1,3,2-OXAZAPHOSPHORINE,2-[BIS(2-CHLOROETHYL)AMINO] TETRAHYDRO-2-OXIDE	10		
50-29-3	DDT	1		
50-29-3	DICHLORO DIPHENYL TRICHLOROETHANE	1		
50-32-8	3,4-BENZOPYRENE	1		
50-32-8	BENZO[A]PYRENE	1		
50-55-5	RESERPINE	5000		

Appendix G: Hazardous Substances (CAS Order)

CAS	CHEMICAL NAME	RQ (LBS)	TPQ (LBS)	NOTES
50-55-5	YOHIMBAN-16-CARBOXYLIC ACID, 11,17-DIMETHOXY-18-[(3,4,5- TRI METHOXYB	5000		
51-28-5	2,4-DINITROPHENOL	10		
51-28-5	PHENOL, 2,4-DINITRO-	10		
51-43-4	1,2-BENZENEDIOL, 4-(1-HYDROXY-2-(METHYLAMINO)ETHYL)-	1000		
51-43-4	EPINEPHRINE	1000		
51-79-6	CARBAMIC ACID, ETHYL ESTER	100		
51-79-6	ETHYL CARBAMATE	100		
51-79-6	URETHANE	100		
52-68-6	TRICHLORFON	100		
52-68-6	TRICHLORFON [PHOSPHONIC ACID,(2,2,2-TRICHLORO-1-HYDROXYETHYL)-,DIMETHY	100		
52-85-7	FAMPHUR	1000		
52-85-7	PHOSPHOROTHIOIC ACID, O,O-DIMETHYL O-[P-[(DIMETHYLAMINO)-SULFONYL]PHEN	1000		
53-70-3	1,2:5,6-DIBENZANTHRACENE	1		
53-70-3	DIBENZO[A,H]ANTHRACENE	1		
53-70-3	DIBENZ[A,H]ANTHRACENE	1		
53-96-3	ACETAMIDE, N-9H-FLUOREN-2-YL-	1		
53-96-3	2-ACETYLAMINOFLUORENE	1		
54-11-5	NICOTINE AND SALTS	100		
54-11-5	PYRIDINE, (S)-3-(1-METHYL-2-PYRROLIDINYL)-, AND SALTS	100		
55-18-5	ETHANAMINE, N-ETHYL-N-NITROSO	1		
55-18-5	N-NITROSODIETHYLAMINE	1		
55-63-0	NITROGLYCERINE	10		
55-63-0	1,2,3-PROPANETRIOL, TRINITRATE-	10		
55-91-4	DIISOPROPYL FLUOROPHOSPHATE	100		
55-91-4	PHOSPHOROFLUORIDIC ACID, BIS(1-METHYLETHYL) ESTER	100		
56-04-2	METHYLTHIOURACIL	10		
56-04-2	4(1H)-PYRIMIDINONE, 2,3-DIHYDRO-6-METHYL-2-THIOXO-	10		
56-23-5	CARBON TETRACHLORIDE	10		
56-23-5	METHANE TETRACHLORO-	10		
56-38-2	PARATHION	10	100	c,d
56-38-2	PARATHION [PHOSPHOROTHIOIC ACID, O,O-DI-ETHYL-0-(4-NITROPHENYL) ESTER]	10	100	c,d
56-38-2	PHOSPHOROTHIOIC ACID, O,O-DIETHYL O-(P-NITROPHENYL) ESTER	10		
56-49-5	BENZ[J]ACEANTHRYLENE, 1,2-DIHYDRO-3-METHYL	10		
56-49-5	3-METHYLCHOLANTHRENE	10		
56-53-1	DIETHYLSTILBESTEROL	1		
56-53-1	4,4'-STILBENEDIOL, ALPHA,ALPHA'DIETHYL-	1		
56-55-3	1,2-BENZANTHRACENE	10		
56-55-3	BENZO[A]ANTHRACENE	10		
56-55-3	BENZ[A]ANTHRACENE	10		
56-72-4	COUMAPHOS	10	100/10000	

Chemical Hazard Communication Guidebook

CAS	CHEMICAL NAME	RQ (LBS)	TPQ (LBS)	NOTES
57-12-5	CYANIDES (SOLUBLE CYANIDE SALTS), NOT ELSEWHERE SPECIFIED	10		
57-12-5	CYANIDES, AS /CN/			
57-14-7	DIMETHYLHYDRAZINE	10	1000	d
57-14-7	1,1-DIMETHYLHYDRAZINE	10		
57-14-7	HYDRAZINE, 1,1-DIMETHYL	10		
57-24-9	STRYCHNIDIN-10-ONE, AND SALTS	10		
57-24-9	STRYCHNINE AND SALTS	10		
57-74-9	CHLORDANE	1	1000	d
57-74-9	CHLORDANE (ALPHA AND GAMMA ISOMERS)	1		
57-74-9	CHLORDANE TECHNICAL	1		
57-74-9	CHLORDANE [4,7-METHANOINDAN,1,2,4,5,6,7,8,8-OCTACHLORO-2,3,3A,4,7,7A-H	1	1000	d
57-74-9	4,7-METHANOINDAN, 1,2,4,5,6,7,8,8-OCTACHLORO-3A,4,7,7A-TETRAHYDRO-	1		
57-97-6	1,2-BENZANTHRACENE, 7,12-DIMETHYL-	1		
57-97-6	7,12-DIMETHYLBENZ[A]ANTHRACENE	1		
58-89-9	GAMMA-BHC	1		
58-89-9	HEXACHLOROCYCLOHEXANE (GAMMA ISOMER)	1		
58-89-9	LINDANE	1	1000/10000	d
58-89-9	LINDANE [CYCLOHEXANE,1,2,3,4,5,6-HEXACHLORO-,(1.ALPHA.,2.ALPHA.,3.BETA	1	1000/10000	d
58-90-2	PHENOL, 2,3,4,6-TETRACHLORO-	10		
58-90-2	2,3,4,6-TETRACHLOROPHENOL	10		
59-50-7	P-CHLORO-M-CRESOL	5000		
59-50-7	PHENOL, 4-CHLORO-3-METHYL-	5000		
60-00-4	ETHYLENEDIAMINE TETRAACETIC ACID	5000		
60-11-7	BENZENAMINE, N,N-DIMETHYL-4-PHENYLAZO-	10		
60-29-7	ETHANE, 1,1'-OXYBIS-	100		
60-29-7	ETHYL ETHER	100		
60-34-4	HYDRAZINE, METHYL	10		
60-34-4	METHYLHYDRAZINE	10	500	
60-51-5	DIMETHOATE	10	500/10000	
60-51-5	PHOSPHORODITHIOIC ACID,O,O-DIMETHYL S-[2(METHYLAMINO)-2-OXOETHYL]ESTER	10		
60-57-1	DIELDRIN	1		
60-57-1	1,2,3,4,10,10-HEXACHLORO- 6,7-EPOXY-1,4,4A,5,6,7,8,8A-OCTAHYDRO-ENDO,EXO-1,4:5, 8-D	1		
61-82-5	AMITROLE	10		
61-82-5	1H-1,2,4-TRIAZOL-3-AMINE	10		
62-38-4	MERCURY, (ACETATO-O)PHENYL-	100		
62-38-4	PHENYLMERCURIC ACETATE	100		
62-44-2	ACETAMIDE, N-(4-ETHOXYPHENYL)-	100		
62-44-2	PHENACETIN	100		
62-50-0	ETHYL METHANESULFONATE	1		
62-50-0	METHANESULFONIC ACID, ETHYL ESTER	1		
62-53-3	ANILINE	5000	1000	d,l
62-53-3	BENZENAMINE	5000		

Appendix G: Hazardous Substances (CAS Order)

CAS	CHEMICAL NAME	RQ (LBS)	TPQ (LBS)	NOTES
62-55-5	ETHANETHIOAMIDE	10		
62-55-5	THIOACETAMIDE	10		
62-56-6	CARBAMIDE, THIO-	10		
62-56-6	THIOUREA	10		
62-73-7	DICHLORVOS	10	1000	
62-73-7	DICHLORVOS [PHOSPHORIC ACID, 2,2-DICHLOROETHENYL DIMETHYL ESTER]	10	1000	
62-74-8	ACETIC ACID, FLUORO-, SODIUM SALT	10		
62-74-8	FLUOROACETIC ACID, SODIUM SALT	10		
62-75-9	DIMETHYLNITROSAMINE	10		
62-75-9	METHAMINE, N-METHYL-N-NITROSO-	10		
62-75-9	N-NITROSODIMETHYLAMINE	10		
63-25-2	CARBARYL	100		
63-25-2	CARBARYL [1-NAPHTHALENOL, METHYLCARBAMATE]	100		
64-18-6	FORMIC ACID	5000		
64-18-6	METHANOIC ACID	5000		
64-19-7	ACETIC ACID	5000		
65-85-0	BENZOIC ACID	5000		
66-75-1	2,4(1H,3H)-PYRIMIDINEDIONE, 5-[BIS(2-CHLOROETHYL)AMINO]-	10		
66-75-1	URACIL 5[BIS(2-CHLOROETHYL) AMINO]-	10		
66-75-1	URACIL MUSTARD	10		
67-56-1	METHANOL	5000		
67-56-1	METHYL ALCOHOL	5000		
67-64-1	ACETONE	5000		
67-64-1	2-PROPANONE	5000		
67-66-3	CHLOROFORM	10	10000	d,l
67-66-3	METHANE TRICHLORO-	10		
67-72-1	ETHANE, 1,1,1,2,2,2-HEXACHLORO-	100		
67-72-1	HEXACHLOROETHANE	100		
70-25-7	GUANIDINE, N-NITROSO-N-METHYL-N'-NITRO-	10		
70-25-7	N-METHYL-N-NITRO-N-NITROSOGUANIDINE	10		
70-30-4	HEXACHLOROPHENE	100		
70-30-4	2,2'-METHYLENEBIS(3,4,6-TRICHLOROPHENOL)	100		
71-36-3	1-BUTANOL	5000		
71-36-3	N-BUTYL ALCOHOL	5000		
71-43-2	BENZENE	10		
71-55-6	METHYL CHLOROFORM	1000		
71-55-6	1,1,1-TRICHLOROETHANE	1000		
72-20-8	2,7:3,6-DIMETHANONAPHTH[2,3-B]OXIRENE, 3,4,5,6,9,9-HEXACHLORO-1A,2,2A,3,6,6A,7	1		
72-20-8	ENDRIN	1	500/10000	
72-20-8	ENDRIN AND METABOLITES	1		
72-43-5	ETHANE 1,1,1-TRICHLORO-2,2-BIS-(P-METHOXYPHENYL)-	1		
72-43-5	METHOXYCHLOR	1		
72-43-5	METHOXYCHLOR [BENZENE, 1,1'-(2,2,2-TRICHLOROETHYLIDENE)BIS[4-METHOXY-	1		

Chemical Hazard Communication Guidebook

CAS	CHEMICAL NAME	RQ (LBS)	TPQ (LBS)	NOTES
72-54-8	DDD	1		
72-54-8	4,4'-DDD	1		
72-54-8	DICHLORODIPHENYL DICHLOROETHANE	1		
72-54-8	TDE	1		
72-55-9	4,4'-DDE	1		
72-57-1	2,7-NAPHTHALENEDISULFONIC ACID, 3,3'-[(3,3'-DIMETHYL-(1,1'-BIPHENYL)-4,4'-	10		
72-57-1	TRYPAN BLUE	10		
74-83-9	METHANE, BROMO-	1000		
74-83-9	METHYL BROMIDE	1000	1000	l
74-87-3	METHANE CHLORO-	100		
74-87-3	METHYL CHLORIDE	100		
74-88-4	METHANE IODO-	100		
74-88-4	METHYL IODIDE	100		
74-89-5	MONOMETHYLAMINE	100		
74-90-8	HYDROCYANIC ACID	10	100	
74-90-8	HYDROGEN CYANIDE	10		
74-93-1	METHANETHIOL	100		
74-93-1	METHYL MERCAPTAN	100	500	
74-93-1	THIOMETHANOL	100		
74-95-3	METHANE DIBROMO-	1000		
74-95-3	METHYLENE BROMIDE	1000		
75-00-3	CHLOROETHANE	100		
75-01-4	ETHENE CHLORO-	1		
75-01-4	VINYL CHLORIDE	1		
75-04-7	MONOETHYLAMINE	100		
75-05-8	ACETONITRILE	5000		
75-05-8	ETHANENITRILE	5000		
75-07-0	ACETALDEHYDE	1000		
75-07-0	ETHANAL	1000		
75-09-2	DICHLOROMETHANE	1000		
75-09-2	METHANE DICHLORO-	1000		
75-09-2	METHYLENE CHLORIDE	1000		
75-15-0	CARBON BISULFIDE	100		
75-15-0	CARBON DISULFIDE	100	10000	l
75-20-7	CALCIUM CARBIDE	10		
75-21-8	ETHYLENE OXIDE	10	1000	d,l
75-21-8	OXIRANE	10		
75-25-2	BROMOFORM	100		
75-25-2	METHANE TRIBROMO-	100		
75-27-4	DICHLOROBROMOMETHANE	5000		
75-34-3	1,1-DICHLOROETHANE	1000		
75-34-3	ETHANE 1,1-DICHLORO-	1000		
75-34-3	ETHYLIDENE DICHLORIDE	1000		
75-35-4	1,1-DICHLOROETHYLENE	100		
75-35-4	ETHENE, 1,1 DICHLORO-	100		
75-35-4	VINYLIDENE CHLORIDE	100		

Appendix G: Hazardous Substances (CAS Order)

CAS	CHEMICAL NAME	RQ (LBS)	TPQ (LBS)	NOTES
75-36-5	ACETYL CHLORIDE	5000		
75-36-5	ETHANOYL CHLORIDE	5000		
75-44-5	CARBONYL CHLORIDE	10		
75-44-5	PHOSGENE	10	10	l
75-50-3	TRIMETHYLAMINE	100		
75-55-8	2-METHYLAZIRIDINE	1		
75-55-8	1,2-PROPYLENIMINE	1		
75-56-9	PROPYLENE OXIDE	100	10000	l
75-60-5	CACODYLIC ACID	1		
75-60-5	HYDROXYDIMETHYLARSINE OXIDE	1		
75-64-9	TERT-BUTYLAMINE	1000		
75-69-4	METHANE TRICHLOROFLUORO-	5000		
75-69-4	TRICHLOROMONOFLUOROMETHANE	5000		
75-71-8	DICHLORODIFLUOROMETHANE	5000		
75-71-8	METHANE, DICHLORODIFLUORO-	5000		
75-86-5	ACETONE CYANOHYDRIN	10	1000	
75-86-5	2-METHYLLACTONITRILE	10		
75-86-5	PROPANENITRILE, 2-HYDROXY-2-METHYL-	10		
75-87-6	ACETALDEHYDE, TRICHLORO-	5000		
75-87-6	CHLORAL	5000		
75-99-0	2,2-DICHLOROPROPIONIC ACID	5000		
76-01-7	ETHANE PENTACHLORO-	10		
76-01-7	PENTACHLOROETHANE	10		d
76-44-8	HEPTACHLOR	1		
76-44-8	HEPTACHLOR [1,4,5,6,7,8,8-HEPTACHLORO-3A,4,7,7A-TETRAHYDRO-4,7-METHANO	1		
76-44-8	4,7-METHANO-1H-INDENE, 1,4,5,6,7,8,8-HEPTACHLORO- 3A,3,7,7A-TETRAHYDRO-	1		
77-47-4	1,3-CYCLOPENTADIENE, 1,2,3,4,5,5-HEXACHLORO-	10		
77-47-4	HEXACHLOROCYCLOPENTADIENE	10	100	d,h
77-78-1	DIMETHYL SULFATE	100	500	d
77-78-1	SULFURIC ACID, DIMETHYL ESTER	100		
78-00-2	PLUMBANE TETRAETHYL-	10		
78-00-2	TETRAETHYL LEAD	10	100	c,d
78-59-1	ISOPHORONE	5000		
78-79-5	ISOPRENE	100		
78-81-9	ISO-BUTYLAMINE	1000		
78-83-1	ISOBUTYL ALCOHOL	5000		
78-83-1	1-PROPANOL, 2-METHYL-	5000		
78-87-5	1,2-DICHLOROPROPANE	1000		
78-87-5	PROPYLENE DICHLORIDE	1000		
78-88-6	2,3-DICHLOROPROPENE (ISOMER)	100		
78-93-3	2-BUTANONE	5000		
78-93-3	METHYL ETHYL KETONE	5000		
78-99-9	1,1-DICHLOROPROPANE	1000		
79-00-5	ETHANE, 1,1,2-TRICHLORO-	100		
79-00-5	1,1,2-TRICHLOROETHANE	100		

Chemical Hazard Communication Guidebook

CAS	CHEMICAL NAME	RQ (LBS)	TPQ (LBS)	NOTES
79-01-6	TRICHLOROETHENE	100		
79-01-6	TRICHLOROETHYLENE	100		
79-06-1	ACRYLAMIDE	5000	1000/10000	d,l
79-06-1	2-PROPENAMIDE	5000		
79-09-4	PROPIONIC ACID	5000		
79-10-7	ACRYLIC ACID	5000		
79-10-7	2-PROPENOIC ACID	5000		
79-19-6	HYDRAZINECARBOTHIOAMIDE	100		
79-19-6	THIOSEMICARBAZIDE	100	100/10000	
79-22-1	CARBONOCHLORIDIC ACID, METHYL ESTER	1000		
79-22-1	METHYL CHLOROCARBONATE	1000		
79-31-2	ISO-BUTYRIC ACID	5000		
79-34-5	ETHANE, 1,1,2,2-TETRACHLORO-	100		
79-34-5	1,1,2,2-TETRACHLOROETHANE	100		
79-44-7	CARBAMOYL CHLORIDE, DIMETHYL-	1		
79-44-7	DIMETHYLCARBOMOYL CHLORIDE	1		
79-46-9	2-NITROPROPANE	10		
79-46-9	PROPANE, 2-NITRO-	10		
80-15-9	ALPHA,ALPHA-DIMETHYLBENZYLHYDROPEROXIDE	10		
80-15-9	HYDROPEROXIDE, 1-METHYL-1-PHENYLETHYL-	10		
80-62-6	METHYL METHACRYLATE	1000		
80-62-6	2-PROPENOIC ACID, 2-METHYL-, METHYL ESTER	1000		
81-07-2	1,2-BENZISOTHIAZOLIN-3-ONE,1,1-DIOXIDE, AND SALTS	100		
81-07-2	SACCHARIN AND SALTS	100		
81-81-2	3-(ALPHA-ACETONYLBENZYL)-4-HYDROXYCOUMARIN AND SALTS	100		
81-81-2	WARFARIN	100	500/10000	
82-68-8	BENZENE, PENTACHLORONITRO-	100		
82-68-8	PENTACHLORONITROBENZENE	100		
82-68-8	QUINTOZENE [PENTACHLORONITROBENZENE]	100		
83-32-9	ACENAPHTHENE	100		
84-66-2	1,2-BENZENEDICARBOXYLIC ACID, DIETHYL ESTER	1000		
84-66-2	DIETHYL PHTHALATE	1000		
84-74-2	1,2-BENZENEDICARBOXYLIC ACID, DIBUTYL ESTER	10		
84-74-2	N-BUTYL PHTHALATE	10		
84-74-2	DI-N-BUTYL PHTHALATE	10		
84-74-2	DIBUTYL PHTHALATE	10		
85-00-7	DIQUAT	1000		
85-01-8	PHENANTHRENE	5000		
85-44-9	1,2-BENZENEDICARBOXYLIC ACID ANHYDRIDE	5000		
85-44-9	PHTHALIC ANHYDRIDE	5000		
85-68-7	BUTYL BENZYL PHTHALATE	100		
86-30-6	N-NITROSODIPHENYLAMINE	100		
86-50-0	GUTHION	1		
86-73-7	FLUORENE	5000		
86-88-4	ALPHA-NAPHTHYL THIOUREA	100		

Appendix G: Hazardous Substances (CAS Order)

CAS	CHEMICAL NAME	RQ (LBS)	TPQ (LBS)	NOTES
86-88-4	THIOUREA, 1-NAPHTHALENYL-	100		
87-65-0	2,6-DICHLOROPHENOL	100		
87-65-0	PHENOL, 2,6-DICHLORO-	100		
87-68-3	1,3-BUTADIENE, 1,1,2,3,4,4-HEXACHLORO-	1		
87-68-3	HEXACHLOROBUTADIENE	1		
87-86-5	PENTACHLOROPHENOL	10		d
87-86-5	PHENOL, PENTACHLORO-	10		
88-06-2	PHENOL, 2,4,6-TRICHLORO-	10		
88-06-2	2,4,6-TRICHLOROPHENOL	10		
88-72-2	O-NITROTOLUENE	1000		
88-75-5	2-NITROPHENOL	100		
88-75-5	O-NITROPHENOL (MIXED)	100		
88-85-7	DINOSEB	1000	100/10000	
88-85-7	PHENOL, 2,4-DINITRO-6-(1-METHYLPROPYL)	1000		
91-08-7	BENZENE, 2,4-DIISOCYANATOMETHYL-	100		
91-08-7	TOLUENE DIISOCYANATE	100		
91-20-3	NAPHTHALENE	100		
91-22-5	QUINOLINE	5000		
91-58-7	BETA-CHLORONAPHTHALENE	5000		
91-58-7	NAPHTHALENE, 2-CHLORO-	5000		
91-59-8	2-NAPHTHALENAMINE	10		
91-59-8	BETA-NAPHTHYLAMINE	10		
91-80-5	METHAPYRILENE	5000		
91-80-5	PYRIDINE, 2-[(2-(DIMETHYLAMINO)ETHYL)-2-THE-NYLA...	5000		
91-94-1	[1,1'-BIPHENYL]-4,4'-DIAMINE, 3,3'-DICHLORO-	1		
91-94-1	3,3'-DICHLOROBENZIDENE	1		
92-87-5	BENZIDINE	1		
92-87-5	(1,1'-BIPHENYL)-4,4'DIAMINE	1		
93-72-1	PROPANOIC ACID, 2(2,4,5-TRICHLOROPHENOXY)-	100		
93-72-1	SILVEX	100		
93-72-1	2,4,5-TP ACID	100		
93-76-5	2,4,5-T	1000		
93-76-5	2,4,5-T ACID	1000		
93-76-5	2,4,5-(TRICHLOROPHENOXY) ACETIC ACID	1000		
93-79-8	2,4,5-T ESTERS	1000		
94-11-1	2,4-D ESTERS	100		
94-58-6	BENZENE, 1,2-METHYLENEDIOXY-4-PROPYL-	10		
94-58-6	1,3-BENZODIOXOLE, 5-PROPYL-	10		
94-58-6	DIHYDROSAFROLE	10		
94-59-7	BENZENE 1,2-METHYLENEDIOXY-4-ALLYL-	100		
94-59-7	SAFROLE	100		
94-75-7	2,4-D ACID	100		
94-75-7	2,4-D SALTS AND ESTERS	100		
94-75-7	2,4-DICHLOROPHENOXYACETIC ACID, SALTS AND ESTERS	100		
94-79-1	2,4-D ESTERS	100		

Chemical Hazard Communication Guidebook

CAS	CHEMICAL NAME	RQ (LBS)	TPQ (LBS)	NOTES
94-80-4	2,4-D ESTERS	100		
95-47-6	O-BENZENE, DIMETHYL	1000		
95-47-6	O-XYLENE	1000		
95-48-7	O-CRESOL	1000	1000/10000	d
95-48-7	O-CRESYLIC ACID	1000		
95-50-1	BENZENE, 1,2-DICHLORO-	100		
95-50-1	1,2-DICHLOROBENZENE	100		
95-53-4	2-AMINO-1-METHYL BENZENE	100		
95-53-4	BENZENAMINE, 2-METHYL-	100		
95-53-4	O-TOLUIDINE	100		
95-57-8	2-CHLOROPHENOL	100		
95-57-8	PHENOL, 2-CHLORO-	100		
95-80-7	TOLUENEDIAMINE	10		
95-94-3	BENZENE, 1,2,4,5-TETRACHLORO-	5000		
95-94-3	1,2,4,5-TETRACHLOROBENZENE	5000		
95-95-4	PHENOL, 2,4,5-TRICHLORO-	10		
95-95-4	2,4,5-TRICHLOROPHENOL	10		
96-12-8	1,2-DIBROMO-3-CHLOROPROPANE	1		
96-12-8	PROPANE, 1,2-DIBROMO-3-CHLORO	1		
96-45-7	ETHYLENE THIOUREA	10		
96-45-7	2-IMIDAZOLIDINETHIONE	10		
97-63-2	ETHYL METHACRYLATE	1000		
97-63-2	2-PROPENOIC ACID, 2-METHYL-, ETHYL ESTER	1000		
98-01-1	2-FURANCARBOXALDEHYDE	5000		
98-01-1	FURFURAL	5000		
98-07-7	BENZENE, TRICHLOROMETHYL	10		
98-07-7	BENZOIC TRICHLORIDE (BENZOTRICHLORIDE)	10	100	d
98-07-7	BENZOTRICHLORIDE	10	100	d
98-09-9	BENZENE SULFONYL CHLORIDE	100		
98-09-9	BENZENESULFONIC ACID CHLORIDE	100		
98-82-8	BENZENE, 1-METHYLETHYL-	5000		
98-82-8	CUMENE	5000		
98-86-2	ACETOPHENONE	5000		
98-86-2	ETHANONE, 1-PHENYL-	5000		
98-87-3	BENZAL CHLORIDE	5000	500	d
98-87-3	BENZENE, DICHLOROMETHYL-	5000		
98-88-4	BENZOYL CHLORIDE	1000		
98-95-3	BENZENE, NITRO	1000		
98-95-3	NITROBENZENE	1000	10000	l
99-08-1	M-NITROTOLUENE	1000		
99-35-4	BENZENE, 1,3,5-TRINITRO-	10		
99-35-4	SYM-TRINITROBENZENE	10		
99-55-8	BENZENAMINE, 2-METHYL-5-NITRO	100		
99-55-8	5-NITRO-O-TOLUIDINE	100		
99-65-0	M-DINITROBENZENE	100		
99-99-0	P-NITROTOLUENE	1000		
100-01-6	BENZENAMINE, 4-NITRO	5000		

Appendix G: Hazardous Substances (CAS Order)

CAS	CHEMICAL NAME	RQ (LBS)	TPQ (LBS)	NOTES
100-01-6	P-NITROANILINE	5000		
100-02-7	4-NITROPHENOL	100		
100-02-7	P-NITROPHENOL (MIXED)	100		
100-02-7	PHENOL, 4-NITRO	100		
100-25-4	P-DINITROBENZENE	100		
100-41-4	ETHYL BENZENE	1000		
100-42-5	STYRENE	1000		
100-44-7	BENZENE, CHLOROMETHYL-	100		
100-44-7	BENZYL CHLORIDE	100	500	d
100-47-0	BENZONITRILE	5000		
100-75-4	N-NITROSOPIPERIDINE	10		
100-75-4	PYRIDINE, HEXAHYDRO-N-NITROSO-	10		
101-14-4	BENZENAMINE, 4,4'-METHYLENEBIS(2-CHLORO-	10		
101-14-4	4,4'-METHYLENE BIS(2-CHLOROANILINE) (MOCA)	10		
101-55-3	BENZENE, 1-BROMO-4-PHENOXY-	100		
101-55-3	4-BROMOPHENYL PHENYL ETHER	100		
103-85-5	THIOUREA, PHENYL-	100		
105-46-4	SEC-BUTYL ACETATE	5000		
105-67-9	2,4-DIMETHYLPHENOL	100		
105-67-9	PHENOL, 2,4-DIMETHYL-	100		
106-42-3	P-BENZENE, DIMETHYL	1000		
106-42-3	P-XYLENE	1000		
106-44-5	P-CRESOL	1000		
106-44-5	P-CRESYLIC ACID	1000		
106-46-7	BENZENE, 1,4-DICHLORO-	100		
106-46-7	P-DICHLOROBENZENE	100		
106-47-8	BENZENAMINE, 4-CHLORO	1000		
106-47-8	P-CHLOROANILINE	1000		
106-49-0	4-AMINO-1-METHYL BENZENE	100		
106-49-0	P-TOLUIDINE	100		
106-51-4	P-BENZOQUINONE	10		
106-51-4	1,4-CYCLOHEXADIENEDIONE	10		
106-89-8	1-CHLORO-2,3-EPOXYPROPANE	100		
106-89-8	EPICHLOROHYDRIN	100	1000	d,l
106-89-8	OXIRANE, 2-(CHLOROMETHYL)-	100		
106-93-4	1,2-DIBROMOETHANE	1		
106-93-4	ETHANE, 1,2-DIBROMO	1		
106-93-4	ETHYLENE DIBROMIDE	1		
107-02-8	ACROLEIN	1	500	
107-02-8	2-PROPENAL	1		
107-05-1	ALLYL CHLORIDE	1000		
107-06-2	1,2-DICHLOROETHANE	100		
107-06-2	ETHANE, 1,2-DICHLORO-	100		
107-06-2	ETHYLENE DICHLORIDE	100		
107-10-8	1-PROPANAMINE	5000		
107-10-8	N-PROPYLAMINE	5000		
107-12-0	ETHYL CYANIDE	10		

Chemical Hazard Communication Guidebook

CAS	CHEMICAL NAME	RQ (LBS)	TPQ (LBS)	NOTES
107-12-0	PROPANENITRILE	10		
107-13-1	ACRYLONITRILE	100	10000	d,l
107-13-1	2-PROPENENITRILE	100		
107-15-3	ETHYLENE DIAMINE	5000	10000	
107-18-6	ALLYL ALCOHOL	100	1000	
107-18-6	2-PROPEN-1-OL	100		
107-19-7	PROPARGYL ALCOHOL	1000		
107-19-7	2-PROPYN-1-OL	1000		
107-20-0	ACETALDEHYDE, CHLORO-	1000		
107-20-0	CHLOROACETALDEHYDE	1000		
107-30-2	CHLOROMETHYL METHYL ETHER	10	100	c,d
107-30-2	METHANE, CHLOROMETHOXY-	10		
107-49-3	PYROPHOSPHORIC ACID, TETRAETHYL ESTER	10		
107-49-3	TETRAETHYL PYROPHOSPHATE	10		
107-92-6	BUTYRIC ACID	5000		
108-05-4	VINYL ACETATE	5000		
108-10-1	METHYL ISOBUTYL KETONE	5000		
108-10-1	4-METHYL-2-PENTANONE	5000		
108-24-7	ACETIC ANHYDRIDE	5000		
108-31-6	2,5-FURANDIONE	5000		
108-31-6	MALEIC ANHYDRIDE	5000		
108-38-3	M-BENZENE, DIMETHYL	1000		
108-38-3	M-XYLENE	1000		
108-39-4	M-CRESOL	1000		
108-39-4	M-CRESYLIC ACID	1000		
108-46-3	1,3-BENZENEDIOL	5000		
108-46-3	RESORCINOL	5000		
108-60-1	BIS(2-CHLOROISOPROPYL) ETHER	1000		
108-60-1	PROPANE, 2,2'-OXYBIS(2-CHLORO-	1000		
108-88-3	BENZENE, METHYL-	1000		
108-88-3	TOLUENE	1000		
108-90-7	BENZENE CHLORO-	100		
108-90-7	CHLOROBENZENE	100		
108-94-1	CYCLOHEXANONE	5000		
108-95-2	BENZENE, HYDROXY-	1000		
108-95-2	PHENOL	1000	500/10000	
108-98-5	BENZENETHIOL	100		
108-98-5	THIOPHENOL	100	500	
109-06-8	2-PICOLINE	5000		
109-06-8	PYRIDINE, 2-METHYL-	5000		
109-73-9	BUTYLAMINE	1000		
109-77-3	MALONONITRILE	1000	500/10000	
109-77-3	PROPANEDINITRILE	1000		
109-89-7	DIETHYLAMINE	100		
109-99-9	FURAN, TETRAHYDRO-	1000		
109-99-9	TETRAHYDROFURAN	1000		
110-00-9	FURAN	100	500	

Appendix G: Hazardous Substances (CAS Order)

CAS	CHEMICAL NAME	RQ (LBS)	TPQ (LBS)	NOTES
110-00-9	FURFURAN	100		
110-16-7	MALEIC ACID	5000		
110-17-8	FUMARIC ACID	5000		
110-19-0	ISO-BUTYL ACETATE	5000		
110-75-8	2-CHLOROETHYL VINYL ETHER	1000		
110-75-8	ETHENE, 2-CHLOROETHOXY-	1000		
110-80-5	2-ETHOXYETHANOL	1000		
110-80-5	ETHYLENE GLYCOL MONOETHYL ETHER	1000		
110-82-7	BENZENE, HEXAHYDRO-	1000		
110-82-7	CYCLOHEXANE	1000		
110-86-1	PYRIDINE	1000		
111-44-4	BIS(2-CHLOROETHYL) ETHER	10		
111-44-4	DICHLOROETHYL ETHER	10	10000	d
111-44-4	ETHANE, 1,1'-OXYBIS(2-CHLORO-	10		
111-54-6	1,2-ETHANEDIYLBISCARBAMODITHIOIC ACID	5000		
111-54-6	ETHYLENEBIS(DITHIOCARBAMIC ACID)	5000		
111-91-1	BIS(2-CHLOROETHOXY) METHANE	1000		
111-91-1	ETHANE, 1,1'-[METHYLENEBIS(OXY)]BIS(2-CHLORO-	1000		
115-02-6	AZASERINE	1		
115-02-6	L-SERINE, DIAZOACETATE (ESTER)	1		
115-29-7	ENDOSULFAN	1	10/10000	
115-29-7	5-NORBORNENE-2,3-DIMETHANOL, 1,4,5,6,7,7-HEXACHLORO, CYCLIC SULFITE	1		
115-32-2	DICOFOL	10		
116-06-3	ALDICARB	1	100/10000	c
116-06-3	PROPANAL, 2-METHYL-2-(METHYLTHIO)-, O-[(METHYLAMINO)CARBONYL]-OXIME	1		
117-80-6	DICHLONE	1		
117-81-7	1,2-BENZENEDICARBOXYLIC ACID, [BIS(2-ETHYLHEXYL)] ESTER	100		
117-81-7	BIS(2-ETHYLHEXYL) PHTHALATE	100		
117-84-0	1,2-BENZENEDICARBOXYLIC ACID, DI-N-OCTYL ESTER	5000		
117-84-0	DI-N-OCTYL PHTHALATE	5000		
118-74-1	BENZENE, HEXACHLORO-	10		
118-74-1	HEXACHLOROBENZENE	10		
119-90-4	(1,1'-BIPHENYL)-4,4'DIAMINE,3,3'DIMETHOXY-	100		
119-90-4	3,3'-DIMETHOXYBENZIDINE	100		
119-93-7	(1,1'-BIPHENYL)-4,4'DIAMINE,3,3'DIMETHYL-	10		
119-93-7	3,3'-DIMETHYLBENZIDINE	10		
120-12-7	ANTHRACENE	5000		
120-58-1	BENZENE, 1,2-METHYLENEDIOXY-4-PROPENYL-	100		
120-58-1	1,3-BENZODIOXOLE, 5-(1-PROPENYL)-	100		
120-58-1	ISOSAFROLE	100		
120-82-1	1,2,4-TRICHLOROBENZENE	100		
120-83-2	2,4-DICHLOROPHENOL	100		
120-83-2	PHENOL, 2,4-DICHLORO-	100		
121-14-2	BENZENE, 1-METHYL-2,4-DINITRO-	10		

Chemical Hazard Communication Guidebook

CAS	CHEMICAL NAME	RQ (LBS)	TPQ (LBS)	NOTES
121-14-2	2,4-DINITROTOLUENE	10		
121-21-1	PYRETHRINS	1		
121-29-9	PYRETHRINS	1		
121-44-8	TRIETHYLAMINE	5000		
121-75-5	MALATHION	100		
122-09-8	ALPHA,ALPHA-DIMETHYL PHENETHYLAMINE	5000		
122-09-8	ETHANAMINE, 1,1-DIMETHYL-2-PHENYL	5000		
122-66-7	1,2-DIPHENYL HYDRAZINE (HYDRAZOBENZENE)	10		
122-66-7	HYDRAZINE, 1,2-DIPHENYL	10		
123-33-1	1,2-DIHYDRO-3,6-PYRIDAZINEDIONE	5000		
123-33-1	MALEIC HYDRAZIDE	5000		
123-63-7	PARALDEHYDE	1000		
123-63-7	1,3,5-TRIOXANE, 2,4,6-TRIMETHYL-	1000		
123-73-9	2-BUTENAL	100		
123-73-9	CROTONALDEHYDE (E)-	100	1000	
123-86-4	BUTYL ACETATE	5000		
123-91-1	1,4-DIETHYLENE DIOXIDE	100		
123-91-1	1,4-DIOXANE	100		
123-92-2	ISO-AMYL ACETATE	5000		
124-04-9	ADIPIC ACID	5000		
124-40-3	DIMETHYLAMINE	1000		
124-40-3	METHANAMINE, N-METHYL-	1000		
124-41-4	SODIUM METHYLATE	1000		
124-48-1	CHLORODIBROMOMETHANE	100		
126-72-7	1-PROPANOL, 2,3-DIBROMO-, PHOSPHATE (3:1)	10		
126-72-7	TRIS(2,3-DIBROMOPROPYL)PHOSPHATE	10		
126-98-7	METHACRYLONITRILE	1000	500	h
126-98-7	2-PROPENENITRILE, 2-METHYL	1000		
127-18-4	ETHENE, 1,1,2,2-TETRACHLORO-	100		
127-18-4	TETRACHLOROETHYLENE	100		
127-82-2	ZINC PHENOLSULFONATE	5000		
129-00-0	PYRENE	5000	1000/10000	c
130-15-4	1,4-NAPHTHALENEDIONE	5000		
130-15-4	1,4-NAPHTHOQUINONE	5000		
131-11-3	1,2-BENZENEDICARBOXYLIC ACID, DIMETHYL ESTER	5000		
131-11-3	DIMETHYL PHTHALATE	5000		
131-74-8	AMMONIUM PICRATE	10		
131-74-8	PHENOL, 2,4,6-TRINITRO- AMMONIUM SALT	10		
131-89-5	4,6-DINITRO-O-CYCLOHEXYLPHENOL	100		
131-89-5	PHENOL 2-CYCLOHEXYL-4-6-DINITRO-	100		
133-06-2	CAPTAN	10		
133-06-2	CAPTAN [1H-ISOINDOLE-1,3(2H)-DIONE,3A,4,7,7A-TETRAHYDRO-2-[(TRICHLOROM	10		
134-32-7	1-NAPHTHYLAMINE	100		
134-32-7	ALPHA-NAPHTHYLAMINE	100		
137-26-8	BIS(DIMETHYLTHIOCARBAMOYL) DISULFIDE	10		
137-26-8	THIRAM	10		

Appendix G: Hazardous Substances (CAS Order)

CAS	CHEMICAL NAME	RQ (LBS)	TPQ (LBS)	NOTES
140-88-5	ETHYL ACRYLATE	1000		
140-88-5	2-PROPENOIC ACID, ETHYL ESTER	1000		
141-78-6	ACETIC ACID, ETHYL ESTER	5000		
141-78-6	ETHYL ACETATE	5000		
142-28-9	1,3-DICHLOROPROPANE	1000		
142-71-2	CUPRIC ACETATE	100		
142-84-7	DIPROPYLAMINE	5000		
142-84-7	1-PROPANAMINE, N-PROPYL	5000		
143-33-9	SODIUM CYANIDE	10	100	b
143-50-0	DECACHLOROOCTAHYDRO-1,3,4-METHENO-2H-CYCLOBUTA[C,D]-PENTALEN-2-ONE	1		
143-50-0	KEPONE	1		
145-73-3	ENDOTHALL	1000		
145-73-3	7-OXABICYCLO[2.2.1]HEPTANE-2,3-DICARBOXYLIC ACID	1000		
148-82-3	ALANINE, 3-[P-BIS(2-CHLOROETHYL)AMINO]PHENYL-, L-	1		
148-82-3	MELPHALAN	1		
151-50-8	CYANIDES			
151-50-8	POTASSIUM CYANIDE	10	100	b
151-56-4	AZIRIDINE	1		
151-56-4	ETHYLENIMINE	1		
152-16-9	DIPHOSPHORAMIDE, OCTAMETHYL-	100	100	
152-16-9	OCTAMETHYLPYROPHOSPHORAMIDE	100		
156-60-5	1,2-TRANS-DICHLOROETHYLENE	1000		
156-60-5	ETHENE, TRANS-1,2,DICHLORO-	1000		
189-55-9	BENZO[RST]PENTAPHENE	10		
189-55-9	1,2:7,8-DIBENZOPYRENE	10		
189-55-9	DIBENZO[A,I]PYRENE	10		
189-55-9	DIBENZ[A,I]PYRENE	10		
191-24-2	BENZO[GHI]PERYLENE	5000		
193-39-5	INDENO [1,2,3-CD]PYRENE	100		
193-39-5	1,10-(1,2-PHENYLENE)PYRENE	100		
205-99-2	BENZO[B]FLUORANTHENE	1		
206-44-0	BENZO[J.K.]FLUORENE	100		
206-44-0	FLUORANTHENE	100		
207-08-9	BENZO[K]FLUORANTHENE	5000		
208-96-8	ACENAPHTHYLENE	5000		
218-01-9	1,2-BENZPHENANTHRENE	100		
218-01-9	CHRYSENE	100		
225-51-4	3,4-BENZACRIDINE	100		
225-51-4	BENZ[C]ACRIDINE	100		
297-97-2	O,O-DIETHYL O-PYRAZINYL PHOSPHOROTHIOATE	100		
297-97-2	PHOSPHOROTHIOIC ACID, O,O-DIETHYL O-PYRAZINYL ESTER	100		
298-00-0	O,O-DIMETHYL O-P-NITROPHENYL PHOSPHOROTHIOATE	100		
298-00-0	METHYL PARATHION	100		

Chemical Hazard Communication Guidebook

CAS	CHEMICAL NAME	RQ (LBS)	TPQ (LBS)	NOTES
298-02-2	PHORATE	10	10	
298-02-2	PHOSPHORODITHIOIC ACID,O,O-DIETHYL S-(ETHYLTHIO) METHYL ESTER	10		
298-04-4	O,O-DIETHYL S-[2-(ETHYLTHIO)ETHYL]PHOSPHORODITHIOATE	1		
298-04-4	DISULFOTON	1	500	
300-76-5	NALED	10		
301-04-2	ACETIC ACID, LEAD SALT	5000		
301-04-2	LEAD ACETATE	5000		
302-01-2	DIAMINE	1		
302-01-2	HYDRAZINE	1	1000	d
303-34-4	2-BUTENOIC ACID, 2-METHYL-, 7-[[2,3-DIHYDROXY-2-(1-METHOXYETHYL)-3-METHY	10		
303-34-4	LASIOCARPINE	10		
305-03-3	BUTANOIC ACID,4-[BIS(2-CHLOROETHYL)-AMINO]BENZENE-	10		
305-03-3	CHLORAMBUCIL	10		
309-00-2	ALDRIN	1	500/10000	d
309-00-2	ALDRIN [1,4:5,8-DIMETHANONAPHTHALENE,1,2,3,4,10,10-HEXACHLORO-1,4,4A,5	1	500/10000	d
309-00-2	1,2,3,4,10,10-HEXACHLORO-1,4,4A,5,8,8A-HEXAHYDRO-1,4,5,8-ENDO,EXO-DIMETHANONAPHTH. ...	1		
311-45-5	DIETHYL-P-NITROPHENYL PHOSPHATE	100		
311-45-5	PHOSPHORIC ACID, DIETHYL P-NITROPHENYL ESTER	100		
315-18-4	MEXACARBATE	1000	500/10000	
319-84-6	ALPHA-BHC	10		
319-85-7	BETA-BHC	1		
319-86-8	DELTA-BHC	1		
329-71-5	2,5-DINITROPHENOL	10		
330-54-1	DIURON	100		
353-50-4	CARBON OXYFLUORIDE	1000		
353-50-4	CARBONYL FLUORIDE	1000		
357-57-3	BRUCINE	100		
357-57-3	STRYCHNIDIN-10-ONE, 2,3-DIMETHOXY-	100		
460-19-5	CYANOGEN	100		
465-73-6	HEXACHLOROHEXAHYDRO-ENDO,ENDO-DIMETHANONAPHTHALENE	1		
465-73-6	1,2,3,4,10,10-HEXACHLORO-1,4,4A,5,8,8A-HEXAHYDRO-1,4,5,8-ENDO,ENDO-DIMETHANONAPHT ...	1		
492-80-8	AURAMINE	100		
492-80-8	BENZENAMINE, 4,4'-CARBONIMIDOYLBIS(N,N-DIMETHYL-	100		
494-03-1	CHLORNAPHAZINE	100		
494-03-1	2-NAPHTHYLAMINE, N,N-BIS(2-CHLOROETHYL)-	100		
496-72-0	DIAMINOTOLUENE	10		
496-72-0	TOLUENEDIAMINE	10		

Appendix G: Hazardous Substances (CAS Order)

CAS	CHEMICAL NAME	RQ (LBS)	TPQ (LBS)	NOTES
504-24-5	4-AMINOPYRIDINE	1000		
504-24-5	4-PYRIDINAMINE	1000		
504-60-9	1-METHYL BUTADIENE	100		
504-60-9	1,3-PENTADIENE	100		
506-61-6	POTASSIUM SILVER CYANIDE	1	500	b
506-64-9	SILVER CYANIDE	1		
506-68-3	BROMINE CYANIDE	1000		
506-68-3	CYANOGEN BROMIDE	1000	500/10000	
506-77-4	CHLORINE CYANIDE	10		
506-77-4	CYANOGEN CHLORIDE	10		
506-87-6	AMMONIUM CARBONATE	5000		
506-96-7	ACETYL BROMIDE	5000		
509-14-8	METHANE TETRANITRO -	10		
509-14-8	TETRANITROMETHANE	10	500	
510-15-6	BENZENEACETIC ACID, 4-CHLORO-ALPHA- (4-CHLOROPHENYL)-ALPHA-HYDROXY-, E	10		
510-15-6	ETHYL 4,4'-DICHLOROBENZILATE	10		
513-49-5	SEC-BUTYLAMINE	1000		
528-29-0	O-DINITROBENZENE	100		
534-52-1	4,6-DINITRO-O-CRESOL AND SALTS	10		
534-52-1	PHENOL, 2,4-DINITRO-6-METHYL- AND SALTS	10		
540-73-8	1,2-DIMETHYLHYDRAZINE	1		
540-73-8	HYDRAZINE, 1,2-DIMETHYL	1		
540-88-5	TERT-BUTYL ACETATE	5000		
541-09-3	URANYL ACETATE	100		
541-53-7	THIOIMIDODICARBONIC DIAMIDE	100		
541-73-1	BENZENE, 1,3-DICHLORO-	100		
541-73-1	1,3-DICHLOROBENZENE	100		
542-62-1	BARIUM CYANIDE	10		
542-75-6	PROPENE, 1,3-DICHLORO	100		
542-76-7	3-CHLOROPROPIONITRILE	1000		
542-76-7	PROPANENITRILE, 3-CHLORO-	1000		
542-88-1	BIS(CHLOROMETHYL) ETHER	10		
542-88-1	METHANE, OXYBIS(CHLORO-	10		
543-90-8	CADMIUM ACETATE	10		
544-18-3	COBALTOUS FORMATE	1000		
544-92-3	COPPER CYANIDE	10		
554-84-7	M-NITROPHENOL (MIXED)	100		
557-19-7	NICKEL CYANIDE	10		
557-19-7	NICKEL(II) CYANIDE	10		
557-21-1	ZINC CYANIDE	10		
557-34-6	ZINC ACETATE	1000		
557-41-5	ZINC FORMATE	1000		
563-12-2	ETHION	10	1000	
563-68-8	ACETIC ACID, THALLIUM (1+) SALT	100		
563-68-8	THALLIUM (I) ACETATE	100		
573-56-8	2,6-DINITROPHENOL	10		

Chemical Hazard Communication Guidebook

CAS	CHEMICAL NAME	RQ (LBS)	TPQ (LBS)	NOTES
584-84-9	BENZENE, 2,4-DIISOCYANATOMETHYL-	100		
584-84-9	TOLUENE DIISOCYANATE	100		
591-08-2	ACETAMIDE, N-(AMINOTHIOXOMETHYL)-	1000		
591-08-2	1-ACETYL-2-THIOUREA	1000		
592-01-3	CALCIUM CYANIDE	10		
592-04-1	MERCURIC CYANIDE	1		
592-85-8	MERCURIC THIOCYANATE	10		
592-87-0	LEAD THIOCYANATE	100		
594-42-3	METHANESULFENYL CHLORIDE, TRICHLORO-	100		
594-42-3	TRICHLOROMETHANESULFENYL CHLORIDE	100		
598-31-2	BROMOACETONE	1000		
598-31-2	2-PROPANONE, 1-BROMO-	1000		
606-20-2	BENZENE, 1-METHYL-2,6-DINITRO-	100		
606-20-2	2,6-DINITROTOLUENE	100		
608-73-1	HEXACHLOROCYCLOHEXANE (ALL ISOMERS)			
608-93-5	BENZENE, PENTACHLORO-	10		
608-93-5	PENTACHLOROBENZENE	10		
609-19-8	3,4,5-TRICHLOROPHENOL	10		
610-39-9	3,4-DINITROTOLUENE	10		
615-53-2	CARBAMIC ACID, METHYLNITROSO-, ETHYL ESTER	1		
615-53-2	N-NITROSO-N-METHYLURETHANE	1		
621-64-7	DI-N-PROPYLNITROSAMINE	10		
621-64-7	N-NITROSODI-N-PROPYLAMINE	10		
624-83-9	ISOCYANIC ACID, METHYL ESTER	1		
624-83-9	METHYL ISOCYANATE	1	500	f
625-16-1	TERT-AMYL ACETATE	5000		
626-38-0	SEC-AMYL ACETATE	5000		
628-63-7	AMYL ACETATE	5000		
628-86-4	FULMINIC ACID, MERCURY(II) SALT	10		
628-86-4	MERCURY FULMINATE	10		
630-10-4	CARBAMIMIDOSELENOIC ACID	1000		
630-10-4	SELENOUREA	1000		
630-20-6	ETHANE 1,1,1,2-TETRACHLORO-	100		
630-20-6	1,1,1,2-TETRACHLOROETHANE	100		
631-61-8	AMMONIUM ACETATE	5000		
636-21-5	BENZENAMINE, 2-METHYL-, HYDROCHLORIDE	100		
636-21-5	O-TOLUIDINE HYDROCHLORIDE	100		
640-19-7	ACETAMIDE, 2-FLUORO-	100		
640-19-7	FLUOROACETAMIDE	100	100/10000	j
684-93-5	CARBAMIDE, N-METHYL-N-NITROSO-	1		
684-93-5	N-NITROSO-N-METHYLUREA	1		
692-42-2	ARSINE, DIETHYL	1		
692-42-2	DIETHYLARSINE	1		
696-28-6	DICHLOROPHENYLARSINE	1		
696-28-6	PHENOL DICHLOROARSINE	1		
696-28-6	PHENYL DICHLOROARSINE	1	500	d,h
757-58-4	HEXAETHYL TETRAPHOSPHATE	100		

Appendix G: Hazardous Substances (CAS Order)

CAS	CHEMICAL NAME	RQ (LBS)	TPQ (LBS)	NOTES
757-58-4	TETRAPHOSPHORIC ACID, HEXAETHYL ESTER	100		
759-73-9	CARBAMIDE, N-ETHYL-N-NITROSO-	1		
759-73-9	N-NITROSO-N-ETHYLUREA	1		
764-41-0	2-BUTENE, 1,4-DICHLORO-	1		
764-41-0	1,4-DICHLORO-2-BUTENE	1		
765-34-4	GLYCIDYLALDEHYDE	10		
765-34-4	1-PROPANAL, 2,3-EPOXY-	10		
765-34-4	1-PROPANOL, 2,3-EPOXY-	10		
815-82-7	CUPRIC TARTRATE	100		
823-40-5	DIAMINOTOLUENE	10		
823-40-5	TOLUENEDIAMINE	10		
924-16-3	1-BUTANAMINE, N-BUTYL-N-NITROSO-	10		
924-16-3	N-NITROSODI-N-BUTYLAMINE	10		
930-55-2	N-NITROSOPYRROLIDINE	1		
930-55-2	PYRROLE, TETRAHYDRO-N-NITROSO-	1		
933-75-5	2,3,6-TRICHLOROPHENOL	10		
933-78-8	2,3,5-TRICHLOROPHENOL	10		
959-98-8	ALPHA-ENDOSULFAN	1		
1024-57-3	HEPTACHLOR EPOXIDE	1		
1031-07-8	ENDOSULFAN SULFATE	1		
1066-30-4	CHROMIC ACETATE	1000		
1066-33-7	AMMONIUM BICARBONATE	5000		
1072-35-1	LEAD STEARATE	5000		
1111-78-0	AMMONIUM CARBAMATE	5000		
1116-54-7	ETHANOL, 2,2'-(NITROSOIMINO)BIS	1		
1116-54-7	N-NITROSODIETHANOLAMINE	1		
1120-71-4	1,2-OXATHIOLANE 2,2-DIOXIDE	10		
1120-71-4	1,3-PROPANE SULTONE	10		
1185-57-5	FERRIC AMMONIUM CITRATE	1000		
1194-65-6	DICHLOBENIL	100		
1300-71-6	XYLENOL	1000		
1303-28-2	ARSENIC PENTOXIDE	1	100/10000	d
1303-28-2	ARSENIC(V) OXIDE	1		
1303-32-8	ARSENIC DISULFIDE	1		
1303-33-9	ARSENIC TRISULFIDE	1		
1309-64-4	ANTIMONY TRIOXIDE	1000		
1310-58-3	POTASSIUM HYDROXIDE	1000		
1310-73-2	SODIUM HYDROXIDE	1000		
1314-32-5	THALLIC OXIDE	100		
1314-32-5	THALLIUM(III) OXIDE	100		
1314-62-1	VANADIUM (V) OXIDE	1000		
1314-62-1	VANADIUM PENTOXIDE	1000	100/10000	
1314-80-3	PHOSPHORUS PENTASULFIDE	100		
1314-80-3	PHOSPHORUS SULFIDE	100		
1314-80-3	SULFUR PHOSPHIDE	100		
1314-84-7	ZINC PHOSPHIDE	100	500	b
1314-87-0	LEAD SULFIDE	5000		

Chemical Hazard Communication Guidebook

CAS	CHEMICAL NAME	RQ (LBS)	TPQ (LBS)	NOTES
1319-72-8	2,4,5-T AMINES	5000		
1319-77-3	CRESOL(S)	1000		
1319-77-3	CRESYLIC ACID	1000		
1320-18-9	2,4-D ESTERS	100		
1321-12-6	NITROTOLUENE	1000		
1327-52-2	ARSENIC ACID	1		
1327-53-3	ARSENIC TRIOXIDE	1		
1327-53-3	ARSENIC(III) OXIDE	1		
1330-20-7	BENZENE, DIMETHYL-	1000		
1330-20-7	XYLENE (MIXED ISOMERS)	1000		
1332-07-6	ZINC BORATE	1000		
1332-21-4	ASBESTOS	1		
1333-83-1	SODIUM BIFLUORIDE	100		
1335-32-6	LEAD SUBACETATE	100		
1335-32-6	LEAD, BIS(ACETATO-O)TETRAHYDROXYTRI-	100		
1336-21-6	AMMONIUM HYDROXIDE	1000		
1336-36-3	POLYCHLORINATED BIPHENYLS	1		
1338-23-4	2-BUTANONE PEROXIDE	10		
1338-23-4	METHYL ETHYL KETONE PEROXIDE	10		
1338-24-5	NAPHTHENIC ACID	100		
1341-49-7	AMMONIUM BIFLUORIDE	100		
1464-53-5	2,2'-BIOXIRANE	10		
1464-53-5	DIEPOXYBUTANE	10	500	d
1464-53-5	1,2:3,4-DIEPOXYBUTANE	10		
1563-66-2	CARBOFURAN	10	10/10000	
1615-80-1	N,N'-DIETHYLHYDRAZINE	10		
1615-80-1	HYDRAZINE 1,2-DIETHYL-	10		
1746-01-6	2,3,7,8-TETRACHLORODIBENZO-P-DIOXIN	1		
1762-95-4	AMMONIUM THIOCYANATE	5000		
1863-63-4	AMMONIUM BENZOATE	5000		
1888-71-7	HEXACHLOROPROPENE	1000		
1888-71-7	1-PROPENE, 1,1,2,3,3,3-HEXACHLORO-	1000		
1918-00-9	DICAMBA	1000		
1928-38-7	2,4-D ESTERS	100		
1928-47-8	2,4,5-T ESTERS	1000		
1928-61-6	2,4-D ESTERS	100		
1929-73-3	2,4-D ESTERS	100		
2008-46-0	2,4,5-T AMINES	5000		
2032-65-7	MERCAPTODIMETHUR	10		
2303-16-4	DIALLATE	100		
2303-16-4	S-(2,3-DICHLOROALLYL) DIISOPROPYLTHIOCARBAMATE	100		
2312-35-8	PROPARGITE	10		
2545-59-7	2,4,5-T ESTERS	1000		
2763-96-4	5-(AMINOMETHYL)-3-ISOXAZOLOL	1000		
2763-96-4	3(2H)-ISOXAZOLONE, 5-(AMINOMETHYL)-	1000		
2764-72-9	DIQUAT	1000		

Appendix G: Hazardous Substances (CAS Order)

CAS	CHEMICAL NAME	RQ (LBS)	TPQ (LBS)	NOTES
2921-88-2	CHLORPYRIFOS	1		
2944-67-4	FERRIC AMMONIUM OXALATE	1000		
2971-38-2	2,4-D ESTERS	100		
3012-65-5	AMMONIUM CITRATE DIBASIC	5000		
3164-29-2	AMMONIUM TARTRATE	5000		
3165-93-3	BENZENAMINE, 4-CHLORO-2-METHYL-, HYDROCHLORIDE	100		
3165-93-3	4-CHLORO-O-TOLUIDINE HYDROCHLORIDE	100		
3251-23-8	CUPRIC NITRATE	100		
3288-58-2	O,O-DIETHYL S-METHYL DITHIOPHOSPHATE	5000		
3288-58-2	PHOSPHORODITHIOIC ACID, O,O-DIETHYL S-METHYLESTER	5000		
3486-35-9	ZINC CARBONATE	1000		
3689-24-5	DITHIOPYROPHOSPHORIC ACID, TETRAETHYL ESTER	100		
3689-24-5	TETRAETHYLDITHIOPYROPHOSPHATE	100		
3813-14-7	2,4,5-T AMINES	5000		
4170-30-3	2-BUTENAL	100		
4170-30-3	CROTONALDEHYDE	100	1000	
4549-40-0	ETHENAMINE, N-METHYL-N-NITROSO-	10		
4549-40-0	N-NITROSOMETHYLVINYLAMINE	10		
5333-41-5	DIAZINON	1		
5344-82-1	1-(O-CHLOROPHENYL)THIOUREA	100		
5344-82-1	THIOUREA, (2-CHLOROPHENYL)-	100	100/10000	
5893-66-3	CUPRIC OXALATE	100		
5972-73-6	AMMONIUM OXALATE	5000		
6009-70-7	AMMONIUM OXALATE	5000		
6369-96-6	2,4,5-T AMINES	5000		
6369-97-7	2,4,5-T AMINES	5000		
6533-73-9	CARBONIC ACID DITHALLIUM (I) SALT	100	100/10000	c,h
6533-73-9	THALLIUM(I) CARBONATE	100		
7005-72-3	4-CHLOROPHENYL PHENYL ETHER	5000		
7421-93-4	ENDRIN ALDEHYDE	1		
7428-48-0	LEAD STEARATE	5000		
7439-92-1	LEAD	1		
7439-97-6	MERCURY	1		
7440-02-0	NICKEL	100		d
7440-22-4	SILVER	1000		
7440-23-5	SODIUM	10		
7440-28-0	THALLIUM	1000		
7440-36-0	ANTIMONY	5000		
7440-38-2	ARSENIC	1		
7440-41-7	BERYLIUM DUST	10		
7440-41-7	BERYLLIUM	10		
7440-41-7	BERYLLIUM DUST	10		
7440-43-9	CADMIUM	10		
7440-47-3	CHROMIUM	5000		
7440-50-8	COPPER	5000		

Chemical Hazard Communication Guidebook

CAS	CHEMICAL NAME	RQ (LBS)	TPQ (LBS)	NOTES
7440-66-6	ZINC	1000		
7446-08-4	SELENIUM DIOXIDE	10		
7446-08-4	SELENIUM DIOXIDE, AS /SEO2/	10		
7446-14-2	LEAD SULFATE	100		
7446-18-6	SULFURIC ACID, THALLIUM(I) SALT	100		
7446-18-6	THALLIUM(I) SULFATE	100		
7446-27-7	LEAD PHOSPHATE	1		
7446-27-7	PHOSPHORIC ACID, LEAD SALT	1		
7447-39-4	CUPRIC CHLORIDE	10		
7488-56-4	SELENIUM DISULFIDE	10		
7488-56-4	SELENIUM SULFIDE	10		
7488-56-4	SULFUR SELENIDE	10		
7558-79-4	SODIUM PHOSPHATE DIBASIC	5000		
7601-54-9	SODIUM PHOSPHATE TRIBASIC	5000		
7631-89-2	SODIUM ARSENATE	1	1000/10000	d
7631-90-5	SODIUM BISULFITE	5000		
7632-00-0	SODIUM NITRITE	100		
7645-25-2	LEAD ARSENATE	1		
7646-85-7	ZINC CHLORIDE	1000		
7647-01-0	HYDROCHLORIC ACID	5000		
7647-18-9	ANTIMONY PENTACHLORIDE	1000		
7664-38-2	PHOSPHORIC ACID	5000		
7664-39-3	HYDROFLUORIC ACID	100		
7664-39-3	HYDROGEN FLUORIDE	100	100	
7664-41-7	AMMONIA	100	500	l
7664-93-9	SULFURIC ACID	1000	1000	
7681-49-4	SODIUM FLUORIDE	1000		
7681-52-9	SODIUM HYPOCHLORITE	100		
7697-37-2	NITRIC ACID	1000	1000	
7699-45-8	ZINC BROMIDE	1000		
7705-08-0	FERRIC CHLORIDE	1000		
7718-54-9	NICKEL CHLORIDE	100		
7719-12-2	PHOSPHORUS TRICHLORIDE	1000	1000	
7720-78-7	FERROUS SULFATE	1000		
7722-64-7	POTASSIUM PERMANGANATE	100		
7723-14-0	PHOSPHORUS	1	100	b,h
7733-02-0	ZINC SULFATE	1000		
7738-94-5	CHROMIC ACID	10		
7758-29-4	SODIUM PHOSPHATE TRIBASIC	5000		
7758-94-3	FERROUS CHLORIDE	100		
7758-95-4	LEAD CHLORIDE	100		
7758-98-7	CUPRIC SULFATE	10		
7761-88-8	SILVER NITRATE	1		
7773-06-0	AMMONIUM SULFAMATE	5000		
7775-11-3	SODIUM CHROMATE	10		
7778-39-4	ARSENIC ACID	1		
7778-44-1	CALCIUM ARSENATE	1	500/10000	d

Appendix G: Hazardous Substances (CAS Order)

CAS	CHEMICAL NAME	RQ (LBS)	TPQ (LBS)	NOTES
7778-50-9	POTASSIUM BICHROMATE	10		
7778-54-3	CALCIUM HYPOCHLORITE	10		
7779-86-4	ZINC HYDROSULFITE	1000		
7779-88-6	ZINC NITRATE	1000		
7782-41-4	FLUORINE	10	500	k
7782-49-2	SELENIUM	100		
7782-50-5	CHLORINE	10	100	
7782-63-0	FERROUS SULFATE	1000		
7782-82-3	SODIUM SELENITE	1000		
7782-86-7	MERCUROUS NITRATE	10		
7783-00-8	SELENIOUS ACID	10	1000/10000	
7783-06-4	HYDROGEN SULFIDE	100	500	l
7783-06-4	HYDROSULFURIC ACID	100		
7783-06-4	SULFUR HYDRIDE	100		
7783-35-9	MERCURIC SULFATE	10		
7783-46-2	LEAD FLUORIDE	100		
7783-49-5	ZINC FLUORIDE	1000		
7783-50-8	FERRIC FLUORIDE	100		
7783-56-4	ANTIMONY TRIFLUORIDE	1000		
7784-34-1	ARSENIC TRICHLORIDE	1		
7784-34-1	ARSENOUS TRICHLORIDE	1	500	d
7784-40-9	LEAD ARSENATE	1		
7784-41-0	POTASSIUM ARSENATE	1		
7784-46-5	SODIUM ARSENITE	1	500/10000	d
7785-84-4	SODIUM PHOSPHATE TRIBASIC	5000		
7786-34-7	MEVINPHOS	10	500	
7786-81-4	NICKEL SULFATE	100		
7787-47-5	BERYLLIUM CHLORIDE	1		
7787-49-7	BERYLLIUM FLUORIDE	1		
7787-55-5	BERYLLIUM NITRATE	1		
7788-98-9	AMMONIUM CHROMATE	10		
7789-00-6	POTASSIUM CHROMATE	10		
7789-06-2	STRONTIUM CHROMATE	10		
7789-09-5	AMMONIUM BICHROMATE	10		
7789-42-6	CADMIUM BROMIDE	10		
7789-43-7	COBALTOUS BROMIDE	1000		
7789-61-9	ANTIMONY TRIBROMIDE	1000		
7790-94-5	CHLOROSULFONIC ACID	1000		
7791-12-0	THALLIUM(I) CHLORIDE	100		
7803-51-2	HYDROGEN PHOSPHIDE	100		
7803-51-2	PHOSPHINE	100	500	
7803-55-6	AMMONIUM VANADATE	1000		
7803-55-6	VANADIC ACID, AMMONIUM SALT	1000		
8001-35-2	CAMPHENE, OCTACHLORO-	1		
8001-35-2	TOXAPHENE	1		
8001-58-9	CREOSOTE	1		

Chemical Hazard Communication Guidebook

CAS	CHEMICAL NAME	RQ (LBS)	TPQ (LBS)	NOTES
8003-19-8	DICHLOROPROPANE - DICHLOROPROPENE (MIXTURE)	100		
8003-34-7	PYRETHRINS	1		
8014-95-7	SULFURIC ACID	1000		
9004-66-4	FERRIC DEXTRAN	5000		
10022-70-5	SODIUM HYPOCHLORITE	100		
10025-87-3	PHOSPHORUS OXYCHLORIDE	1000	500	d
10025-91-9	ANTIMONY TRICHLORIDE	1000		
10026-11-6	ZIRCONIUM TETRACHLORIDE	5000		
10028-22-5	FERRIC SULFATE	1000		
10031-59-1	SULFURIC ACID, THALLIUM(I) SALT	100		
10031-59-1	THALLIUM(I) SULFATE	100		
10039-32-4	SODIUM PHOSPHATE DIBASIC	5000		
10043-01-3	ALUMINUM SULFATE	5000		
10045-89-3	FERROUS AMMONIUM SULFATE	1000		
10045-94-0	MERCURIC NITRATE	10		
10049-05-5	CHROMOUS CHLORIDE	1000		
10099-74-8	LEAD NITRATE	100		
10101-53-8	CHROMIC SULFATE	1000		
10101-63-0	LEAD IODIDE	100		
10101-89-0	SODIUM PHOSPHATE TRIBASIC	5000		
10102-06-4	URANYL NITRATE	100		
10102-18-8	SODIUM SELENITE	100	100/10000	h
10102-43-9	NITRIC OXIDE	10	100	c
10102-43-9	NITROGEN (II) OXIDE	10		
10102-44-0	NITROGEN (IV) OXIDE	10		
10102-44-0	NITROGEN DIOXIDE	10	100	
10102-45-1	THALLIUM(I) NITRATE	100		
10102-48-4	LEAD ARSENATE	1		
10108-64-2	CADMIUM CHLORIDE	10		
10124-50-2	POTASSIUM ARSENITE	1	500/10000	d
10124-56-8	SODIUM PHOSPHATE, TRIBASIC	5000		
10140-65-5	SODIUM PHOSPHATE, DIBASIC	5000		
10192-30-0	AMMONIUM BISULFITE	5000		
10196-04-0	AMMONIUM SULFITE	5000		
10361-89-4	SODIUM PHOSPHATE TRIBASIC	5000		
10380-29-7	CUPRIC SULFATE AMMONIATED	100		
10415-75-5	MERCUROUS NITRATE	10		
10421-48-4	FERRIC NITRATE	1000		
10544-72-6	NITROGEN (IV) OXIDE	10		
10544-72-6	NITROGEN DIOXIDE	10		
10588-01-9	SODIUM BICHROMATE	10		
11096-82-5	AROCLOR 1260	1		
11096-82-5	POLYCHLORINATED BIPHENYLS	1		
11097-69-1	AROCLOR 1254	1		
11097-69-1	POLYCHLORINATED BIPHENYLS	1		
11104-28-2	AROCLOR 1221	1		

Appendix G: Hazardous Substances (CAS Order)

CAS	CHEMICAL NAME	RQ (LBS)	TPQ (LBS)	NOTES
11104-28-2	POLYCHLORINATED BIPHENYLS	1		
11115-74-5	CHROMIC ACID	10		
11141-16-5	AROCLOR 1232	1		
11141-16-5	POLYCHLORINATED BIPHENYLS	1		
12002-03-8	CUPRIC ACETOARSENITE	1		
12039-52-0	THALLIUM(I) SELENIDE	1000		
12054-48-7	NICKEL HYDROXIDE	10		
12125-01-8	AMMONIUM FLUORIDE	100		
12125-02-9	AMMONIUM CHLORIDE	5000		
12135-76-1	AMMONIUM SULFIDE	100		
12672-29-6	AROCLOR 1248	1		
12672-29-6	POLYCHLORINATED BIPHENYLS	1		
12674-11-2	AROCLOR 1016	1		
12674-11-2	POLYCHLORINATED BIPHENYLS	1		
12771-08-3	SULFUR MONOCHLORIDE	1000		
13463-39-3	NICKEL CARBONYL	10	1	d
13463-39-3	NICKEL TETRACARBONYL	10		
13560-99-1	2,4,5-T SALTS	1000		
13597-99-4	BERYLLIUM NITRATE	1		
13746-89-9	ZIRCONIUM NITRATE	5000		
13765-19-0	CALCIUM CHROMATE	10		
13765-19-0	CHROMIC ACID, AS /H2CRO4/, CALCIUM SALT (1:1)	10		
13765-19-0	CHROMIC ACID, CALCIUM SALT	10		
13814-96-5	LEAD FLUOBORATE	100		
13826-83-0	AMMONIUM FLUOBORATE	5000		
13952-84-6	SEC-BUTYLAMINE	1000		
14017-41-5	COBALTOUS SULFAMATE	1000		
14216-75-2	NICKEL NITRATE	100		
14258-49-2	AMMONIUM OXALATE	5000		
14307-35-8	LITHIUM CHROMATE	10		
14307-43-8	AMMONIUM TARTRATE	5000		
14639-97-5	ZINC AMMONIUM CHLORIDE	5000		
14639-98-6	ZINC AMMONIUM CHLORIDE	5000		
14644-61-2	ZIRCONIUM SULFATE	5000		
15699-18-0	NICKEL AMMONIUM SULFATE	100		
15739-80-7	LEAD SULFATE	100		
15950-66-0	2,3,4-TRICHLOROPHENOL	10		
16721-80-5	SODIUM HYDROSULFIDE	5000		
16752-77-5	ACETIMIDIC ACID, N-(METHYLCARBAMOYL)OXY-LTHIO-, METHYL ESTER	100		
16752-77-5	METHOMYL	100	500/10000	h
16871-71-9	ZINC SILICOFLUORIDE	5000		
16919-19-0	AMMONIUM SILICOFLUORIDE	1000		
16923-95-8	ZIRCONIUM POTASSIUM FLUORIDE	1000		
18883-66-4	D-GLUCOPYRANOSE, 2-DEOXY-2-(3-METHYL-3-NITROSOUREIDO)-	1		
18883-66-4	STREPTOZOTOCIN	1		

Chemical Hazard Communication Guidebook

CAS	CHEMICAL NAME	RQ (LBS)	TPQ (LBS)	NOTES
20816-12-0	OSMIUM OXIDE	1000		
20816-12-0	OSMIUM TETROXIDE	1000		
20830-81-3	DAUNOMYCIN	10		
20830-81-3	5,12-NAPHTHACENEDIONE, (8S-CIS)- 8-ACETYL-10-(3-AMINO-2,3,6 TRIDEOXY-ALPHA-	10		
20859-73-8	ALUMINUM PHOSPHIDE	100	500	b
23950-58-5	3,5-DICHLORO-N-(1,1-DIMETHYL-2-PROPYNYL)BENZAMIDE	5000		
23950-58-5	PRONAMIDE	5000		
25154-54-5	DINITROBENZENE (MIXED)	100		
25154-55-6	NITROPHENOL (MIXED)	100		
25155-30-0	SODIUM DODECYLBENZENE SULFONATE	1000		
25167-82-2	TRICHLOROPHENOL	10		
25168-15-4	2,4,5-T ESTERS	1000		
25168-26-7	2,4-D ESTERS	100		
25321-14-6	DINITROTOLUENE	10		
25321-22-6	DICHLOROBENZENE (MIXED)	100		
25376-45-8	DIAMINOTOLUENE	10		
25376-45-8	TOLUENEDIAMINE	10		
25550-58-7	DINITROPHENOL	10		
26264-06-2	CALCIUM DODECYLBENZENE SULFONATE	1000		
26471-62-5	BENZENE, 2,4-DIISOCYANATOMETHYL-	100		
26471-62-5	TOLUENE DIISOCYANATE	100		
26628-22-8	SODIUM AZIDE	1000	500	b
26638-19-7	DICHLOROPROPANE	1000		
26952-23-8	DICHLOROPROPENE	100		
26952-23-8	DICHLOROPROPENE, N.O.S.	100		
27176-87-0	DODECYLBENZENE-SULFONIC ACID	1000		
27323-41-7	TRIETHANOLAMINE DODECYLBENZENESULFONATE	1000		
27774-13-6	VANADYL SULFATE	1000		
28300-74-5	ANTIMONY POTASSIUM TARTRATE	100		
30525-89-4	PARAFORMALDEHYDE	1000		
32534-95-5	2,4,5-TP ACID ESTERS	100		
33213-65-9	BETA-ENDOSULFAN	1		
36478-76-9	URANYL NITRATE	100		
37211-05-5	NICKEL CHLORIDE	100		
39196-18-4	3,3-DIMETHYL -1-(METHYLTHIO)-2-BUTANONE,O-[(METHYLAMINO)CARBONYL] OXIME	100		
39196-18-4	THIOFANOX	100	100/10000	
42504-46-1	ISOPROPANOLAMINE DODECYLBENZENESULFONATE	1000		
52628-25-8	ZINC AMMONIUM CHLORIDE	1000		
52652-59-2	LEAD STEARATE	5000		
53467-11-1	2,4-D ESTERS	100		
53469-21-9	AROCLOR 1242	1		
53469-21-9	POLYCHLORINATED BIPHENYLS	1		
55488-87-4	FERRIC AMMONIUM OXALATE	1000		
56189-09-4	LEAD STEARATE	5000		

Appendix G: Hazardous Substances (CAS Order)

CAS	CHEMICAL NAME	RQ (LBS)	TPQ (LBS)	NOTES
61792-07-2	2,4,5-T ESTERS		1000	

NOTES

a. This chemical does not meet acute toxicity criteria. Its TPQ is set at 10,000 pounds.

b. This material is a reactive solid. The Threshold Planning Quantity does not default to 10,000 pounds for non-powder, non-molten, non-solution form.

c. EPA changed the calculated Threshold Planning Quantity and the reader is referred to the *Federal Register* of April 22, 1987 for further details.

d. EPA has indicated that the reportable quantity is likely to change when the assessment of potential carcinogenicity and chronic toxicity is completed.

e. Statutory reportable quantity for purposes of emergency notification under Section 304(a)(2) of the Emergency Planning and Community Right-to-Know Act.

f. EPA has indicated that the statutory one pound reportable quantity for methyl isocyanate may be adjusted in a future rulemaking.

g. New chemicals added that were not part of the original list of 402 substances.

h. Revised TPQ based on new or re-evaluated toxicity data.

j. TPQ is revised to its calculated value and does not change due to technical review as in proposed rule.

k. The TPQ was revised after proposal due to calculation error.

l. Chemicals on the original list that do not meet the toxicity criteria but because of their high production volume and recognized toxicity are considered chemicals of concern.

m. Hydrogen chloride is regulated as an extremely hazardous substance for the gas only.